COST ORIENTED AUTOMATION 2004
(COA 2004)

A Proceedings volume from the 7th IFAC Symposium,
Gatineau, Québec, Canada, 6–9 June 2004

Edited by

MAREK ZAREMBA
Département d'informatique, Université du Québec en Outaousi,
Québec, Canada

JUREK SASIADEK
Department of Mechanical and Aerospace Engineering,
Carlton University, Ottawa, Ontario, Canada

and

HEINZ H. ERBE
Technische Universität Berlin,
Center of Human-Machine Systems,
Berlin, Germany

Published for the

INTERNATIONAL FEDERATION OF AUTOMATIC CONTROL

by

ELSEVIER LIMITED

ELSEVIER Ltd
The Boulevard, Langford Lane
Kidlington, Oxford OX5 1GB, UK

Elsevier Internet Homepage
http://www.elsevier.com

Consult the Elsevier Homepage for full catalogue information on all books, journals and electronic products and services.

IFAC Publications Internet Homepage
http://www.elsevier.com/locate/ifac

Consult the IFAC Publications Homepage for full details on the preparation of IFAC meeting papers, published/forthcoming IFAC books, and information about the IFAC Journals and affiliated journals.

First edition 2005

Library of Congress Cataloging in Publication Data

A catalogue record for this book is available from the Library of Congress

British Library Cataloguing in Publication Data

A catalogue record for this book is available from the British Library

ISBN 0080443095

9 780080 443096

Transferred to digital print 2008
Printed and bound by CPI Antony Rowe, Eastbourne

To Contact the Publisher

Elsevier welcomes enquiries concerning publishing proposals: books, journal special issues, conference proceedings, etc. All formats and media can be considered. Should you have a publishing proposal you wish to discuss, please contact, without obligation, the publisher responsible for Elsevier's industrial and control engineering publishing programme:

Christopher Greenwell
Publishing Editor
Elsevier Ltd
The Boulevard, Langford Lane   Phone: +44 1865 843230
Kidlington, Oxford   Fax: +44 1865 843920
OX5 1GB, UK   E.mail: c.greenwell@elsevier.com

General enquiries, including placing orders, should be directed to Elsevier's Regional Sales Offices – please access the Elsevier homepage for full contact details (homepage details at the top of this page).

7th IFAC SYMPOSIUM ON COST ORIENTED AUTOMATION

Sponsored by
International Federation of Automatic Control (IFAC)
Technical Committee on Cost Oriented Automation

Co-sponsored by
International Federation of Automatic Control (IFAC)
Technical Committees on:
- Computers for Control
- Components & Instruments
- Robotics
- Human-Machine-Systems
- Manufacturing Plant Control
- Enterprise Integration and Networking
- Mining, Mineral and Metal Processing
- Social Impact of Automation
- Developing Countries
IFIP TC 5 - Computer Application in Technology

International Programme Committee (IPC)
H.-H. Erbe (D) - chair
G. Olling (USA) - vice-chair (industry)

Albertos, P. (ES)
Asano, K. (JP)
Bayart, M. (F)
Benyo, Z. (HU)
Boverie, S. (F)
Bernhardt, R. (D)
Camarinha-Matos, L. (P)
Carelli, R. (RA)
Cernetic, J. (SLO)
Cipriano, A. (RCH)
Craig, I. (ZA)
Dimirovski, G. (MK)
Dinibütün, T. (TR)
Jämsä-Jounela, S. (FI)
John, J. (CZ)

Kopacek, P. (A)
Miyagi, P. (BR)
Molina, A. (MX)
Morel, G. (F)
Nof, S. (USA)
Nunes, U. (P)
Ollero, A. (ES)
Rojek, R. (PL)
Sanz, R. (ES)
Sasiadek, J. (CA)
Stahre, J. (SE)
Wang, W. (PRC)
Won, S. (KR)
Zaremba, M. (CA)
Zuehlke, D. (D)

National Organizing Committee (NOC)
J. Vaillancourt   Chair
J. Dale   Vice-Chair (Industry)
P. Bergeron   Coordinator

J. O'Shea
J. Sasiadek
M. Zaremba

PREFACE

This symposium on Cost Oriented Automation brought together researchers and industrialists in plenary and regular sessions, and industrial workshops as well. The organizers intended to focus on theoretical and practical solutions, and the applications of cost effective automation systems. Despite possible expensive components, the life cycle of systems will be considered regarding design, production, implementation, operating, maintenance, reconfiguration, and recycling.

Cost aspects are always considered when designing automation systems. Industry is constantly looking for intelligent solutions and engineering strategies for saving cost, but also maintaining a high and secure performance. Field robots in several domains such as manufacturing plants, buildings, offices, agriculture and mining are candidates for reducing the cost. Enterprise Integration and support for networked enterprises are considered as cost saving strategies. Human-Machine collaboration is a new technological challenge, and promises more than cooperation. Last but not least, condition monitoring of machines to reduce the maintenance cost, and to avoid standstill of machines and equipment, if even possible, is also a new challenge, and promotes e-maintenance and e-service.

All these themes have been discussed within the symposium and particularly with the panelists of the two industrial workshops.

The symposium is the 7th in a sequence of successful symposia on Low Cost Automation (1986-2001).The Technical Committee on Cost Oriented Automation is a cross section of theory, technology, and application. Therefore this symposium was highly co-sponsored by other TC´s of the IFAC community.

Heinz-H. Erbe
Chair of the International Program Committee
Chair of the IFAC TC on Cost Oriented Automation

CONTENTS

PLENARY PAPERS

COST REDUCING ENGINEERING STRATEGIES AND COST IMPACTS OF INTERNATIONAL STANDARDS ON AUTOMATION

SENSOR AND DATA-FUSION

ARCHITECTURES FOR ENTERPRISE INTEGRATION AND NETWORKING

HUMAN COOPERATION WITH AUTOMATION SYSTEMS

PLANT AND BUILDING AUTOMATION

COST REDUCTION WITH E-MAINTENANCE SYSTEMS IN MANUFACTURING

SME-ORIENTED DECISION SUPPORT SYSTEMS

NETWORK RELIABILITY

PANEL DISCUSSIONS

ELSEVIER
IFAC
PUBLICATIONS
www.elsevier.com/locate/ifac

COST-EFFECTIVE PRODUCT RELIZATION: SERVICE-ORIENTED ARCHITECTURE FOR INTEGRATED PRODUCT LIFE-CYCLE MANAGEMENT

Bart O. Nnaji, Yan Wang, Kyoung-Yun Kim

NSF Center for e-Design
University of Pittsburgh, PA, USA
nnaji@engr.pitt.edu

Abstract: The worldwide availability of technology, capital, information, and labor makes today's manufacturing enterprises global. Information incompleteness, inconsistency, and improccessability are problems that collaborative design groups are facing. There is a need to improve the current product development process by allowing distributed and real-time collaboration, endorsing knowledge sharing, and assisting better decision making in product life-cycle management. This will fundamentally transform product development to achieve significant reduction in time to market and cost. This paper addresses a new integration environment that enables the evolution of collaborative *e*-design paradigm. This design paradigm aims at seamless and dynamic integration of distributed design objects and engineering tools over the Internet. A service-oriented architecture is introduced to ensure design collaboration, good interoperability, scalability, extensibility, and portability of the engineering tools. The *e*-design environment transparently integrates overall product life-cycle management activities, including conceptualization, detailed design, virtual simulation and testing, and supply chain management. *Copyright © 2004 IFAC*

Keywords: e-design, e-service, collaborative e-work, product life-cycle management, cost-reducing strategies

1. INTRODUCTION

Today, manufacturing enterprises are globalized with the world-wide availability of technology, capital, information, and labor. Faster change in market demand drives faster obsolescence of established products. Global marketing competition makes manufacturers more conscious of quality, cost, and time-to-market. This distributed economic and technological environment poses a challenge of how to manage collaborative engineering, that is, how to let engineers collaborate globally during the product development period. In recent years, the Internet has evolved rapidly and has made enormous impact on the whole spectrum of industries. The application of network technologies in manufacturing is indispensable because manufacturers face numerous challenges in the practice of collaborative design: lack of information from suppliers and working partners; incompleteness and inconsistency of product information/knowledge within the collaborating group; incapability of processing information/data

from other parties due to the problem of interoperability. Hence, collaborative design tools are needed to improve collaboration among distributed design groups, enhance knowledge sharing, and assist in better decision making. There now exist many computational tools in those different areas. However, there are many problems that inhibit them from working together transparently and seamlessly without human intervention. Problems mostly come from the lack of common communication protocols, such as different CAD data formats, different computer operating systems, and different programming languages.

This paper addresses a new integration environment that enables the evolution of collaborative e-design paradigm. This design paradigm aims at seamless and dynamic integration of distributed design objects and engineering tools over the Internet. e-Design involves conceptualizing, designing and realizing a product using tools that allow for interoperability of remote and heterogeneous systems, collaboration among remote supply chain and multidisciplinary enterprise product design team stakeholders, and virtual testing and validation of a product in a secure Internet based information infrastructure.

An e-design system is an integrated product development environment that allows for customers, suppliers, engineers, sales personnel, and other stakeholders to participate in product lifecycle management, while shortening product development time and cost. This integration should be realized by using a service-oriented infrastructure. Service provides functional use for a person, an application program, or another service in the system, which is the core for integration of engineering tools. Various computational engineering tools make certain services available to other design participants in a network-based distributed environment. Instead of traditional client/server relations, peer-to-peer relations exist among service providers. The services that are provided by different engineering tools are published by a service manager, and are available within this distributed environment.

2. BACKGROUND

The advent of the Internet and World Wide Web (WWW) introduced a new wave of research on collaborative product development environment. There are two major research areas in this field. One is the research on how to manage product life-cycle information effectively within a distributed enterprise environment. The other is on network-centric concurrent design and manufacturing, which concentrates on new product design and manufacturing methodology facilitated by network technologies.

In the first research area, research topics comprise of the integration of product and process information temporally and spatially. The product information for the whole life cycle needs to be stored and retrieved enterprise-wide. The accessibility, security, and integrity of information are of the major concerns. By merging the processes of design documentation and design data management through linking CAD drawings with external network-accessible relational databases, integrated geometric information and related documentation can be shared enterprise-wide (Maxfield et al., 1995; Dong and Agogino, 1998; Roy and Kodkani, 1999; Huang and Mak, 1999; Kan et al., 2001). This group of research utilizes existing network protocols to achieve enterprise-wide communication. Some research focuses on agent-based communication methodology over networks. Those researchers (Kumar et al., 1994; Sriram and Logcher, 1993; Huang and Mak, 2000) considered the following research issues of collaborative design system: multimedia engineering documentation, messages and annotations organization, negotiation/constraint management, design, visualization, interfaces, and web communication and navigation among agents.

In the second research area, research is more focused on the feasibility for product design and manufacturing collaboration by the aids of networked computers in a distributed environment. The importance of design collaboration has gained attention of industry (NSF Workshop, 2000; FIPER, 2003; OneSpace; Windchill). Meanwhile, the possibility of distributed environment for product designers and manufacturers has been studied by several academic research groups (Chui and Wright, 1999; Wagner et al., 1997; Kao and Lin, 1998; Larson and Cheng, 2000; Qiang et al., 2001). Some research utilizes middleware technologies for communication in the areas of feature modeling, feature recognition, and design composition (Han and Requicha, 1998; Abrahamson et al., 2000; Lee et al., 1999).

Instead of looking at various engineering tools from traditional computation viewpoint, as above researchers have done, e-design focuses on the engineering implication of those tools from a more abstract level. This approach assures good openness of collaborative engineering systems. The Internet is no longer a simple network of computers. From an application perspective, the Internet is a network of potential services. Functional views of services need to be clearly defined during the design of an Internet-based distributed engineering system. The information supply chain (from customers to product vendors and makers) is required to deal with several information issues including life cycle needs; protection and security; seamless sharing of information across international boundaries; storage/retrieval and data mining strategies; creation of a knowledge depository; classification in the depository (proprietary, public and shared) along with the means to deposit information and knowledge into corporate memory; maintaining and representing the interpretation of information for use by down stream applications and processes; information interpretation for consistency; and record of reasoning process.

In this paper, the concept of service-oriented product engineering architecture is presented to make e-design system cost- and time-effective. By using service-oriented e-design system, customers, designers, manufacturers, suppliers, and other stakeholders can participate in early stage of product design so as to reduce the new product development cycle time and cost.

3. SERVICE-ORIENTED ARCHITECTURE

Integrated product engineering requires collaboration of various engineering tools from various disciplines, such as aesthetic, drafting, material, manufacturing, quality, marketing, maintenance, and government regulations. The Internet provides an opportunity for these engineering tools to work together and utilize these services optimally. To connect these "islands of automation" transparently, universally accepted protocols should be defined at different levels.

In service-oriented architecture, any service can be integrated and shared with architecture components in a legitimate network. To utilize this architecture, services should be specified from the functional aspect of service providers. To make an existing tool available online or to build a new tool for such a system, services associated with this tool should be defined explicitly. The service transaction among service providers, service consumers, and the service manager within e-design system is illustrated in Figure 1. Once a service is registered at a central administrative manager, it is then available within the legitimate domain. This process is service publication. When a service consumer within the system needs a service, it will request a lookup service from the service manager. This process is service lookup. If the service is available, the service consumer can request the service from the service provider by the aid of the service manager. Most importantly, this service triangular relationship should be built at run-time. The service consumer (client) does not know the name, the location, or even the way to invoke the service from the service provider (server). The collaboration between engineering tools is established and executed based on the characteristics of services that can be provided.

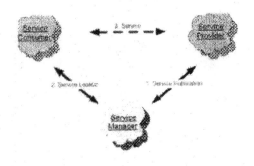

Figure 1: Service triangular relationship

Service publication and lookup are the primary services provided by the Service Manager. As depicted in Figure 2, service publication for service providers includes name publication, catalog publication, and implementation publication. Name publication service is similar to the white-page service provided by telephone companies, by which the name of the service provider is published. Catalog publication service is similar to the yellow-page service: the name and the functional description of the service are published. Implementation publication service is the procedure by which the service provider makes its implementation and invocation of services public so that clients can invoke the service at run-time. Service lookup for service consumers includes name lookup, catalog lookup, and interface lookup. Name lookup service is provided so that consumers can locate the service providers based on the service names. Catalog lookup service is for those consumers who need certain services according to their needs

and specifications but do not know the names of the services. Interface lookup service is to provide a way such that consumers can check the protocols of how to invoke the service.

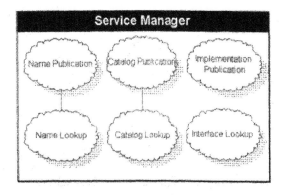

Figure 2: Services provided by Service Manager

A cost-effective collaborative design system should consider: security (which includes access control, identification, authentication, and auditing); concurrency and consistency; heterogeneity and transparency (which includes transparency of access, location, performance, and scaling); inter-process communication; naming (which separates physical and logical names to preserve scalability and transparency); scalability; and resource sharing and management. It is also desirable to reduce the coupling and dependency of data, control information, and administrative information.

4. SERVICE-ORIENTED E-DESIGN INFRASTRUCTURE

In order to shorten product development life-cycle, thus reduce overall cost, an open system for ease of collaboration is needed. The openness provides required extensibility, portability, interoperability, and scalability. In this paper, a service-oriented architecture is employed to conceptualize a future collaborative development environment (i.e., e-design). In a service-oriented e-design architecture, service providers that provide different services such as drafting, assembly, manufacturing, analysis, optimization, procurement, and ergonomics can be developed independently. As showed in Figure 3, servers that provide different engineering services (which are represented by nodes) are linked by the Internet. Each node in this network may require or provide certain engineering services. Thus, it could be a client or a server for different services depending on whether it is the recipient or the provider of such a service. The client/server relationship is determined at run-time. The system is open for the future expansion and extension, in case that more services are available. Plug-and-Play (PnP) is an important consideration of this structure.

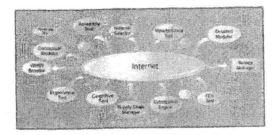

Figure 3: e-Design System Architecture

architecture. In addition, the architecture can realize enterprise-to-enterprise (i.e., inter-enterprise) collaboration. This service transaction chain should be transparent to the stakeholders. The following sub-sections explain the components of the e-design infrastructure in more detail.

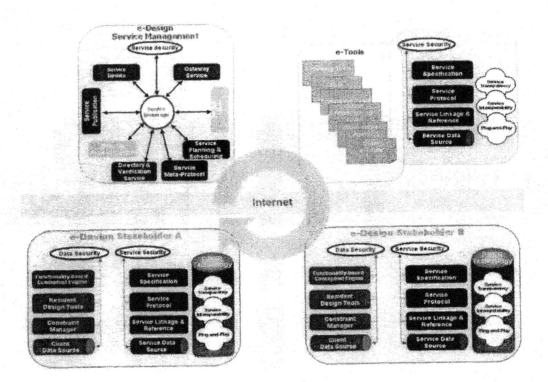

Figure 4: Service-oriented e-Design Information Infrastructure

Figure 4 illustrates the service-oriented e-design information infrastructure. The demonstration illustrated in this figure is named Pegasus Designer System. This infrastructure includes three major components, (i.e., e-design service management, e-tools, and participating e-design stakeholders). These components are integrated through the Internet and share their resources. The e-design service management components provide administrative services (that is brokerage service). Each e-tool has its own services defined. For example, the FEA tool provides the services of product finite element analysis to analyze engineering properties. As another illustration, the ergonomics tool provides product ergonomics analysis. During service transaction, an e-tool may request multiple transactions in collaboration with other e-tools. The e-design stakeholder, (i.e., an enterprise) can have its own collaboration network. The intra-enterprise collaboration can be managed by the e-design

4.1 e-Design service management

The service management builds a bridge between stakeholders and e-tools, in which engineering services are defined, queried, dispatched, and protected according to real-time requirements. The management tasks can be performed in a central service provider as well as distributed service managers. The required services provided by this management role include:

- *Service brokerage*: It allows a transparent and extensible service transaction between stakeholders and e-tools, where distributed and specialized computational tools can be developed and interconnected in a modular way and dedicated e-tools can serve more effectively.
- *Service publication*: It allows for e-tools or third-party agents to publish engineering services for legitimate clients so that open information flow can assist community communication.

4

- *Service lookup*: It is a directory service provided for stakeholders or third-party agents to query and retrieve engineering service meta information.
- *Service subscription*: It provides different levels of client access to service and allows ease of service customization.
- *Security and trust management*: It ensures secured system for service brokerage and secured information transaction so that collaborators can have a trustable and accountable environment.
- *Service certification*: It ensures security and quality of services by an independent certifier for e-tools as well as brokers.
- *Service planning and scheduling*: It handles distributed resource management (such as differentiation of services and distribution of jobs among service providers with identical services, and throughput and service cycle-time).
- *Service update*: Service version control which maintains consistency for interface and implementation.
- *Protocol publication* (service meta-protocol): Service provided for service brokers. It provides meta-information about service protocols such that brokers can lookup and update as necessary.
- *Financial accounting management*: It allows financial compensation transactions for e-tools being monitored and managed in the pay-per-use service model.

4.2 e-Tool

e-Tools can be any hardware and software resources providing engineering services. Example can be engineering solvers (e.g., finite element analysis (FEA) and computational fluid dynamics (CFD) solvers), information database/knowledge base (e.g., material library, part library), web servers (e.g., ontology server), intelligent agents (e.g., design and assembly advisors), as well as computing servers (e.g., high-performance computing group, supercomputers). To be seamlessly utilized as e-Tools, the tools should be implemented considering following issues.

- *Service specification*: Detail definition of service, such as name, type, function, metrics, and version, etc.
- *Service protocol*: Interfacing protocol, input and output parameters, detailed implementation, and brokerage requirement (such as implementation requirement).
- *Service linkage and reference*: Other service providers/third parties, which are required for the service, should be specified and referred.
- *Service data source*: necessary data and information for all above functions.

4.3 Enterprise Collaboration

In this paper, design collaboration is categorized based on different aspects including the types of requests, the scope of collaboration, and the characteristics of transaction. Collaboration types should be first considered when building collaboration relations.

When the types of service requests are considered, two design collaboration cases can be found. A client may request services directly or service providers themselves based on the requirements of performance, complexity of the service, as well as the frequency of the service.

Service request: For a complex or rare service, the client submits a service request and receives a result from available service providers.

Service provider request: For a simple or recurrent service, tools (such as a plug-in for Excel) are downloaded for a local use.

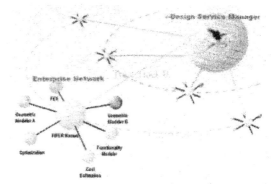

Figure 5: Intra- and Inter-enterprise collaboration

In context of the scope of collaboration, the design collaboration can be classified into two cases. A client may request services within the enterprise or across the boundary of the enterprise based on the availability of service.

- *Intra-enterprise collaboration*: Collaboration within an enterprise where firm's collaboration policy is easier to embody. As shown in Figure 5, a collaboration network can exist within an enterprise.
- *Inter-enterprise collaboration*: Collaboration among enterprises where collaboration policy should be strictly followed and trust and contract management is required. Thus, more overhead is involved. Figure 5 illustrates the concept of collaboration among multiple enterprises (inter-enterprise collaboration). This collaboration can be realized based on a legitimate trust track.

A client may request services with different patterns of service cycles, either single cycle or multiple cycles. This affects the specifications, protocols, and performance requirements of services when collaboration is defined.

- *Single cycle transaction*: By one cycle, service final results can be obtained.

- *Multiple cycle transaction*: It requires multiple cycles and may use service manager multiple times.

The service-oriented e-design environment should be able to allow those design collaboration cases.

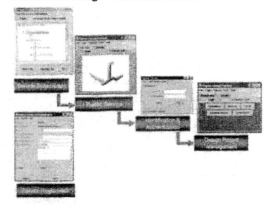

Figure 6: e-Design Service Set-up

4.4 e-Design Infrastructure Requirements

To fully realize the service-oriented e-design environment, the e-design infrastructure should meet the requirements from various design and analysis aspects in collaborative design, as listed in Table 1 (Annex 1).

5. COLLABORATION SCENARIOS

To satisfy the openness requirements, such as extensibility, portability, interoperability, and scalability, a service-oriented architecture is a core concept of e-design systems. In the service-oriented architecture, various engineering services should be available to e-design stakeholder. Figure 6 illustrates a procedure to make the services available for clients. When a company wants to make its service available for e-design stakeholders, the company needs to register its service by providing various information (e.g., company information, service name and category, and transaction protocol). An e-design participant or stakeholder needs to subscribe service. Through this service subscription process, services are classified into visible or invisible services. A set of services can be formed as a "bundle," such as conceptual design tools. All e-design participants can access public service, which provides fundamental services, such as graphic viewer. However, to use the subscribed services, the participants must go through an identification and authentication process. The subscribed service will be displayed through a common interface, called *e-design gateway*.

The e-design gateway provides an environment customized for a participant's design project. Each enterprise or participant may require different design environment due to the different characteristics of their project. For example, if a design project is generative in nature and is to conceptualize design specifications without having a previous design, the project will require conceptual design tools, such as a functionality-based design tool. But, if the project is based on an existing design, this project will require CAD tools compatible to the existing design. Especially, in case where the designs are from different CAD systems, the relevant interoperable environment must be set up. Figure 7 illustrates different design projects of multiple enterprises and examples of customized e-design gateway.

Figure 7: e-Design gateway and customization

Figure 8 shows a procedure of e-design project customization. By specifying project type (e.g., new or existing project), collaboration type (e.g., enterprise-wide or enterprise-to-enterprise collaboration), existence of an initial model, and system environment (e.g., CAD systems of collaborators), proper e-design system environment including interoperability among CAD tools, analysis tools, data, can be customized. It should be noted that additional tools and functions can be appended by client request.

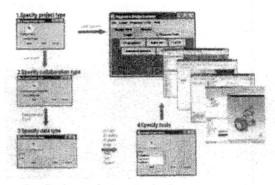

Figure 8: e-Design project customization

For different collaboration scenarios, the e-Design environment should provide transparent views to both stakeholders and e-tools. That is, e-Design stakeholders have uniform functional views of e-Tools without notification of physical locations. And e-Tools can provide services without imposing system-dependent restrictions to stakeholders. Services are based on interoperable network and application protocols without implementation restriction during the system's evolution. Information flow process within the e-Design environment should be designed from the viewpoints of efficiency and security besides transparency. Lean information exchange should be supported among heterogeneous tools considering the limitation of communication bandwidth. Protection of Intellectual Property is vital to build a trustable cyberspace for product development.

To illustrate the service architecture of Pegasus system, a demonstration is described in this section. Figure 9 illustrates a typical assembly design cooperation in the Pegasus service architecture between the Pegasus service manager, a geometric modeler, an analysis service provider, and an engineering material service provider. The various Pegasus component implementations for Figure 9 are shown in Figures 10 to 13.

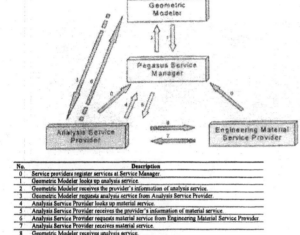

No.	Description
0	Service providers register services at Service Manager.
1	Geometric Modeler looks up analysis service.
2	Geometric Modeler receives the provider's information of analysis service.
3	Geometric Modeler requests analysis service from Analysis Service Provider.
4	Analysis Service Provider looks up material service.
5	Analysis Service Provider receives the provider's information of material service.
6	Analysis Service Provider requests material service from Engineering Material Service Provider
7	Analysis Service Provider receives material service.
8	Geometric Modeler receives analysis service.

Figure 9. Assembly design cooperation

In this assembly design example, the Geometric modeler (Figure 10-a) provides detailed design services (such as sketching, feature generation, assembly design, etc.) for design engineers. Assuming the designer want to join the an exhaust manifold and an engine case by mechanical fastening, the designer should know the physical effects, such as deformation, bucking, stress/strain distribution on the assembly. Once the designer indicates a certain joining method (in this case a mechanical fastener), a virtual assembly analysis tool (Figure 10-b), which is a tool of the Geometric modeler, is automatically triggered and it generates analysis parameters based on the assembly design information. The assembly design information has been imposed on the design model by the system and the designer does not need to know any details about this translation. The virtual assembly analysis tool is a simple interface to show the analysis model and to be confirmed by the designer, and is linked to the assembly design cooperation network (service-oriented architecture). The physical effect can be simulated by use of mechanical analysis service. This mechanical analysis service, which is provided by analysis service providers, such as ANSYS, ADINA, and ABAQUS, can be invoked remotely through the

Pegasus service manager (Figure 11). To accomplish this analysis, the Analysis service provider (Figure 12) needs the joining parameters and material properties. The material properties, which are sometimes but essential for engineering design, can be provided by engineering material service provider. An Engineering Material service provider (Figure 13) offers engineering material lookup services.

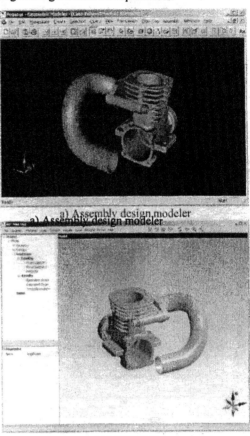

a) Assembly design modeler

b) Virtual assembly analysis tool
Figure 10. Geometric modeler

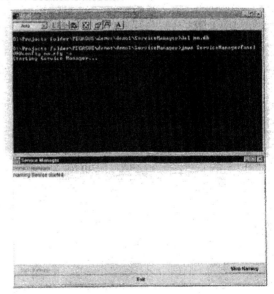

Figure 11. Service manager

Figure 12. Analysis service provider

Figure 13. Engineering Material service provider

In order to perform a realistic simulation of the physical effects of the joining operations, the analysis service provider (transparent to the Geometric modeler) looks up and acquires the material information on the specified material type from the Engineering Material Service Provider. Remark that the Geometric modeler provides only material type and design configuration. When the analysis is completed, the Analysis service provider returns the analysis results (e.g., output files, animation movies) to the virtual assembly analysis tool (Figure 14).

The locations of various service providers are not known until at run-time. The relation between the service consumers and the service providers is built dynamically. This relation can be viewed as a service chain, which connects service provider with client/server affiliation, as illustrated in Figure 15.

The Service manager plays an important role in this service chain management. It allocates service resources according to service consumers' demand and service providers' capability and capacity. In a more matured system, it is possible that several providers can offer same services. Thus when Service Manager makes decisions, some other factors (geographical location, capacity, compatibility level, etc.) will be considered.

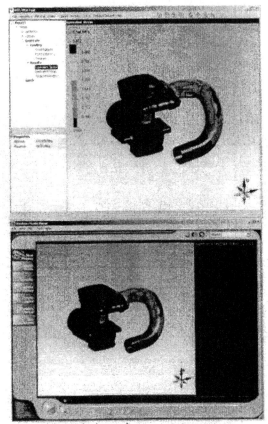

Figure 14. Analysis service response

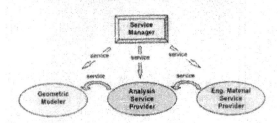

Figure 15. Service chain relations

6. CONCLUSION

This paper addresses the concept of a new design paradigm, e-design, enabling seamless and dynamic integration of distributed design objects and engineering tools over the Internet. In order to shorten product development life-cycle, and consequently reduce overall cost, an open system for ease of

8

collaboration is needed. The openness provides required extensibility, portability, interoperability, and scalability. In this paper, a service-oriented open architecture is introduced to conceptualize a future e-design environment, which allows cost-effective distributed design collaboration. The *e*-design environment transparently integrates overall product life-cycle management activities, including conceptualization, detailed design, virtual simulation and testing, and supply chain management. Various requirements to realize future e-design systems for different collaboration scenarios are described.

REFERENCES

Abrahamson, S., Wallace, D., Senin, N. and Sferro, P. (2000) Integrated Design in a Service Marketplace, *Computer-Aided Design*, **32** (2), 97-107

Chui, W.H. and Wright, P.K. (1999) A WWW computer integrated manufacturing environment for rapid prototyping and education, *International Journal of Computer Integrated Manufacturing*, **12** (1), 54-60

Dong, A and Agogino, A.M. (1998) Managing design information in enterprise-wide CAD using 'smart drawings', *Computer-Aided Design*, **30** (6), 425-435

FIPER (2003) 2003 FIPER Workshop and National Institute of Standard Technology Final Review on FIPER project, Cincinnati, OH, December 17-19, 2003

Han, J.H. and Requicha, A.A.G. (1998) Modeler-independent feature recognition in a distributed environment, *Computer-Aided Design*, **30** (6), 453-463

Huang, G.Q. and Mak, K.L. (1999) Design for manufacturing and assembly on the Internet, *Computers in Industry*, **38** (1), 17-30

Huang, G.Q. and Mak, K.L. (2000) WeBid: A Web-based Framework to Support Early Supplier Involvement in New Product Development, *Robotics and Computer Integrated Manufacturing*, **16** (2-3), 169-179

Kan, H.Y., Duffy, V.G. and Su, C.J. (2001) An Internet Virtual Reality Collaborative Environment for Effective Product Design, *Computers In Industry*, **45** (2), 197-213

Kao, Y.C. and Lin, G.C.I. (1998) Development of a collaborative CAD/CAM system, *Robotics and Computer-Integrated Manufacturing*, **14** (1), 55-68

Kumar, V., Glicksman, J. and Kramer, G.A. (1994) A SHAREd Web To Support Design Teams, in *IEEE Proceedings of the Third workshop on Enabling Technologies: Infrastructure for collaborative Enterprises*, April 17-19, Morgantown, West Virginia, pp.178-182

Larson, J. and Cheng, H.H. (2000) Object-Oriented Cam Design Through the Internet, *Journal of Intelligent Manufacturing*, **11** (6), 515-534

Lee, J.Y., Han, S.B., Kim, H. and Park, S.B. (1999) Network-Centric Feature-Based Modeling, in *IEEE Proceedings of The Seventh Pacific Conference on Computer Graphics and Applications*, October 5-7, Seoul, Korea, pp.280-289

Maxfield, J., Fernando, T. and Dew, P. (1995) A Distributed Virtual Environment for Concurrent Engineering", in *IEEE Proceedings on Virtual Reality Annual International Symposium*, March 1-15, Research Triangle Park, North Carolina, pp.162-170

NSF Workshop (2000) National Science Foundation Workshop on e-Product Design and Realization for Mechanically Engineered Products, University of Pittsburgh, Pittsburgh, PA, October 19-20, 2000

OneSpace, CoCreate Corporate, http://www.cocreate.com

Qiang, L., Zhang, Y.F. and Nee, A.Y.C. (2001) A Distributive and Collaborative Concurrent Product Design System through the WWW/Internet, *International Journal of Advanced Manufacturing Technology*, **17** (5), 315-322

Roy, U. and Kodkani, S.S. (1999) Product modeling within the framework of the World Wide Web, *IIE Transactions*, **31** (7), 667-677

Sriram, D. and Logcher, R. (1993) The MIT Dice Project, *IEEE Computer*, **26** (1), 64-65

Wagner, R., Castanotto, G. and Goldberg, K. (1997) FixtureNet: Interactive Computer-Aided Design via the World Wide Web, *International Journal of Human-Computer Studies*, **46**, 773-788

Windchill, Parametric Technology Corporate, http://www.ptc.com

Annex 1:

Table 1 The requirements of e-Design infrastructure for collaborative design

Interoperability from CAD to CAD, from CAD to CAE, from CAE to CAE	Multidisciplinary design in Product Lifecycle Management	Multidisciplinary constraints & preferences capture	Lean product data management, instant access & visualization	Proactive analysis & transparency	Virtual simulation & prototyping	Security and trust management
• Standards and protocols supplement • Parameters and constraints representation • Non-geometric constraints capturing design intents	• System engineering approach • Conceptual design (functionality-based, ergonomics & cognitive-based)Direct constraint imposition • Conflict resolution & management • Design activity based cost modeling • Multi-attributes decision models • Reliability matrix • Computationally efficient, high-fidelity predictive models	• Multidisciplinary constraints & preferences representation and management • Multidisciplinary constraints representation & multi-objective decision making • Conflict resolution • Material representation methodology	• Lean product data modeling • Subscription-based hybrid data modeling • Distributed data linkage mode • Multi-views with different levels of details	• Capacity for process of analysis, etc, "behind the scene" at remote locations without setup • High level modeling knowledge capturing in Engineering Analysis Modeling • Domain specific analysis knowledge modeling & ontology Model reusability, adaptability, and interoperability • Open system architecture	• Development of Physics-based models • Virtual reality, product, environment models for system level simulation • Models for simulation-based design under uncertainty • Virtual collaboration and sharing • Virtual assembly analysis, design, and knowledge capturing • Simulation-based acquisition • Real-time visualization	• Security and trust modeling • Trust for service for distributed, enterprise-wide e-Business networks • Trust issues concerning honestly, openness, reliability, competence and benevolence • Trust-support infrastructure

Annex 2:

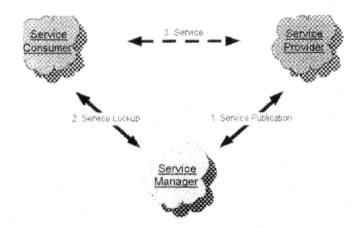

Figure 1: Service triangular relationship

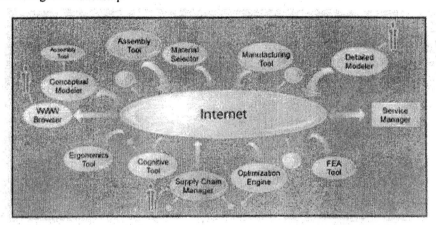

Figure 3: e-Design System Architecture

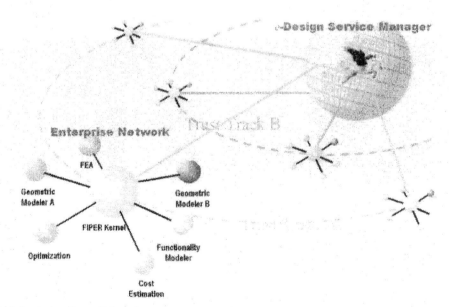

Figure 5: Intra- and Inter- enterprise collaboration

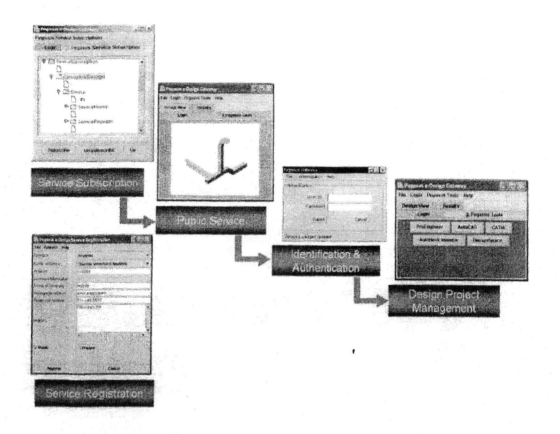

Figure 6: e-Design Service Set up

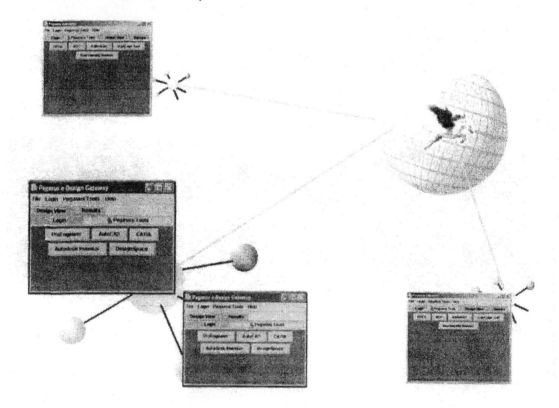

Figure 7: e-Design Gateway and Customization

PROOF-ORIENTED FAULT-TOLERANT SYSTEMS ENGINEERING: RATIONALES, EXPERIMENTS AND OPEN ISSUES

Gérard MOREL[1], Dominique MERY[2],
Jean Baptiste LEGER[3], Thierry LECOMTE[4]

[1]*gerard.morel@cran.uhp-nancy.fr,* [2]*dominique.mery@loria.fr,*
[3]*jean-baptiste.leger@predict.fr,* [4]*thierry.lecomte@clearsy.com*

Abstract: Proving system properties such as fail-safety is a challenge for systems engineering since industrial automation is nowadays embedding intensive on-site and remote infotronics components engineered with increasing intuitive ease-of-use techniques. Since a formal proof of the complete safe-behaviour of the resulting ad-hoc system is not possible, this paper argues that *Proof Oriented Systems Engineering* formal techniques should bridge the gap with *Fault Tolerant Systems Engineering* practical techniques in order to mathematically check the proof of fail-safety. Rationales, experiments and open issues are addressed on combining the formal B event-based method using the B proof assistant with a technical-safety modelling formalized-framework. *Copyright © 2004 IFAC*

Keywords: Fault-Tolerant Systems Engineering, Proof-Oriented Fail-Safe Systems Engineering, Formal Methods, Automation Engineering, Safe-Process.

1. INTRODUCTION

Previous works (Morel *et al.*, 2001b) stressed the consideration of the performance of the complete system as a compromise between end-use system goals, such as CRAM (Cost, Reliability, Availability, Maintainability, and Productivity) parameters in maintenance, rather than the control performance only (fig. 1) as addressed currently in System Theory and Automatic Control (Cassandras and Lafortune, 1999).

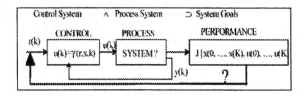

Figure 1: A closed-loop system with system performance optimization rather than control performance optimization (Morel *et al*, in Erbe, 2003).

Others rationales and works addressed in automation engineering the impact of the human designer (Lhote et al., 1999) on the system design and, consequently, on the operational performance with regards to the ideal and reachable ones (fig. 2).

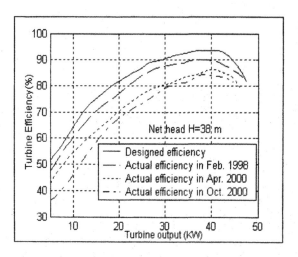

Figure 2: Operational efficiency evolvement of a hydro turbine from the designed efficiency (Liu *et al.*, 2003)

Since the trend is nowadays to merge on-site and remote intensive infotronics technologies (Lee, in Pereira *et al.*, 2004) in order to recover expected performances as well as to face increasing demands on performance (agility, interoperability, reactivity, configurability, security certification, failure recovery, ...), the threat could be that software dependability may limit further automation progress at the Business-to-Manufacturing level (fig.3).

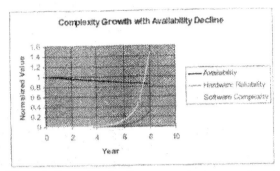

Figure 3: Scenario of software complexity growth and system availability decline (Johnson, in Pereira *et al.*, 2004)

The difficulty to balance availability and dependability during runtime is strengthened by the ease-of-use trend of standardised automation-objects for intuitive engineering (Vyatkin, in Pereira *et al.*, 2004) infering a lack of sound foundations for the required completeness, consistency, unambiguousness and correctness of the system model (fig. 4).

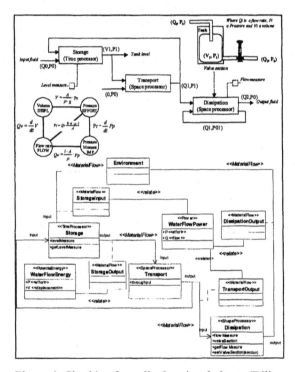

Figure 4: Checking formally functional charts (Féliot, 1997) and UML diagrams (Panetto *et al.*, 2003) from basic constructs (Paynter, 1961).

Many rationales and experiments emphasize that the compromise between safety, security, dependability and availability should be carried out with a system point of view all along the engineering life-cycle (Moik, 2003) in order to bridge the gap between traditional techniques focusing on particular system performance parameters (reliability analysis, cost evaluation, multi-criteria optimisation,...) and engineering-related formal methods (fig. 5).

Since a complete mathematical proof of high-level safety-integrity levels[1] is not possible, this paper deals with *proof-oriented fail-safe systems engineering* combining formal-related software-engineering techniques and fault-tolerant automation-engineering techniques.

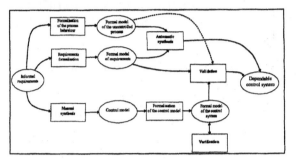

Figure 5: Role of formal methods to improve automation software dependability (Faure, in Pereira *et al.*, 2004).

Section 2 addresses Computing Science rationales for *Proof Oriented System Development (POSD)*. These rationales emphasize technical statements about *POSD* in section 3 before to strengthen the efficiency of the *B methodology* in section 4. Perspectives on *Proof Oriented Systems Engineering (POSE)* state in section 5 that the system goals should be deeply founded before to split the resulting system model into hardware/software sub-models and, beyond, into process/control systems for automation purposes. Moving from *POSD* and *POSE* to *Proof-Oriented Fault-Tolerant Systems Engineering (POFTSE)* is illustrated in section 6 by an industrial experiment aiming to formally check a high-level of fail-safety. Rationales and open issues to make *POSE* and *POFTSE* efficient in practice are discussed in conclusion.

2. PROOF-ORIENTED SYSTEM-DEVELOPMENT RATIONALES

The idea of refinement is not an original invention of computing science but is fundamental to the scientific approach to engineering. Traditional engineering disciplines translate a given physical problem into a mathematical model. From this conceptual mathematical model, a solution is derived by applying engineering methods that embody scientific knowledge. Typically, the solution can not be obtained in one step; intermediate models serve to identify sub-problems, sufficient approximations etc. Every step in the development is justified by mathematical proof (e.g., calculation). The abstraction gained by considering mathematical models instead of physical, chemical or biological reality helps to concentrate on the relevant aspects of the problem and (for the scientist) to state laws that are universally applicable. Validation is still necessary, e.g. by conducting experiments on scale models, to ensure that all relevant aspects have been accounted for in

---

[1] IEC 61508

14

the abstraction. Applying this view to the development of computer-based systems, the refinement enables a rational approach to system development, separating the engineering invention from the mechanical aspects related to proofs.

Moreover, system designers should be able to follow standard patterns that are generic to well-identified problem classes and whose correctness has been established once and for all. These Capability Maturity Models (CMM) provided by technical societies[2] aim to aid engineers to combine prescriptive and descriptive models of system parts in order to bring the system into being without any proof on the resulting ad-hoc system model (fig. 6).

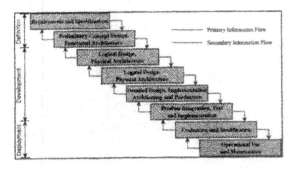

Figure 6: A Systems Engineering life-cycle (Boland et al., 2001)[3]

POSD aims to reinforce standard methods for the design of computer-based systems by formal description and analysis techniques that can help to ensure higher levels of reliability and correctness. Based on a precise mathematical semantics, it offers powerful techniques for the validation and analysis of system models, including comprehensive testing and verification that accompany and guide the development process.

Although formal development techniques have been studied since several decades and have had a substantial impact on the understanding of the development activity as well as on language design, they have not yet found widespread acceptance in the development community. Individual techniques such as model checking, formal approaches to testing, and semantic analysis have been successfully applied in industry, following some spectacular failures (e.g., Therac~25, Ariane~5, Pentium bug, …) as well as certification requirements for certain products that contain software components. The mastering of quality and safety requirements of computer-based systems has thus become a challenge for system developers.

Clearly, much work remains to be done to make such an approach practical and applicable to complex systems. We believe that it should be integrated with standard methods of system design that have proven in practice to represent useful views of systems. In

this respect, the UML has over the past years established itself as the de-facto framework for object-oriented software design, offering several types of diagrams to represent static, dynamic, and architectural views of systems, at different phases of the development cycle. Formal methods should provide the necessary semantics to justify such transitions between development phases as valid refinement steps.

3. PROOF-ORIENTED SYSTEM-DEVELOPMENT TECHNIQUE

We focus our research on systems that may include hardware and software parts; the proof technique plays a central role in our approach, since it helps in validating the consistency of the developed models. Our systems are generally called reactive systems and are very close to the control systems for which *proof-oriented modelling* consists in giving a mathematical meaning to the operators of the automation paradigm into a logical framework (fig. 7), as previously addressed by (Fusaoka et al., 1983) according to :

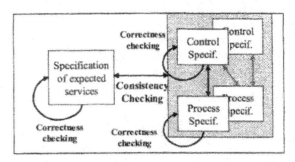

Figure 7: Verification of the automation paradigm (Morel et al., 2001a)

We found our work on the seminal framework of (Back, 1979), namely actions systems, and the refinement relationship defined between models or action systems. Other scientists worked on the same topics like (Dijkstra, 1976), (Morgan, 1990), (Lamport, 1994), (Abrial, 1996) and we apply those techniques in the development of proved systems; the *proof of a system* is always given with respect to *goals of system*. We should be careful and be precise in defining what is to be proved and what is the relationship between the system and goals.

First at all, the object (built in a given notation) is a model that should be verified and validated. The adequacy of the model with respect to requirements should be checked in a way. The traceability of requirements allows us to identify formal elements and informal ones; the informal elements are clearly simpler to understand by the so called customer. The integration of formal techniques is mandatory and is considered as a crucial point to ensure the quality of the developed systems. The design of formal models for a given system may be either based on an *a posteriori approach* using automatic techniques like

[2] www.incose.org; www.isa.org;
[3] www.lar.ee.upatras.gr/;

model checking (Clarke *et al.*, 2000) or abstract interpretation-based ones and allowing to establish the link between the formal model and required properties (goals), or based on an *a priori approach* allowing to derive systematically a sequence of formal models by preserving the link between formal models and required properties.

In the second approach, the models are related by a *refinement relationship* that is able to preserve properties through transformations of models. A transformation of a model into another one consists in concretizing it or in making it more precise; the rule of the game is to start by a simple high level model with simple properties to check and to gradually add detail by preserving previous properties. We summarize the approach as follows:

(M1,G1) refined by (M2,G2)
refined by .... refined by (Mn,Gn)

Mi is the ith model and it satisfies the goal *Gi* and the goals of smaller number; the *refined by* relationship ensures the preservation of goals but the proof of refinement should be given. It means that if a new model is derived from *(Mn, Gn)*, we should prove that the new model refines the previous one and that the new properties are true. The verification is much simpler than to verify every goal in one step and we can build the system like an onion starting from a very high level view of the system into a very concrete view. The initial pair *(M1, G1)* should be validated and we should prove that *M1* satisfies *G1* and since *M1* is supposed to be simple or as abstract as possible, the proof is simple. However, the key point of this approach is to start by a simple abstract view or model of the system and to add step by step details: it is a very creative point that leads to a simplification of the proof-based development. The a posteriori approach checks that the pair *(M,G)* is correct: *M* satisfies *G* and the problem is that *M* may be complex and details or choices of design are forgotten and useful for proof. It is why the automatic techniques for proving this one step development are used on approximation or abstractions of the real system or on finite states models (model checking). Both approaches can be also combined and we have proposed integration formalism based on a graphical notation called predicate diagrams; we define a refinement between diagrams and we provide tools for interactive and automatic analysis. In fact, the concepts of abstraction and refinement are dual; the incrementality of the proof-based development is crucial and requires the use of a proof assistant which helps in discharging conditions over models and goals called proof obligations. Since the proof assistant can not in the general case prove every theorem, the incrementality of the approach helps the assistant prover in the proof task. The user should interact and interactions can provide information on the current model: make the life of proof assistant as much as possible simpler.

4. PROOF ORIENTED SYSTEM DEVELOPMENT USING THE B EVENT-BASED APPROACH

Case studies are central in our work and the *B event-based method* provides a framework to develop formal models:

- Distributed systems like the IEEE 1394 tree identification protocol (Abrial *et al.*, 2002)
- Cryptographic protocols (Cansell *et al.*, 2002)
- Electronic circuits (Cansell *et al.*, 2003)
- Access control systems (Pétin et al., 1998).

The benefits of an approach based on refinement are numerous: from the point of view of the system developer, system requirements can be addressed in several steps (or cycles) of system development and feedback on properties of the current model of the system or on design errors is obtained quite early. From the point of view of the verifier, the burden of proof is spread over the development process, and the preservation of key properties such as safety, security or availability is guaranteed. The presence of intermediate system models both reduces the complexity of the proof obligations (allowing for a higher degree of automation) and produces a trace of *milestones* produced during the development of a system, documenting the design. Validation techniques such as tests and simulation can be applied to the intermediary models and enable the early detection of design faults.

Classical B is a state-based method developed by (Abrial, 1996) for specifying, designing and coding software systems. It is based on Zermelo-Fraenkel set theory with the axiom of choice. *Sets* are used for data modelling, *generalised substitutions* are used to describe state modifications, the *refinement calculus* is used to relate models at varying *levels of abstraction*, and there are a number of *structuring mechanisms* (model refinement, implementation) which are used in the organisation of a development. The first version of the B method is extensively described in the B-Book. It is supported by the Atelier B tool. Central to the classical B approach is the idea of a software operation which will perform according to a given specification if called within a given *pre-condition*. Subsequent to the formulation of the classical approach, (Abrial, 2000) and others have developed a more general approach in which the notion of *event* is fundamental. An event has a firing condition (a *guard*) as opposed to a pre-condition. It may fire when its guard is true. *Event based models* have proved useful in requirement analysis, modelling distributed systems and in the discovery/design of both distributed and sequential programming algorithms. After extensive experience with B, current work by (Abrial, 2001) is proposing the formulation of a second version of the method. This distils experience gained with the event based approach and provides a general framework for the development of *Discrete Systems*. Although this widens the scope of the method, the mathematical foundations of both versions of the method are the same.

A model (fig. 8) has a name *m*; the clause *sets* contains definitions of sets of the problem; the clause *constants* allows one to introduce information related to the mathematical structure of the problem to solve and the clause *properties* contains the effective definitions of constants: it is very important to list carefully properties of constants in a way that can be easily used by the tool. Another point is the fact that sets and constants can be considered like parameters and extensions of the B method exploit this aspect to introduce parameterization techniques in the development process of B models. The second part of the model defines dynamic aspects of state variables and properties over variables using the invariant called generally inductive invariant and using assertions called generally safety properties. The *invariant I(x)* types the *variable x*, which is assumed to be initialized with respect to the initial conditions and which is preserved by events (or transitions) of the list of events. Conditions of verification called proof obligations are generated from the text of the model using the first part for defining the mathematical theory and the second part is used to generate proof obligations for the preservation of the invariant and proof obligations stating the correctness of *safety properties* with respect to the invariant.

Model m
Sets s
Constants c
Properties P(s,c)
Variables x
Invariant I(x)
Safety A(x)
Initialisation substitution
events e1,e2,e3, …, en
End

Figure 8: B formal model

A model is valid or correct when the following proof obligations hold; they are checked by the proof assistant with possible interactions.

(INV1) Init(x) => I(x)
(INV2) for each event e of the system,
I(x) ∧ BA(e)(x,x') => I(x')
(DEAD) I(x) => (grd(e1) ∨ … ∨ grd(en))
(FIS) for each event e of the system,
I(x) ∧ grd(e)(x) => exists x'. BA(e)(x,x')

Proof obligations for a model are generated by the proof-obligations generator of the B environment; the sequent calculus is used to state the validity of the proof obligations in the current mathematical environment defined by constants, properties. Several proof techniques are available but the proof tool is not able to prove automatically every proof obligation and interactions with the prover should lead to prove every generated proof obligation. We say that the model is internally consistent when every proof obligation is proved.

5. PROOF-ORIENTED SYSTEMS-ENGINEERING PERSPECTIVES

System development integrates formal languages which should be completed by formal techniques for validating resulting models. It is clear that the main problem is to separate or to assign an objective value to models of system. The mathematics and the engineering sciences are based on theories built from axioms using rules of deduction to infer new properties. System development is in essence an engineering science that should be based on mathematically and logically based techniques. The construction of a model for a given system can be decomposed into several steps and the very initial one is the elicitation of a very first model generally written in a semi-formal language but using a formalized notation like UML (fig. 9).

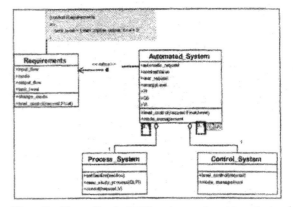

Fig. 9: Formalized UML automation descriptive paradigm (Panetto *et al.*, 2003)

A document is derived from the notation and we call it *SD (System Documentation)*. Basically, the problem is to provide a list of requirements for the system. A second step leads to list properties of the system and properties can be functional ones or non-functional ones: safety, security, integrity, liveness, real time properties, fairness, availability. It is clear that a property language is required and we can use any expressive language suitable for our task of expressing properties.

In our previous experiments, we have used a language based on ZF set theory and predicate calculus, namely B, for expressing invariance properties and safety properties; we have stated more complex properties like liveness, fairness or eventuality properties using temporal logic like TLA of (Lamport, 1994). Remember that we want to develop a system from a very abstract level which allows us only a simple list of properties like a simple invariant with safety properties. Hence, we have a first list of properties for the system and we called the list *SP* (for *System Properties*). SP is built from SD by analysing the document called *SD*.

Now, the proof-based development can start and we should extract a very abstract model *M1*, which is defined by a list of (state) *variables x* and a list of *events*. Variables should satisfy a very simple invariant *I1(x)* and the first model should be consistent: proof obligations must be true. The invariant expresses that nothing bad will happen and that variables are in a given set of possible values but it is a symbolic view; the invariant is clearly extracted from the list of properties namely *SP* and safety properties *SA1* can be found in the list of properties but we should prove that they are proved under the invariant. Since the first model is abstract and simple, the proofs are simple and not too complex for the proof assistant. Now, the current state of our development is characterized by the following relation:

M1 satisfies I1(x) and SA1

M1 includes both events of the process to control and events of the control; the partition is not yet defined, since the model is still abstract, but the refinement will help in making views more and more precise. Our model *M1* contains either the (abstract) environment or plant *MP1* or the (abstract) control system *MC1*.

MP1, MC1 satisfies I1(x) and SA1

The next step of the development process is the refinement step, which is a way to concretize state variables or to introduce new events (refining skip or invisible at the abstract level) or to introduce new variables. The refinement should be correct or consistent; it means properties holding in the abstract model should remain true in the concrete or refinement model. The proof of refinement is established by the proof of proof obligations for the refinement. Intuitively, a model *M1* is refined into a model *M2*, if, when any event of *M2* is observed, then an abstract event is also observed at the abstract level; the refinement model adds new constraints on the invariant but the previous invariant remains an invariant. This technique allows us to obtain a stratified view of the system modelling. Each step is carefully and completely proved.

M1 satisfies I1(x) and SA1

is refined into

M2 satisfies I2(x) and SA2

The refinement step improves the invariant by adding constraints that are mentioned in the requirements and we should add a simple new property or new events. The process goes ahead and is over, when one has a sufficiently detailed view of the system (properties, events). The final model can be then decomposed into two parts for automation purposes: the process system model and the control system model. Our works extended the Fusaoka's predicate to automation engineering in order to consider first the global specification of the system services required for process control before to design and to integrate

hardware-software process-control components as a system (fig. 10).

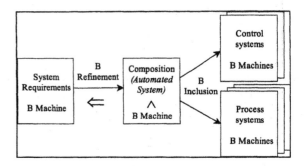

Figure 10: B formal automation prescriptive paradigm (Lamboley, 2001)

6. PROOF-ORIENTED FAULT-TOLERANT SYSTEMS ENGINEERING EXPERIMENT

Proof of fail-safety should be checked and certified according to standards and guidelines of safety-relevant systems which recommend the use of formal methods but without defining how they can be applied.

Previous works (Leger *et al.*, 2001) have formalized into a logical framework a global methodology based on industrially well-established qualitative methods - HAZOP (hazard operability analysis), Hardware and Software FMECA (failure mode effects and criticality analysis), ...- in order to improve the completeness, the consistency, the unambiguousness and the correctness of both the normal and abnormal system behaviour.

Although this formalized technical-safety engineering framework allows to identify the critical processes or equipments and to model the system properties degradation in order to design the fault-tolerant processes - monitoring, diagnosis, prognosis, compensation, accommodation...-, only the use of proof-oriented formal techniques allows meeting higher-level of safety-integrity according to:

Process System ∧ Control System ⟹ Fail-Safe System Goals

The industrial application of such a proof-oriented safety-related formal approach is based on the design of an embedded system used for railway transport in conformance with IEC 61508 standard in order to satisfy a *Safety Integrity Level of 4 (SIL4)*. The role of the embedded system (fig. 11) is to alert the train driver when a problem occurs on the railway signalization and to stop the train if necessary. This electronic device aims to control a safety relay (output: switch on/off signal) according to measures (input: analogical current signal), states (inputs: on/off switch and day-night switch) and parameters (input: static parameters) as well as to alert at distance operators (output: numerical remote maintenance).

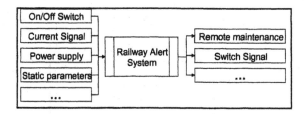

Figure 11: Embedded system to be fail-safe proven

Integral safety engineering is first based on both a technological and a methodological approach. Technology-related approach consists in two redundant actuation, processing and measurement hardware systems coming from different manufacturers. Methodology-related approach consists in independent software co-design by two teams with different tools. Integral safety engineering is secondly based on the proof-design and the safety analysis jointly made by PREDICT and CLEARSY companies in order to improve the first approach towards legacy certification. Main *SIL4* to be proven implies high-level human-safety with regards to train traffic operational availability. Although only parts of models are described because of confidentiality concerns, this section deals with the main steps of the proposed proof-oriented fail-safe engineering experiment.

First *POFTSE* step consists in a formalized description (Leger *et al.*, 1998) of a Domain Process and of its related *ALERTMD-DP* system-part functioning (fig. 12)

$$DP = \left\{ \begin{array}{l} NameDP, De\ fDP, ObjDP, CstDP, D\ RDP, \\ FIDP, CIDP, RIDP, FOD\ P, CODP, R\ ODP, \\ PRSDP, EvD\ P, Employs\ DP, Where-\ UsedDP \end{array} \right\}$$

Name$_{DP}$ = {ALERTMD}
Def$_{DP}$ = {Alert the train driver when a default is detected on the railway}
Obj$_{DP}$ = {Detect problem on the railway, Alert in real-time the train driver, Alert the Office by tele-maintenance}
FI$_{DP}$ = {Current Signal, On/Off Switch}
CI$_{DP}$ = {Parameters}
RI$_{DP}$ = {Power Supply}
FO$_{DP}$ = {Switch Signal}
Employs$_{DP}$ = {MEASUREMENT, TREATMENT, ACTUATION}
Where-Used$_{DP}$ = {RAILWAY SYSTEM}

Figure 12: Formalized ALERTMD-DP functional descriptive model

The related high-level abstract-model (fig. 13) aims then to formally express the interactions between the system and its environment with simpler properties to be checked all along the *POFTSE* life-cycle.

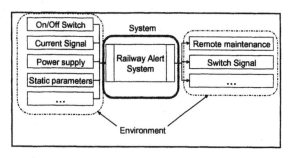

Figure 13: High-abstract B model of the embedded system.

For example, a system-level safety-property to be always (at each time) verified is:

INVARIANT: Switch signal is up
⇒ Railway Alert System is Nominal

meaning that output switch-signal can be up only in nominal mode.

Safety-related architecture-redundancy implies the refinement of this initial abstract model into two parallel processing chains (fig. 14) with result an improvement of the above system-level safety-property:

INVARIANT: Switch signal is up
⇒ Railway Alert System 1 and 2 are Nominal

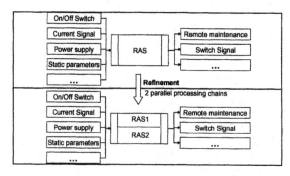

Figure 14: Refined B model of the redundant architecture

Stepwise refinement (fig. 15) allows to gradually add details (components, sequence …), to exhibit synchronisation points (data should be exchanged and be available before executing some action) and to specify required interface (co-design).

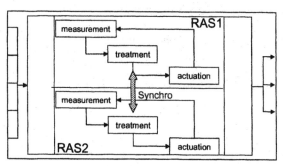

Figure 15: Detailed B model of the refined redundant architecture

Each of the redundant chains is duplicated into a *formal B model*. (fig.16):

SUBSYSTEM RAS1	SUBSYSTEM RAS2
(...)	(...)
EnterActuation =	**EnterActuation =**
SELECT	SELECT
Mode=Nominal &	Mode=Nominal &
State=Treatment	State=Treatment
THEN	THEN
State := Actuation	State := Actuation
END;	END;
ActuateUp =	**ActuateUp =**
SELECT	SELECT
Mode=Nominal &	Mode=Nominal &
State=Actuate &	State=Actuate &
ComputedOrder=Up	ComputedOrder=Up
THEN	THEN
Output:=Up	Output:=Up
END;	END;
(...)	(...)

Figure 16: Formal B prescriptive model of the redundant architecture

Design correctness is automatically verified by proof. For example, exchange of orders allows testing the synchronization correctness between the two chains in nominal mode (fig. 17) and to check the redundant behaviour correctness all along *POFTSE* by adding details on synchronization.

SUBSYSTEM RAS1	SUBSYSTEM RAS2
(...)	(...)
EnterActuation =	**EnterActuation =**
SELECT	SELECT
Mode=Nominal &	Mode=Nominal &
State=Treatment	State=Treatment
THEN	THEN
State := Actuation	State := Actuation
END;	END;
ActuateUp =	**ActuateUp =**
SELECT	SELECT
Mode=Nominal &	Mode=Nominal &
State=Actuate &	State=Actuate &
ComputedOrder=Up &	ComputedOrder=Up &
OrderReceivedFrom	**OrderReceivedFrom**
RAS2=Up	RAS1=Up
THEN	THEN
Output:=Up	Output:=Up
END;	END;
(...)	(...)

Figure 17: Formal checking of the redundant behaviour correctness

Causal relationships between system dysfunctioning and external or internal causes are formalized with predicate calculus ($\wedge$) and S5 modal operators ($\blacksquare$ necessity, of depth n+1, implying that the hypothesis is true all the times before) and consistently with an industrially well-established *FMECA* graphical description. This systematic modelling is made for each level of sub-model –Alertmd (fig.18), Measurement, Treatment, Actuation...- from a software and hardware points of view and formalizes *FMECA* for signal, information and workflow as well as for hardware and software.
FMECA are completed with a consistency *SD* report allowing identifying if both model and sub-models are analysed and if the input/output (signal and data) are studied.

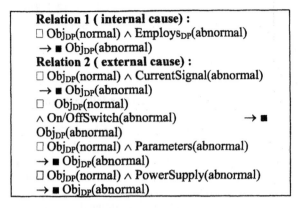

Figure 18: Formalized ALERTMD dysfunctional descriptive model

This formalized knowledge on system malfunctioning allows overloading the event-driven functional B model with events related to faults according to:

<event name> = If <Gard (or condition)> then <Substitution (or action)>

Fault test consists in generating events derived from functional-model events by considering external faults (inputs) or internal faults occurring on conditions (guards) different from initial specification and/or acting differently (substitution). For example, a wrong actuation not under control is checked acting on process availability (fig. 19).

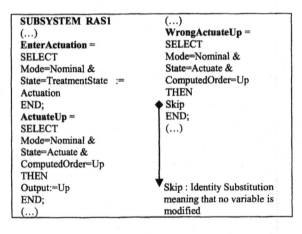

Figure 19: Proven fault actuation processing

Adding *FT* details on *PO* models consists in deriving formalized *DPs* from reference ones before to prove fail-safe properties by playing test scenario.

In order to check the global ALERTMD system malfunctioning model from a hardware/software point of view, *ALERTMD-DP* integrates *DPs* dedicated to Monitoring DP_{moni} and Diagnosis DP_{diag} (fig. 20) in order respectively to monitor PowerSupply and to diagnose latent faults on SwitchSignal.

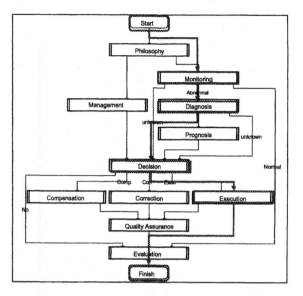

Figure 20: Formalized Fault-Tolerant DP descriptive reference model

PowerSupplyMonitor DP (fig. 21) is implemented with electronic components in order to detect all the variation of the voltage and current of the 24 VDC power supply.

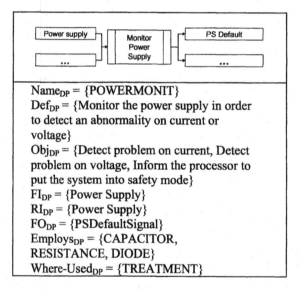

Name$_{DP}$ = {POWERMONIT}
Def$_{DP}$ = {Monitor the power supply in order to detect an abnormality on current or voltage}
Obj$_{DP}$ = {Detect problem on current, Detect problem on voltage, Inform the processor to put the system into safety mode}
FI$_{DP}$ = {Power Supply}
RI$_{DP}$ = {Power Supply}
FO$_{DP}$ = {PSDefaultSignal}
Employs$_{DP}$ = {CAPACITOR, RESISTANCE, DIODE}
Where-Used$_{DP}$ = {TREATMENT}

Figure 21: Formalized Fault-Tolerant PowerSupplyMonitor DP descriptive model

LatentFaultDiagnosis DP (fig. 22) is implemented within software modules in order to diagnose a latent fault on one of the two parallel chains such as the abnormal dynamic activation of the output because of a relay sticking.

These formalized descriptions of DPs are compliant with the triple *{{BM, COMP}, CXT, OBS}* where:

- BM corresponds to the description *{Def$_{DP}$}*, that is to say the gathering of the normal description *{Defn$_{DP}$}* and the abnormal one *{Defabn$_{DP}$}* related to the causality,
- COMP is the structural description of the system *{Employs$_{DP}$}*,

- CXT is an empty set because all the input objects and the input flows of the activity are considered as potential causes,
- OBS represents the observations to be made to implement the diagnosis.

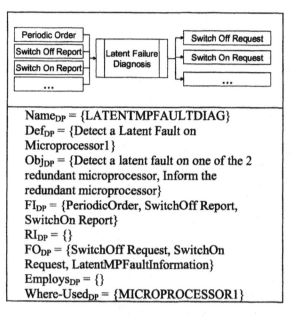

Name$_{DP}$ = {LATENTMPFAULTDIAG}
Def$_{DP}$ = {Detect a Latent Fault on Microprocessor1}
Obj$_{DP}$ = {Detect a latent fault on one of the 2 redundant microprocessor, Inform the redundant microprocessor}
FI$_{DP}$ = {PeriodicOrder, SwitchOff Report, SwitchOn Report}
RI$_{DP}$ = {}
FO$_{DP}$ = {SwitchOff Request, SwitchOn Request, LatentMPFaultInformation}
Employs$_{DP}$ = {}
Where-Used$_{DP}$ = {MICROPROCESSOR1}

Figure 22: Formalized Fault-Tolerant LatentFaultDiagnosis DP descriptive model

This *POFTSE* iterative process between *POSE* and *FTSE* should run until all the required fail-safe properties are proven by:

- adapting the functional model properties in order to take into account the introduction of degradations,
- specifying a new control enabling to verify the safe behaviour of the system.

Despite the efficiency of the above *POFTSE* experimentation to assist the *SIL4* certification process of the studied embedded system, this approach combining formalized descriptive models and formal prescriptive ones remains hardly pragmatic.

For example, the consistency between the malfunctioning of ALERTMD system-model and sub-models (Measurement, Treatment, Actuation...) is only related to the input/output coherence between level structuring FIDP, CI$_{DP}$, RI$_{DP}$, FO$_{DP}$. Relationships between malfunctioning model and malfunctioning sub-models should be formalized to check the global ALERTMD system malfunctioning model according to:

Alertmd(abnormal) = f[Measurement(abnormal), Treatment(abnormal), Actuation(abnormal)]

Also, overloading *POSE* by *FTSE* is not automated and consequently the fail-safe resulting models not a priori proven. Next efforts should be made to ensure the automatic translation between descriptive/prescriptive models, to discharge the proof assistant by identifying pertinent FT testscenarii, to improve the reproducibility of the results of the iterative design process, ..., in order formally proof the fail-safe behaviour of a system..

7. CONCLUSION AND OPEN ISSUES

Proof Oriented System Engineering (POSE) should ideally refine the atomic relationship defining the customer's requirements of a product (good and service), from which emerges the *system (model)* throughout an engineering life-cycle (fig. 23).

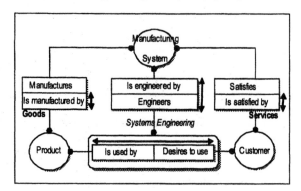

Figure 23: 'Emergence' of a manufacturing system from an atomic Product/Customer interaction (Lavigne, *et al.*, 2003).

In practice, *POSE* consists currently in satisfying a set of relationships ensuring the checking of a set of correct *sub-systems (models)* leading in fine to a consistent *ad-hoc system (model)*.

Among many key elements in making *POSE* practicable are the combination of formal *POSD* approaches such as B methodology using theorem-proving constructs and set-theory notation with formalized technical-oriented approaches such as *FTSE* in order to check from the early phase of a project life-cycle the completeness, the unambiguousness, the consistency and the correctness between all the various requirements and specifications of different functions of the system (Shell, 2001). These prescriptive approaches focusing on the system and on proper and comprehensive understanding of its goals are the critical link in bridging the gap with the UML-based normative descriptive approaches (CMM) that focus primarily on the project and the qualitative and quantitative management of the engineer's activities.

Although much works remain to be developed (Papadopoulos, in Pereira *et al.*, 2004) to make such formal *POSE* practicable to complex systems, time is ripe, as addressed by (Jonhson, in Pereira *et al.*, 2004), to improve automation dependability by using formal methods in order to check the increasing impact of infotronics technologies.

This *POFTSE* should challenge both Computing Science and Automatic Control R&D communities for practical experiments as presented in this paper.

REFERENCES

Abrial J.R. (1996), Assigning Programs to Meanings, *The B book,* Cambridge University Press.

Abrial J.R. (2000), Event driven sequential program construction., *Internal Note Version 17,* Consultant

Abrial J.R. (2001), Event driven distributed program construction, *Internal Note Version 5,* Consultant.

Abrial J.R., Cansell D., Méry D. (2002), A mechanically proved and incremental development of the IEEE 1394 Tree Identify Protocol, *Formal Aspects for Computing,* Digital Equipment Corporation.

Back R.J.R. (1979), On correct refinement of programs, *Journal of Computer and System Sciences,* **23/1,** 49-68.

Boland D., Groumpos P.P., Jagdev H.S., Karcanias N. (Eds) (2001), Technolgical roadmap, *Network Of Excellence in Intelligent Control and Integrated Manufacturing Systems,* ESPRIT Project 23447.

Cansell D., Gopalakrishnan G., Jones M., Méry D., Weinzoepflen A. (2002), Incremental Proof of the Producer/Consumer Property for the PCI Protocol, *Lectures Notes in Computer Science,* **2272,** 22-41.

Cansell D., Tanougast C., Berviller Y., Méry D., Proch C., Rabah H., Weber S. (2003), Proof-based design of a microelectronic architecture for MPEG-2 bit-rate measurement, *Forum on specification and Design Languages,* September, Frankfurt, Germany.

Cassandras C.G., Lafortune S., (1999), Introduction to discrete event systems, *Book,* Kluwer Academic Publishers.

Clarke E.M., Grunberg O., Peled D.A. (2000), Model Checking, *Book,* MIT Press.

Dijkstra E.W. (1976), A Discipline of Programming, *Prentice Hall International Series in Computer Science.*

Féliot C. (1997), Complex Systems Modelling: Models Integration and Formalisation, *PhDThesis,* University of Lille 1, 1997 (in French, at LAIL).

Fusaoka A., Seki H., Takahashi K. (1983), A description and reasoning of plant controllers in temporal logic, *Proceedings of the International Joint Conference on Artificial Intelligence,* 405-408, August 8-12, Karlsruhe, Germany.

Lamboley P. (2001), Production Systems Automation Formal Method Proposal, *PhDThesis,* University Henri Poincaré, Nancy 1, 2001 (in French, at CRAN).

Lamport L. (1994), A temporal logic of actions, *Transactions on Programming Language and Systems,* **16/3,** 872–923.

Lavigne J.P., Mayer F., Lhoste P. (2003), Category Theory Based Approach for IMS Modelling, *Proceedings of the 7th IFAC workshop on Intelligent Manufacturing Systems,* April 6-8, Budapest, Hungary.

Leger J-B., Iung B. (1998), Methodological approach to modelling of degradation detection and failure diagnosis in complex production systems, *Proceedings of the 9th international workshop on principles of Diagnosis*, NASA Ames Research Center (eds.), 209-216, USA.

Leger J.B., Morel G. (2001), Integration of maintenance in the enterprise towards an enterprise modelling-based framework compliant with proactive maintenance strategy, *Production Planning & Control*, **12/2**, 176-187.

Lhote F., Chazelet Ph. and Dulmet M., (1999), The extension of principles of cybernetics towards engineering and manufacturing, *Annual Reviews in Control*, **23/1**, 139-148.

Liu Y., Ye L., Iung B., Liu X., Cheng Y., Morel G., Fu C. (2003), Economic performance evaluation method for hydroelectric generating Unit, *Journal of Energy Conversion and Management*, **44/6**, 797-808.

Moik A. (2003), Engineering-related formal methods for the development of safe industrial automation systems, *Automation Technology in Practice*, **1**, 45-53.

Morel G., Pétin J.F., Lamboley P. (2001a), Formal specification for manufacturing systems automation, *Procceedings of the 10th IFAC symposium on INformation COntrol problems in Manufacturing*, September 20-22, Vienna, Austria.

Morel G., Iung B., Suhner M., Leger J-B. (2001b), Maintenance holistique framework for optimising the cost/availability compromise of manufacturing systems, *Proceedings of the 6th IFAC symposium on Low Cost Automation*, October 8-9, Berlin, Germany.

Morgan C. (1990), Programming from Specifications, *Prentice Hall International Series in Computer Science*.

Panetto H., Pétin J.F. (2003), Setting up UML stereotypes for production systems modelling, *Proceedings of the ISPE conference on Concurrent Engineering*, 1, 747-754, July 26-30, Madeira, Portugal.

Paynter M. (1961), Analysis and design of engineering systems, *M.I.T. Press*, Cambridge.

Pereira C.E., Morel G., Kopacek P. (Eds.) (2004), *Proceedings of the 11th IFAC symposium on INformation COntrol in Manufacturing*, April 5-7, Salvador da Bahia, Brazil.

Petin J.F., Morel G., Méry D., Lamboley P. (1998), Process control engineering: contribution to a formal structuring framework with the B method, *Lectures notes in Computer Science*, **1393**, 198-209.

Shell T. (2001). Systems functions implementation and behavioural modelling: system theoretic approach, *International Journal of Systems Engineering*, **4/1**.

HIGH TECHNOLOGY IN LOW COST MANUFACTURING - WITH SPECIAL EMPHASIS ON SME'S

Peter Kopacek

Institute of Handling Devices and Robotics
Vienna University of Technology
Favoritenstr. 9–11, A–1040 Vienna
Tel: +43-1- 58801 31801, FAX: +43-1- 58801 31899
e-mail: kopacek@ihrt.tuwien.ac.at

Abstract: AI methods were introduced more and more in Computer Integrated Manufacturing (CIM) Systems. This results in intelligent CIM components - ICAD, ICAP, ICAM, ICAQ - and intelligent CIM systems (ICIM) or in Intelligent Manufacturing Systems (IMS). This new philosophy requires a lot of prerequisites and research. „ICIM" or „IMS" is partially introduced in the industry but mainly for large companies. AI in form of knowledge based and expert systems is ready to be introduced in an efficient way in CIM components. The implementation of AI methods in „low cost" CIM systems depends from the availability of AI software packages for this hardware configuration. In this paper a short outline will be given and some examples for low cost, industrial solutions will be discussed. *Copyright © 2004 IFAC*

Keywords: Computer Aided Manufacturing, Intelligent Manufacturing, "Low-Cost" Automation, Intelligent CIM, Robots, Assembly, Disassembly

1. INTRODUCTION

The industry in some European countries – especially the new members of the EC - is and will be dominated by „Small and Medium Sized Companies – SME's" - mostly defined as companies with less than 500 employees. Under the influence of a future common market these companies have special demands to automation technologies. For such companies automation could be a prerequisite to survive in a common European market. For this market highest flexibility is necessary. One possibility was 25 years ago „Computer integrated manufacturing – CIM". The main idea beyond was a fully computer controlled production including all the necessary auxiliary tasks. In the last time, methods of artificial intelligence are more and more introduced in technical applications. This yields to „intelligent" automation – in the special case of production systems to „intelligent manufacturing systems - IMS".

2. CIM FOR SME'S

Especially SME's had a lot of difficulties to introduce CIM or parts of CIM. Today only in few European SME's a total CIM concept is realized. In most cases only some CIM components (CAD, CAP, CAM, CAQ, PPS) or parts of these are realized. European and especially Austrian SME's need more time to get familiar with the CIM philosophy and to introduce more components or parts.

Software packages for the CIM components are available commercially from various sellers. Usually these packages are only partially suitable for SME's. They offer a lot of features, often not necessary for the demands of SME's – on the other hand important features are missing. Adding some of these takes a long time and it is usually very costly and time consuming. Furthermore most of these packages require a cost intensive computer hardware.

Especially for the demands of SME's a flexible, modular „low cost" CIM concept was developed 20 years ago. From the side of hardware, the basic philosophy was to use PC's connected by a local area network (LAN) with a host computer for database tasks.

The modular "low cost" CIM concept is shown in Fig.1. It uses two types of computers: A UNIX

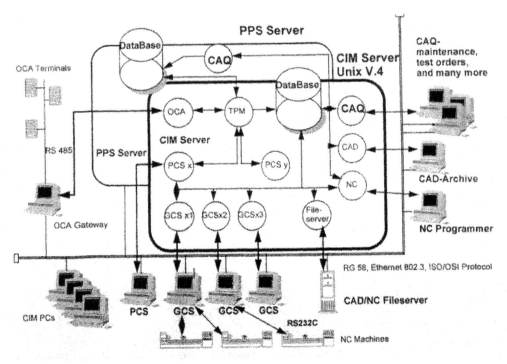

Fig.1.   "Low cost" CIM concept (Kopacek, 1995)

machine serves as a database and network server, various PC's serving as network stations with different tasks. For OCA (Operating Characteristics Aquisition) special terminals are used. The database-server and all workstations are connected by a local area network (Ethernet). A second network connects the OCA-terminals with the OCA-server, which works as an intelligent gateway between the two networks.

The various control systems recognize on their own, if the connection to the host server is possible and/or succeeded. They build up the connections self-acting and adjust the data between the different databases automatically.

Each workstation has its own server process running on the network server. These server processes handle the communication between workstations, processes and databases. The software is completely modular. "Modules" could be commercially available or individually written according to specific demands of the user. These modules can be combined with a minimum of interface problems – as usual in CIM systems today. The programming language is C or C++.

This concept has following advantages:

- low cost
- possibility of stepwise realization
- easy combination of software modules for a specific solution

- possibility to include AI methods in some modules

3. INTELLIGENT CIM

Artificial Intelligence (AI) has been an extensive research field for some decades already. The first knowledge based or expert systems were available in the late seventies – especially for narrow fields in medicine and not for technical purposes. One reason being the computer hardware as AI methods require very high computing power. In recent years the computer capacity has increased dramatically – especially those from PC's – and therefore some methods of Artificial Intelligence can be applied in an efficient way for real industrial processes. For such purposes, programming languages and software tools are available for "low cost" industrial applications.

Therefore, AI methods are introduced more and more in production automation or CIM systems. This leads to intelligent CIM components – ICAD, ICAP, ICAM, ICAQ – and to intelligent CIM systems (ICIM) or intelligent manufacturing systems (IMS).

The implementation of AI methods in "low cost" CIM systems as knowledge based and expert systems depends on the availability of AI software packages for a distinct hardware configuration.

Expert or knowledge based systems are used in CIM systems for the following areas:

- Design

26

- Manufacturing
- Process planning, shop planning, layout planning
- Production planning and scheduling:
  PPS systems, shop control, scheduling, logistics
- Operation:
  assembly, maintenance
- Quality assurance

4. IMS FOR SME'S

Relevant influences for further developments in production automation are:

- local area networks (LAN's)
- new computer architectures
- transferable real-time languages
- hierarchical structures
- modular design of hardware and software
- "intelligent" (embedded) sensors
- CIM concepts and components

SME's in Austria as well as in Europe had some problems to introduce CIM or some parts of CIM in their production.

The main reasons were:

- less experiences
- restricted investment possibilities
- no market forecast for a longer time
- no specialists
- no company-own R&D departments
- employees are not familiar with new technologies

At begin of the 21th century these companies – just confronted with CIM – should be attracted to introduce the next higher step – IMS. The IMS test cases and its results were definitely not appropriate for immediate use in SME's. This knowledge had to be selected, concentrated and making applicable for such companies.

5. EXAMPLES

In the following some examples for introducing AI methods on a very low and cheap level in SME's will be shortly described and discussed. These solutions were developed with and are realised in companies.

5.1. Selection of components for robotized assembly and disassembly cells (Kopacek et al., 1992)

One important component of flexible (dis-)assembly systems is a robotized (dis-)assembly cell. It could be a stand alone unit with one or two robots, or it can be integrated in an assembly line. The task of the planner is to find the optimal configuration of the robotized assembly cell. The assembly robot, grippers, assembly tools, and auxiliary components have to be selected and placed in the robots working

area. Accomplishing this task is time consuming, and special knowledge as well as experience is necessary. Several iterations are often necessary in order to find an optimal solution. Therefore computer aided systems for planning of assembly systems have been developed to support the planner in this time-consuming job and finding the optimal solution. It is known, that the development of such systems – including the building of the integrated data base, clear definitions of planning goals and planning steps, choosing the appropriate programming tools and environment is a complex task.

The computer aided planning systems ROBPLAN and LASIMCO (Kopacek, Noe, 2001) are based on the supposition, that an assembly cell is built for the set of assembly operations and their sequencing and that the assembly cell is a compound of modules such as the robot, clamping system, assembly tools, feeding components, transport system, and others. Determination of assembly operations and their sequencing, component selecting and layout of system with evaluation of assembly time and system costs in variants are the tasks of the planning process.

One of the main parts of the system was creating the appropriate component data base and component selecting procedures. Within the boundary conditions imposed by the customer, such as product to be assembled, assembly operations sequences, workspace size and shape, the component properties, the selection of components must be optimised with respect to a minimum assembly cell cost or minimizing time. Based on these conditions and the particular point of view, a group of components, that satisfy proposed constrains, can be selected. To catch a final optimal solution, two selecting approaches can be used: a multiple attribute decisions making method and developing the selecting systems where expert systems technology is used. Both forced us to build the appropriate component database and to determinate the selecting criteria.

5.2. An expert system for fault detection in assembly cells (Beneder, Kopacek, 1994)

A medium sized company typical for Austria is producing welding transformers. In cooperation with University institutes a CIM concept and an automatic assembly cell were developed. For the improvement of error handling in this work cell a diagnosis system was provided additionally.

For automation of the assembly process of six different types of welding transformers a robotized work cell consisting of two robots, tools, parts and tools storage devices, a screen printing machine and a transportation system was developed. Due to the market conditions smaller lots and therefore more variations of the six basic types are necessary. Today approximately 1300 variations with an average lot size of 10 are produced in the cell.

The introduced assembly plant can be supervised by a single person, who is responsible for the refill of the part magazines and for the handling of arising errors.

A system for diagnosis of errors in the plant was provided in addition of the control programme. Since the diagnosis programme is a separate tool, the actual state of the work cell is written to temporary files for information transfer.

In case of machine failure the error detection module can be called by a function key. For maximum security, leaving the control programme causes both robots to stop in their momentous position. In that case the control programme stops, all actual parameters are stored and the error diagnosis programme starts. In the diagnosis session the errors are identified and corrected by the user in interaction with the system.

The next step following the use of the diagnosis programme is the elimination of the error reason by selecting suitable actions out of the menus and by manual intervention (removing defective parts, replacing screws, refilling the magazins, ...). When the production process is restarted, the control programme is continued at the same point it was stopped.

5.3. A consulting system for production planning and manufacturing control (Kopacek et al., 1990)

A consulting system was developed which is suitable for the computer-aided operative production planning and manufacturing control of process consisting for the most part of workshop manufacturing processes (discrete manufacturing sections in a well-defined sequence at different work stations).

In contrast to a number of other methods developed in recent years in case of which job release takes place after a load analysis, the consulting system described above is based on an OPT conception (optimized production technology) in which the actually stochastic character of the manufacturing process is decisively determined by exact planning and the consideration of all important disturbances that can be predicted. The core of the simulation model is a time table in which the utilization of the installations is entered with a fixed cycle time. The job arrangement is carried out in accordance with the respective technique in consideration of the side conditions mentioned above. Thus, a partial solution of the optimization problem will be approached for a gliding time horizon which is smaller than the planning horizon.

Although the production plan on the whole is not optimal, it represents a satisfactory solution which had not been obtained with the former planning methods.

5.4. Disassembling of minidisks

Sony DADC Austria is one of the largest producers of optical storage units – Mini Discs as their main product. As it is typical for every industrial production, some of the produced Mini Discs do not satisfy the desired high quality standard. Due to the rising waste disposal costs and the high costs of human work, an automatic recycling of Mini Discs was the key aspect of this project. There are two different types of Mini Discs in production: a playback-only and a recordable.

An assembled playback-only Mini Disc consists of the following parts- Upper Cartridge, Label, Disk, Clamping Plate, Shutter Lock, Shutter, and Lower Cartridge. The recordable one has a Shutter on both sides.

The clamping plate is made of a special type of magnetic steel - the Label is made of paper. The Upper and Lower Cartridge as well as the Disc consist of Polycarbonat, the Shutter Lock and the Shutter of Ployoxymethylen.

The disassembly cell consists of two main components: a feeding system and the disassembly system itself. Furthermore there is the cell control unit, a transportation unit between the two components, and sensors to control the operation.

The feeding system takes the Mini Discs from a container. A recognition or inspection of the particular Mini Disc is not necessary, because every disk is the same and there is no wear, pollution or damage. The transportation system is equipped with sensors to orientate and align the Mini Discs.

Afterwards the Mini Disc is taken to the disassembly system. It will be fixed and cracked with wedges from the side separating the Upper and the Lower Cartridge. A vacuum gripper picks the Upper Cartridge and puts it into the special container. An optical sensor controls that the Upper Cartridge reaches its destination. At the next stations of the system the Clamping Plate, the Shutter Lock, and finally the Shutter are removed with special tools. The proper courses of these operations are also controlled by sensors. Every part is given into a specific container, where they are stored for further processes.

5.5. Disassembly cell for printed circuit boards (Kopacek, 2001; Drinek et al. 2001)

On old or new PCBs there are several re-useable parts. These chips can be soldered in old or new technique or socketed. The task was to develop a semi automatized disassembly cell for both kinds of chips.

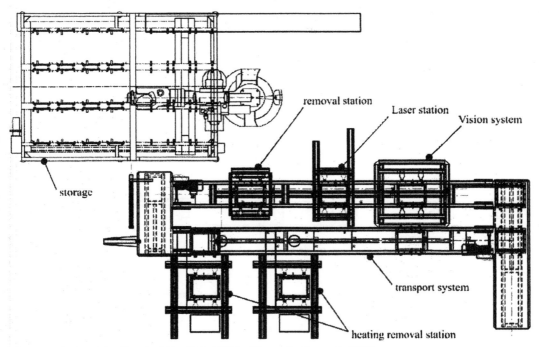

Fig. 2. Layout of the disassembly cell for PCB's (Kopacek, Kopacek, 2002)

The layout of the cell is shown in Fig. 2. The basis of a disassembly cell is a very stiff frame construction developed from commercially available profiles. In a manual feeding station the PCBs with a maximum size of 300 x 220 mm are attached on special work holding device. The disassembly cell consists of 4 stations:

- Vision system
- Laser desoldering system
- Removal station
- Heating removal station

The vision system has several tasks. It has to recognize the re-useable parts by means of a data base containing the data (kind, production company, assigned, dimensions). The vision system has to detect the re-useable part and to determine the position, the size and the centre of inertia. Furthermore it has to classify the useable parts to be desoldered or removed from sockets.

The desoldering station consists of a cross table – two linear axes – controlled to reach every point (centre of inertia) on the PCB. The desoldering process is carried out by laser technology. The desoldered parts are put on a distinct area outside the laser from which they are removed by the industrial robot and to put into the appropriate magazines.

The third station is the removal station for socket parts. An industrial robot equipped with special grippers as well as external sensors carries out process. The robot removes these parts and puts them

also in the right magazines. A prototype of this disassembly cell is now in the test phase.

5.6. A semiautomatized Disassembly Cell for Mobile Phones (Kopacek, Kopacek, 2003)

After a detailed analysis of used mobile phones concerning the parts as well as the assembly technology and tests for disassembly with the most frequent mobile phones the following concept for the disassembly cell was created (Fig. 3.). It consists on five automated stations plus a manual feeding and removal station:

- Feeding and removal station
- Drilling and milling station
- Removal station for the covers
- Drilling station
- Circuit board removal station
- Drilling station

For disassembly of the mobile phones they were fixed on a pallet in a distinct position. These pallets are moving around on a transportation system. According to the necessary disassembly operations the pallets with the mobile phones to be disassembled are stopped, lifted and fitted in a distinct station.

Before the mobile phone is fixed on a pallet the power supply will be removed and the type of the handy will be recognized by a barcode reader manually. Now the control computer knows exactly the type of the handy. The main dimensions of the handy are stored in a database of the host computer.

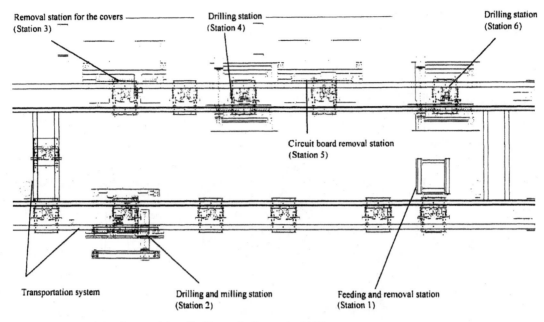

Fig. 3.   Layout of the disassembly cell. (Kopacek, Kopacek, 2003)

In the drilling and milling station (no. 2) the upper part of the handy will be cut off from the lower part and the screws – usually between 4 and 17 – are removed by a simple drilling mechanism. The dust content is removed by air from the pallet.

In the third station – the cover removal station – the cover as well as the keyboard of the handy will be removed by pneumatic sucks. These two parts are separated in a storage device. In the next station – drilling station; no. 4 – the screws which connect the printed circuit board on the lower part of the housing are removed. In the printed circuit removal station various other parts will be removed from the handy and separated in special storage devices. Because some mobile phones have additional parts connected with the power part of the housing of the handy the remaining screws will be removed in the last drilling station – station 6. Finally the lower part of the handy will be removed in the fixing and removal station.

As a development of this semi-automated disassembly cell for used mobile phones some previous tests were necessary. For the milling in the drilling and milling station (no. 2) it was necessary to make tests with grinding wheels, with different saws and with milling devices. Finally a milling device was chosen as the right tool for this task.

Further extensive tests were carried out for the removal of the screws. From the literature there are very high sophisticated, complicated and therefore very expensive and heavy devices known. We found a very simple and very cheap method for the removal of the screws.

5.7. A 'Tool Kit' for mobile robots

The basis of a modular concept for mobile robots is the Mobile Robot Platform (MRP) which can be described as a multi-use mobile robot, developed in its basic configuration.

These platforms can be divided in some basic systems:

- Locomotion system
- Driving system
- Main control system
- Communication system

The locomotion system can be realized on different principles, like wheel-, chain-, walking- or special-locomotion. All these principles have different characteristics in regard to costs, weight and efficiency in various terrains.

The driving system consists of: power sources, actuators and transmissions, which enable the platform to perform the necessary movements.

The main control system is usually microprocessor based and responsible for the actions executed by the robot. It should be powerful enough to meet the requirements of all current tasks and possible future tasks.

The navigation system normally consists of an array of, possibly different, sensors for the perception of the robots environment and of course of some kind of software which designates the rules for the robot-movement.

The communication system connects all the systems in the platform and the platform with the environment.

Modular Mobile Robot is intelligent (low, medium or high degree of intelligence) semi or fully autonomous vehicle (wheel, legged, chain, crawling, climbing or special locomotion) with all its systems (locomotion, driving, control, navigation and communication) build on a modular principle, able to carry peripheral systems (robot arms etc.) or tools (conventional or special) for transporting of loads or executing different industrial or service operations in world coordinates.

The mobile robot platform has to be equipped by a mechanical or other system able to manipulate tools or grippers and operate with them. Simple arms, lifts, fork mechanisms etc. are needed to perform some simple tasks. Though these tasks could be performed by a dexterous powerful arm as well, it would be simply too expensive.

Grippers are designed to imitate the human hand and operations which are fulfilled by it. It is highly complicated to imitate the complex motion sequences of a human hand. Therefore most of the grippers are only a simplified copy with less DOF.

The mobile robot platform can be upgraded and modified by adding a number of peripheral systems and tools for the performance of different tasks or functions. There is a large variety of tools, which can be used.

Basically these tools can be divided into two major categories:

- Conventional tools
- Special tools

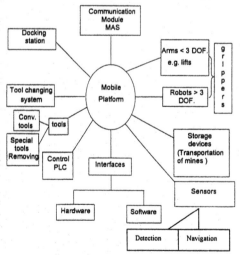

Fig. 4.   Modular Robot System (Shivarov, 2001)

Conventional tools (screw drivers, drilling tools, polishing tools, etc) are similar in regard to their

function to conventional hand-held tools for manual operations. The difference is in their design, since they have to be fixed on the mobile robot platform, and actuation.

Special tools installed onboard of a mobile robot platform changes the same to a specialized mobile robot system. When special tools are lightweight constructed the manipulation system can be more flexible and with wider reach. Heavier tools cannot be very flexible. They need more rigid and strong manipulation systems. So there is often only one degree of freedom applied, and the other DOFs are realized by the mobility of the platform.

Installing a tool changing system enables the robot to achieve a wide variety of performable operations. Tool changing systems are normally placed at the end of a robot arm. They have to be light, simple and very reliable.

The basic configuration of each mobile robot platform has its integrated sensors. The navigation system makes excessive use of sensor for position determination and collision avoidance. But there are numerous possibilities to upgrade the system with additional sensors for some special applications or to extend its abilities.

In many mobile robot applications transportation is an important part of the overall task. To transport different items mobile robot platforms have to be upgraded with another type of peripheral devices: special storage systems or devices.

Although mobile robot platforms are normally equipped with a communication system it could be necessary to use some special communication systems. Especially in multi agent systems (MAS) where a team of robots acts together is communication between the team members of importance.

This system was applied in edutainment (robot-soccer) and will be used the for humanitarian demining actions.

6. CONCLUSION

SME's have to use more and more "High Technology – Low cost" manufacturing systems. Manufacturing automation guaranties flexibility necessary to survive in a common, global market.

Starting from a classical, low cost CIM concept some realized „mosaic stones" developed and introduced in SME's are shortly described and discussed.

To develop cost oriented solutions for SME's the headlines are modularity and flexibility as well as "buy and adapt" and compensate hardware inaccuracies by software. Furthermore we try to use new and cheap technologies. If there are commercial solutions available for some of the presented examples the cost reduction is approximately 50 to 70%.

Currently and in the future "IMS" methods are and will be introduced more frequently in SME's. Further projects are going on in the field of "IMS" and we have to look on the results to make these applicable for SME's.

REFERENCES

Beneder M., P. Kopacek (1994). An Expert System for Fault Detection in Assembly Cells. In: *Proceedings of the First World Automation Congress - WAC'94*, 14.-17.8.1994, Maui, Hawaii, Vol. 1, p. 19-22.

Gschwendtner, G.; Kopacek, P.; Prucker, K.: Robotized Disassembling of MiniDiscs. In: *Proceedings of the 26th International Symposium on Industrial Robots ISIR'95*, October 4-6, 1995, Singapore, p. 463-468.

Drinek, M., K. Daichendt and H. Zebedin (2001): A new Concept for a fully automatized Disassembly Cell. *Proceedings of the 10th International Workshop "Robotics in Alpe Adria Danube Region – RAAD'01"*, Vienna, May 16. – 18. 2001, p.383-388.

Kopacek,B. and P.Kopacek (2002): A robotized disassembly cell for printed circuit boards. In: *Proceedings of the 33rd International Symposium on Robotics*, October 7–11, 2002, Stockholm, Schweden, p. 641 – 643.

Kopacek, P., P. Otto, and J. Wernstedt (1990). A Consulting System for the Computer-Aided Operative Production Planning and Manufacturing Control. In: *Proceedings of the IFAC Symposium on Skill Based Automated Production – SBAP'89*, Vienna, Pergamon Press, p.167-170.

Kopacek, P., R. Probst, D. Noe, R. Stangl (1993). CAP-AC: Computer Aided Planning of Assembly Cells. In: *Proceedings of the 12th IFAC World Congress*, Sydney, Australia, Vol.10, pp. 257-260.

Kopacek, P. (1995). CIM for Small and Medium Sized Companies. In: *Preprints of the 4th Symposium on Low Cost Automation – LCA'95*, 13.-15.9.1995, Buenos Aires, Argentinia, p.188-191.

Kopacek, P., B. Kopacek and H. Zebedin (2000): Hierarchical Control of Disassembly Cells, In: *Preprints of 4th IFAC Symposium on „Intelligent Components and Instruments for Control Applications – SICICA'00"*; Buenos Aires; September 2000, pp. 265-272.

Kopacek,P. and D.Noe (2001): Intelligent, flexible assembly and disassembly. In: *Proceedings of the IFAC Workshop „Intelligent Assembly and Disassembly – IAD'01"*, p.1–10.

Kopacek,P. and B.Kopacek (2003): Robotized disassembly of mobile phones. In: *Preprints of the IFAC Workshop „Intelligent assembly and disassembly – IAD'03"*, Bucharest, p.142-144.

Shivarov, N. (2001): A 'Tool Kit' for modular, intelligent, mobile robots, PhD. Thesis, Vienna University of Technology, 2001.

ELSEVIER
IFAC
PUBLICATIONS
www.elsevier.com/locate/ifac

UBIQUITOUS COMPUTING AND NEW FRONTIERS OF AUTOMATION

Wilhelm Bruns

Bremen University, Germany
Research Center for Work, Environment and Technology (artec)
bruns@artec.uni-bremen.de

Abstract: *Ubiquitous Computing*, a vision of invisible computing integrated in our everyday surroundings, first introduced by M. Weiser and his group at Xerox PARC in 1988, is still in its early infancy and far from leaving the laboratory stage. Nevertheless there are foreseeable applications in specific areas like automotive automation, health care, home automation, advanced manufacturing. Some aspects of *Ubiquitous Computing* from an automatic control perspective and its relation to mixed reality, augmented reality and pervasive computing will be covered. Several problems like geometric representation, extensible computing, scalability, movable interactions, integration of various network-technologies, user interface design for multi-modality, design methodologies and evaluation techniques, security/privacy issues and enabling software concepts are touched and it will be speculated about how Ubiquitous Computing might influence Low Cost Advanced Manufacturing and how experiences from the automation and control field might influence the emerging community. *Copyright © 2004 IFAC*

Keywords: ubiquitous computing, intelligent environments, invisible computing, pervasive computing

Endlich sind alle Glühbirnen auf Erden miteinander verbunden.
Die Vorstellungswelt des Paranoikers, seit jeher beherrscht von universeller Vernetzung,
liegt uns nun vor als Spielfeld.
Seine Software erwies sich als die vorteilhafteste Investition in die Zukunft
Nun ist seine Welt nicht mehr vorstellbar
Und wir alle befinden uns auf der Suche nach einer neuen Irrealität
Der Geist arbeitet auf der Höhe seines Könnensbewusstseins
Halb Chip, halb Tiefe

Botho Strauß (1999)

Finally all electric bulbs are connected.
The imagination of the paranoid, at all times governed by universal networking,
now lies in front of us as a playground
His software proved to be the most advantageous investment into the future.
Now his world is not imaginable any more
and we all find ourselves on a search for a new unreality.
The mind is working on the summit of its consciousness of ability
Half chip - half depth
(Translation by WB)

1. INTRODUCTION

The above text of a famous dramatist of German language has been used as an entrance to a theatre performance *Theatre of the Machines* presented by a Bremen project of computer science students at the International Theatre Festival in Stuttgart-Germany. The main focus was on the question: who is controlling whom in a networked world of computers, marionettes, avatars, robots, sensors and a musician (Fig. 1).

Ubiquitous Computing or computing presence everywhere and at the same time, is an omnipotent vision, not completely unknown to control engineers and scientists of automation, animating the lifeless. But knowing some of the organisers and participants, I hope this conference of the IFAC, to be in the tradition of N. Wiener's concern in "Cybernetics: or Control and Communication in the Animal and the Machine" (p. 28), writing about the first industrial revolution as the devaluation of the human arm and the second industrial revolution as the devaluation of the brain: "However, taking the second revolution as accomplished, the average human being of mediocre attainments or less has nothing to sell that is worth anyone's money to buy. The answer, of course, is to have a society based on human values other than buying or selling." Therefore beyond the title of this conference "Cost oriented Automation" I am also looking for aspects of human-centred-ness in ubiquitous computing.

Some motivations of Mark Weiser, when he introduced UbiComp, obviously were influenced by the anthropological studies of work life of Lucy Suchman (1985) and Jean Lave (1991) who "teach us that people primarily work in a world of shared situations and unexamined technological skills. However the computer today is isolated and isolating from the overall situation, and fails to get out of the way of work." (Weiser, 1993). Ideas of the "intimate computer" (Kay 1991), "rather like a human assistant" (Tesler 1991) or "Rechner im Rücken" (Bruns 1993) were all around and found some condensed presentation in a Communication of the ACM Volume "Back to the real World".

Fig. 1. Theatre of the Machines: A struggle between Man, Machines and Nature. Performance at the Stuttgart Youth Theatre Festival 2000. (Robot-Man controlling virtual reality and a marionette or vice versa?)

There were mainly three approaches to face this re-discovery of the real reality connected to the virtual reality from a sensory and control theoretical point of view:

- sensors and actuators being integrated into real objects, recognising the intention of the user and the environment (reaction model),
- sensors and actuators being at the fingertips of the user, manipulating the real world by grasping and signing, recognised by the computer (action model),
- independent third perspective of sensing and manipulating the world.

Embedding computation in physical artefacts and spreading them throughout our environment was the first approach. A behaviour construction kit (Resnick, 1993) allowed to build models out of computerized LEGO pieces with electronic sensors which could be programmed using LEGO/Logo, able to interact with users or physical objects on a low cost level. This sensorisation of physical objects required the implementation of a model of how to react on changes, these bricks experience from the user. A different approach was to use the hands of a user, equipped with sensors, and implement a model about how to change the computational repre-sentation of the environment. Roughly speaking, in the first approach, the computer has information about tracked physical objects and a model of the user to relate changes of the physical world to actions and intentions of the user. In the second approach the computer has information about the tracked hands and body of a user and a world model to relate them to changes of its internal world representation. Two reasons for us to follow the second approach were *costs* and *controllability*. In certain application areas like modelling a production plant or a control circuit it is less expensive to sense the modellers hand than to sense all objects (figure 2-3). Furthermore, a user wearing sensors has control over them, she can turn them on or off, as opposed to a sensory world, in which she has no influence about being tracked. Of course, the most powerful method but also most expensive way is to combine these approaches and there the latter problem is in again.

Whereas these early days of *Ubicomp* were concentrated on a more receptive view of the world, looking for new input-output-devices and how to handle **information**, later years more and more brought up the idea of an activated world with small invisible devices, sensing **and acting** into the world around us. Therefore we may see ubicomp in a narrow way as ubiquitous computing for information services (the sensory coffee-cub or refrigerator to inform about the state of real objects) or in a more broader sense as ubiquitous sensing and acting, as pervading automation. Or, as Mitchell (2003) proposes, the "trial separation" of bits and atoms is over. With increasing frequency, events in physical space reflect events in cyberspace, and vice versa. This latter view is the challenge for control people. Beside science fiction authors speculating about material compiler and matter transmitter, like Neal Stephenson (1995) with his Young Lady's illustrated Primer, a serious computer-graphics pioneer, E. Sutherland, early mentioned a vision of the ultimate display, a device able to project computer internal information into the real world like a bullet, hitting careless people. "The ultimate display would, of course, be a room within which the computer can control the existence of matter. A chair displayed in such a room would be good enough to sit in. Handcuffs displayed in such a room would be confining, and a bullet displayed in such a room would be fatal. With appropriate programming such display could literally be the Wonderland into which Alice walked." (Sutherland 1965, p. 508). However, even this far reaching view is still a display-view, a bullet is shot and then let alone. From a control perspective, closed loops are aimed at.

Therefore, let me only shortly cover some information technology issues, as this is done at the moment on many congresses, and concentrate instead on control issues, not so often covered. For other perspectives on Ubiquitous Computing see i.e. Strassner & Schoch (2002) and Pervasive Computing conferences.

Fig. 2. Concrete Modelling of a Control System with Sensory Hand

Fig. 3. Specifying dynamic systems by concrete demonstration with sensory hand

2. CHALLENGING PROBLEMS OF UBIQUITOUS COMPUTING

Considering actual ubiquitous computing research, we find that it is still dominated by the visual and sensory technology. Augmented reality, overlapping with ubiquitous computing, is mainly discussed as an augmentation of reality through computer generated image projection into reality or into video-images of reality, but not as an augmentation by actuators and actions.

The information related view, of course, opens up enough interesting applications and problems, like

- wireless media access with appropriate physics and protocols
- input-output devices with image, gesture, speech recognition and synthesis
- architectures for distributed services
- design environments for ubiquitous computing
- geometric representation, discovery/recognition
- information-technological scalability, how to manage and select devices in open networks

- movable interactions, how do they move fluidly with the user
- machine learning for pattern recognition of behaviour, optimising performance, customisation, failure prediction
- privacy, data safety and security
- GRID computing and massive parallelism

Besides general information-technological problems of distributed systems (Eckert 2004) and real-time performance (Diethers et al. 2004) especially the geometric representation problem of one unique correspondence between a distributed real and virtual world has to be solved (Brumitt et al 2000). Among these are: How to handle contradictory facts about the physical world from multiple sensors, how to combine sensors to improve discovering the physical relationship between entities in the world, how to select appropriate devices. New design methods for not only useful but sometimes playful and highly experience oriented environments are necessary (Pokahr et al 2002, Mueller 1998).

The control related view rises other issues. Ubiquitous computing is represented by a variety of different shapes and sizes, from nano-technical devices, smart cards, paper like notepads, wall sized projections, mixed reality cave elements. Weiser, had the vision, that for each person in an office there would be one or two boards, tens of pads, hundreds of tabs, and one could add, thousands of nano-objects. In some sense, this is not yet true for the office, but consider everyday surroundings and the statement of Intel-Chairman A. Grove, that the average US-American encounters 70 microprocessors each day before lunch, mainly in consumer electronics at home, in elevators and security systems of public spaces and in vehicles and traffic control systems! I could not count them this morning, because they aught to be invisible, but I expect near a hundred. However, the increasing pervasion or **ubiquitous computing and acting**, opens up old and new frontiers: technological and social (pedagogical, psychological, ethical). Here, I will cover mainly technological issues.

Maurer (2004) foresees the PC in 10 years as being one almost on a credit card with no hard-disk, keyboard and screen, but instead distributed over users body with projection glasses, tiny cameras, voice transmitting collar, head-tracking and body localisation devices, full global network integration. Again, this vision is mainly an information view, missing an action support. Exoskeleton devices and helpful physical ghosts could be around. The ultimate penetration of the physical and the information world would not be if all objects have a RFID Chip, as sometimes stated, but if they also have some means of remote action support.

M. Weiser had this far reaching perspective, he distinguished ubiquitous computing from PDAs and autonomous software agents using as example the lifting of a heavy object. "You call in your strong assistant to lift it for you, or you can be yourself made effortlessly, unconsciously stronger and just lift it … ubiquitous computing aims at the latter" (Weiser, 1993). This could be understood metaphorically, but I would like to interpret it in a concrete way. Then we face the problem of how to handle effort and flow of energy in a distributed human-machine related way. CoBots in an assembly line, production factory, the household or an intelligent building, as elucidated by Erbe (2004), represent good examples of ubiquitous automation. This close coupling of physical objects and information is a domain, control people are very familiar with. Therefore, ubiquitous automation might be an important field of future work of IFAC.

Weiser's example leads me to a related research field with similar, sometimes inverse problems: *Humanoids, Telepresence* and *Intelligent Buildings* research (Hamel 2003, Becher et al 2003, Hirche et al 2003, Hirche & Buss 2003, Morpha 2003, QRIO 2003, Streitz et al 1999).

How can autonomous robots recognise and act in a changing natural environment and how can they collaborate with surrounding humans and other CoBots? Humanoids share their workspace with humans. They could take over the heavy load of a task, letting the human fine adjust and control the work (Takubo et al 2000). Distributing this functionality into visible and invisible surrounding objects yields to ubiquitous computing and acting. To experience problems and possibilities on a low cost experimental level, with off-the shell toys and tools, we need some well developed similarity theory, as it is known from engineering disciplines, like fluid dynamics scaling in relation to Reynolds Number (length of object * velocity of flow / cinematic viscosity of the medium). There is a strong theoretical background in control theory, but no clear criteria and evaluation scheme to compare different control schemes. Melchiori (2003) contributed to close this gap, and presented an overview and criteria for the analysis and comparison of control schemes for tele-manipulation systems. He also found that there is a need for laws of proper scaling even for force/velocity systems in higher dimensions. But ubiquitous computing and acting requires more than a representation and application of just forces. All physical phenomena could be candidates for mediation. This means, the relating laws of control, criteria of stability, inertia and damping, tracking, stiffness, applied to forces and velocities of robot arms should be generalised to effort/flow phenomena in the sense of Bond-Graph theory to cover full ubiquitous computing. This is one important theoretical challenge I see. *Bond-Graph Theory* still has a missing link to computer science. Graph-transformations, as they are used in language processing (compiler building, theorem proof etc) offer powerful means to handle graph representations, graph grammars and replacement algorithms. As there is a strong theoretical and experimental background in more or less disjoint disciplines, like electrical, thermodynamic, mechanical and fluid-dynamic engineering, the existing unified concepts, like Petri-Nets and Bond-Graphs, could improve the transfer and merge of practical system solutions for ubiquitous computing and acting. Unified views of information and energy flow, the merging of discrete and continuous system behaviour, self-similarity and hierarchical composition of large distributed systems are still posing non-trivial problems.

For cost oriented ubiquitous computing and acting, we need good simulation and reality integration. Wollherr & Buss (2004) demonstrate how a MAT-LAB Realtime Workshop in connection with Simulink on a Realtime (RT) Linux can be used to combine virtual reality with realtime reality control of a SCARA robot balancing an inverted pendulum and a car prevented to roll over using impedance control. This combination of real and virtual representations proved not only to be useful for the human-system interaction, but also for students and their understanding of the adequate or inadequate reduction of simulation models (i.e. not modelled system properties such as friction). In our mixed reality learning environment (see below) we found similar benefits, however we also missed the inadequate possibilities of powerful modern simulators to allow modelling and simulation on the fly. Justified by a tradition of well designed simulation modelling separated from experimentation: first build a model, then compile it to run efficiently and then perform series of reproducible experiments, then draw consequences for reality or the model, this procedure is still dominating the simulation community but inadequate for interactive worlds. When we faced the problem to work with pneumatic cylinders, valves, pressure sources and connecting tubes to build electro-pneumatic systems interactively in virtuality and wanted to experience a similar behaviour as in reality already during the building process, there was a problem. Simulators do not support this modelling with components "already under physics", on the fly. To connect a tube under pressure to a cylinder should immediately drive out the piston and a tube with one open end should whip around.

Learning system behaviour by demonstration is a further challenge to be faced, if ubiquitous "intelligent" devices want to adapt itself to a changing environment and the habits of users. In recent projects supported by the DFG (German Science Foundation) we investigated possibilities to generate abstract behaviour descriptions for robot tasks and plc-algorithms from concrete movements of sensory hands (Fig. 3) (Schäfer & Bruns 2001). We were able to generate simple plc-programs this way, but much more research has to be done, to be able to handle open situations in ubiquitous computing and acting. According to Moore's Law, the cost of a system on a chip including input and output controller is falling to a few cents. One important question is, how to integrate the expectable manifold of devices into a controllable system. Some research counts on self-organizational principles to handle these "Spray Computers" (Mamei & Zambonelli, 2003), clouds of microcomputers to be sprayed into the environment With specific smart sensing and effecting functionality they imagine to go to a local store and buy a "pipe repairing" spray of a cloud of microcomputers. In a recently started joint initiative between industry and German Universities a similar focus is on *Organic-Computing*: learning from nature to adapt and emulate self-organizational principles and architectures to handle complexity. Envisioned applications of these controller/observer driven systems are cars, intelligent buildings, intelligent autonomous robots and vehicles and production systems (OC, 2004).

User-Interface design will completely change. Driven by classical ergonomics, software ergonomics first was concentrated on physiological and psychological "objective" performance criteria of man-(soft)- machine or task relations. Scandinavian approaches towards cooperative design (Ehn, 1988), Useware design (Zühlke & Wahl 1999), interaction design (Caroll 1991, Cooper 1999) and experience design (Laurel 1986, Hagita 2003, Paulos 2003) more and more shift towards a playful emotional useless and useful relation to an everyday environment. This means, that new aesthetic forms will emerge and have to be supported in the design process. Fishwick (2002) and others investigate means to express functionality and behaviour in alternative aesthetic styles with a broad variety of user oriented materials, symbols and rules. To support translations between these notations is an emerging research area.

To sum up, we need

- Frameworks of Control Models (Melchiorri, 2003)
- Unified models of mapping physics to information and vice versa (Bondgraphs, Paynter, 1996 and Petri-Nets)
- Physical scalability
- Advanced sensor-actuator interfaces (Hyper-Bonds, s.b.)
- Mixed reality frameworks of distributed model-view-control
- Mixed reality multi user environments (who owns the real process?)
- Tactile interfaces (tele-manipulation)
- Abstractions from recorded activities and events, specifying by demonstration
- Low power devices and integrating and alternating mechatronic perspectives on form, function and behaviour

- More consideration of the importance of enjoyment, bridging perspectives of toys and tools

Two concepts, recently developed in my group, may be a modest contribution to ubiquitous computing: *complex objects* and *hyper-bonds*. Complex objects are objects having one real part and several virtual representations closely coupled.

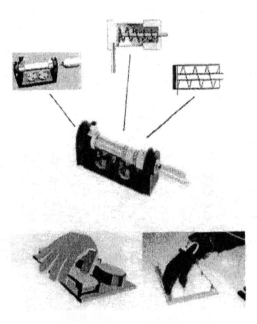

Fig. 4. Complex Objects with real tangible parts and various digital representations

In Fig. 4, two different kinds of complex objects are presented, one for a pneumatic cylinder and one for a conveyor belt. Computer based links between real and virtual parts ensure the synchronization of their states. They can be realized by video-image-recognition or, as shown in figure 3, by data glove tracking. Starting from a reference situation, changes of state are sensed by a graspable user interface and used to update the complementary part (Bruns, 1993). The term *complex object* is an allusion to the mathematical notion of complex numbers. Similar to complex numbers, having a real and an imaginary part, the complex object contains an abstract, virtual object with enriched possibility of mathematical treatment and behavior (algorithms, data-structures) and the controlled automation device as its projection into reality. With construction kits, containing sets of these complex objects for specific application areas, it is possible to construct a system in reality and synchronously generate a corresponding virtual model, which can be tested, analyzed and transmitted to remote places.

The concept of Hyper-Bonds is an electro-mechanical interface mechanism to sense and generate physical effort and flow phenomena and to relate this to the well known theory of bond-graphs. A first implementation of this concept has been demonstrated in a European project DERIVE – Distributed real and virtual Learning Environment for Mechatronics and Tele-Service (Fig. 5). However, the implementation is for non-time-critical electro-pneumatic systems of state-automata type, with discrete pressure and voltage. This allows a distribution via low band-width Internet. To integrate and distribute hard real-time analogue processes, like remote force-feedback, is still heavily restricted by the quality of the Internet, by process control sensors and actuators and by theoretical issues of merging and cutting bond-graphs and Petri-net simulations.

Fig. 5. Distributed real and virtual world, mixing physical and information phenomena

In two running projects Lab@Future and MARVEL we are developing perspectives and prototypes about future laboratory work. Many of the above questions have to be considered there. In these projects we extended Milgram & Colquhoun's (1999) Mixed Reality Taxonomy by emphasizing the focus of attention and real experiences, Fig. 6. Three dimensions are (Fig. 7)
- focus of attention (reality – virtuality)
- centeredness as location of our experience with sensing and acting (ergo-exocentric)
- directness of the action-reaction control circuit (direct-indirect).

Centeredness is egocentric, if we see, hear, feel, as if we were in the center. Exocentric is a perspective if we consider the world from a distant third-person perspective. Directness of control would be in the sense of our known control models, together with cognitive directness (orientation, time-delay, multi-sensory reception). This taxonomy could help to systemize the design and evaluation of mixed reality applications.

In DERIVE we faced a low scale distributed problem with a central server-client architecture. The architecture consists of several modules that together build the system (Fig. 8).

This will certainly be inadequate for massively parallel systems. The challenge will be to combine many of these types of server, recording and handling the states of and accesses for overlapping realities.

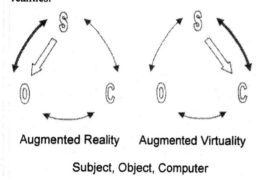

Augmented Reality   Augmented Virtuality

Subject, Object, Computer

⟹ **Attention**   ↔ **Sensing-Acting**

Fig. 6: Relations in Mixed Reality

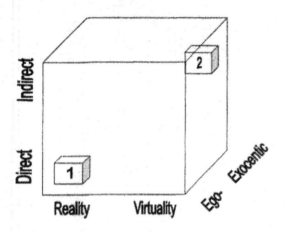

Fig. 7. Taxonomy of Mixed Reality (Control, Location, Perspective)

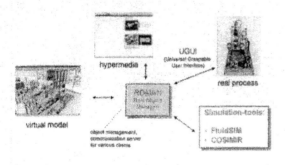

Fig. 8. System Architecture Overview

3. SOCIAL ASPECTS

Tracking the life-cycle of perishable food, workflow-systems, personal access control, identification of objects and persons in public and work places might be a promising perspective, but the arising problems are immense. The right of informational self-determination will be strongly attacked by these developments. In a recent case in Germany it was discovered that a food-store distributed active smart-cards among their customers being able to collect data about their custom preferences and use behaviour. Ubiquitous Computing rises in a new way the question of transparency. One up-to-date aim of computer-design is transparency, making the system invisible, independent of size, location, physical implementation: computing power everywhere at every time. If we do not see the all surrounding system, the possibility of misuse increases, be in the name of home security or public safety.

I have no solutions for these problems, but the more we reflect about them and the more sensibility we develop for these questions, the higher the chance, that we are not overrun by technology.

Wenn der Schein wild wird nach Gestalt, wird er den Spiegel zum Bersten bringen

Botho Strauß
When display is getting wild for form, it will bring the mirror to burst
(translation WB)

I am not as pessimistic as the dramatist, who fears, that we are entering an era of collective trance, but I believe that we have a lot to do, to be aware of the possibilities and dangers of a world where computing is everywhere.

In another student project we animated a rose-garden through a nightly installation *Sensory Garden*, where visitors could experience the virtual duplication of a lifeless stone statue, flowers and humans in a mixed reality presence (Fig. 9). These performances are some modest low cost tries to face the above questions (Richard 2004).

4. CONCLUSIONS

Some aspects have been presented, how and why ubiquitous computing may be developed in the direction of ubiquitous *computing and action* and how this is related to automation technology. Automation Control theory and practice might provide a fruitful contribution to the emerging field of ubiquitous computing and action. Some small examples have been given.

Fig. 9. Statue Aegina and her double on a projection wall

REFERENCES

Becher, R., P. Steinhaus and R. Dillmann (2003). The Collaborative Research Center 588: „Humanoid Robots – Learning and Cooperating Multimodal Robots". *Proc. Int. Con. Humanoids 2003*, Karlsruhe & München, Germany, 1.-3. Oct. 2003.

Brumitt, B., J. Krumm, B. Meyers and S. Shafer (2000). Ubiquitous Computing & The Role of Geometry. *IEEE Personal Communications*, **vol. 7, No. 5**, Oct. 2000, pp. 41-43.

Bruns, F. W. (1993). Zur Rückgewinnung von Sinnlichkeit – Eine neue Form des Umgangs mit Rechnern. *Technische Rundschau* **Nr. 29/30**, pp. 14-18

Bruns, F. W. & V. Brauer (1996). Bridging the Gap between Real and Virtual Modeling - A new Approach to Human-Computer-Interaction - *IFIP5.10-Workshop „Virtual Prototyping"*, May 6-8, Arlington, Texas, 1996 (also artec Paper Nr. 46, Universität Bremen)

Bruns, F. W. (1999). Complex Construction Kits for Coupled Real and Virtual Engineering Workspaces. In: *Cooperative Buildings – Integrating Information, Organizations and Architecture*. Second Int. Workshop, CoBuild'99. (Streitz, N. A., J. Siegel, V. Hartkopf and S. Konomi, (Ed.)). Lecture Notes of Computer Science **1670**, pp. 55-68. Springer, Heidelberg.

Bruns, F. W. (2000). Complex Objects and anthropocentric Systems Design. In: *Advances in Networked Enterprises* (L. M. Camarinha-Matos, H. Afsarmanesh, H.-H. Erbe (Ed.)), pp. 249-258. Kluwer, Boston.

Bruns, W., H. Ernst, M. Faust, P. G. Amaral, H. Gathmann, S. Grund, I. Hadfield, J. Huyer, U. Karras, R. Pundt and K. Schmudlach (2000). Distributed real and virtual learning environment for mechatronics and tele-service.

Final Report. *Artec-Paper* **102**, Bremen University.

Carroll, J. M. (1991). *Designing Interaction*. Cambridge University Press, Cambridge.

Diethers, K., B. Finkemeyer an N. Kohn (2004). Middelware zur Realisierung offener Steuerungssoftware für hochdynamische Prozesse, *it* **1/2004**, pp. 39-47.

Eckert, C., K. Bayarou ad S. Rohr, (2004). NGN, All-IP, B3G: Enabler für das Future Net?!. *Informatik Spektrum* **Vol 27, No 1**, Feb. 2004, pp. 12-34.

Ehn, P. (1988). Work-Oriented Design of Computer Artifacts. *Arbeitslivscentrum*, Stockholm.

Erbe, H. (2004). On Human Robot Collaboration. *Proc. Int. IFAC COA*, Ottawa, Canada

Fishwick, P. (2002). Aesthetic Programming. *Leonardo* **35(4)**, MIT Press.

Fitzmaurice, G. W., H. Ishii and W. Buxton (1995). Bricks: Laying the Foundations for Graspable User Interfaces. *ACM CHI'95 Mosaic of Creativity*, pp. 442-449.

Hagita, N. (2003). Introduction to "Ubiquitous Experience Media". *ATR Workshop on Ubiquitous Experience Media 2003*, Keihanna Science City, Kyoto, Japan

Hamel, W. R. (2003). Fundamental Issues in Telerobotocs. *Proc. Int. Con. Humanoids 2003*, Karlsruhe & München, Germany, 1.-3. Oct. 2003

Hirche, S., B. Stanczyk and M. Buss (2003). Transparent Exploration of Remote Environments by Internet Telepresence. *Proceedings of Int. Workshop on High-Fidelity Telepresence and Teleaction jointly with the conference HUMANOIDS*, München, 2003.

Hirche, S. and M. Buss (2003). Study of Teleoperation using Realtime Communication Network Emulation. *Proceedings of IEEE/ASME International*

Conference on Advanced Intelligent Mechatronics, Kobe, Japan.

Ishii, H. and B. Ullmer (1997). Tangible Bits: Toward Seamless Interfaces between People, Bits and Atoms. *CHI 1997*, ACM, Atlanta, Georgia.

Kang, S. B. and K. Ikeuchi, (1994). Grasp Recognition and Manipulative Motion Characterization from Human Hand Motion Sequences. *Proc. of IEEE Int. Conf. on Robotics and Automation*, **Vol 2**, pp. 1759-1764. San Diego, Cal.

Karnopp, D. C., D. L. Margolis and R. C. Rosenberg (1990). *System Dynamics – A unified Approach*. John Wiley, New York.

Laurel, B. (1993): Computers as Theater. Addison-Wesley, Reading, MA.

Lave, J. (1991). *Situated learning: legitimate peripheral participation.* Cambridge University Press. Cambridge. New York, NY.

Maurer, H. (2004): Der PC in zehn Jahren. *Informatik Spektrum* **Vol 27, No 1**, Feb. 2004, pp. 44-50.

Mamei, M. and F. Zambonelli (2003). Spray Computers: Frontiers of Self-Organization for Pervasive Computing. Zeus.elet.polimi.it/is-manet/documenti/pop-dismi-10.pdf, see also Tutorial 1 at *PerCom 2003, IEEE Int. Conf. on Pervasive Computing and Communications*, Dallas-Fort Worth, Texas.

Melchiorri, C. (2003). Robotic Telemanipulation Systems: An Overview on Control Aspects. *Proc. 7th IFAC Sympos. on Robot Control*, Sept. 1-3, Wroclaw, Poland

Milgram, F. and H. Coquhoun (1999). A Taxonomy of Real and Virtual World Display Integration. In: *Mixed Reality – Merging Real and Virtual Worlds.* (Ohta, Y. and H. Tamura. (Ed.)), New York

Mitchell, W. J. (2003). *Me++: The Cyborg Self and the Networked City*, MIT Press.

Morpha (2003). Kommunikation, Interaktion und Kooperation zwischen Menschen und intelligenten anthropomorphen Assistenzsystemen.
http://www.morpha.de/php_d/index.php3

Mueller, D. (1998). *Simulation und Erfahrung – Konzeption und Gestaltung rechnergestützter Simulatoren für die technische Bildung.* Dissertation. Bremen University

Murray, H. J. (2001). *Hamlet on the Holodeck – The Future of Narrative in Cyberspace*, MIT Press, Cambridge, MA.

OC (2004). Organic-Computing-Initiative position paper. http://www.sra.uni-hannover.de/forschung/forschung.htm

Ohta, Y. And H. Tamura (1999). *Mixed Reality – Merging Real and Virtual Worlds*. New-York.

Paulos, E., B. Brown, B. Gaver, M. Smith and N. Wakeford (2003). Mobile Play: Blogging, Tagging, and Messaging. *5th Int. Conf. on Ubiquitous Computing*. Panel Discussion. Seattle, Washington.

Paynter, H. M. (1961). *Analysis and Design of Engineering Systems*. MIT Press, Cambridge, MA.

Pokahr, A., L. Braubach, A. Bartelt, D. Moldt and W. Lamersdorf (2002). Vesuf, eine modellbasierte User Interface Entwicklungsumgebung für das Ubiquitous Computing. In: *Mensch & Computer: Vom interaktiven Werkzeug zu kooperativen Arbeit- und Lernwelten.* (M. Herczeg et al (Ed.)), p. 185-194. Teubner, Stuttgart.

QRIO (2003). Sony Dream Robot. http:www.sony.net/SonyInfo/QRIO/story/

Resnick, M. (1993). Behavior Construction Kits. *Communications of the ACM*. **36(7)**, pp. 64-71.

Richard, J. (2004). Theatrum Mundi, das Bilboquet, die Maus und die Avatare. Geschichte und Geschichten der Spielgesellschaft. In: Reisen – Erkunden – Erzählen. Bilder aus der europäischen Ethnologie und Literatur. (M. Nagel (Ed.)), pp. 125-142. edition lumière, Bremen.

Schäfer, K. and F. W. Bruns (2001). PLC-Programming by Demonstration with Graspable Models, *Proceedings of 6th IFAC Symposium on Cost Oriented Automation*, Berlin, 8.-9.10., pp. 88-93.

Stephensen, N. (1995). *The Diamon Age or, A Young Lady's Illustrated Primer*, Bantam Books, New York

Strauss, B. (1999). *Der Aufstand gegen die sekundäre Welt – Bemerkungen zu einer Ästhetik der Anwesenheit.* pp. 93-105. Hanser Verlag München.

Strassner, M. and T. Schoch (2002). Today's Impact of Ubiquitous Computing on Business Processes. In: Mattern, F. and M. Naghshineh (Eds.). *Pervasice Computing. First Int. Conf. Pervasive Computing 2002*, Zürich, Switzerland.

Streitz, N. A., J. Siegel, V. Hartkopf and S. Konomi, (1999). Cooperative Buildings – Integrating Information, Organizations and Architecture. Second Int. Wrokshop, CoBuild'99. *Lecture Notes of Computer Science* **1670**, Springer, Heidelberg.

Suchman, L. A. (1985). Plans and Situated Actions: The problem of human-machine commuinication. *Xerox PARC Technical report* **ISL-6**, Feb 1985.

Sutherland, Ivan E. (1965). The Ultimate Display. *Proc. IFIP Congress*, **Vol. 1**, p. 508. Washington.

Takubo, T., H. Arai and K. Tanie (2000). Virtual Nonholonomic Constraint for Human-Robot Cooperation in 3-D Space. *Proc. 2000*

IEEE/RSJ Int. Conf. on Intelligent Robots and Systems (IROS2000), pp.300-305.

Weiser, M. (1993). Some Computer Science Issues in Ubiquitous Computing. *Communications of the ACM*, **36/7**, July 1993. http://www.ubiq.com/hypertext/weiser/UbiC ACM.html

Wellner, P., W. Mackay and R. Gold (1993). Computer-Augmented Environments: Back to the Real World. *Communications of the ACM*, **36, 7**, pp. 24ff.

Wiener, N. (1948). *Cybernetics: or Control and Communication in the Animal and the Machine*. MIT Press paperback edition (1965).

Wollherr, D. and M. Buss (2004). Cost Oriented Virtual Reality and Realtime Control System Architecture.

Zühlke, D. and M. Wahl (1999). Hardware, Software – Useware. Maschinenbedienung in vernetzten Produktionssystemen. *Elektronik* **23**. pp. 54-62.

Virtual Automation System Standards

Rolf Bernhardt, Alexander Sabov, Cornelius Willnow

Fraunhofer Institute Production Systems and Design Technology (IPK)
Pascalstr. 8-9
10 587 Berlin
Germany
cornelius.willnow@ipk.fhg.de

Abstract: Computer based models have become an indispensable means for the planning, optimisation and operation of industrial manufacturing systems. The usability and significance of the models strongly depends on their accuracy, availability and functional extent. Today, suitable models for components of automation systems are available only for some types of components. Functionality for simulation purposes is often limited and integration costs are often still considerable. These problems were widely solved for industrial robots in the 'Realistic Robot Simulation' (RRS) projects. There, world-wide standards for Virtual Robot Controllers (VRCs) were created. Now, this success should be transferred to other components of automation systems like programmable logic controllers (PLCs), human-machine interface (HMI) systems, bus systems and effector sensor systems. *Copyright © 2004 IFAC*

Keywords: Automation, Concurrent engineering, Industrial robots, Industrial control systems, Interfaces, Modelling, Robotics, Simulation, Simulators, Virtual Reality

1. INTRODUCTION

Computer-based models and simulation have become an indispensable means for planning, optimisation and operation of industrial manufacturing systems. The usability of simulation results depends highly not only on the availability and accuracy of the models, but also on the ability to integrate them into different simulation systems and into the overall data flow during engineering and operation of installations.

World-wide standards exist for simulating robot controllers, and numerous compatible products are on the market, as will be outlined below.

For PLC simulation, several manufacturers provide simulation models (Siemens 2003, Cape Software), and a number of simulator manufacturers have integrated them (Cosimir 2004, e4engineering, Tecnomatix). These models already provide high simulation accuracy since they include the original PLC software. Data and program exchange with the real installation has already reached a moderately good state, since programs and data can be transferred directly.

Integration with higher level planning data, however, is still limited. Furthermore, the integrations rely on proprietary interfaces or they are based on standards like OPC (OLE for Process Control). But proprietary interfaces restrict users in combining controller models with simulators, and integration has to be repeated for each manufacturer. OPC is at a first glance promising, exp. since it is in widespread use. However, OPC is designed for efficient real-time data access. For this reason, it does not support the synchronization mechanisms required for precise simulation of time behaviour and controlled data consistency between shop floor and planning level.

In the area of HMI-systems, integration via OPC is common praxis, but shows disadvantages similar to the ones of PLC integration. Today, simulating effector and sensor systems relies almost completely on the generic functionality provided by simulators. Incorporating communication busses into simulation based planning and programming is an almost untouched area.

Systematics for integrating automation system components into simulators are outlined for example in (Danielsson 2002).

In summary, precise models of automation system components are provided as commercially available products only for a limited number of components. Except for robot controller models, integration takes place by proprietary interfaces, or by using existing standards that, however, show disadvantages in compatibility and performance.

2. APPROACH TO A SOLUTION

For these reasons, suitable interface standards for integrating models of automation system components into simulators must be defined.

To enable an easy integration of the models into simulation systems, standard interfaces are required for each type of control system component. Each manufacturer of a component can then provide a virtual model of its component that is equipped with the standard interface. In turn, each manufacturer of a simulation platform provides the standard interfaces. The user of a simulation system can then buy the required simulation models, plug them into the simulation system and run precise simulation scenarios for low efforts.

Figure 1 illustrates the coupling of virtual automation system components with simulation system.

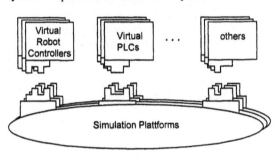

Fig. 1. Crosswise coupling of any virtual model with any simulator via standard interfaces

This will result in a new generation of highly precise simulation technology. It will create a new market for virtual components. The new products are compatible, precise, supplier-made models with high availability.

In addition to increased performance, the approach brings considerable advantages to all concerned types of companies:
- Component manufacturers must realize the standard interfaces only once. The models of their components can then be run in different simulators for different purposes. This enables optimal application, which adds value to the components.
- Simulator manufacturers are relieved from modelling components or adapting their system to different integration methods. This saves costs and enables them to concentrate on the core business.
- End users can buy component models commercially and plug them into their preferred simulator without depending on special and often costly developments.

3. PRECEDING DEVELOPMENTS

To achieve this, the proven approach of the 'Realistic Robot Simulation' (RRS) projects can be adopted. These projects were initiated by major automotive companies and included major robot manufacturers, simulator manufacturers, and system integrators as illustrated in Figure 2.

Fig. 2. The RRS-Consortium

In 1991, the RRS-1 project by the consortium. It defined the 'Robot Controller Simulation' (RCS) interface standard, which was published in 1994. Today it is the world-wide de-facto standard for precise simulation of robot motion behaviour (RRS-Owners 1997 (Version 1.3)). All major robot and simulator manufacturers support this standard.

Based on the experience gained, the consortium started the RRS-2 project in 1998 to develop the next generation standard. The resulting 'Virtual Robot Controller' (VRC) interface standard was published in 2001 (Bernhardt, Schreck, Willnow 2000; VRC-Specification Owners 2004 (Version 1.1)). It covers almost the complete robot controller software, including language system, technology control and user interface functionality.

The VRC-interface also includes a variety of additional features such as user activity models, automatic program generation, consistent data flow between planning and shop floor, as well as efficient distribution and integration methods. Furthermore, it supports data and model exchange between the different companies that are involved in the development of installations.

4. TECHNICAL APPROACHES

The VRC-Interface standard was designed to be the fundamental solution for the next decades. Its functionality is carefully balanced for achieving long term generality and for enabling a wide area of applications. Two canonical situations cover the interface functionality:
- running simulation scenarios for different purposes, and
- supporting continuous data flow between planning and production.

4.1 Interface Functionality for Running Simulation Scenarios

The VRC-Interface functionality for running simulation scenarios consists of the interface areas illustrated in Figure 3.

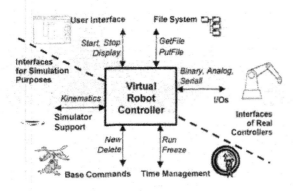

Fig. 3. VRC-Interface Areas for Running Simulations

The three interface areas at the top and at the right reflect interfaces of real controllers:
- I/Os models the I/O-lines of real controllers, i.e. binary, analogous and serial inputs and outputs.
- File System allows for transferring single files and complete file systems.
- User Interface allows for handling user interfaces, e.g., for editing, handling and debugging programs, machine data, etc.

The three interface areas at the bottom and left are required for handling VRCs in simulation environments:
- Base Commands include creating, booting, deleting and listing available VRCs.
- Virtual Time Management controls the simulation equivalent of real time (Willnow, 2001). In connection with I/Os, this is of high significance for, e.g.,. cycle time prediction.
- Simulator Support simplifies several tasks of simulators like kinematics transformations.

These interface areas can be used in application scenarios reaching from cell layout, detailed programming, shop floor maintenance and operator training.

4.2 Support for Continuous Data Flow between Planning and Shop Floor

The support for continuous data flow foresees
- dataflow between planning systems and virtual controller models, as well as
- dataflow between virtual controller models and physical shop floor production systems.

Both is illustrated in Figure 4.

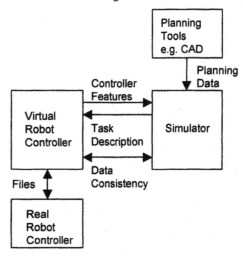

Fig. 4. VRC-Interface support for continuous data flow

During the initialisation phase, the specific Controller Features are passed to the simulation system, e.g. the number and type of I/Os, the available command set and geometry information about controlled mechanics. This is vital for plug-and-playability, since it enables an automatic adaptation of simulators to a given VRC (Bernhardt R., Schreck G., Willnow C., 2001).

In order to convert data in planning systems to robot task programs, the planning system sends these data as a Task Description to the VRC. From this, the VRC creates a robot task program in the robot specific task programming language (code generation, post-processing). The robot task programs can be edited and tested in the VRC, and can then be transferred to the real robot controller in the shop floor.

Data Consistency mechanisms allow for the exchange of, e.g., geometry data like motion targets. This enables the reuse and update of data between the product model and the task program. The mechanisms are synchronized by protocols and consistency states that support different updating procedures.

Data exchange with the shop floor is implemented by the described file system functionality. It allows for updating program and machine data files in both directions. Complete consistency between VRC and the shop floor can be obtained by exchanging complete file systems.

Altogether, these mechanisms enable bi-directional consistency. For example, geometry data of spot welding points in CAD systems can be transferred via simulators and VRCs to the shop floor robot controllers. On the other hand, modifications performed at the shop floor can be up-loaded from the shop floor, via VRCs and simulators to the product model in the CAD-system.

4.3 Inter-Company Data Flow

The mechanisms for continuous data flow are also used for obtaining data exchange between the companies involved in the construction of installations. For example, layouts with task descriptions can be sent from end user companies to system integrators. There, the detailed robot task programs are developed in a virtual environment. Then they are returned to the end user for transferring them to the installation.

During maintenance tasks, the precise virtual models of installations are exchanged in the same way.

Overall, these efficient methods for data exchange enable high consistency between suppliers, users and related information systems. This saves expenditures and time spent in engineering, maintenance and service. Development and integration efforts are reduced and virtual training scenarios are easily created. This leads to faster developments, more efficient installations, better usage of resources and improved training.

5. FURTHER REQUIRED DEVELOPMENTS

RRS is geared mainly towards robots. The basic concepts and a number of mechanisms can, however, be transferred to further components of automation systems. A typical scenario for an automation system is shown in Figure 5.

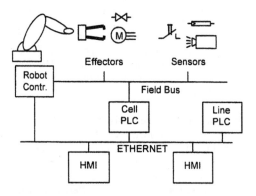

Fig. 5. Typical scenario for an automation system in parts production industries

To transfer the achievements of RRS and to complete such scenarios, the following interface standards are to be developed:

- Virtual Programmable Logic Controllers(PLCs),
- Virtual Human-Machine-Interface Systems (HMIs),
- Virtual Bus-Systems, and
- Virtual Effector and Sensor Systems.

The development of these additional standards results in a family of standards that share common data formats and mechanism, as illustrated in Figure 6.

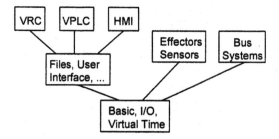

Fig. 6. Family of consistent and interoperable standards that share common interface areas

In addition to reusing concepts of the VRC-interface, the consistent family of standards will also ensure the required compatibility of the virtual models. Interfacing models with different interface data formats and exchange and synchronization mechanisms would require interface adaptors that typically lead to conversion losses. The interoperability of a family of standards, however enables the aimed-at high precision in simulation.

6. EXTENSION TO FURTHER INDUSTRIAL SECTORS

The developments of RRS started with robot application in automotive industries. But the scenario for automation systems in Figure 5 holds also for a variety of other industrial sectors. These include

- chemical, pharmaceutical, petrochemical,
- energy, oil, gas,
- supply, white goods,
- aviation, space,
- steel, plastics,
- maritime industries,
- printing, food and health care.

In these industrial sectors, the use of simulation is growing and the need for significant simulation results is increasing. For this reason, the development of the new simulation technology is of high interest for these industrial sectors and should also regard their specific needs.

Many of the companies in these sectors are large enterprises. The number of suppliers and system integrators that are small and medium enterprises (SMEs) however increases (Leupold, Voges, 2003). For this reason special attention has also to be paid to SME-needs.

7. APPLICATION AREAS DURING THE INSTALLATION LIFE-CYCLE

In addition to the range of major automation system components and the range of industrial sectors where these components are applied in installations, the different application areas of virtual scenarios throughout the life-cycle of installations must also be covered.

The major areas of application include
- planning of new installations,
- programming of controllers,
- commissioning installations,
- education and training of operator and maintenance personnel,
- maintenance and optimisation of running installations, and
- re-construction, modification and extension of installations.

An overview on the use of virtual models during the different life-cycle phases is given in figure 7.

Life-cycle Phase	Use of Virtual Models
Pre-productive Phases	
Planning	- Design Optimisation - Design Verification
Construction	- Programming in Parallel - Program Test
Commissioning	- Testing - Optimisation - Operator Training
Productive Phases	
Optimisation	- Analysis of the Installation - Performance Tests - Verifying Modifications
Product Change	- Feasibility Tests - Program Preparation - Program Tests
Production Re-design	- Combinations of the Above

Fig. 7. Activities with virtual models during the life-cycle phases of an installation

Systematic usage of the virtual models throughout these phases also brings synergetic effects that include:
- the consistency of model significance during the different phases and their application areas, and
- the reuse of models created in previous phases during subsequent phases, which increases cost efficiency of model creation by model reuse.

8. COST REDUCTION EFFECTS

Cost reduction results directly from the use of the virtual models during the different phases and is based on the following major effects:
- Virtual models make installations available even before they are physically realized.

- Virtual models allow for early testing and verification before errors appear in the real installation.
- Virtual models allow for optimisation and testing without involving the real system.
- Virtual models forgive errors, while testing in real installations may harm machines or even persons.
- Virtual models enable work on a running installation without interfering with production.
- Virtual models can be easily copied for low costs, which enables more work in parallel.

An overview of cost reduction effects during the life-cycle phases is given in Figure 8 (Baumgartner., Bernhardt, Schreck, Willnow 2001).

Life-cycle Phase	Costs Reduction by Models
Pre-productive Phases	
Planning	- Verified Concepts and Designs - Optimised Layouts - Parallel Work
Construction	- Programming in Parallel - Reduced Risk of Damage
Commissioning	- Higher Availability of Equipment - Reduced Risk of Damage - Verified Designs - Less Late Repair
Productive Phases	
Optimisation	- Maintaining Production - Improved Productivity - Verification before Performance
Product Change	- Work in Parallel to Production - Reduced Risk of Damage - Verification before Execution - Reduced Down-times
Production Re-design	- Combination of the Above

Fig. 8. Economisation using virtual models during the life-cycle of an installation

Precise figures for the economisation potential are difficult to obtain and estimates are rare. This is also due to the fact that comparative values for 'with-simulation' and 'without-simulation' are not available.

For automotive industries, estimates are based on the costs for the development of a car model variant of some 100 Mio. € and the costs for, e.g., a body-in-white line of about 140 to 200 Mio. €. For these investments, interest accumulate until start of production.

With advanced simulation technologies, production can be started several days or even weeks earlier. Taken alone, the daily payment of interest represents considerable potential for economisation. In addition, with each day of earlier start of production, some 1000 more cars can be produced.

For chemical and petrochemical industries, a survey on operator training simulation (Kroll 2003) reports estimated economisations achieved by improved operation of installations. Figures were given for 5 installations. Reported economisations reach from 1.6 Mio. Euros to 4 Mio. US$ per year. In one case gains of about 4 Mio. US$ were reported from improved ramp-up during the commissioning phase.

9. FURTHER PROCEEDING

For developing the required family of standards for virtual automation systems, a number of related international projects must be performed. Each of the projects should concentrate on one type of automation system, i.e. PLCs, HMI systems, bus-systems and sensor and effector systems.

Each of the projects has to cover all states in knowledge creation from research institutes over component and simulator manufacturers, via system integrator companies to final value creation by end user companies.

Furthermore, the projects must include system integrators and end users from different industrial sectors like automotive, supply, chemical, pharmaceutical, energy, steel and maritime industries, for covering their specific requirements and needs.

There is also a need for the participation of manufacturers for the most relevant types of simulation systems. This concerns, on one hand, different manufacturing processes like parts production, as in automotive and supply industries, but also continuous and flow production as in chemical industries. On the other hand, the scale of the user and system integrator companies should be regarded, and include simulation systems for large enterprises and for SMEs.

Finally, the different application areas for virtual models have to be covered, reaching from planning via construction and programming to training simulators.

10. SUMMARY AND CONCLUSIONS

The paper presents concepts for the creation of a new generation of high performance simulation models for components of automation systems, available on the market. The methodology is based on the successes of the RRS-consortium that created the world-wide Virtual Robot Controller (VRC) interface standard for precise and cost efficient simulation of robot controllers.

To complete the virtual scenarios for automation systems, a number of related projects for defining compatible standards for Virtual PLCs, Virtual HMI-systems, virtual bus-systems and virtual effector and sensor systems are proposed. The framework conditions for successful project performance and expected results for Virtual Automation Systems were shown.

REFERENCES

Baumgartner A., Bernhardt R., Schreck G., Willnow C. (2001), *Realistic Robot Simulation in Concurrent Engineering of Manufacturing Lines in Automotive Industries*, Proceedings of the 8th International Conference on Concurrent Engineering; West Coast Anaheim Hotel, California, USA; July 29 - August 1, 2001

Bernhardt R., Schreck G., Willnow C. (2000), *The Virtual Robot Controller Interface*, Proceedings of the ISATA 2000, Track Simulation and Virtual Realitiy, Dublin, Ireland, September 25-27, 2000

Bernhardt R., Schreck G., Willnow C., *Virtual Robot Controllers as Simulation Agents*, Proceedings of the 2nd Workshop on Agent Based Simulation, Passau, Germany, April 2-4 2001

Cape Software, *Cape Software_s VPLink for Allen Bradley PLC-5 Series*, www.capesoftware.com/Docs/AB5web.pdf

Cosimir, *Simulation of S5/S7 copatible Programmle Logic Controllers (PLC)*, http://www.cosimir.com/COSIMIR_MP/CD/English/Professional/PLCSimulation/PLCSimulation.htm, 2004

Danielsson F., *Off-line Programming, Verification and Optimisation of Industrial Control Systems*, Ph.D. thesis, Faculty of Computing Sciences and Engineering, De Montfort University, UK, November 2002

e4engineering, http://www.e4engineering.com/item.asp?id=41022&type=product%20notes&pub=ims

Kroll A. *Trainingssimulation für die Prozess-industrien: Status, Trends und Ausblick - Teil 1* Automatisierungstechnische Praxis Heft 2, 2003

Leupold M., Voges S., *Zukunftsschmiede "Digitale Fabrik"*, Computer&Automation, 9/03

RRS-Owners (1997), *RRS-Interface Specification*, Version 1.3, available through Fraunhofer IPK, September 23, 1997

Siemens, *Engineering Software, PLC Simulation - SIMATIC S7-PLCSIM, Product Sheer*, 04/2003

Willnow C. (2001) *Virtual Time Management of Virtual Robot Controllers*, Proceedings of the 15th European Simulation Muiltconference 2001 (EMS 2001), June 6-9, 2001, Prague, Czech Republic

VRC-Specification Owners (2001), *VRC Interface Specification*, Version 1.1, available through Fraunhofer IPK, June 2001

ELSEVIER
IFAC
PUBLICATIONS
www.elsevier.com/locate/ifac

A FRAMEWORK FOR INTEGRATING CONTROL ENGINEERING KNOWLEDGE IN COMPUTER SYSTEMS

**Isaías García, José R. Villar, Carmen Benavides, Francisco J. Rodríguez
and Ángel Alonso**

[dieigr, diejvf, diecbc, diefrs, dieaaa]@unileon.es
Systems Engineering and Control Group
Dept. of Electrical and Electronic Engineering, University of León
Edificio Tecnológico, Campus de Vegazana s/n, 24071 León (SPAIN)

Abstract: One of the most important issues when talking about cost in industry is cost regarding software applications' development and use. This paper shows new directions related to this discipline, aimed at the improvement of the different technologies and applications used today in every area of the product life cycle, and stress the need for a more conceptual view needed in order this improvement can be really useful. The combination of techniques appeared within software engineering, Artificial Intelligence and Knowledge Engineering will be shown to be the ones that can achieve this objective. A common framework for representing the knowledge involved in the domain of control engineering will be shown and some of the tools and methodologies needed to implement it will be presented. *Copyright © 2004 IFAC.*

Keywords: Control Engineering Software, Software Engineering, Software Integration, Knowledge Engineering, Engineering Ontologies, Knowledge-Based Systems.

1. INTRODUCTION

This paper shows new directions on the use of the computer in the field of control engineering (and the industry in general). The aim is to draw a general picture of how the knowledge of this domain can be represented in a computer system in an integrated and conceptually-enriched framework. The need for integration has arisen due to the characteristics shown by the current situation of the technology and its relationships with social and economical issues. Enterprises need today to manage not only the increase in complexity in the technological artifacts, but also the relationship with social and economical aspects that are present along the product life cycle.

The flow of information related to or found within the manufacturing processes is gaining more and more importance over the flow of energy.

Information processing and communications are the two key concepts present in every area of the modern manufacturing and instrumentation issues as is reported in (Ollero *et. al.*, 2002) and new architectures to efficiently deal with this information and able to cooperate in a distributed environment are needed.

Regarding control software, some of the needs for new techniques and approaches to be used are summarized in (Heck *et. al.*, 2003):

- Component-based architectures, stressing the use of object-oriented structures for reusing.
- Building distributed architectures that make use of the network communications technologies.
- Researching about the real-time issue, that is, finding ways of assuring or specifying a

time interval within which the system is able to respond.

But the problems regarding the use of computers in the industry goes beyond the kind of problems and needs stated in the former paragraph. As was discussed at the 15th IFAC World Congress held in Barcelona in 2002 (Verbruggen *et. al.*, 2002), the use of computers in control face a number of problems, ranging from the numeric or "algorithmic level" to the "cognitive level" where the term "knowledge management" brings up a lot of new concepts to deal with (ontological issues, knowledge acquisition, knowledge representation).

Software applications are key elements in today's industry, and an important issue when talking about cost. Investment in software is today a rather critical decision and extensibility and maintenance issues are often a source of extra and unpredicted cost.

Few effort has been devoted so far to the task of diminishing costs by means of improving the creation and the use of software in engineering, but there are several ways in which an industrial company could reduce costs by means of software. To achieve this objective, software developers should:

- Creating software able to manage the increasing complexity in the processes where some kind of automation is involved, improving the insight of the system and allowing to build applications that could be useful to simulate, analyze or design new control schemas or build training systems for the operators.
- Creating software systems that embrace and integrate the different areas of the enterprise activities by improving the communication processes among the different agents involved in the product life cycle and thus reducing the manufacturing costs.
- Using software modeling approaches that allow the ease in the software developing process, facilitating the reuse of structures across different applications and domains, and thus also reducing the software design and development costs.

The first two facts are conceptually very close to the domain of automation and control engineering while the last one is more related to the software engineering and knowledge engineering field. This way, while the software engineering has been searching for solutions to the processes of building and maintaining large software applications with the aim of reducing costs, the same cannot be said about the industry and automation field in the same manner.

The rest of the paper is organized as follows: Section 2 gives a brief summary of the different uses of the computer for control engineering, posing the different problems that the traditional approach involves in terms of costs, Section 3 outlines the research aimed at the integration and improvement of the isolated applications in different activities and introduces the disciplines to be used for solving the problems formerly mentioned, showing some ongoing work related with the ideas presented in this paper. In Section 4, a brief discussion is presented.

2. THE COMPUTER IN THE CONTROL ENGINEERING DOMAIN

Computers have been used mainly as a numeric processing machines when talking about control systems. As well as discrete time control, modeling and simulation are some of the fields from which more benefits have been obtained. MATLAB (and SIMULINK) are the prototypical examples of the numeric processing paradigm. These tools, being valuable don't offer an insight of the system at hand because they work at the "mathematical layer", that is, in a very low level.

Intelligent control is the name given to the use of the computer to build artificial intelligence artifacts that help in the control or supervision of systems. Expert Systems (mostly rule based), artificial neural networks or fuzzy control are some of the technologies used. Once more, the key aspects are the data processing in the computer and the need for "working in real time".

Computer software has also been devoted to tasks like computer aided design (CAD). These tools are graphical applications that help in the research and development of components and products, but once more, they are numeric processing applications whose greater advantage is the ability for rendering complex structures with a great flexibility.

All these systems have shifted the control and automation activities toward the computer. The human worker gets separated from the process and the engineer finds it difficult to handle the information provided by the computer because the great majority is only found in machine readable format or cannot be used from one application to another.

All these facts result in the existence of valuable computer applications that resolve punctual problems in a very efficient way but pose also great problems in terms of cost.

The first problem arises because of the mentioned low conceptual level of representation of the information used in these applications, what supposes a great effort when using and interpreting the results and thus leaves the use of these programs to the skilled engineer. This implies that the use of a piece of software application usually needs a training period specific to that application and thus an investment directed only to one person and one specific application.

A second problem is found in the isolation of the different applications, that makes almost impossible any kind of joint work among them. In order to work with different applications to achieve a given task, multiple specialists must work in isolation with their known programs and when the data exit one application and enter another one, both specialists must communicate in order the receiving application is able to understand the data obtained in the previous phase. This is also an important source of cost in industry.

The current software packages also lack the ability to handle large or complex systems, in particular those that cannot be grasped by the traditional mathematical representation used by the mentioned tools. This fact implies that great parts of these kind of complex systems are not introduced in the computer, and are only present in the mind of the expert. Then, when trying to "reason" about that kind system at a high level, a great effort (and thus cost) must be done in order to join the results of the different applications to the concepts and knowledge inside the mind of the expert.

3. THE NATURE OF THE PROPOSED ENGINEERING SOFTWARE

Today, there's a shift toward integration of all the different aspects of automation in industry into holonic systems. Some approaches to integrate the design and assembly tasks (Zha, 2000) or design and manufacturing (Su and Wakelam, 1998) have been proposed. The integration will eventually span over the entire life cycle of the product, as pointed in (Morel and Grabot, 2003).

Those approaches tend to join the existing stand-alone applications in a networked and distributed environment, but lack a common representation framework for the knowledge they are sharing.

Moreover, those approaches promote the integration by means of standardization of the information exchange (like, for example, the IEC 62264 Enterprise-control system integration -ISO/TC 184/SC 5 Enterprise-) but they don't improve the level of representation of the information in the computer.

As well as for implementing the previously mentioned integration, the approach outlined in this paper proposes the use of computers in an augmented way, representing higher conceptual levels of the knowledge of the field at hand.

3.1 A New Framework for Control Engineering Software: Joint use of Software Engineering and Knowledge Engineering

Enterprises have been working in this kind of use of computers described in the previous paragraph for some years, storing their "business knowledge" into the so-called "knowledge management" systems. The complexity of today's global economy has led to the need of developing such kind of complex computer systems that reflect the entire business processes and all the information relevant for the activities in the company. The tools used to build such systems come from the fields of software engineering and knowledge engineering (which in turn comes from Artificial Intelligence).

- software engineering is devoted to search for and build new methodologies, formalisms and tools to make the process of developing software in an easier and more flexible way in order to cope with the complexity that the last generation software applications must have.

- knowledge engineering is an important research area within the broad field of Artificial Intelligence. This engineering was born because of the need for representing the knowledge of a domain as an essential step in the building of the so-called knowledge-based systems (the software that results from the knowledge engineering process). The building of the domain models in knowledge engineering is based on the modeling of the concepts and the tasks that are found in that domain. This conceptualization is done without thinking about the later representation formalism that will be used when coding the application into the computer, as opposed to the software engineering, where the final coding is present in the conceptualization phase, conditioning it.

It could be said that software engineering, that was born to solve the problem in the software developing can be perfectly combined and used along with knowledge engineering, that was born to solve the knowledge representation problem. As is represented in figure 1, software engineering is near the "code or symbol level" while knowledge engineering is close to the "knowledge level" (Newell, 1982). In fact, both disciplines tend to merge and interchange knowledge and research.

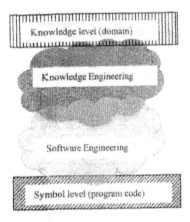

Figure 1. The knowledge and the symbol levels

The problem when building software using these two disciplines is the asymmetry in their applicability. Software engineering, being so tied to the objective of creating code for the applications necessarily produces useful tools that have been used for years while knowledge engineering, having a more abstract aim (working in the knowledge level), has been almost a pure research discipline for many years.

3.2 Current Trends in the Software Engineering Field (Object-oriented, Component-based and Distributed)

The current approach to software developing is the object orientation. The object is a software structure that encapsulates some kind of functionality and so provides a level of abstraction over the traditional function or code statement.

Another important paradigm in software engineering is reuse. Reuse is the use of the same code structures across different software applications. The piece of code that is reused is called "component" and has led to the so called component-based architectures (Heineman and Councill, 2001).

The use of components in a networked environment leads to the component distribution or distributed object computing (Object Management Group, 2001) a key concept in the building of complex software systems where heterogeneous pieces of code coming from different platforms can interoperate in a transparent fashion.

Modeling tools coming from software engineering are focused on this coding schema and so the world is described in terms of objects, properties, methods, interfaces, etc. The Unified Modeling Language (UML) (Booch *et. al.*, 1999) has become the de-facto standard for the modeling of object oriented like applications.

Object orientation is extensively used today in the software applications. In the control engineering

community object orientation has been used to build applications that provide a higher abstraction level respect to the pure numeric tools like MATLAB or SIMULINK. MODELICA (Mattsson *et. al.*, 1997), uses an object-oriented paradigm to represent systems modeled by means of bond graphs (Karnopp *et. al.*, 2000) in an effort to offer some insight of the system at hand. Real Time Innovations (RTI) has developed a tool where UML is used as a modeling language for the specification of system design, these specifications are then converted and linked to the SIMULINK notation in order to be simulated and analysed (Real Time Innovations, 2002).

3.3 Current Trends in the Knowledge Engineering Field (Ontologies and Problem Solving Methods)

The great majority of the current research in the knowledge engineering domain is the study of the methodologies for the representation of the tasks and concepts (along with their properties, relations and constraints) of a domain in a conceptual (knowledge level) structure called an ontology (or application ontology) (Gruber, 1993). An ontology is an specification of a conceptualization, something similar to an UML schema, but with much greater representation capabilities.

The tasks so defined in the ontology should be mapped to one or several generic problem solving methods (PSM), that may be used in a great variety of different problems (configuration, classification, design, etc) (Musen, 1998).

Here, the focus is also on reusability, as was the case with software engineering. The difference is that reusability in this case is stated at the knowledge level and not at the symbol or code level.

The main concern about knowledge engineering is the need of methodologies to build ontologies and problem solving methods. This is a very active field of research nowadays, as is the verification and validation of the structures obtained.

It then can be said that the building of knowledge-based system today consists in building an ontology of the domain knowledge and finding and using (or building) the problem-solving methods that fulfill all the functionality of the domain's processes. This approach is accepted by the most actual, prominent methodologies to build knowledge-based systems like Protégé (Gennari *et. al.*, 2002) or CommonKADS (Schreiber *et al.*, 1993).

Some good surveys on ontologies and problem-solving methods can be found at (Corcho *et al.*, 2001; Gómez-Pérez and Benjamins, 1999).

Few research has been devoted to apply knowledge engineering to the engineering disciplines, most of them aimed at building engineering ontologies.

The main task when building an ontology is finding the common concepts and processes that are usually involved in the field to be represented. In the engineering domain, some of these concepts may be:

- Description of physical systems, what implies description of mechanical systems with spatial and compositional relations among them.
- Description of the functionality of the different parts of a system.
- Description of casual relations among components.
- Description of the mathematical representation of the behavior of the system.
- Description of processes like design, modeling, analysis, etc.

For example in (Gruber et al., 1995) is described an ontology about mathematics for engineering called EngMath. This ontology has conceptual foundations for scalar, vector and tensor quantities, physical dimensions, units of measure, etc.

One of the most important engineering ontologies built so far is PhysSys (Borst, 1997). This ontology aims to represent the domain of engineering modeling, simulation and design. It has some very interesting ideas about ontology construction in general and about engineering ontologies in particular.

The modeling and design of engineering systems is one of the more promising fields of application of engineering ontologies. In (Borst et. al., 1997) a system comprising an ontology for engineering design is shown. The modeling of a system is viewed there in three conceptually different and separated levels: the component level, the process level and the mathematical level. This way, models can be built from components stored in a library and tested by using simulation at the mathematical level. The connection among the different levels makes it easy to redesign or modify the system very quickly.

4. DISCUSSION

It is clear that an improved and augmented representation of the knowledge of (control) engineering in computers should be a great advance in the use of the software in industry, leading to a better performance of the applications used and thus to reducing costs.

Some approaches, led to the integration of manufacturing systems into the enterprise have been carried out, but these systems lack a powerful representation of the knowledge of the domain.
Some studies have been done in order to bring the knowledge engineering structures to the engineering software, but these systems are based on the mere representation of usually low-level concepts in ontologies, as has been seen.

The framework proposed here is based on the representation of the domain at every level of abstraction, from the low level concepts (components, signals, etc) to the high level ones (design, analyze, etc). These concepts would be built into what is called a "knowledge based system" and will be the starting point to build different kind of applications. Moreover, as well as representing "static" concepts, actions should also be modeled.

The main problem facing this approach is the lack of knowledge of the control engineering domain by researchers in software engineering and also the lack of knowledge of the software and knowledge engineering by the researchers in control engineering. In order to break this vicious circle, our approach tries to build the kind of software described in this paper from the control engineering community. In order to do it, an intelligent tutoring (or training) system to help in the process of designing a lead-lag compensator is going to be developed. Building this system by means of the mentioned framework supposes to represent in the computer:

- Knowledge about the basic primitives like component, connection, etc.
- Knowledge of the analysis process (for example, frequency-domain analysis techniques) and the design process (what would suppose to represent heuristic knowledge in some cases).
- Knowledge of the tracking of activities of the human doing the design (actions, changes, etc).
- Knowledge about "what causes what" in order to give an explanation about why something has happened and how should things be done.

This development is still in its early stages, but the results obtained so far are very promising.

5. CONCLUSIONS

In this paper, a new framework for integrating the different software technologies and applications devoted to the (control) engineering field has been presented. This approach is based on the joint use of the software engineering and knowledge engineering fields in order to build applications that use knowledge of the domain at a higher level of abstraction.

REFERENCES

Booch, G., Rumbaugh, J. and Jacobson, I. (1999) The Unified Modeling Language user guide. Reading, MA, Adisson-Wesley

Borst, W. N. (1997) Construction of engineering ontologies. Ph. D-Thesis, University of Twente, NL-Centre for Telematics and Information Technology.

Borst, W. N.; Akkermans, J. M.. and Top, J. L. (1997) Engineering Ontologies. International Journal of Human-Computer Studies, 46, Academic Press, pp. 365-406

Corcho, O., Fernández-López, M., and A. Gómez-Pérez (2001) Ontoweb technical roadmap v1.0. Deliverable 1.1. Available online at: http://www.ontoweb.org/download/deliverables/D11_v1_0.pdf [active link on 23th, Jan, 2003]

Gennari, J., Musen, M. A., Fergerson, R. W., Grosso, W. E., Crubézy, M., Eriksson, H., Noy, N. F., and Tu, S. W. (2002) The Evolution of Protégé: An Environment for Knowledge-Based Systems Development. Technical Report SMI-2002-0943, SMI, Stanford University

Gómez-Perez, A., and V. R. Benjamins (1999) Overview of knowledge sharing and reuse components: Ontologies and problem-solving methods. Workshop on Ontologies and Problem-Solving Methods: Lessons Learned and Future Trends. (IJCAI99). Stockholm

Gruber, T. R. (1993) A translation approach to portable ontology specifications. Knowledge Adquisition, 5, pp. 199-220

Gruber, T. (1995) An ontology for engineering mathematics. In Doyle, J., Torasso, P. and E. Sandewall, Eds., Fourth International Conference on Principles of Knowledge Representation and Reasoning, Gustav Stresemann Institut, Bonn, Germany, Morgan Kaufmann.

Heck, B. S., Wills, L. M., and Vachtsevanos, G. J. (2003) Software Technology for Implementing Reusable, Distributed Control Systems, IEEE Control Systems Magazine, pp. 21-35

Heineman, G. T. and Councill, W. T. (2001) Component-Based software engineering: Putting the pieces together. Reading, MA, Addisson-Wesley

Karnopp, D. C., Margolis, D. L. and Rosenberg, R. C. (2000) System Dynamics: Modeling and Simulation of Mechatronic Systems, John Wiley and Sons

Mattsson, S. E., Elmqvist, H. E. and Broenink, J. F. (1997) Modelica™ – An international effort to design the next generation modeling language, Journal on Automatic Control 38 (3), pp. 16 – 19

Morel, G. and Grabot, B. (2003) Editorial of the special issue on manufacturing, Engineering Applications of Artificial Intelligence, 16, Elsevier, pp. 271-275

Musen, M. A. (1998) Modern Architectures for Intelligent Systems: Reusable Ontologies and Problem-Solving Methods. In C.G. Chute, Ed., AMIA Annual Symposium, Orlando, FL, pp. 46-52.

Newell, A. (1982) The Knowledge Level. Artificial Intelligence, 18, pp. 87-127

Object Management Group. (2001) The Common Object Request Broker: Architecture and Specification. Object Management Group, 2.5 edition, September 2001

Ollero, A., Morel, G., Bernus, P., Nof, S. Y., Sasiadek, J., Boverie, S., Erbe, H. and Goodall, R. (2002). Milestone Report of the Manufacturing and Instrumentation Coordinating Committee: From MEMS to Enterprise Systems. Annual Reviews in Control, 26, Elsevier, pp. 151-162

Real-Time Innovations (RTI) (2002), Cosntellation and Simulink, the Complete Controls Software Platform, White Paper from Real-Time Innovations.

Schreiber, A. Th., Wielinga, B. J., and J. A. Breuker eds. (1993) KADS: A Principled approach to knowledge-based system development, vol. 11 of Knowledge-based systems books series. Academic Press, London.

Su, D. and Wakelam, M. (1998) Intelligent hybrid system for integration in design and manufacture, Journal of Materials Processing Technology, 76, Elsevier, pp. 23-28

Verbruggen, H. B., Park, J., Halang, W., Irwin, G. and Zalewski, J. (2002) Milestone Report Coordinating Committee on Computer Control.

Zha, X. F. (2000) An object-oriented knowledge based Petri net approach to intelligent integration of design and assembly planning, Artificial Intelligence in Engineering, 14, Elsevier, pp. 83-112

OPERATION CYCLE OPTIMISATION OF THE LAROX PRESSURE FILTER

Jämsä-Jounela S-L. [1], Vermasvuori M. [1], Kämpe J. [1], and Koskela, K. [2]

1) Helsinki University of Technology
Department of Chemical Technology
Laboratory of Process Control and Automation
Kemistintie 1. FIN-02150 Espoo
FINLAND
E-mail:Sirkka-l@hut.fi
2) Larox Corporation, P.O. Box 29, 53101 Lappeenranta, FINLAND

Abstract: Artificial intelligence methods such as expert systems, fuzzy systems, neural networks and combinations of these, have become invaluable tools in helping operators to monitor and control processes. These methods can also be used to run processes in a more economically effective way and, in the case of equipment malfunction, they can propose appropriate corrective measures. In this paper a system for operation cycle optimisation of the Larox pressure filter is presented and some test results are discussed. *Copyright © 2004 IFAC*

Keywords: Expert systems, Optimisation problem, Fault detection, Computer applications and Models.

1. INTRODUCTION

Artificial intelligence methods have become more common in a wide range of scientific and engineering fields. The number of applications is increasing rapidly and successful results have been reported. Artificial intelligence methods such as expert systems, fuzzy systems, neural networks and combinations of these, have proved to be excellent tools for the control of mineral processes (Jämsä-Jounela et. al., 1996)

During the last 25 years many advanced, model-based fault diagnostic systems have been developed (Isermann, 1997, Himmelblau, 1978). Estimation methods for evaluating changes in parameters and states have been presented in Isermann (1993). There have also been reviews on different fault diagnosis methods and their current status and the future trends (Isermann and Ballé, 1997, and Jämsä-Jounela, S-L., 2001).

This paper presents an intelligent operation support and optimisation system for a Larox pressure filter. The system is programmed using Java, a platform-independent object-oriented programming language. The structure of the system is modular, which makes it updateable and expandable. The system consists of classification, modelling, optimisation, economical, fault diagnostic and remote support modules. The aim of the system is to maximise the capacity of the filter by optimising the cycle stage times and to find optimal operation parameters so as to ensure that the end product meets the quality criteria. Economical efficiency is improved not only by increasing capacity, but also by lowering the amount of energy consumed and by using fault diagnostic and remote support modules to decrease process downtime.

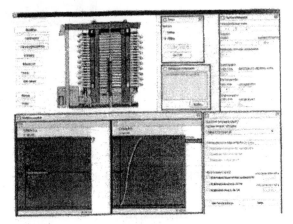

Figure 1. Screen capture of the support system software.

2. THE STRUCTURE OF THE SUPPORT SYSTEM

2.1 Classification module

The aim of the classification module is to determine what kind of slurry is being filtered. The optimal operation parameters depend on the type of feed, and the parameters are optimised separately for every type of feed. The classification of the feed type is performed on the basis of measurements that describe the physical properties of the slurry, and it is implemented by means of an artificial neural network, the Kohonen Self-organizing map (SOMs) (Kohonen 1990, Jämsä-Jounela, 1998).

2.2 Modelling module

The modelling module is used to monitor the progress of the feeding stage of the filtration cycle. A model has been derived for a single chamber, constant-pressure filter. The starting point for the model is Darcy's equation (1).

$$Q = k \frac{A \cdot \Delta p}{\mu \cdot l} \qquad (1)$$

where
Q = the accumulation rate of filtrate [m³/s],
k = a constant, permeability of the cake [m²],
A = the area of filtration [m²],
Δp = total filtration pressure difference [Pa],
μ = viscosity of the filtrate, and
l = thickness of the cake

The achieved model is presented in (2)

$$t = aV^2 + bV \qquad (2)$$

where
V = the volume of the filtrate [m³] and dummy variables a and b are:

$$a = \frac{\mu \cdot \alpha \cdot c}{A^2 \cdot \Delta p} \qquad b = \frac{\mu \cdot R_m}{A \cdot \Delta p} \qquad (3)$$

The duration of the feeding stage is then a function of the filtrate volume and, when we know the desired volume, the time taken to reach this volume can be determined if the parameters a and b are known. The parameters a and b are identified using the recursive least squares method proposed in Åström et al., 1984. The detailed derivation of the model is presented in Jämsä-Jounela and Oja (2000).

2.3 Optimisation module

The optimisation module optimises the feeding, pressing and drying stages of the filtration cycle. The individual stages are optimised independently.

The feeding stage. In the feeding stage the goal is to continue feeding long enough for the chambers of the filter to become filled with slurry. The stage should not last longer than this because overfilling of the chambers causes problems during the next operation cycle. The optimal increase in the mass of the filter during the feeding stage is determined on the basis of information about successful operation cycles in the past. The time when the mass of the filter reaches this optimal value and the feeding can be stopped is predicted using the recursive model described earlier.

The pressing stage. The aim in the pressing stage is to get the cake dense enough for the drying stage. If the cake is not pressed hard enough, it may break during the drying stage. If the cake cracks, the pressurised air used for drying flows mainly through the cracks and the cake will not dry properly. If this happens then the desired moisture level of the cake cannot be achieved. The pressing stage can be divided into two separate sub-stages (Kämpe, 1999). At first, the excess liquid on top of the cake is pressed through the cake and filtering medium. The cake is then further pressed in order to achieve the density required in the drying stage. During the second sub-stage the amount of filtrate coming from the filter is much smaller than during the first sub-stage. As a result, the transition from the first sub-stage to the second can be seen from the rate at which filtrate accumulates.

The drying stage. In the drying stage the object is to dry the cake to the desired moisture level as fast as possible. As for the pressing stage, the drying stage can also be divided to two sub-stages: in the first stage the liquid between the particles is blown out, and in the second sub-stage the air flow removes moisture from the surface of the particles. Again, the transition between sub-stages can be seen from the change in the rate of accumulating filtrate.

2.4 Economical module

The main purpose of this module is to support the decision making of the optimisation module and to motivate the operator to run the process in an economical way by monitoring and displaying the operating costs of the filtering process. If the same quality and amount of final product can be achieved with different operating strategies, the one with the lowest costs is chosen. The operating costs are calculated once per cycle using information from all process stages.

2.5 Fault diagnostic module

The fault diagnostic module uses parameter values a and b of the modelling module to monitor the behaviour of the filter. During the normal feeding stage, the parameters level off at certain near-constant values. In the case of a specific problem, the model does not describe the situation, which is indicated by the continuously changing values of the parameters. Normal process measurements are monitored and, if they behave abnormally, symptoms are generated. The symptoms are then compared against known symptom combinations of different malfunctions in order to identify the abnormal operating condition.

2.6 Remote support module

The purpose of the remote support module is to provide a secure and reliable way of transferring data from the filter to experts, who can then analyse the data if malfunction occurs. The remote support module consists of an SQL database server that is connected to a Larox pressure filter, a Microsoft Internet Information Server (IIS), workstations, and a LAN network. The workstations are connected to each other either over the Internet or via modems. For security reasons the Internet connection has been secured by means of a Virtual Private Network (VPN) and access to the local network is restricted by a firewall. This structure is presented in figure 2.

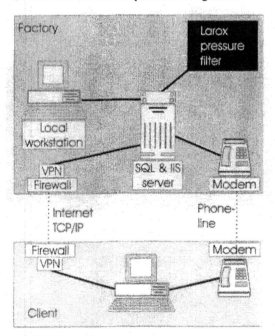

Figure 2. The structure of the remote support module.

3. TESTING OF THE SUPPORT SYSTEM

The support system software has been tested off-line with data collected from a multiple chamber Larox pressure filter (model Powerex PF60/60), which has 10 chambers 60 mm in height and a total pressing area of 60 m². The test data have been collected in February 2003 using the remote support module. The encrypted data were sent over the Internet and the connection was secured using VPN.

The system has also been tested on-line with two different kinds of pressure filter: a single chamber Larox PF variable volume filter with a 1.6 m² filtration area, and the Larox Powerex PF60/60.

3.1 Testing with a single chamber filter

For the feeding stage the modelling module used the model in (13). Values of a and b gradually level off to constant values, as can be seen in figure 3. When the parameter values no longer change, the model can be used to accurately predict the time when enough slurry has been fed to the chamber and the feeding stage should be stopped. The model works very well, and there is only minimal difference between the measured and predicted values in the cumulative amount of filtrate, as can be seen in figure 4.

The fault diagnostic module was tested with different kinds of problem in the process: missing the control signal of the valve, problems in measuring the feeding pressure, low feeding pressure of the slurry, significant and rapid changes in the composition of

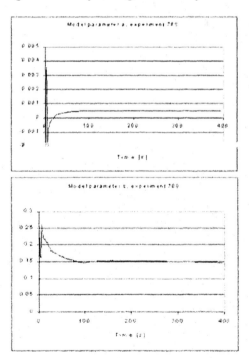

Figure 3. The evolution of the model parameters a (top) and b (bottom).

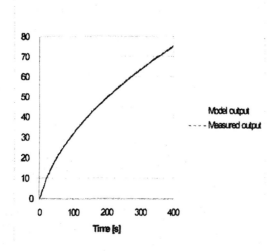

Figure 4. Modelling the change in filtrate volume.

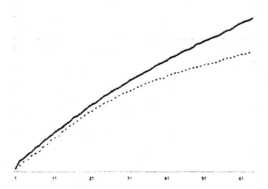

Figure 5. Development of the amount of filtrate under normal operating conditions (solid line) and when the pump pressure is low (dashed line).

the slurry, and clogging of the filtering medium.

In one experiment the pump pressure was lower than that in normal operation. Compared to the normal feeding stage, the amount of filtrate coming through the filter increases more slowly. This natural consequence of a low feeding pressure is shown in figure 5. In addition, parameters a and b behave rather differently compared to the normal situation. Their values do not level off to any single value, which indicates that there is a problem in the process. The evolution of parameter values in normal operation and in this low pressure experiment are illustrated in figures 6 and 7.

3.2. Testing with a Multi chamber filter

The model derived for a constant pressure, single chamber filter cannot be used as such with the Powerex PF60/60 filter as can be seen in figure 8. The

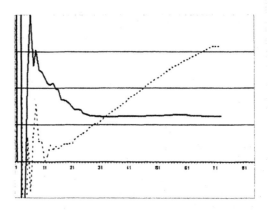

Figure 6. The behaviour of parameter a during the normal feeding stage (solid line) and in the low pump pressure experiment (dashed line).

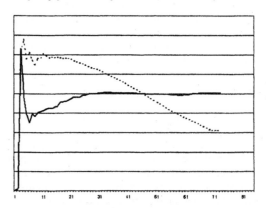

Figure 7. The behaviour of parameter b during the normal feeding stage (solid line) and in the low pump pressure experiment (dashed line).

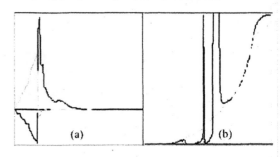

Figure 8. (a) Poor behaviour of parameters a (grey) and b (black) and (b) predicted (black) vs. measured (grey) filtration on the right.

main problem is that in a real industrial size filter the feeding pressure is not constant, but is raised stepwise to its final value. The change in feeding pressure is shown in figure 9.

To solve this problem the model was modified to use the actual pressure difference according to pressure measurements instead of a constant value.

$$t = \frac{a}{\Delta p}V^2 + \frac{b}{\Delta p}V \qquad (4)$$

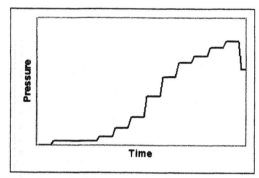

Figure 9. Gradual increase in feeding pressure during the feeding stage.

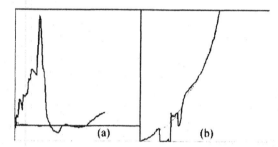

Figure 10. (a) The development of parameters a (grey) and b (black) and (b) the predicted (black) vs. measured (grey) amount of filtrate on the right.

Much better results were obtained using model (4) instead of (2), as can be seen in figure 10. One case dependent problem remains; the parameter values are still fluctuating in the beginning of the stage and during the relatively short feeding stage in this case, the parameter values do not have enough time to level off at near-constant values.

Because there is some delay between the filter and the tank where the filtrate is collected and weighed, the weight of the filter itself was used with the modelling module. The weight initially increases rapidly as the chambers are filled, but then the cake formation obstructs the flow filtrate and the rate of change gradually decreases. Much better behaviour of the parameter values was achieved using this delay-free measurement and the prediction of the amount of filtrate is more accurate, as can be seen in figure 11.

3.2.1 On-line tests with the Powerex PF60/60

The system was tested on site with an industrial-size pressure filter, Powerex PF60/60, in June 2003. The moisture content of the cake was measured on 42 cycles during which the feeding, pressing and drying times were altered. The feeding stage time varied from 65 to 75 seconds, pressing time from 150 to 480 seconds and drying time from 270 to 330 seconds.

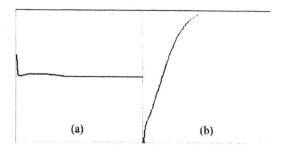

Figure 11. (a) Parameters following the desired behaviour and (b) prediction of the amount of filtrate determined using the mass of the filter with the modelling module.

Feeding stage. The feeding time was kept at a constant 65 seconds during most of the experiments since the predictions of the modelling module could be examined without making changes to the feeding time. During the testing period the modelling module suggested feeding times from 68 to 72 seconds. These values are good estimates for the optimal stage time since, with a 65 second feeding time, the cakes were not at their maximum thickness.

Pressing stage. The results of for pressing stages of different duration are shown in figure 12. During these tests the drying time was kept constant at 300 seconds. It seems that, if the pressing time is sufficiently long for the cake to become solid enough, further pressing will actually increase the moisture content of the cake. So, instead of trying to press the moisture out of the cake, pressing should be used only to make the cake solid and give it the right form, and then use compressed air to remove excess liquid.

Drying stage. The effect of drying time on cake moisture is shown in figure 13. During these tests the pressing time was kept constant at 200 seconds. After a certain time in the drying stage, the mass of the filter decreases almost linearly, as can be seen in figures 14 and 15. This weight loss is completely due to mass loss of the drying cakes, i.e. removal of moisture. This information can be used to predict how much longer the drying stage should last.

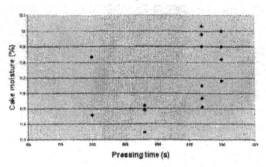

Figure 12. The effect of pressing time on cake moisture.

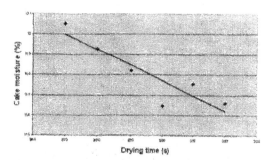

Figure 13. The effect of drying time on cake moisture.

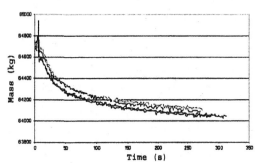

Figure 14. The weight of the filter during a number of drying stages.

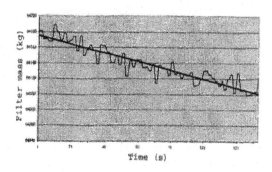

Figure 15. Mass of the filter during part of a drying stage and its linear approximation.

Whole cycle. During the brief testing period the system was able to reduce the filtration cycle time by over 12 percent, which means a considerable increase in the production rate.

4. CONCLUSION

There have been very few reported attempts to optimize the operation cycle of pressure filters. In this paper an operator support system, which performs the optimisation, was introduced.

The support system software seems to work well and the modelling and optimisation algorithms produce good results. In addition, the fault diagnosis module can be used in abnormal operation situations to alert operators and the data describing anomalies can be sent via a remote support module to experts for further analysis.

ACKNOWLEDGEMENT

We would like to thank TEKES and Larox Corporation for funding this research project and providing technical help and real industrial data from the pressure filtering process.

REFERENCES

Himmelblau, D. M., (1978) Fault Detection and Diagnosis in Chemical and Petrochemical Processes. Elsevier, Amsterdam 1978, 414 p.

Isermann, R., (1993) Fault Diagnosis of machines via parameter estimation and knowledge processsing - tutorial paper. *Automatica* **29** (1993) pp 815-835.

Isermann, R., and Ballé, P., (1997) Trends in the application of model-based fault detection and diagnosis of technical processes. *Control Engineering Practice* **5** (1997) pp 709-719.

Jämsä-Jounela, S-L. Ore Type based Expert Systems in Mineral Processing Plants, *Particle & Particle Systems Characterization* **4** (1998).

Jämsä-Jounela, S-L., (2001) Current Status and future trends in the automation of mineral and metal processing. *Control Engineering Practice* **9** (2001) pp 1021-1035.

Jämsä-Jounela, S-L., Laine, S. and Ruokonen, E. Recent Developments in Designing Ore Type Based Expert System at the Hitura and Kemi concentrators. Proceedings of IFAC 13th World Congress, San Francisco. (1996)

Jämsä-Jounela, S-L., Oja, M., Modelling module of the intelligent control system for the variable volume pressure filter, *Filtration and Separation*, **37**(2), 2000, pp 39-49.

Kohonen, T. The Self Organizing Map, Proceedings of the IEEE **78** (9), 1990, pp 1464-1480.

Kämpe, J., (1999) Fault diagnostic system for a pressure filter. Master of Science (Tech.) Thesis, HUT.

Åström, K. J. and Wittenmark, B. (1984). Computer-Controlled systems-Theory Kemi concentrators. Proceedings of IFAC 13th World Congress, San Francisco.

ELSEVIER

IFAC

PUBLICATIONS
www.elsevier.com/locate/ifac

AUTOMATION SYSTEM DESIGN THROUGH COST OPTIMISATION UNDER DEPENDABILITY CONSTRAINTS

Blaise CONRARD, Mireille BAYART

Laboratoire d'Automatique, Génie Informatique & Signal
(LAGIS UMR CNRS 8146)
USTL, Cité scientifique, Bât. Polytech-Lille
59 655 Villeneuve d'Ascq Cedex - FRANCE
Blaise.Conrard@polytech-lille.fr, Mireille.Bayart@univ-lille1.fr

Abstract: This paper deals with control system design and optimisation according to a cost criterion and dependability constraints related to the number of tolerated faults. The proposed method is designed to be used with a system model that describes both functional and dysfunctional behaviour and takes the different hardware architecture options into consideration. This model is derived through a functional decomposition of the system, completed by dependability data.

From this model, we apply an optimisation algorithm based on a branch and bound method that can select the best system architecture, one that meets a set of dependability constraints while maintaining the lowest cost. This paper also presents the different techniques for optimising our results in order to provide an optimal solution for rather large systems. *Copyright © 2004 IFAC*

Keywords: Computer-aided control systems design, Structural optimization, Decision trees, Safety, System reliability, Global optimization, Structural constraints.

1. INTRODUCTION

In terms of production systems, automation systems (or control systems) are ordered sets of standard components, such as sensors, actuators and processing units, interconnected in point-to-point lines or fieldbuses. Designing the architecture for such systems boils down to determining the best assembly of the different components. Thus, for each process, designers must first identify the basic components required and choose those that are best suited to the needs of the particular system from the diverse component types available—standard, safety or intelligent (often called smart). At the same time that they choose the specific components that will be used, they must also determine which to make redundant and how to organise the various elements to achieve the best result.

Basically, this means selecting from those components available on the market the ones that will significantly improve system reliability and safety without increasing costs. Since architectural choices have a significant impact on dependability,

independent of cost, the appropriate distribution of sensors and actuators on each processing unit can actually increase the capacity of the system to be fault tolerant (Gehin, 1999).

This paper is organised as follows. Section 2 presents and explains the criteria and constraints needed to optimise the cost of a control system. Sections 3 and 4 deal with the optimisation method: the first concerns the system modelling and the second, the optimisation algorithm. Finally, section 5 provides a sample application of the method.

2. OPTIMISATION OF AN AUTOMATION SYSTEM

Designing an automation system consists of determining the appropriate architecture for the system components. This requires comparing the different possible solutions according to certain criteria. When dealing with production systems, the main and unique criterion is Life Cycle Cost (LCC). By definition, LCC includes the cost of development, acquisition, operation, support (including manpower), and where applicable, disposal. In automation system design, LCC can be reduced to two elements: a cost criterion and a dependability constraint. The second can have quite a large impact on LCC, because both system unavailability (Laprie, 1999), with its resulting production stoppages, and system failure, with its decreased or poor quality production, induce profit loss.

Consequently, the problem of automation system design can be reduced to creating a system whose level of dependability meets the required standards for a given process and whose component cost is minimal. There are several ways to characterize dependability level. This article characterises it as the minimal number of failure components that can cause an undesirable system event. Using such semi-quantitative dependability constraints provides certain advantages to the designer. First, the modelling task requires less data and the dependability assessment is simplified, thus making the method more suitable for the optimisation of large system architectures. Moreover, this characterisation of dependability resolves a number of contemporary design problems linked to the difficulty of obtaining quantitative information and feedback about the robustness and reliability of new components, which are increasingly electronic.

The second criterion can lead to particular problems. Indeed, setting dependability level according to the number of failed components is sometimes inappropriate for comparing components with varying levels of robustness. For an elementary function performed by a single component, several component types (standard, safety or intelligent) are possible, each with varying costs and varying levels of robustness. To compare the different possibilities, the faults must be quantified in terms of their relative failure probability. For instance, if the failure mode of a standard component is equal to a coefficient 1, the coefficient of the same failure mode for a safety component would be related to how much less likely it was to occur (maybe 2, 10 or 100 depending on the number of times that the probability of its occurrence is reduced).

The problem then becomes a question of designing a system with a minimal cost that ensures a set of dependability constraints. Each constraint is expressed in the following form: "For a given failure mode of the system, there can be no combination of failed components whose failure mode coefficients total more than the limit set for inducing its occurrence".

3. SYSTEM MODELLING FOR SYSTEM OPTIMISATION

3.1 Automation system modelling

Optimisation methods are used to select the best solution from the set of potential solutions to system problems. The first step in optimising a system is,

thus, to build a system model that adequately represents the various implementation possibilities and that will allow the various criteria to be evaluated and the constraints, checked.

The following model, based on a functional decomposition of the system to be designed, fulfils the requirements described in the previous paragraph. An automation system usually has several secondary missions to accomplish in addition to its primary mission. For each of these missions, a functional decomposition can be established, by breaking each mission down into its various functions, which can in turn be further broken down into sub-functions, and so on. The final level of the hierarchy is reached when the decomposition produces functions that are elementary enough to be performed by one standard automation component.

Such modelling makes it possible to envisage several control architectures. This method requires defining a variety of possible implementations for each elementary function, using a set of components that each have different costs and different reliability levels. The method also offers the possibility of non-implementation, in which case, the associated component cost is zero, and the function is considered to be perpetually unavailable. Non-implementation is permitted only for redundant or optional functions whose only role is to improve the system, or for those functions that can be replaced by another function.

With this model, the optimisation process entails examining each elementary function in the decomposition hierarchy and deciding whether or not to implement it. If the decision is made to implement, the executing component must be chosen from the set of possibilities. If the model is developed enough, meaning that all the complex functions have been broken down into their various elementary functions, many different architectures can be evaluated during the optimisation phase, and the one that is the most appropriate to the designer's goals can be chosen.

3.2 Behavioural description

The problems related to dependability constraint verification can be solved entirely only if the model integrates a description of system behaviour in terms of the failure state of its components. Basically, there is a relationship between the state of the individual components (either in an operational mode or in a failure mode) and the state of the entire system. This is a boolean relationship that characterises the combination of sub-function failure modes that cause each subsequent failure mode to occur. The functional decomposition described in section 3.1 can be used to establish the state of each function in terms of the state of its sub-functions, for every level of the decomposition hierarchy. Globally, each failure mode of the system can be described in terms of the set of combinations of faulty elementary functions that induce it. The dependability assessment is done by applying the dependability coefficients of selected components to each corresponding elementary function in order to evaluate the minimal sum of coefficients that can induce each undesirable event.

3.3 Reusing of partial solutions

One of the particularities of automation systems is that, as an assembly of standard components, their design is essentially an application of the reuse principle. This principle, often called COTS ("Component Off The Shelf" (Simpson 1996)), consists of using (or reusing) components designed for a general purpose. This same principle can be used to create the functional decomposition model described above. Reusing classic subsets not only simplifies the modelling task and thus decreases the time spent doing it, it also makes it unnecessary to redefine the relationship between the failure of a function and that of its sub-functions.

3.4 Two levels of modelling

The model discussed in the previous sections describes the components that can be used to accomplish each mission. Depending on the time available to the designer and the desired accuracy of the results, the model can be more or less elaborate. Two levels of elaboration are possible. The first, the simplest and thus the most rapid to construct, takes into account only the instrumentation (sensors and actuators). The second, more detailed, integrates the processing unit and the communication media.

For the first level, the search for a suitable architecture is done in two steps because a reduced model of the system is used. The first step consists of determining the optimal instrumentation, one that respects the dependability constraints while maintaining the lowest cost. The second consists of distributing the instruments among several processing units; this choice is made with respect to the dependability constraints. The second refinement level requires only the first step, but it is applied to a larger and more complex model of system.

The first level of elaboration is well suited to required standard automation architecture, corresponding as it does to the control system architecture proposed by component suppliers (figure 1). This architecture has two levels. The lowest level, composed of sensors and actuators, constitutes a process interface; the second one is composed of processing units. Each of these units has its own sensors and actuators, which are interconnected via point-to-point lines or an I/O fieldbus. In order to coordinate their actions, these units are linked through a federate (field)bus. When intelligent instruments (Staroswiecki, 1996) are used, their enhanced communication and their processing capabilities allow them to be directly connected to the federate network. Thus, with intelligent instruments, distribution on the different processing units is not necessary.

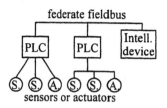

Fig 1. Standard control architecture

4. OPTIMISATION

The optimisation algorithm is a two-part procedure. First, the solution space is scrutinised to determine a set of possible solutions. Second, the cost and the dependability of each potential solution are assessed in order to retain the best solution.

4.1 Scrutinising the solution space

4.1.1 Generalities

Due to the nature of optimisation problem, each solution can be represented as a vector. Each parameter of the vector is then associated to an elementary function, and the parameter value indicates the component used for its implementation, or non-implementation. The search for the optimal solution becomes a question of scrutinising the solution space (i.e. the set of possible vectors) to determine which solution has the smallest cost and respects the dependability constraints imposed by the designer.

An automation system requires a large number of elementary functions, each of them involving several components. Thus, the number of possible solutions is very large (the number of possible components × the number of possible functions). Given this large number, the optimisation algorithm can rarely scrutinise all possible solutions, and so studies only the potentially interesting solutions. In this paper, we propose a method based on the Branch-and-Bound algorithm (Chen, 1990), as well as several techniques that can speed up the search for the optimum solution.

4.1.2 Principle of the algorithm

The optimisation algorithm scrutinises solutions by testing the different possible values for each parameter of the solution vectors. This search can be modelled as a tree, in which each node corresponds to a vector parameter and each of the branches corresponds to one of the different values of this parameter. To find the optimum solution, it is only necessary to determine which leaf on the tree respects the constraints and has the minimal associated cost. Figure 2 illustrates this approach.

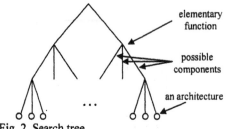

elementary function

possible components

an architecture

Fig. 2. Search tree

4.1.3 Improving the research

In order to avoid scrutinising every leaf on the tree (which would be a gargantuan task), several techniques are possible.

1) Branch sorting
Since the goal of the algorithm is to find a minimum, sorting the branches according to increasing cost would speed up the process. By putting the solutions with the lowest costs first, this technique would increase the probability of finding the optimal solution quickly.

2) Parameter assessment order
Evaluating the cost criterion before the dependability criterion would also improve the algorithm's speed. Since the cost evaluation can be done more quickly than the dependability evaluation, interesting solutions can be compared in terms of cost, with only the less expensive solutions being retained for further evaluation in terms of dependability. This technique allows time-consuming, but ultimately non-productive dependability evaluations to be avoided.

In addition, by determining which vector parameters induce greater costs, the algorithm can skip over those solutions that are obviously not the optimum. This is possible only if an initial solution is found and a maximal cost is determined. If the new potential solution has a greater cost, the assessment algorithm can identify which parameters induce this greater cost. Knowing these parameters, the scrutinising algorithm can select a new potential solution with new values for these parameters. In such a way, the algorithm avoids testing other solutions that have the same combination of bad parameter values. Indirectly, a bound is made in scrutinising the solution tree, and the algorithm saves time overall in the search for the optimal.

3) Use of dependability criteria information
The faulty parameters in vector solutions that do not satisfy the dependability constraints can be identified and used in the same way that unsatisfactory cost criteria were used above. Knowing these parameters allows the next envisaged potential solution to be one with another combination of parameter values. As was true for the optimisation technique, a bound is made in scrutinising the solutions tree, allowing a significant number of unsatisfactory solutions to be ignored. This technique, which allows whole groups

of solutions to be eliminated, is essential to the efficiency of algorithm.

4) Splitting the solution space
Studying the constraints before starting the optimisation process can reveal independent subsets of variables. (Two variable subsets are considered independent if the subsets have no common members.) In such cases, the optimisation algorithm can be applied to each subset independently, thus saving a lot of time. The global optimum is the union of the optimal solutions for each of the sub-sets.

Cases involving independent subsets of variables are rather common when dealing with automation systems because the automation process is often a string of independent subsystems (ex.: assembly line or production line...). Given configurations in which the subsystems are independent, a failure generally effects only the subsystem to which it belongs, and the solutions generally require only local reconfigurations (within the same task) rather than reconfigurations of whole systems. Given such independence, applying fault tolerance constraints to the whole system amounts to applying those constraints to each sub-system.

4.2 Assessing the solution

Two aspects of the solutions proposed during the scrutinising phase of the algorithm have to be studied: the cost criterion and the dependability constraints. The first refers to system cost. This can be evaluated by doing a simple sum of the cost of the components selected for the potential solution.

The second aspect is more complex. One way to evaluate dependability to establish a priori all the combinations of elementary faults that induce the occurrence of an undesirable event. But decision graphs (based on binary decision diagrams, BDD (Rauzy, 1996)) seem to be the most appropriate method for coding this sets of fault combinations when their number is sufficiently important. Indeed, the verification process proceeds quickly to verify whether a potential solution satisfies the constraints once the initial diagrams are drawn.

However the combinations are coded, verifying that a defined number of component faults do not cause the occurrence of an undesirable event consists of checking the order of the minimal cuts associated to this event. A cut is a combination of failed components that induces the failure of the system. A cut is minimal if the system returns to an operating state when one of its components goes back to a normal state. The order of the cut refers to the number of components belonging to the cut in question. For each potential solution (i.e. an architecture) and for each given event, the associated constraint is satisfied only if the order of all the minimal cuts for this event reaches the required value. In this way, verifying a constraint consisting of considering each cut composed of elementary functions and extracted from the previously determined set of combinations. For each cut, the elementary functions are replaced by the corresponding selected component, or removed, in the case of non-implementation. Finally, the order of the new cut is revaluated and compared to the required value in order to determiner if the constraint has been satisfied or not. During this revaluation, the robustness coefficient can be used to assess the order.

With the use of a decision diagram, all minimal cuts (of elementary functions) can be reorganized into a more compact format better suited to data processing. Each node of this diagram is associated to the failure mode of an elementary function, and its branches correspond to the function's presence or absence in a cut. The terminal leaves of the diagram are labelled and define whether or not the elements present on the path constitute a cut. If a leaf corresponds to a cut, the dependability level (i.e. the number of tolerated faults) is associated to it. With such a diagram, all the cuts of a particular solution can be studied simultaneously, by scrutinising the diagram from its root toward its leaves. At each node, the minimal number of faults needed to reach this node is evaluated iteratively by considering the component associated to the elementary function and increasing the number of faults from node to node according to this component's reliability coefficient. (This coefficient is described in section 2.) Scrutinising the whole diagram yields the number of faults needed to reach each leaf. The respect of the constraints is verified by comparing this number with the minimum required for each leaf.

5. APPLICATION

In the proposed application, our method is applied to an average large system composed of a reactor in which two fluids are mixed and heated for a set period. At the end of this period, the tank is drained and the product is provided to the final user.

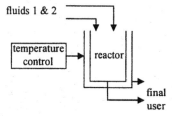

Fig.3. Process for which a control system must be designed

First, the system must be modelled through a process of hierarchical functional decomposition. This system has three main functions: filling, heating and draining. The function 'filling' requires several channels from the supply tanks to the reactor. (One of the goals of optimisation is to determine the optimal number of required channels.) Each channel is composed of three elements (valve, pump and flowmeter), which comprise the elementary

functions. Each function can be implemented in several different ways. For example, three possibilities are offered for the elementary function 'pump': non-implementation, standard pump and robust pump. Each possibility has an associated financial cost: 0, 10, 20, respectively. (These values are arbitrary.) The function 'valve' offers two cases of non-implementation: unimplemented with a blocked channel and unimplemented with an open channel. The optimisation process has to determine whether or not the valve must be implemented and whether or not implementation contributes to making system safer.

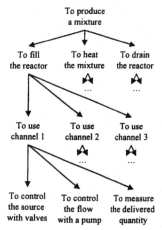

Fig.4. Hierarchical functional decomposition

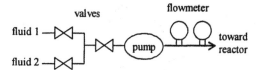

Fig.5. A channel for supplying reactor

A table for each elementary function supplies the robustness of each potentially usable component. For each possible failure mode and for each implementation case, this table provides a dependability coefficient that can be compared to a number of standard failures to obtain a failure mode.

Table 1 & 2 Example of proposed implementation

Implementation case for function 'pump'	failure mode	
	unavailable and stop	continuous pumping
unimplemented case	0	∞ (=never)
standard pump	1	∞ *
robust pump	2	∞ *

*: only due to the control system

Implementation case for 'valve' function	failure mode	
	unavailable and blocked	unavailable and opened
Unimplemented and open case	∞	0
Unimplemented and blocked case	0	∞
Standard pump	1	1
Robust pump	2	2

The second step involves defining the failure mode for each function and the relationship between these modes. In the case of the system function "to produce the mixture", these failure modes are "unavailable for nominal functioning" (when the reactor is filled with the two fluids simultaneously), "unavailable for degraded functioning" (when both fluids arrive simultaneously, but one fluid is pumped more slowly), "poorly produced mixture" (for example, a wrong ratio between the two fluids, or the incorrect heating temperature), for an accident (the overflow of the reactor, for example). Similar modes can be attributed to the other functions and sub-functions and the relationship between these modes can be established. For example, the unavailability of the function "filling", "heating" or "draining" causes the unavailability of the system prevents its mission "to produce the mixture" from being fulfilled.

Using these definitions, the optimisation of the instrumentation can be done according to the cost of the components (valve:5, pump:10, flowmeter:3). Table 2 sums up the results for the filling function tin terms of the various desired dependability levels.

Table 3 Different results for various constraints

Failure mode	Desired dependability level			
Unavailable in nominal mode	0	1	1	2
Unavailable in degraded mode	1	1	2	2
Bad mixture	2	2	3	3
Accident	2	2	3	3
Minimal obtained cost :	13	26	42	58

Finally, using the proposed instrument architecture, the number of control units can be determined from the second step of the design method that consists of distributing instruments to several control (or processing) units. For the application, the required number of control units is equal to the limit set for the failure "Unavailable in degraded mode".

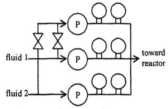

fluid 1

fluid 2

toward reactor

Fig.6. A possible architecture

At the same time, the components are allocated to these control units to yield the final control system architecture. Due to the relative simplicity of the process, one control unit per channel is enough to ensure both the control and safety of the system. Figure 6 shows the instrument architecture proposed for last column of the table 3. It is composed by 3 channels; two of them are dedicated to pumping each fluid and the last can pump one of two fluids via a net of valves. With this architecture, the failure of one component still permits the reactor to be filled with the two fluids simultaneously. Moreover, even if 2 components fail, the reactor can be filled, albeit with one fluid at a time, and so, the system stays available although in a degraded mode.

CONCLUSION

The method proposed in this article is designed to optimise automation system architecture using a cost criterion and dependability constraints. The system is modelled from a set of elementary functions, and several components are proposed for the support of each one. The optimisation consists of selecting the best combination of components that will ensure the required dependability level while maintaining the lowest cost. The dependability constraints are expressed from a semi-quantitative point of view. They define for each undesirable event a given minimal number of failed components that can induce the occurrence of this event. This formulation of the design problem requires first establishing the relationship between the system failures and those of the elementary functions or components. These relationships are defined through a functional decomposition of the system.

Because of the nature of the problems, using an algorithm based on the Branch-and-Bound method and the Binary Decision Diagram allow the optimal solution for systems composed of one, or several hundred, components to be obtained within a suitable timeframe. The proposed method can be used very early in the design phase. It provides a preliminary evaluation both of the system cost and of the hardware architecture that seems to be the most suitable. Moreover, considering dependability from a semi-quantitative point of view responds to current designer problems, including the difficulty in finding quantitative information about the reliability of new components, due to lack of feedback and the complex behaviour of their electronic parts.

REFERENCES

Gehin, A.L. and M. Staroswiecki (1999). A formal approach to reconfigurability analysis application to the three tank benchmark. In: proceedings of European Control Conference, ECC'99, Karlsruhe, Germany.

Laprie J.C. (1999), *Dependability of software-based critical systems, Dependable network computing*, Ed. DR.Avresky, Kluwer Academic Publishers.

Rauzy A. (1996) An Brief Introduction to Binary Decision Diagrams, In : *RAIRO-APII-JESA 30(8)*, pp. 1033-1051, Hermes, Paris

Staroswiecki M., Bayart. M. (1996), Models and Languages for the Interoperability of Smart Instruments In: *Automatica*, Vol. 32, No. 6.

Chen, G.H., & Yur, J-S. (1990). A branch and bound with underestimates algorithm for the task assignment problem with precedence constraint. In : *Proc. 10th International Conference On Distributed Computing Systems*, pp. 494-501, Paris, France.

ELSEVIER
IFAC
PUBLICATIONS
www.elsevier.com/locate/ifac

DEPENDABILITY STUDY IN DISTRIBUTED CONTROL SYSTEMS INTEGRATING SMART DEVICES

Pavol Barger*, Jean-Marc Thiriet*, Michel Robert*, Jean-François Aubry**

* Université Henri Poincaré Nancy 1
Centre de Recherche en Automatique de Nancy / Nancy Research Centre for Automatic Control CNRS UMR 7039
2, rue Jean Lamour, 54 519 VANDOEUVRE les Nancy cedex
France, {firstname.familyname}@esstin.uhp-nancy.fr

**Institut National Polytechnique de Lorraine (Polytechnical National Institute of Lorraine)
Centre de Recherche en Automatique de Nancy / Nancy Research Centre for Automatic Control CNRS UMR 7039
2 avenue de la Forêt-de-Haye, 54516 Vandoeuvre, France, Jean-Francois.Aubry@ensem.inpl-nancy.fr

Abstract: The modern control systems are composed of several intelligent or smart devices, that is components with calculation capabilities and communication functions; these components are generally organized around a communication network. The dependability and safety evaluation of such systems or architectures is generally not easy to quantify, notably regarding the IEC 61508 standards. The paper focuses on the evaluation of some dependability parameters, on a pilot system represented as a Colored Petri Net model, taking into account dynamic aspects, thanks to Monte-Carlo simulation. *Copyright © 2004 IFAC*

Keywords: Intelligent devices, Communication network, Component failures, Scenarios, Failure rate estimation, Colored Petri nets

1. INTRODUCTION

Recent advances in informatics and electronics have changed the nature of control system classical architecture (a closed loop) basically composed of sensors, controllers and actuators. Ever more electronic sub-systems are present in the closed loop and the components contain a certain amount of intelligence (smart devices). By intelligence we mean the capacity 1.) to treat the data and 2.) to communicate with other components. So these systems became hybrid integrating continuous/discrete (time-driven) and event-driven components.

To reduce the cabling cost and to increase the communication capacity in the system the communication networks (e.g. fieldbus) become also integrated in the closed loop (Walsh, *et al.*, 2001) In general such networks can be characterized with a variable transmission delay and a jitter (delay distributed around a certain value).

This phenomenon can be associated with the study of delayed systems. In some cases a communication frame can be lost and thus not be available for the receiver at the moment needed. This case is addressed in the irregular sampling theory.

Adding all mentioned components and their embedded electronics and intelligence together induces a great complexity of these systems. The mutual influences of different subsystems are difficult to estimate.

In principle every component and its sub-element can present one or more malfunctions which can be of different nature and whose characteristics and consequences may not be known at the system design phase. Thus the system can be said to be in a number of functioning modes.

No dependability method at the moment enables a consideration of all given points within one single study. A proposal of one such method is the main goal of this communication.

2. DISTRIBUTED SYSTEM

Distributed Control Systems (DCS) are mainly composed of sensors, decisions elements (controllers), final elements (actuators) and some means of communication. Two basic communication schemes can be distinguished: 1.) an analogue peer-to-peer connection and 2.) a shared communication network. Safety criteria evaluation in both cases is a challenge and its complexity increases in the case of the communication network presence.

The main problem in the dependability studies of DCS are mutual influences of events present in various times on various components. The temporal propagation of these events is not easy to specify and can greatly modify the system behavior. Let's illustrate this with an example.

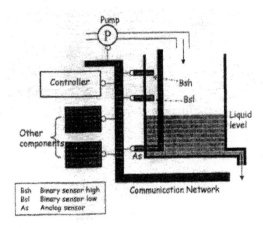

Figure 1: Control system example

On the figure 1 a simple networked control system is represented. Its goal is to fill the tank and maintain the liquid at a desired level. Each component in the example can fail. It is evident that some failures will have more influence on the system dependability than the others. For example, the low binary sensor may not be crucial to the system functioning in the juxtaposition to the controller failure which can to a great extent modify the system behavior. But these 'evident' classifications are not really that evident.

The importance of a communication frame can be very variable. These changes are highly dynamic. One frame can for example contain a security alarm and the next one a data not essential to the system functioning. Thus the importance of a communication frame is a function of the contained data. It is also a function of the system history. In fact, if the alarm frame is redundant to another alarm frame, its corruption does not necessarily imply the same consequences. Some other attributes (general

system state, communication protocol) influence as well the importance of communication frames.

3. METHOD DESCRIPTION

A Colored Petri Net (Jensen, 1997) model of the figure 1 was realized with Design/CPN (Design) tool (Son, et al., 2003). Some parts of the model are presented in (Barger, et al., 2003). This model represents at the same time the functional and the malfunctional behavior and characteristics of every component.

The reasons for the choice of Petri Nets for the modeling and analysis are: 1.) their capacity to model time- and event-triggered components (hybrid systems), 2.) their general acceptance for performance and dependability studies (Meunier, et al., 2000, Juanole, et al., 2002) and 3.) the modular model conception which makes the model modification and evolution very simple.

Colored Petri Nets are used, since they allow a concise representation of the phenomena as the liquid level, which is far from being Boolean (binary) and the communication frames composed of various non Boolean arrays. An example of a binary sensor used in the present study is shown on the figure 2.

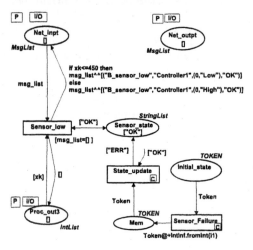

Figure 2: Sensor CPN model

Monte Carlo simulation (Labeau, et al., 2002, Moncelet, et al., 1998) is used for the analysis. Its basic principle is the repeated simulation of the system functioning cycle. The results are then subject to standard statistical analysis which gives us the information on the mean values and their distributions.

4. APPLICATION

The studied system, represented on the figure 1, is composed of:
− a pump,
− a controller,

- 3 sensors (2 binary and 1 analog),
- a communication network.

8 possible component failures and errors are considered:

Code	Event name	Mean occurrence rate
1	Sensor_low_failure	1/1800 sec
2	Sensor_high_failure	1/1800 sec
3	Sensor_analog_failure	1/1800 sec
4	Controller_failure	1/1800 sec
5	Actuator_blocked,	1/100 starts
6	Actuator_wearout,	1/100 sec
7	Frame_alternation.	1/50 frames
8	Frame_loss,	1/25 frames

Table 1: Failure events

Events 1 to 6 are permanent failures. Their repair is not considered. Events 7 and 8 are network related errors. They are instantaneous and affect one communication frame only. In the real world, they appear for example at the moment of a short electro-magnetic perturbation.

Events 1-4 and 6 are time functions and events 5, 7, 8 are event functions. Their failure probabilities are much higher than the real-life values, but are acceptable as the test values for the proposed method in order to verify the method feasibility. It has to be understood that if even if the method suites well a study with these parameters, the real life values can make its use very difficult.

Two more events are also safety related:

Code	Event name
9	Alarm_activated
10	Alarm_applied

Table 2: Safety related events

These two cases represent the safety mechanism functioning. The alarm is activated, when the information of the high level (delivered by the binary high sensor) arrives to the controller. The alarm value is applied when the safety command reaches the actuator.

Two system missions are taken into account for the analysis of the system: 1.) fill and 2.) maintain. Consequently, we can identify two system failures:
- failed to fill,

- failed to maintain.

The system functions as follows:

This study is limited only to the filling phase. It starts always with an empty tank and is defined as finished when $\sum_{k=1}^{n} abs(y_k - \bar{y}) < \varepsilon$ where y_k is the liquid level at the k instant, $\bar{y}$ and ε is a user-defined threshold. Under the nominal functioning the analog sensor measures are used to determine the appropriate control value from the controller to the actuator. If the liquid level becomes higher than the binary high sensor, the later sends information to the controller which recognizes it as an alarm and orders an immediate closure of the filling pump. The pumping may not be restarted as long as the binary high sensor reports the high level. If by any mean the level continues to rise and reaches the value Max, the tank is considered as overflown and the mission is considered as failed. The mission is also considered as unsuccessful if after 10 sampling periods the liquid level is still lower than 200 height units. This failure is called Not_started.

The simulation ends if:
- The filling phase is stabilized,
- The tank level overflows (liquid level higher than Max),
- The liquid level is lower than 200 height units at the 10th sampling period[1].

Table 3 summarizes different final states.

State number	State name	Condition	Comment
1	Successful	400< level < Max	desired
2	Started_not_finished	level < 400	undesired, safe
3	Overflow	level > Max	undesired, dangerous
4	Not_started	level (4*)Ts <20	undesired, safe
5	Forced_stop	not stabilised	?, safe

Table 3: Simulation final states

5. STUDY PRESENTATION

A question has to be asked about the stabilization success. In fact not all stabilized simulations mean that the filling is successful. For example the filling may start correctly but later a failure may appear and the system stabilizes at another place than the desired level neighborhood. This acceptable interval in this study is defined as (400, Max). According to the definition, if the final value is equal to or greater than Max, the tank overflows. If it is lower than 400

[1] In this communication the term 'sampling period' is used in reference to the time-triggered sampling of the tank and of the controller, which are in fact simultaneous.

height units, the failure is called Started_not_finished.

The first question concerns the safety mechanism effectiveness, that is: What is the safety mechanism influence on the system reliability. The answer to this question is fairly simple. It can be responded by determining the percentage of successful scenarios in the Monte Carlo simulation. Without the safety mechanism 78% simulations finished as successful. With the safety function it was 83%. That means that the safety mechanism improves the system reliability by 5 points. This difference is achieved mainly trough a decrease in failed scenarios, which is slightly less than 5 points.

Total scenario/simulation count	10000
Successful scenario count	8218
Unsuccessful scenario count	1782
Started_not_finished scenario count	150
Overflow scenario count	1412
Not_started scenario count	156
Forced_stop scenario count	64
No event present in scenarios (empty scenarios):	94

Table 4: Monte Carlo results with safety function

Even though the safety mechanism improves the system characteristics, it is far from making the system absolutely safe. For example: more than 6% of failed scenarios have at least once activated the safety mechanism.

This brings us to the second question on the causes of this failure. This question is partially answered in the following section.

6. FAILED SCENARIOS

In this section we work exclusively on the set of scenarios leading to the Overflow state. According to the table 4 there are 1412 such scenarios. The next table shows the length[2] distribution of these failed scenarios.

Scenario length	Scenarios count	%
1	466	33.00
2	422	29.89
3	289	20.47
4	131	9.28
5	49	3.47
6	15	1.06
7	8	0.57
8	10	0.71
9	4	0.28
10	4	0.28
11	2	0.14
12	6	0.42
13	1	0.07
14	2	0.14
15	0	0.00
16	1	0.07
17	0	0.00
18	1	0.07
19	0	0.00
20	0	0.00
21	1	0.07

Table 5: Length distribution of failed scenarios

One third of all failed scenarios contains only one event and almost another third is composed of scenarios recording 2 events. Only very few scenarios are longer than 6 events.

Let's now look closer at those failed scenarios composed of a unique event. There are 466 of them. One logically asks what event is it that precedes the overflow. Table 6 synthesizes these scenarios.

Single-event & failed scenarios	Count
Sensor_low_failure	0
Sensor_high_failure	0
Sensor_analog_failure	2
Controller_failure	1
Actuator_blocked	0
Actuator_wearout	5
Frame_alternation	143
Frame_loss	315
Alarm_activated	0
Alarm_applied	0
total count	466

Table 6: Event distribution of failed one-event scenarios

Before looking some further for the events causing the Overflow we'll present a look at others single-event scenarios. There are 866 altogether and their final states are as follows:

---

[2] Under scenario length we understand the number of events composing the scenario.

Final state	count
Successful	345
Started_not_finished	2
Overflow	466
Not_started	53
Forced_stop	0
total count	*866*

Table 7: Final states of one-event scenarios

The count of successful one-event scenarios is comparable to the one of the Overflow scenarios. Table 8 presents the events participating in the successful scenarios.

Single-event & successful scenarios	Count
Sensor_low_failure	1
Sensor_high_failure	1
Sensor_analog_failure	0
Controller_failure	2
Actuator_blocked	1
Actuator_wearout	6
Frame_alternation	113
Frame_loss	221
Alarm_activated	0
Alarm_applied	0
total count	*345*

Table 8: Event distribution of successful one-event scenarios

It is evident that both Successful and Overflow scenarios, the network errors are the most present. At this point it is impossible to recognize which network error leads potentially to the Overflow and which one not. This is why, the corrupted communication frames were recorded during the Monte Carlo simulation.

334 single-event scenarios composed exclusively of network errors have finished as successful and 458 such scenarios finished as failed. Thus the next step consists in creating classes of similar communication frames and observing whether these classes lead only to failed final state or if they can also lead to the successful final state.

The basic information for frames classification is: 1.) sender and receiver, 2.) time of emission and 3.) data transported. The frames in failed scenarios are exchanged principally between analog sensor and the controller. The exchanges between the controller and the actuator are less frequent.

The frames sent from the analog sensor to the controller are sent in the 2nd, 3rd and 4th sampling period of the process. The data that they contain are different according to the emission time. Searching for similar frames in the set of one-event successful scenarios has shown that no similar frame exists.

The simulation is a little different for the second group of frames (sent from the controller to the actuator). The main difference is that if this network error appears in the first two sampling periods then the Overflow follows. If it appears in the third period, then the Overflow does not follow. And if it occurs in the 4th or 5th period then the Overflow can follow.

These results are fundamental. They emphasize the main difficulty in the dependability evaluation of smart systems. In fact, no event can be analyzed by itself. Its context has to be taken into account. In this particular case, it means to consider not only the network error, but its occurrence instant as well as the corrupted frame content. It is only the set of these elements that allows distinguishing whether a network error is likely to be followed by the tank Overflow or not.

The question whether the results obtained from single-event scenarios can be considered as general is addressed in the following part.

7. RESULTS PORTABILITY

After having obtained the two groups of corrupted frames from single-event scenarios it is important to examine whether these results are valid also in other scenarios or not. This verification is done on the following example: 130 single-event scenarios leading to the Overflow are composed of loss or modification of the frame send from the Analog sensor to the Controller in the 2nd sampling period. The measures communicated varied from 155 to 200 height units. As 466 single-event scenarios followed by the Overflow were recorded so these 130 scenarios represent more the 1/3 of total single-event failed scenarios.

The loss or modification of this frame was observed 565 times (table 9).

Final state	Event occurrence
Successful	239
Started_not_finished	2
Overflow	323
Not_started	0
Forced_stop	1
total count	*565*

Table 9: One particular event occurrence and its final state

From the given table, it is evident, that the corruption of the frame given above does not imply that the scenario will finish as Overflow. Only a little more 57% of this event is present in Overflow scenarios. So even though this frame corruption is greatly present in the single-event failed scenarios and completely absent from the group of single-event successful scenarios, in general it cannot be concluded that the considered event is a cause of the Overflow.

So it is obvious that the results from single-event scenario study cannot be generalized. In fact we have found no event that can be by itself considered as the Overflow cause. To identify the real causes a sequence of more than one event has to be taken into consideration. Our first approach to the study of event sequences is presented in (Barger, *et al.*, 2003).

8. OTHER SCENARIOS

8.1 Overflow scenarios

In the simulation set without security 1412 scenarios finished as Overflow. Out of them 1391 have overflow in less or equal than 120 time units. Only 21 scenarios lasted longer than 120 time units. The instant of 120 time units is chosen as the limit since it separates two types of failures: immediate overshooting (liquid level raise is monotone) and overshooting where a certain decrease in the liquid level can be observed. 6 (28%) out of 21 scenarios contained only network errors. 15 (72%) contained other events as well. This means that the case of slow overshooting is more dependent on other events than only the network errors.

On the other hand, if the fast overshooting (1391 cases) 1210 (87%) overflows are exclusively caused by network errors and only 181 (13%) remaining cases contain other event than network errors.

This confirms clearly our statement in (Barger, *et al.*, 2003), that the influence of network errors is the most important at the beginning of the functioning and its importance decreases with the time.

8.2 Not_started scenarios

The first result in the simulation concerns the Not_started scenarios. They appeared 156 times in 10000 simulations. The analysis has shown that in every one of these scenarios at least one component failure (1-6) is present. A consequence of this is that if the system comes to this state, the problems have to be searched outside of the communication network since the later has not the potential to cause this particular failure mode.

9. CONCLUSION

This work is a contribution in studying the safety criteria of the distributed intelligent systems through scenario analysis and Monte Carlo simulation.

The main application domain of this method is the integration of the safety criteria (such as a network failure rate) directly into the design process of Distributed Control Systems. It allows estimating the global failure rate when only individual component failure rates are known and their correlations are not specified. It permits also a dynamical classification of events according to their influence on the safety criteria while depending on the global system state. Thus this method is a contribution to the common failure causes detection.

REFERENCES

Barger, P., J.M. Thiriet and M. Robert (2003). Safety analysis and reliability estimation of a networked control system, *IFAC-Safeprocess*, Washington (USA)

Design Design/CPN software homepage: http://www.daimi.aau.dk/designCPN.

Jensen, K. (1997). Coloured Petri Nets. Basic Concepts, Analysis Methods and Practical Use. *Monographs in Theoretical Computer Science*, Springer-Verlag.

Juanole, G. (2002). Quality of service of communication networks and distributed automation: models and performances, *Invited paper: 15th Triennial World Congress of the IFAC*, Barcelona, Spain.

Labeau, P.E. and E. Zio (2002). Procedures of Monte Carlo transport simulation for applications in system engineering, *Reliability Engineering and System Safety*, 77 (2002), pp. 217–228.

Meunier, P., B. Denis and J.J. Lesage (2000). Safety Analysis During the Control Architecture Design of Automated Systems, *Safeprocess 2000 (IFAC)*, Labeau Budapest, Hungary,

Moncelet, G. (1998). Application des réseaux de Petri à l'évaluation de la sûreté de fonctionnement des systèmes mécatroniques du monde automobile, *PhD. thesis, Université Paul Sabatier, Toulouse, France*.

Son, H.S. and P.H. Seong (2003). Development of a safety critical software requirements verification method with combined CPN and PVS: a nuclear power plant protection system application, *Reliability Engineering and System Safety*, 80 (2003), pp. 19–32.

Walsh, G.C. and H. Ye (2001). Scheduling of networked control systems, *IEEE control systems magazine*, February 2001, pp. 57-65.

ELSEVIER

IFAC

PUBLICATIONS
www.elsevier.com/locate/ifac

A low cost Stereovision Sensor for

Non-Contact Measurement Purpose

M. Zayed[*+], J. Boonaert[*] and M. Bayart[+]

[*] Laboratoire des Systèmes Industriels Distribués, Ecole des Mines de Douai
941 rue Charles Bourseul, 59508 Douai - France
Tél. : +(33) 327 712121   Fax. : +(33) 327 712980

[+] Laboratoire d'Automatique, de Génie Informatique et Signal,
UMR 8146,Bât Polytech-Lille, 59565 Villeneuve d'Ascq Cédex - France
Tél. : +(33) 320 434565   Fax. : +(33) 320 337189

Abstract: In this paper, we propose a cheap stereovision sensor. It is made of two
Webcams based on low cost CMOS technology imaging sensors. Computation and
images' acquisitions are performed by a standard Computer (PC). We first focus on the
advantages of such a system on CCD-based technologies, and then we explain how the
tri-dimensional measurements can be obtained. Before concluding, we give some
experimental results expressing the effectiveness of the approach. *Copyright © 2004
IFAC*

Keywords: Stereovision; Imaging technology; Flexible calibration; Localisation; Non-
contact measurements; Low cost devices.

1. INTRODUCTION

The stereovision is one of the measurement methods
that makes possible to obtain the 3D coordinates of a
point from two (or more) of its images. If we apply
this process to several points of a given object, it is
easy to compute its actual dimensions and position in
the Euclidian space, without any kind of contact with
the object. This last condition is required for
industrial processes needing dimensional
measurements while contact with the process parts is
strictly forbidden (nuclear plants, hazardous
environments, etc).

Unfortunately, the use of stereovision sensor suffers
from two major disadvantages. At first, the price of
the sensor itself represents a considerable percentage
of the global system cost. In the second place, the
lake of flexibility of many industrial realizations
makes them not easily transportable.

To cope with this problem, we propose a flexible
stereoscopic sensor, containing low cost cameras
(modified WebCams) and that does not require any
other additional hardware devices than a nowadays
PC (or a laptop PC). This approach greatly reduces
the cost of the global measurement system.

This article presents the different steps that took
place in the set up process of our system and details
the algorithms we applied to extract dimensional data
from the observed scene.

In this article, we first present our CMOS system
and compare it with what can be found on the market
of Vision-based measurements (CMOS vs. CCD
devices). Advantages and drawbacks of each of these
technologies will be discussed. The second part of
this work more especially focus on the preliminary
steps needed to efficiently exploit a stereoscopic
sensor. Finally, experimental results highlight the
robustness of the system that appears to be efficient
and flexible, regardless of its low price.

2. STEREOVISION SENSOR

Stereovision systems (fig.1) conquers nowadays
more & more new application fields such as
automotive (Zayed and Boonaert, 2003b), Robotics
(Lobo *et al.,* 2003) thanks to its ability to extract 3D
world space information from its projection.

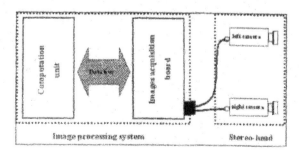

Fig.1.   A standard stereovision system.

In many cases, the two visual sensors are industrial CCD cameras. This leads to an expensive solution. Then, although it implies to solve some specific problems, changing the type of sensors is a good way to greatly reduce the global cost of the system. Webcams, based on low cost CMOS imaging technology, are certainly the cheapest kind of camera it is possible to find. For this reason, our stereoscopic system is made of two webcams that have been modified to fit the constraints of a measurement application. Because the poor quality affecting images taken from a webcam is mainly a consequence of the poor optical quality of the featured lens, the first adaptation we made consisted in replacing this one by C-type mounting screw lenses that provides (much) better optical characteristics while remaining cheap. Of course, this change of lens needed the sensors to be completely "repackaged". This gave us the opportunity to design simple and robust low cost cameras, suited for outdoor applications. In the next section, we expose the reasons why CMOS technology can now be considered as a valuable alternative solution, regarding the CCD imaging sensors.

3. IMAGING TECHNOLOGIES

The use of CCD camera in the field of computer Vision (and consequently concerning 3D measurements) is predominant. Due to special features (e.g. Random pixel access, characteristic curve) CMOS sensors with similar resolution were used in other fields (e.g. Interferometers, digital holography, …). The goal of this section is to draw up a comparison between the characteristics of both sensor types and their influence on vision-based measurements.

3.1 CMOS vs. CCD

Both CCD and CMOS image sensors are manufactured in a silicon foundry and the base materials (silicon, silicone oxide, polysilicon,…) and equipments are the same. However CCD processes and imagers have been specifically optimised for imaging applications for more than 3 decades. As a result, they present today excellent performances and image quality, thanks to extremely low noise, low dark current, high quantum efficiency and fill factor (Blanc, 2001).

3.1.1 Technical aspects

In what follows, the practical importance of some of the two sensors characteristics are discussed.

a) Dark current and Noise

Dark current mainly arises from thermal energy existing within the silicon lattice. Independently from the light falling on the detectors' area, electrons are created over time (Magnan, 2003). The occurrence of this phenomenon can be described in statistical terms, known as "dark current noise", owing to the fact that this kind of fluctuations still exists with no light falling on the detector. Notice that the dark current

would be only a problem for long exposure times (in the range of a few seconds).

b) Pattern Noise

In contrast to CCD sensors, CMOS sensors are active pixel sensors where every pixel has its own sense node and amplifier (Janesick, 2002a; Janesick, 2002b; Eastman 2000). Because of the parametric dispersion affecting each of these components during the manufacturing process, the output grey values vary from pixel to pixel even under homogeneous sensor illumination. The corresponding sensor's "pattern noise" can be measured once for different values of grey levels. The data can be stored in a matrix and then can be used for the correction of pixel intensity (Helmers and Schellenberg, 2003). Furthermore, reducing the pattern noise is possible by correlated double sampling (Janesick, 2002b; Kavadis, 2000).

c) Quantum efficiency

Quantum efficiency (QE) is the measurement of the effectiveness of an imager to produce electronic charge from incident light.

The physical quantum efficiency is defined as the ratio of the number of collected electrons per the number of incident photons, in a sensor area that is entirely photosensitive. For CCD and CMOS images, the quantum efficiency is typically excellent in the visible range (400 – 700 nm) (Blanc, 2001). However, CMOS imagers are generally less sensitive in the blue region (300 – 400 nm) than CCD ones.

d) Fill factor

The fill factor, defined as the ratio of light-sensitive area to the total pixel surface, also determines the maximum achievable sensitivity. Due to their architecture and their operating mode, the fill factor of full-frame-transfer CCD sensors is close to 100%, while the fill factor of the interline-transfer CCD and CMOS sensors is about 30 %. It is worth to say here that it is possible to increase the fill factor by keeping the pixel size not too small but decreasing the space used by the transistor found in each individual pixel.

e) Low Power

CCD cameras require numerous "companion" chips (such as drivers and signal conditioners) to be connected to the sensors. In the opposite, the standard CMOS mixed signal technologies allows the integration of all the needed functions (e.g. timing, exposure control, …) on one single piece of silicon. Consequently, CMOS sensors consume much less power (at least 10 time less) than their CCD counterparts.

3.1.2 Economical aspects

a) Lower Cost

Due to the wide use of the CMOS technology for numerous types of integrated circuits, these chips benefit from manufacturing scale saving that drive down defect density and production costs. Higher yields and less susceptibility to fail make CMOS a lower cost technology than CCD for image sensors. Fewer parts, smaller form factor, and higher reliability in the end product system mean cost savings to the systems manufacturer.

It is worth noting at the end of this section that CCD technology has been able to provide top-level performances for imaging sensors, thanks to designers' experience and process specialisation. However, it is also clear that CMOS sensors exhibit intrinsic advantages: low power consumption, integration capability, random pixel access, etc. Taking advantage of the large amount of works that have been done in the CCD domain, several techniques are now applied to CMOS: micro-lenses are widely used to compensate the loss of fill-factor (Lobo *et al.*, 2003).

As the cost and the possibility to develop our system using standard tools and operating system were key constraints of our project, two WebCams containing low-cost CMOS sensors are used to build a stereovision sensor for tri-dimensional measurements. The following sections are aimed to recall the principles of stereovision exploited by this sensor.

4. THE STEREOVISION SYSTEM GEOMETRY

4.1 The geometrical camera model

In what follows, we consider a pinhole camera model. This model connects a point of the 3D space to its perspective projection (i.e. its image) on the cameras' retina. This transformation is linear in the projective space and is described by the intrinsic and extrinsic parameters. The intrinsic camera parameters, such as the focal length (f), the pixels size (k_u, k_v), and the image centre coordinates $(u_0 \; v_0)^T$ describe the transformation between the camera and the image references. While the extrinsic parameters describe the rigid transformation between the world reference and that of the camera, this transformation is entirely defined by a 3x3-rotation matrix and 3x1-translation vector.

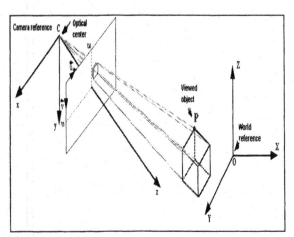

Fig. 2. Pinhole camera model

Using the pinhole camera model, the relation between the image coordinates of the projection of a point P on the image can be expressed using its coordinates in the scene reference. Using homogeneous coordinates in a projective space, we get:

$$\begin{pmatrix} x \\ y \\ s \end{pmatrix} = \underbrace{\begin{pmatrix} \alpha_u & 0 & u_0 & 0 \\ 0 & \alpha_v & v_0 & 0 \\ 0 & 0 & 1 & 0 \end{pmatrix}}_{A} \begin{pmatrix} R & T \\ 0^T & 1 \end{pmatrix} \begin{pmatrix} X \\ Y \\ Z \\ 1 \end{pmatrix} \quad (1)$$

Where:

- $(x \; y \; s)^T$ is the homogeneous coordinates of the projection of point P on the image.

- α_u, α_v Are defined as follows:

$\alpha_u = \dfrac{f}{k_u}$ and $\alpha_v = \dfrac{f}{k_v}$. with k_u and k_v : the pixels width and height.

- R is the rotation matrix, expressing the rotation between the scene reference and the camera reference.

- $(X \; Y \; Z \; 1)^T$ are the homogeneous coordinates of point P in the scene reference.

- T is the translation vector expressing translation between the scene reference and the camera's reference.

- A is the perspective projection matrix. It expresses the transformation between the camera coordinates system and that of the image.

The model above describes a projective transformation from 3D space (Euclidian space) to 2D space (camera retina plan). A 3x4-projection matrix, which depends on intrinsic and extrinsic camera parameters, describes this transformation. The operation that enables us to extract these parameters is called "camera calibration", and is detailed in the next section.

4.2 Camera calibration

Extracting 3D information from a stereoscopic pair of 2D images needs the cameras to be calibrated. Some of the first efficient tools were developed by the community of photogrametry (Brown, 1971; Faig, 1975), while further major evolutions can be credited to the particular field of computer Vision. Wherever they come from, these techniques can be divided into three categories: *photogrametric calibration, self –calibration* and finally the *hybrid calibration.*

4.2.1 Photogrametric calibration.

The *photogrametric calibration*, also called the "classical" calibration, allows a very high degree of accuracy in the cameras parameters estimation. Its major defect lies in its lack of flexibility, owing to the fact it requires a 3D calibration pattern with well-known dimensions.

4.2.2 Self calibration.

On the other hand, the *self – calibration* proposes an outstanding flexibility compared with the other two techniques, since no particular calibration pattern is needed, but this method still misses maturity (Bougnoux, 1998).

4.2.3 Hybrid calibration.

Finally, the *hybrid calibration* technique tries to take advantages from both of the two preceding approaches; this method fills the lack of robustness of the self–calibration while allowing a rather high degree of flexibility (Zhang, 2000).

Fig. 3. Some images taken by a WebCam for the checkerboard used as a calibration pattern for *hybrid calibration*.

Because we need our system to be easily set up, this kind of calibration process has been preferred to the other ones. More precisely, we are calibrating our cameras thanks to the ZhangYou Zhang technique (for more details see(Zhang, 2000)). This method requires that the camera observes a 2D well-known dimensions calibration pattern (see fig.3) at least two times with different orientations. From these pictures, we are able to compute intrinsic and extrinsic parameters. This procedure can be done in real time without operator intervention, thanks to the checkerboard properties.

4.3 The stereopsis

Locating a point in 3D space needs this one to be viewed from at least two points of view. This requirement can be met either by viewing the scene point using two different cameras (fig. 4) or by moving a camera around. Of course, the motion between the two cameras positions must be known in both cases. However, the stereovision system (made of two cameras at a fixed location from each other) is more suitable for a measurement task.

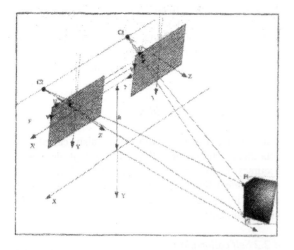

Fig. 4. Description of the stereovision system geometry.

Let P_1 be a scene point observed by the two cameras. From (1), we can write:

$$P_{il} = R_l P_i + T_l \qquad (2)$$
$$P_{ir} = R_r P_i + T_r \qquad (3)$$

where:

- i : is the point index.
- P_{il} and P_{ir} : Coordinates of a scene point expressed in left and right cameras coordinates system respectively.
- R_l and R_r : Rotation matrix expressing rotations between the scene reference and the left and right cameras references, respectively.
- P_i : Coordinates of the scene point P_i in the scene reference.
- T_l, T_r : The translations vectors, expressing translations between the scene reference and the left and right cameras references, respectively.

From (2) and (3), we can express the coordinates of a scene point P_i in the left camera reference, such as:

$$P_{il} = \mathbf{R}^T P_r + \mathbf{T} \qquad (4)$$

with:

$$\mathbf{R} = R_r R_l^T$$
$$\mathbf{T} = T_l - \mathbf{R}^T T_r$$

Equation (4) allows us to express the known coordinates of a point from one camera reference to the other one, which is primordial for the 3D points localisation.

4.4 3D points localization

From fig. 4 we can see that the location of a given point P_i in the 3D space is the intersection between lines $(C_1 p_{il})$ and $(C_2 p_{ir})$, with C_1, C_2 the cameras' centres and p_{il}, p_{ir} the projections of P_i in the left and right camera references, respectively.

In fact, due to the noise affecting the detection of p_{il} and p_{ir}, the two lines don't actually intersect and many techniques were developed to cope with this problem (for more information, see (Hartley and Sturm, 1997)). In the particular context of our application, we exploit the mid-point technique, which is less sensitive to noise. It consists in computing the mid-point coordinates of the shortest line segment joining $(C_1 p_{il})$ and $(C_2 p_{ir})$, satisfying the following equation:

$$a p_{il} + c w = \mathbf{T} + b \mathbf{R}^T p_{ir} \qquad (5)$$

where: $a, b, c \in \Re$, $a p_{il}$: is the line passing through C_1 and p_{il}, expressed in left camera reference, $\mathbf{T} + b \mathbf{R}^T p_{ir}$ defines the line passing through C_2 and p_{ir}, expressed in left camera reference, $w = [p_{il}]_x \mathbf{R}^T p_{ir}$ is an orthogonal vector to both lines ($[p_{il}]_x$ is the skew symmetric matrix of p_{il}).

By solving the linear system (5) for (a,b,c), one can compute the coordinates of the two extremities of the shortest segment joining the two lines. Of course, iterating such a process for multiple points of one given object allows us to determine some of the object's dimensions.

5. EXPERIMENTAL RESULTS

After the calibration procedure described above, we are able to test our stereovision sensor. All of the tests have been performed with the system you can see on the picture below (figure 5):

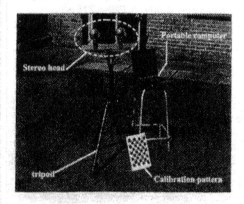

Fig. 5. The complete stereovision system.

Figure 6 is aimed to give details concerning the stereo head itself, and especially the kind of lens that can be adapted to the new cameras' form factor.

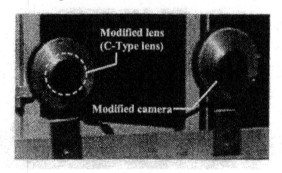

Fig. 6. Detailed view of the stereo head.

The first experiment we present here consists in estimating the displacements of a cube on the ground. The points we took into account are these delimiting the front bottom edge of the cube. Figures 7-a and 7-b illustrate the displacements' range. Figures 8 and 9 represent the estimated values versus the actual ones.

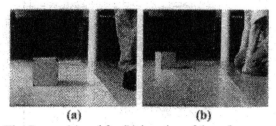

(a)   (b)

Fig. 7. near (a) and far (b) location of the cube.

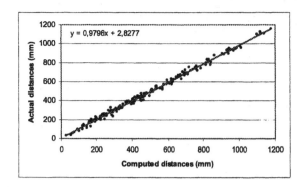

Fig .8. Estimated distance versus actual (first serial).

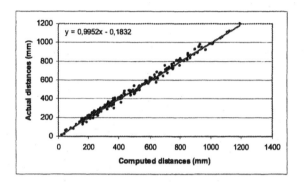

Fig.9. Estimated distance versus actual (2nd serial).

As expected, the estimated displacements are very close to the actual ones. The stereo sensor's characteristic is linear with a very good approximation (owing to the high correlation between actual and computed data and the low standard deviation affecting the differences between actual data and the linear regression line). Notice that the estimation error seems not to be correlated to the distance itself. This first example shows that it is possible to measure displacements on a small area (within the range 0 to 2 meters) with a sufficiently good precision (a few millimetres) for many applications (objects location on an automated working area, accident or crime scenes, etc).

The second application presented here consists in estimating the size of outdoors elements and characteristic lengths of building parts (see figure 10-a and 10-b below).

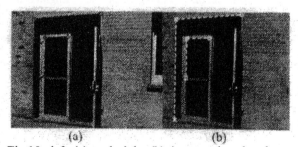

(a)   (b)

Fig.10. left (a) and right (b) image taken by the sensor.

In the context of this application, points in the left image and their matches in the right image were taken by hand, but this matching procedure can be automated (for more information, see (Zayed and Boonaert, 2003a; Zhang et al., 1994). The actual width of the wall aperture in which the door is located is 1,800 meter, and the estimated width is 1,790 meter. The actual height of this element is 2,400 meter and the estimated height is 2,404 meter. Concerning the left door's upper window, the actual dimensions are 0,630 m x 0,945 m while the system gives 0,661 m x 0,943 m. Again, estimated and actual dimensions are close enough for this kind of application. Notice that the adaptation of the stereovision system configuration (the relative position of the two cameras) to a specified measurement task can considerably improve the precision of the measurements.

6. CONCLUSION

In this article, we present a low cost and efficient stereoscopic sensor. The precision we get is good, although, of course, it still can be improved. The key point here is that our system shows that low costs materials can now give satisfactory results, in "out of the lab" applications. The global cost of our stereo sensor (if we except the time we spent making it a real thing...) does not exceed 1500 euros. Industrial solutions providing the same degree of flexibility and robustness often cost about ten times this price. Another aspect to be put the stress on is that the CMOS technology exhibits enough maturity to be viewed as a valuable competitor to CCD-based Vision system, owing to its intrinsic low cost and its increasing imaging performances.

Our further work will consist in improving the software modules we developed in order to make this stereo sensor as easy to use as a simple camcorder.

REFERENCES

Blanc, N., (2001) *"CCD versus CMOS – has CCD imaging come to an end ?"*. p. 131 – 137.

Bougnoux, S. (1998), from projective to Euclidian space under any practical situation, a criticism of self-calibration. *In Proceedings of the 6th International Conference on Computer Vision,* pages 790-796.

Brown, D.C. (1971), Close-range camera calibration. *Photogrammetric Engineering,* 37(8):855-866.

Faig, W. (1975), Calibration of close-range photogrammetry systems: Mathematical formulation. *Photogrammetric Engineering,* 41(12): 1479-1486.

Hartley, R., Sturm P. (1997), *" Triangulation", Computer Vision and Image Understanding* number 2 **vol. 68** p.146-157.

Helmers, H., Schellenberg M. (2003), *"CMOS vs. CCD sensors in speckle interferometry". Optics & Laser Technology,* **vol. 35** p.587 – 595.

Janesick, JR. (2002a), *"Dueling detectors – CMOS or CCD ?" SPIE's OE Magazine,* p. 30 – 3.

Janesick, J. (2002b), *"Lux transfer: complementary metal oxide semicondictors versus charge-coupled devices". Opt Eng 2002*; 41(6):1203 – 15.

Kavadis, S., Dierickx B., Scheffer D., Alaerts A., Waerts D.U and, Bogaerts J. (2000), *"A logarithmic response CMOS image sensor with on-chip calibration". IEEE Solid-State Circuits* 2000;35(8):1146 – 52.

Eastman Kodak Company (2000). *"Kodak CMOS image sensors – white paper"*, Rochester.

Lobo, J., Queiroz C. and Dias J. (2003), *" World feature detection and mapping using stereovision and inertial sensors ", Robotics and Autonomuous Systems* **vol. 44** p. 69-81.

Magnan, P. (2003), "Detection of visible photons in CCD and CMOS: A comparative view". *Nuclear Instruments and Methods in Physics Research A 504,* p.199 – 212.

Zayed, M., Boonaert J. (2003a), *"An effective Stereo-head points matching approach".* The Fourth International Conference AIAI.

Zayed, M., Boonaert J. (2003b), *"Obstacles detection from disparity properties in a particular stereovision system configuration". The 6th IEEE on Intelligent Transportation Systems.* p 311-316.

Zhang, Z., Deriche R., Faugeras O. and Luong Q.-T. (1994), *"Robust technique for matching two uncalibrated images through the recovery of the unknown epipolar geometry" . INRIA* RR – N° 2273.

Zhang, Z. (2000), *"A Flexible New Technique for Camera Calibration". IEEE Transactions on Pattern Analysis and Machine Intelligence.* 22(11):1330-1334.

OPTIMAL SENSOR PLACEMENT USING
BOND-GRAPH MODEL FOR FDI DESIGN

M. Khemliche, B. Ould Bouamama and H. Haffaf

LAGIS, UMR CNRS 8146021, Ecole Polytechnique de Lille,
Cite scientifique
59655 Villeneuve d'Ascq Cedex, France
Tel.: (+33)320337139, Fax: (+33)320337189
E-mail: belkacem.bouamama@univ-lille1.fr

Abstract: The sensor placement for monotorability specifications using model based approach requires model and analytical redundancy relations. The innovative interest of the present paper is the use of only one representation: the bond graph tool for modelling and sensor placement. The methodology is well suited not only for modelling but also for sensor location because of it's causal, graphical, and structural properties. The physical sensor position is explicitly displayed in the graphical model, and the ARRs (which are used for monitorability analysis) are directly and systematically generated from a model depending on the supposed instrumentation architecture using covering causal path. The developed methodology is illustrated using a pedagogical example. *Copyright © 2004 IFAC*

Keywords: Sensors placement, bond-graph, causal path, detectability, monitorability, process engineering.

1. INTRODUCTION

In the past, automated production systems have aided operators in controlling the process in order to improve the quality of the finished product, the safety and the industrial units efficiency. Recently, another challenge has appeared: it concerns the diagnosis procedures automation using an intelligent control. The improvement of the process safety is essentially based on the Fault Detection and Isolation (FDI) procedures. The FDI algorithms are based on the same principle: the comparison between the real behavior of the process provided by a set of sensors and a reference behavior provided by a model under normal operation. The monitorability (ability to detect and to isolate faults) of the system depends mainly on the implemented instrumentation architecture.

Two sensor placement methodologies have been developed, depending on the kind of knowledge

used to describe the process : model based (the model is given under analytical form), and the non model-based (the knowledge is given under rules, tables, pattern recognition, ...).

Among the non-based model approaches, we can cite neuronal approaches (RN), simulated annealing algorithm (SA), the iterative algorithm of insertion/deletion (I/D) (Worden and Burrows, 2001) and genetic algorithms (GA) (Holland, 1975). The main drawback of these methods is the lack of physical knowledge. Some of them need pattern recognition. The model based approach bears on Analytical Redundancy Relations (ARRs) for which is applied the sensor placement algorithm. The used analytical model can be given under structural (Carpentier and Litwak, 1996) or state space equation form (J. Donald-Chmielewski, May 2002). For the cited methods the sensor location can not be defined explicitly. Furthermore in addition to the modelling step problem, ARRs generation is not trivial and needs complicated unknown variables theory elimination (Cox *et al.*, 1992).

The innovative interest of the present paper is the use of only one representation: the bond graph tool for modelling and sensor placement. Indeed, this methodology is well suited not only for modelling as shown elsewhere (Paynter, 1961), (Thoma and Ould Bouamama, 2000) but also for sensor location because of it's causal and structural properties (Ould Bouamama *et al.*, 2000a). The physical sensor position is explicitly displayed in the graphical model, and the ARRs (which are used for monitorability analysis) are directly and systematically generated from a model depending on the supposed instrumentation architecture using covering causal path .

After a brief presentation of bond graph tool, the first part of the paper presents the optimal sensor placement strategy based on a bond graph modelling. The developed methodology is illustrated using a pedagogical example. Finally the third part concludes the paper.

2. OPTIMAL SENSOR PLACEMENT STRATEGY IN VIEW OF COMPONENTS MONITORABILITY

2.1 *Bond-graph element basis*

The bond graph modelling as unified language is well developed elsewhere (for survey see (Karnopp *et al.*, 1990), (Thoma and Ould Bouamama, 2000). During the ten last years, a bond graph is used not only for modelling but for control and monitoring analysis because of it's structural and causal properties.

The key to bond graph modelling is the representation (by a bond) of power as the product of efforts and flows, with elements acting between these variables and junction structures to put the system together. As shown in Figure 1(a), the power exchanged between two systems A and B indicated by a bond is the product of two variables - a potential variable (*e.g.* pressure, electrical potential, temperature, chemical potential, force, etc.) called effort (e) and a current variable (*e.g.* volume flow, current, entropy flow, velocity, molar flow, etc.) referred to as flow (f).

From the FDI point of view, one important structural property of the bond graph is its causality concept. Indeed, the determination of causes and effects in the system is directly deduced from the graphical representation. In the bond graph, it is denoted by the cross-stroke on the right indicating that the effort acts to the right, the side of the cross-stroke while the flow is in the reverse direction. As example in Figure 1(a), assigned causality means that system A imposes efforts on B. In the corresponding block diagram given by the Figure 1(b), the direction of action is indicated by an arrow on each connection as illustrated. Independently of the causality, the direction of the positive power is indicated by the half-arrow on the bond.

A bond graph model can be considered as a graph which describes power exchange in physical systems. The vertices of the graph are of two kinds: basic elements (I, C, R, TF, GY) namely inertial, capacity, resistive, transformer and gyrator elements and junction nodes $(0, 1)$ which represent energy conservative laws .

The computational aspect is represented by causal stork to impose which of the two variables is known. Contrary to the classical description using equation model based, the sensor location corresponding to a physical placement on the process is explicitly displayed on the bond graph model. Indeed, sensor placement on "0" and "1" junction corresponds to a physical component (tank, pump,...) where occurs the energy conservation. In bond graph, two types of sensors are used : the effort detectors De (pressure sensor, temperature sensor, ...) and the flow detectors Df (mass flow, volume flow). The power on sensors is the product of effort and flow variables in true bond graph. In the pseudo bond graph (well suited for process engineering modelling) (Karnopp and Azerbaijani, 1981) the power is provided by the flow variable.

An effort detector De is placed in the "0" junction and measures the energy (displacement) stored by the physical component associated with this junction. A flow detector Df is located in "1" junction while it measures the flow through this junction.

(a   (b

Fig. 1. Bond Graph representation (a) and causality (b)

2.2 *Technical specification for sensor placement*

The goal is to provide an optimal sensor placement on the bond graph model in order to make all components monitorable. We assume as constraints that the faults are not multiple and may affect only components.

Let given a bond graph model obtained from physical process. We suppose that the sensors are not placed yet on bond graph model.

Let x_i and y_j the boolean variables corresponding to the sensor placement such as :

$$
\begin{aligned}
&x_i = 1 \text{ if the } i^{th} \text{ sensor is placed} \\
&\text{on the } i^{th} \text{ "0" junction} \\
&x_i = 0 \text{ otherwise} \\
&y_j = 1 \text{ if the } j^{th} \text{ sensor is placed} \\
&\text{on the } j^{th} \text{ "1" junction} \\
&y_j = 0 \text{ otherwise}
\end{aligned} \tag{1}
$$

Let :

- N_0 number of "0" junctions
- N_1 number of "1" junctions
- n_i number of bonds attached to the ith "0" junction $(i = 1, N_0)$
- m_j number of bonds attached to the jth "1" junction $(j = 1, N_1)$

- K the set of the known variables. The sub-

set K of known variables contains the control variables u, the variables whose value is measured by sensors (De and Df), the controlled (MSe, MSf) and perturbation (Se, Sf) sources

$$
K = MSe \cup MSf \cup De \cup Df \cup Se \cup Sf \cup u. \tag{2}
$$

- X the set of the unknown variables. The unknown variables are the power variables (flow and effort) which label the bonds. The vector X containing all the power variables is

$$
\begin{aligned}
&X(t) = \{e_1(t), f_1(t)\} \ldots \cup \{e_{ne}(t), f_{ne}(t)\} \\
&X \in \Re^{2 * n_e},
\end{aligned}
$$

n_e is the number of bond graph R,C, and I elements

$$\tag{3}$$

Based on causal properties of the bond graph modelling, the unknown variables can be calculated using covering causal paths (this methodology is developed in (Busson, 2002)). For the "0" and the "1" junction (fig 2), the unknown variable (based on fixed causality) is calculated as follows:

$$
\left\{
\begin{aligned}
&f_{Ci} = \Phi_{Ci}[s\{(1 - x_i)e_{Ci} + x_i De_i\}], \ i = 1, N_0 \\
&e_{Ci} = \frac{1}{s}(1 - x_i)\Phi_{Ci}^{-1}(f_{Ci}) + x_i De_i
\end{aligned}
\right. \tag{4}
$$

$$
\left\{
\begin{aligned}
&e_{Rj} = \Phi_{Rj}[(1 - y_j)f_{Rj} + y_j Df_j]), \ j = 1, N_1 \\
&f_{Rj} = (1 - y_j)\Phi_{Rj}^{-1}(e_{Rj}) + y_j Df_j
\end{aligned}
\right. \tag{5}
$$

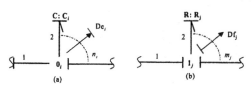

Fig. 2. Junction structure constraints

2.3 Monitorability conditions

The set of *ARRs* generates a binary table. The columns of this table are called failure signatures (in our case are considered only components faults which may affect C and R bond graph elements). A "1" entry in the *ith* row and the *jth* column of the table indicates that the residual r_i is sensitive to the *jth* fault.

The structure sr of a residual "r" is constituted by the variables occurring in this residual. Each kth variable to be monitored is associated with its Φ_k relation; it can be defined by :

$$
sr(r, \Phi_k) = \left\{
\begin{array}{ll}
1 & \text{if and only if } \Phi_k \text{ exists in ARR } r \\
0 & \text{otherwise}
\end{array}
\right. \tag{6}
$$

where Φ is the set of constraints deduced from a bond graph model (for more details see (Ould Bouamama *et al.*, 2000b)) $\Phi = \{\Phi_i\}_{i=1}^{N_0} \cup \{\Phi_j\}_{j=1}^{N_1}$

The vector of failure signatures sd_i associated at the relation Φ_i is defined by :

$$
sd_i = [sr(r_1, \Phi_i) \ldots sr(r_m, \Phi_i)]^t \tag{7}
$$

$$
sr(r_k, \Phi_i) = \left\{
\begin{array}{ll}
1 & \text{if the residual } r_k \text{ use the relation } \Phi_i \\
0 & \text{otherwise}
\end{array}
\right. \tag{8}
$$

The failure which may affect the component i is detectable if and only if the vector sd_i is different of zero, and is isolable if and only if it is detectable and

$$
\forall \Phi_j \in \Phi / i \neq j \ \exists r_k \in R \ / \ sr(r_k, \Phi_i) \oplus sr(r_k, \Phi_j) = 1.
$$

The methods used until now test the binary vectors using Hamming distance (Carpentier, 1999).

In the next section is studied through an example, the combinatorial problematic of sensor placement. It is shown that it can be considered as a binary integer program where each variable can be assigned to $0 - 1$ value.

3. APPLICATION

Let a scheme of thermofluid system in figure 3.

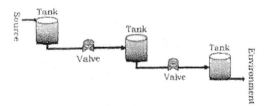

Fig. 3. Thermofluid process

The corresponding bond-graph model is shown by figure 4.

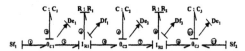

Fig. 4. Bond graph model with virtual detectors

This model is composed by :

3 junctions 0: 0_{C1}, 0_{C2} and 0_{C3}

2 junctions 1: 1_{R1} and 1_{R2}

5 components: C_1, R_1, C_2, R_2 and C_3

2 sources : Sf and Se

The problematic is that we want to place a minimal number of sensors to monitor 5 components. We suppose that sensors and sources are not affected by faults. The set of known variables is K={Sf$_1$,Sf$_2$,De$_1$,Df$_1$,De$_2$,Df$_2$,De$_3$} and the set of unknown variables is $X = \{e_2, f_2, e_4, f_4, e_6, f_6, e_8, f_8, e_{10}, f_{10}\}$

In this example, the equations in junctions are given by :

For 0_{C1} junction we have :

$$\begin{cases} f_1-f_2-f_3=0 \text{ and } e_1=e_2=e_3 \\ f_1=Sf \text{ and} \\ f_{C1}=f_2=\Phi_{C1}[s\{(1-x_1)e_2+x_1De_1\}] \\ e_2=e_{C1}=\dfrac{1}{s}(1-x_1)\Phi_{C1}^{-1}(f_2)+x_1De_1 \end{cases} \quad (9)$$

For 1_{R1} junction we have :

$$\begin{cases} e_3-e_4-e_5=0 \text{ and } f_3=f_4=f_5 \\ e_{R1}=e_4=\Phi_{R1}[(1-y_1)f_4+y_1Df_1] \\ f_4=f_{R1}=(1-y_1)\Phi_{R1}^{-1}(e_4)+y_1Df_1 \end{cases} \quad (10)$$

For 0_{C2} junction we have :

$$\begin{cases} f_5-f_6-f_7=0 \text{ and } e_5=e_6=e_7 \\ f_{C2}=f_6=\Phi_{C2}[s\{(1-x_2)e_6+x_2De_2\}] \\ e_6=e_{C2}=\dfrac{1}{s}(1-x_2)\Phi_{C2}^{-1}(f_6)+x_2De_2 \end{cases} \quad (11)$$

For 1_{R2} junction we have :

$$\begin{cases} e_7-e_8-e_9=0 \text{ and } f_7=f_8=f_9 \\ e_{R2}=e_8=\Phi_{R2}[(1-y_2)f_8+y_2Df_2] \\ f_8=f_{R2}=(1-y_2)\Phi_{R2}^{-1}(e_8)+y_2Df_2 \end{cases} \quad (12)$$

For 0_{C3} junction we have :

$$\begin{cases} f_9-f_{10}+f_{11}=0 \text{ and } e_9=e_{10}=e_{11} \\ f_{C3}=f_{10}=\Phi_{C3}[s\{(1-x_3)e_{10}+x_3De_3\}] \\ e_{10}=\dfrac{1}{s}(1-x_3)\Phi_{C3}^{-1}(f_{10})+x_3De_3 \end{cases} \quad (13)$$

The set of equations (9)-(13) leads to the following system :

$$\begin{cases} Sf_1-\Phi_{C1}[s\{(1-x_1)e_2+x_1De_1\}] \\ -(1-y_1)\Phi_{R1}^{-1}(e_4)-y_1Df_1=0 \\ \dfrac{1}{s}(1-x_1)\Phi_{C1}^{-1}(f_2)+x_1De_1 \\ -\Phi_{R1}[(1-y_1)f_4+y_1Df_1] \\ -\dfrac{1}{s}(1-x_2)\Phi_{C2}^{-1}(f_6) \\ -x_2De_2=0 \\ (1-y_1)\Phi_{R1}^{-1}(e_4)+y_1Df_1- \\ \Phi_{C2}[s\{(1-x_2)e_6+x_2De_2\}] \\ -(1-y_2)\Phi_{R2}^{-1}(e_8)-y_2Df_2=0 \\ \dfrac{1}{s}(1-x_2)\Phi_{C2}^{-1}(f_6)+x_2De_2 \\ -\Phi_{R2}[(1-y_2)f_8+y_2Df_2] \\ -\dfrac{1}{s}(1-x_3)\Phi_{C3}^{-1}(f_{10})-x_3De_3=0 \\ (1-y_2)\Phi_{R2}^{-1}(e_8)+y_2Df_2- \\ \Phi_{C3}[s\{(1-x_3)e_{10}+x_3De_3\}]+Sf_2=0 \end{cases} \quad (14)$$

From the binary variables x_i $(i = 1, 2, 3)$ and y_j $(j = 1, 2)$ we can determine the final structure of monitorable system. Three combinations with a placement of 4 sensors and one combination with a placement of 3 sensors provide the monitorability of the 5 components.

For $[x_1 \ y_1 \ x_2 \ y_2 \ x_3]^t = [1 \ 1 \ 0 \ 1 \ 1]^t$ the obtained residuals after elimination of unknown variables are:

$$r_1=Sf_1-\Phi_{C1}[sDe_1]-Df_1 \quad (15)$$

$$r_2=De_1-\Phi_{R1}[Df_1]-\frac{1}{s}\Phi_{C2}^{-1}(Df_1-Df_2) \quad (16)$$

$$r_3=De_3+\Phi_{R2}[Df_2]-\frac{1}{s}\Phi_{C2}^{-1}(Df_1-Df_2) \quad (17)$$

$$r_4 = Df_2 - \Phi_{C3}[sDe_3] + Sf_2 \qquad (18)$$

The structured residuals are given by the fault signature matrix 1:

	Φ_{C1}	Φ_{R1}	Φ_{C2}	Φ_{R2}	Φ_{C3}
r_1	1	0	0	0	0
r_2	0	1	1	0	0
r_3	0	0	1	1	0
r_4	0	0	0	0	1

Table 1. Signature faults using four sensors

The fault signatures are different from each other and not equal to zero, then components C_1, R_1, C_2, R_2 and C_3 are monitorable. The same result is found for the combinations $[1\ 1\ 1\ 0\ 1]^t$ and $[1\ 0\ 1\ 1\ 1]^t$. By exploring all combinations, see table 3, we remark that only the following combination with 3 sensors provide the monitorability of the 5 components. The optimal case corresponds to the $[1\ 0\ 1\ 0\ 1]^t$ combination and the resulting residuals are :

$$r_1 = Sf_1 - \Phi_{C1}[sDe_1] - \Phi_{R1}^{-1}(De_1 - De_2) \quad (19)$$

$$r_2 = \Phi_{C2}[sDe_2] - \Phi_{R1}^{-1}(De_1 - De_2) + \Phi_{R2}^{-1}(De_2 - De_3) \qquad (20)$$

$$r_3 = \Phi_{R2}^{-1}(De_2 - De_3) - \Phi_{C3}[sDe_3] + Sf_2 \quad (21)$$

The signature fault table is given by table 2:

	Φ_{C1}	Φ_{R1}	Φ_{C2}	Φ_{R2}	Φ_{C3}
r_1	1	1	0	0	0
r_2	0	1	1	1	0
r_3	0	0	0	1	1

Table 2. Signature faults using three sensors

All results for placement of 4 and 3 sensors are resumed on the following table 3 :

number of sensors	Comb.	Monitorable	Detectable
4	11011	C_1, R_1, C_2, R_2, C_3	ϕ
4	11101	C_1, R_1, C_2, R_2, C_3	ϕ
4	10111	C_1, R_1, C_2, R_2, C_3	ϕ
3	10101	C_1, R_1, C_2, R_2, C_3	ϕ
3	10110	C_1, R_1, C_2	R_2, C_3
4	11110	C_1, R_1, C_2	R_2, C_3
4	01111	C_2, R_2, C_3	C_1, R_1
3	01011	C_2, R_2, C_3	C_1, R_1
3	01101	C_2, R_2, C_3	C_1, R_1
3	11010	C_2, R_2, C_3	C_1, R_1
3	10011	R_1, C_2, C_3	C_1, R_2
3	11001	C_1, C_2, R_2	R_1, C_3
3	00111	R_2, C_3	C_1, R_1, C_2
3	11100	C_1, R_1	C_2, R_2, C_3
3	01110	C_2	C_1, R_1, R_2, C_3

Table 3. Sensor placement resul

3.1 Method of placement

The innovative interest of the presented work consists in assigning detectors on junction nodes of bond graph without generating ARRs.

From table3, the following strategy of sensors placement on mono-energy bond graph is deduced. Intuitively, this method is like heuristic method which includes a set of rules applied to the combinations. Suppose in the first stage that the model does not contain causality loops.

It is useless to generate the analytical redundancy relations because the vector of sensors placement combination leads directly to the structure of residuals $[\Phi_{C1}, \Phi_{R1}, \Phi_{C2}, \Phi_{R2}, \Phi_{C3}]$.

This method is as follows:

For each '1' in the combination, we associate residual structure by applying following rules :

1. When we have two adjacent 1, we place 1 on the position where sensors are placed and all the other positions get "0" :

$[1\ 1\ z\ z\ z] \rightarrow [1\ 0\ 0\ 0\ 0]$ where $z = 0$ or 1

2. If 1 is adjacent to 0, we conserve the position of 1 and we replace 0 by 1:

$[1\ 0..] \rightarrow [1\ 1..]$

$[0\ 1..] \rightarrow [1\ 1..]$

3. If 1 is followed by two consecutive 0 then we conserve the 1 place and we replace 0 by 1

$[1\ 0\ 0] \rightarrow [1\ 1\ 1]$

$[0\ 0\ 1] \rightarrow [1\ 1\ 1]$

4. If we have two 0 followed by 1, the last 0 becomes 1 (the last '1' is conserved) :

$[1\ 0\ 0\ 0\ 1] \rightarrow [1\ 0\ 0\ 1\ 1]$

These rules indicate in fact how to get ARRs from the bond graph model using the causal paths properties. If a variable can not be eliminated at a junction node, then we use the causal path to go to the next junction and then the component attached to this junction will appear in the structure of the residual. The 0 position in the combination shows that the component is attached to the junction node but we have no detector to measure the flow (or the effort) variable at the this node. Thus, this variable is still unknown until another junction with a sensor is reached through the causal path.

4. CONCLUSION

The sensor placement for monitorability specifications using model based approach requires model from which analytical redundancy relations are deduced. In this paper it is shown how the bond graph methodology can easily (because of its graphical aspect) provide powerful procedure to insure monitorability of component failures with any need of calculation. This method is based on set of transformation rules allowing to avoid ARRs computational process handling. The proposed heuristic rules are efficient ones: this method circumvents the solution set exploration and operates transformations directly on the sensor placement combination to obtain residual structure. The method might be generalized to more complex bond graph models (multi-ports models, causal loops, ...) describing process engineering systems.

5. REFERENCES

Busson, F. (2002). Modèlisation et Surveillance Des Processus En Génie Des Procédés À L'aide Des Bond Graphs Multiénergies. Thése de doctorat. Université des Sciences et Technologies de lille1. Villeneuve d'Ascq (France).

Carpentier, T. (1999). Placement de Capteurs Pour la Surveillance Des Processus Industriels Complexes. PhD thesis. Université des sciences et Technologies de Lille (France).

Carpentier, T. and R. Litwak (1996). Algorithms and criteria for sensors location in view of supervision. In: *CESA'96 IMACS Multiconference*. Lille, France. pp. 684–689.

Cox, D., J. Little and D. O'Shea (1992). *Ideals, Varieties, and Algorithms, Undergraduate Texts in Mathematics*. Springer-Verlag.

Holland, J.H. (1975). Adaptation in natural and artificial systems. *MIT Press*.

J. Donald-Chmielewski, P. Tasha, Vasilios M (May 2002). On the theory of optimal sensor placement. *AIChE Journal* **Vol. 48, nř 5**, 1001–1012.

Karnopp, D.C. and S. Azerbaijani (1981). Pseudo bond graphs for generalised comportmental models in engineering and physiology. *Journal of The Franklin Institute* **312**(2), 95–108.

Karnopp, D.C., D. Margolis and R. Rosenberg (1990). *Systems Dynamics: A Unified approach*. second ed.. John Wiley. New York.

Ould Bouamama, B., F. Busson, G. Dauphin-Tanguy and M. Staroswiecki (2000a). Analysis of structural properties of thermodynamic systems. In: *4th IFAC Symposium on Fault Detection Supervision and Safety for Technical Processes*. Vol. 2. IFAC. Budapest. pp. 1068–1073.

Ould Bouamama, B., G. Dauphin-Tanguy, M. Staroswiecki and D. Amo-Bravo (2000b). Bond graph analysis of structural FDI properties in mechatronic systems. In: *1st IFAC Conference on Mechatronic Systems*. Vol. 3. IFAC. Darmsdadt, Germany. pp. 1057–1062.

Paynter, H.M. (1961). *Analysis and design of Engineering Systems*. M.I.T. Press.

Thoma, J. U and B. Ould Bouamama (2000). *Modelling and Simulation in Thermal and Chemical Engineering. Bond Graph Approach*. Springer Verlag.

Worden, K. and A. Burrows (2001). Optimal sensor placement for fault detection. *Engineering Structures* **23**, 885–901.

VIRTUAL SENSORS UNDER DELAYED SCARCE MEASUREMENTS

Albertos P. * Peñarrocha I. * García P. *

* *Departamento de Ingeniería de Sistemas y Automática*
Universidad Politécnica de Valencia
Apdo. 22012, E-46071, Valencia, Spain.
e-mail: pedro@aii.upv.es, pggil@isa.upv.es, igpeaal@doctor.upv.es

Abstract: In this work, the output estimation under delayed and scarce measurements conditions is addressed. The process model is used to estimate the measurements that are not available. The available measurements are used to correct the estimated measurements by means of a complete regression vector update. In order to estimate the actual process output (delay-free), a model based predictor is applied. Copyright ©2004 IFAC

Keywords: Measurement delays, samples post-processing, scarce measurements.

1. INTRODUCTION

The data gathered from a process through measurements gives information about the state of the process and the disturbances acting on it. In some applications, due to hardware or availability constraints the number of measurements is reduced. This situation arises when dealing with chemical analysers to determine the concentration of the outcome of a process or, in general, in those measurement conditions where the measurement device cannot be placed at the required position.

On the other hand, the actuators, motors or valves, may be updated at a fast rate. Thus, the sampling rate is lower than the input updating rate.

Also, in many control applications, the data acquisition system involves a pre-processing of the sample. This results in a delay, with respect to the sampling instant, and probably, some changes in

their own value. This delay may be due to sample preparation or transportation, sample processing or even computation time.

The different nature of these delays implies that, in some cases, the delay can be greater than the sampling period. That is, a number of samples are under the processing line before the output is computed or estimated. This situation may be combined with missing data or irregular sampling, if the samples are not taken equally spaced on time, or some samples are not being processed.

A typical application is depicted in figure 1. In order to control the quality of the paper, the moisture (and the basis weight) of the paper web at the output of the headbox should be measured. But the measurement of the moisture can be only achieved down in the paper flow, well after the wet end and the dryer. As a result, the web humidity changes, due to the wet end and drying processes, and the measurement is taken after a transportation delay. If these processes are assumed to be time invariant, it is acceptable to assume a constant relationship between the

[1] This project has been partially granted by the CICYT project number DPI2002-04432, and through FPI grant number CTBPRB/2002/245 from Technology and Science Office from Generalitat Valenciana.

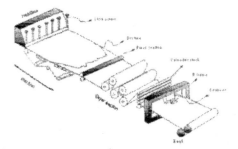

Fig. 1. Paper making machine

moisture at the headbox exit and the measured moisture, some time later.

A simplified model of the measurement system can be derived as follows: The web, wide L, runs at a constant speed $v(t)$ in the machine direction. The sensor frame is located m web meters down stream. Thus, a transport time delay $\tau = m/v$ appears. The sensor is transversely flying at speed $v'(t)$, taken samples at different cross-directional positions, giving an averaged moisture of the web each period $T(= L/v')$. The goal is to determine the moisture at the headbox exit, $y(t)$, based on modified (reduced) moisture measurements $y_r(t)$ taken at the sensor position:

$$y_r(t+\tau) = \gamma\, y(t) \qquad (1)$$

where $\gamma < 1$ is the humidity reduction factor due to the wet-end and dryer processes. The model is discretized with sampling period T. As an additional simplification the time delay τ is usually assumed to be a multiple of the sampling period T.

Some motivations and tools for investigating time-delay systems are presented in (Richard, 2003). In several works, a Smith predictor has been used for predicting the current output and afterwards a standard controller is applied. In (Lozano et al., 2003) an internal representation to observe the state of an unstable system and then predict the state to implement a state feedback control is used. In these works, only delayed measurements have been considered, but no scarce ones.

In works like (Albertos and Goodwin, 2002) and (Sanchis and Albertos, 2002) an internal model of the system has been used to estimate the missing outputs with the use of a Kalman filter with high complexity and computational cost. In these works, it is assumed that the available measurements have no time-delay.

In this contribution, different working conditions are analysed and some solutions to predict the plant output (all them are based on the availability of a plant model) are proposed. First, the plant is described. Then, the scarce sampling pattern is considered and an output estimator is proposed. The unavoidable delay is counteracted by using a

model-based predictor, and a general schema for the control of plants with delayed and scarce measurements is proposed. The approach is applied to some illustrative examples and its suitability is also discussed.

2. PROBLEM STATEMENT

Consider the system shown in figure 2 that consists of a linear process $G(s)$ whose input is updated at period T by a computer with a zero-order hold. The process output is measured with a sensor. This measurement system can be a dynamical system presenting also a time-delay.

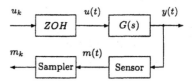

Fig. 2. Process with a general measurement system

2.1 The plant

The process $G(s)$ is modelled by its discrete ZOH equivalent at period T, defined by the difference equation:

$$y_k = -\boldsymbol{\theta}_a^{\mathsf{T}} Y_{k-1} + \boldsymbol{\theta}_b^{\mathsf{T}} U_{k-1} \qquad (2)$$

where $\boldsymbol{\theta}_a = [a_1 \cdots a_n]^{\mathsf{T}}$ and $\boldsymbol{\theta}_b = [b_1 \cdots b_n]^{\mathsf{T}}$ are the parameter vectors, $Y_{k-1} = [y_{k-1} \cdots y_{k-n}]^{\mathsf{T}}$ is the output vector and $U_{k-1} = [u_{k-1} \cdots u_{k-n}]^{\mathsf{T}}$ is the input vector (with $y_i = y(iT)$, $u_i = u(iT)$). Equation (2) can be rewritten as:

$$\begin{aligned} Y_k &= \begin{bmatrix} -\boldsymbol{\theta}_a^{\mathsf{T}} \\ \boldsymbol{I}_{(n-1)\times(n)} \end{bmatrix} Y_{k-1} + \begin{bmatrix} \boldsymbol{\theta}_b^{\mathsf{T}} \\ \boldsymbol{O}_{(n-1)\times(n)} \end{bmatrix} U_{k-1} \\ &= \boldsymbol{A}\, Y_{k-1} + \boldsymbol{B}\, U_{k-1} \end{aligned} \qquad (3)$$

where $\boldsymbol{I}_{(n-1)\times(n)}$ are the $n-1$ first rows and n first columns of the identity matrix, and $\boldsymbol{O}_{(n-1)\times(n)}$ is a null matrix of order $(n-1) \times (n)$.

One of the most common scenarios is the use of a chemical analyser that needs some time to evaluate some property. The chemical measurements will not be taken equally spaced on time, that is equivalent to a scarce and irregular pattern of measurements. That means that there will be only a few measurements available and they will have a delay because of the analysis process and maybe because of the communication channel linking to the measurement device.

In the following, the sensor dynamics is assumed to be too much faster than the process one, so, the sensor can be modelled as a pure time-delay.

3. OUTPUT PREDICTION. PROPOSED APPROACH

Consider the scheme shown in figure 3, where the process output $y(t)$ is measured with a sensor with a delay of dT seconds, and affected by a noise $n(t)$. The resulting measured signal is sampled synchronously with the input update at a larger period T_0. If the measurement pattern is regular, then the sampling period will be an integer multiple of the input update period ($T_0 = NT$), leading to m_k. The output estimation $\hat{y}_k$ is given at a period T by taking the inputs at period T and the measured signal sampled at the lower rate.

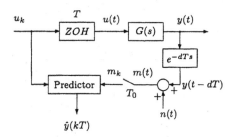

Fig. 3. Time-delay measurement system with scarce sampling

The estimation of the actual process output (y_k) can be separated into two subproblems. First, the available measured outputs (m_k when possible) are used to estimate the actual measurement ($\hat{m}_k$ every sampling time) to solve the problem of sporadic data gathering (see 3.1). Then, this virtual measurements are used to run the model in order to rectify the delay between output and measurement (see 3.2).

3.1 Measurement estimation

The estimator uses the process model (2), all the previous inputs and the actual or estimated measurements of the output. When there is no measurement available, the measurement is estimated running the model in open loop and taking into account the time-delay of the measurement system (e^{-dTs} equivalent to a delay of d samples), just transferring the delay at the process input. This leads to:

$$\hat{m}_{k|k-1} = -\boldsymbol{\theta}_a^{\mathsf{T}} \hat{M}_{k-1} + \boldsymbol{\theta}_b^{\mathsf{T}} U_{k-1-d} \qquad (4a)$$

$$\hat{M}_{k|k-1} = \left[\hat{m}_{k|k-1}\ \hat{m}_{k-1|k-1}\ \cdots\ \hat{m}_{k-n+1|k-1} \right]^{\mathsf{T}}$$
$$= \left[\hat{m}_{k|k-1}\ \hat{M}_{k-1,(1:n-1)}^{\mathsf{T}} \right]^{\mathsf{T}} \qquad (4b)$$

where $\hat{m}_{k-i|k-j}$ is the estimation of the measurement at instant $k-i$ computed with the data until instant $k-j$, $\hat{M}_k = [\hat{m}_{k|k}\ \cdots\ \hat{m}_{k-n+1|k}]^{\mathsf{T}}$

is the measurement regression vector, and $\hat{M}_{k|k-1}$ its initial estimation. By using the notation in (3), it can be expressed by

$$\hat{M}_{k|k-1} = \boldsymbol{A}\,\hat{M}_{k-1} + \boldsymbol{B}\,U_{k-1-d} \qquad (5)$$

If there is no measurement (i.e. m_k is not available), the best actual measurement estimation ($\hat{m}_k \equiv \hat{m}_{k|k}$) and regression vector ($\hat{M}_k$) are defined by:

$$\hat{m}_k = \hat{m}_{k|k-1} \qquad (6a)$$

$$\hat{M}_k = \hat{M}_{k|k-1} \qquad (6b)$$

Note that, in this case

$$\hat{m}_k = \boldsymbol{h}^{\mathsf{T}} \hat{M}_k = \boldsymbol{h}^{\mathsf{T}} \hat{M}_{k|k-1} \qquad (7)$$

where $\boldsymbol{h} = [1\ 0 \cdots 0]^{\mathsf{T}}$. When there is a new measurement (i.e. m_k is available), the pre-estimated measurement and regression vector calculated by equations (4) are updated by (Sanchis and Albertos, 2002), (see the references therein for details):

$$\hat{m}_k = \hat{m}_{k|k-1} + l_1(m_k - \hat{m}_{k|k-1}) \qquad (8a)$$

$$\hat{m}_{k-i|k} = \hat{m}_{k-i|k-1} + l_{i+1}(m_k - \hat{m}_{k|k-1});$$
$$i = 1, \ldots, n-1 \qquad (8b)$$

where the gain vector $\boldsymbol{l} = [l_1 \cdots l_n]^{\mathsf{T}}$ can be designed, for instance by pole placement, to achieve a desired predictor dynamic behaviour.

These equations can be written as

$$\hat{m}_k = (1 - l_1)\,\hat{m}_{k|k-1} + l_1\,m_k \qquad (9a)$$

$$\hat{M}_k = (\boldsymbol{I} - \boldsymbol{l}\,\boldsymbol{h}^{\mathsf{T}})\,\hat{M}_{k|k-1} + \boldsymbol{l}\,\boldsymbol{h}^{\mathsf{T}}\,M_k \qquad (9b)$$

where $M_k = [m_k\ \cdots\ m_{k-n+1}]^{\mathsf{T}}$ is the complete vector of measurements, available or not.

Let us now introduce δ_k, the availability factor of measurements ($\delta_k = 1$ if there is one measurement available, at instants $k = jN$, and $\delta_k = 0$ if not). Then, combining equations (5), (6) and (9), the measurement estimation can be computed at every sampling instant by:

$$\hat{M}_k = (\boldsymbol{I} - \boldsymbol{l}\,\boldsymbol{h}^{\mathsf{T}}\,\delta_k)\left(\boldsymbol{A}\,\hat{M}_{k-1} + \boldsymbol{B}\,U_{k-1-d}\right)$$
$$+ \boldsymbol{l}\,\boldsymbol{h}^{\mathsf{T}}\,M_k\,\delta_k \qquad (10)$$
$$\hat{m}_k = \boldsymbol{h}^{\mathsf{T}}\,\hat{M}_k$$

The dynamics of the measurement estimation error ($E_k = M_k - \hat{M}_k$) under a regular sampling ($T_0 = NT$) is defined by (see (Sanchis, 1999) for the proof):

$$E_{k+N} = \left(\boldsymbol{I} - \boldsymbol{l}\,\boldsymbol{h}^{\mathsf{T}}\right)\boldsymbol{A}^N\,E_k \qquad (11)$$

3.2 Current output prediction

As the measurement system has a delay (see figure 3), the output must be estimated from the

87

available measurements m_k. For that purpose, an output predictor has been added to the system. This predictor estimates the actual process output from the available vector of parameters, inputs and measurement regression vector. The process output and the measurement are considered to be related with a single delay:

$$m_k = y_{k-d} + n_k \qquad (12)$$

where n_k is the sampled noise signal. If there is no delay on the system ($d = 0$), the output is evaluated with the measurement estimation given by (10):

$$\hat{y}_k = \hat{m}_k \qquad (13)$$

If the measurement has one sample-delay ($d = 1$), $\hat{Y}_{k-1} = \hat{M}_k$, the output is estimated running the process model (3), in open loop:

$$\hat{Y}_k = A\,\hat{M}_k + B\,U_{k-1} \qquad (14)$$

If the measurement system has a delay of $d = 2$ samples, the model has to be run twice in order to estimate the output, leading to the equation:

$$\hat{Y}_k = A^2\,\hat{Y}_{k-2} + A\,B\,U_{k-2} + B\,U_{k-1} \qquad (15a)$$

$$\hat{Y}_k = A^2\,\hat{M}_k + A\,B\,U_{k-2} + B\,U_{k-1} \qquad (15b)$$

In general, the output under a delay d is estimated as:

$$\hat{Y}_k = A^d\,\hat{M}_k + \sum_{j=1}^{d} A^{j-1}\,B\,U_{k-j} \qquad (16)$$

In any case, the current output is computed as

$$\hat{y}_k = h^\top\,\hat{Y}_k \qquad (17)$$

Combining equations (10) and (16), the output predictor is:

$$\hat{M}_k = (I - lh^\top\delta_k)\left(A\hat{M}_{k-1} + BU_{k-1-d}\right) + lh^\top M_k\delta_k$$

$$\hat{y}_k = h^\top\left(A^d\,\hat{M}_k + \sum_{j=1}^{d} A^{j-1}\,B\,U_{k-j}\right) \qquad (18)$$

4. SIMULATION RESULTS

One of the most useful applications of this prediction scheme shown in figure 3 can be the design of a control system with delayed and scarce measurements, as shown in figure 4. In that scheme, a standard digital controller $C(z)$, designed for the discrete time plant model $G(z)$, needs the sequence of outputs at period T, but the process output is only measured at period $T_0 = N\,T$ (if regular sampling is considered), and with a delay of d samples. Therefore, the $N - 1$ missing inter-sampling measurements have to be estimated, and the actual output must be "predicted" from the estimated measurements (which are delayed). For that purpose, a model based output predictor has been added in the closed loop.

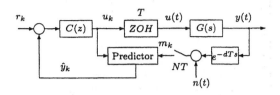

Fig. 4. Control system with measurement delay under scarce measurement

This approach is a simple way to estimate the outputs of a process with delay and with scarce measurements. Here, an input-output model has been used to estimate the (scarce) measurements, and then to predict the output.

The proposed output predictor gives solution to closing the loop with a standard controller that is designed taking into account the delay-free system ($G(s)$). The effects of the measurement processing and delay can be checked in real systems like the one illustrated in figure 1, the paper making machine. Nevertheless, the predictor is also useful in really complex measurement scenarios, dealing with unstable plants.

4.1 Example

In order to emphasize the delay effect, let us consider an unstable system defined by the transfer function:

$$G(s) = \frac{1}{(s-1)\,(s+3)}$$

with a sampling period of $T = 0.1$ seconds. The equivalent discrete transfer function for a zero-order-hold is:

$$G(z) = \frac{4.69\,z^{-1} + 4.39\,z^{-2}}{1 - 1.846\,z^{-1} + 0.8187\,z^{-2}}\,10^{-3}$$

The sensor is modelled as a single delay of 1 second, equivalent to a delay of $d = 10$ samples. Moreover, the measurements are only available every $T_0 = 1.2$ seconds ($N = 12$).

Running the model in open loop ($l = 0$) with estimation of missing data and output prediction results in an unstable behaviour of the prediction. This can be checked by computing the eigenvalues of the prediction error, given by:

$$\lambda\left((I - l\,h^\top)\,A^N\right) = \lambda\left(A^N\right) = \{3.32,\ 0.0273\}$$

A modelling error, a measurement noise and a wrong initial estimation of the outputs has been added to test the predictor. A predictor has been designed following (18). The updating vector l is designed for stable eigenvalues of the estimator $(0.3, 0.3)$, leading to $l = [0.0079\ \ -0.3606]^\top$. This predictor is added in a closed loop system as shown in figure 4, using a simple controller $C(z) = 4$, just to see the effectiveness of including

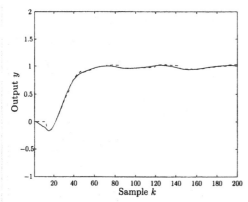

Fig. 5. Prediction error in an unstable system with delay $d = 10$ and scarce measurements ($N = 12$). Output prediction (dashed line) and process output (full line) for the closed loop predictor. Process output with open loop predictor $l = 0$ (dotted line).

the proposed predictor. The simulation results are shown in figure 5, showing how the closed loop scheme is unstable if an open loop predictor is used ($l = 0$ dotted line). If the proposed closed loop predictor is used, the prediction (dashed line) and the output (full line) are stable.

5. DISCUSSION AND CONCLUSIONS

In this paper the problem of output prediction of a process under delayed and scarce measurements has been addressed.

In order to estimate the measurements which are not available at the input update instants, a model based measurement estimator has been added. This calculation uses the available measurements to correct the estimations so the error has a stable dynamical behaviour. With the estimated measurements, the model is computed in open loop as many times as needed (the number of delay samples) to predict the process output.

As a difference with respect to the typical Smith predictor to overcome the time delays, the open-loop process stability is not required.

Some possible real applications have been commented and one example illustrates the validity of the proposed prediction of an unstable system under a delayed and scarce measurements.

The approach can be extended to consider other measurements conditions. Let us consider that the measurement vector m_k is obtained from the output sequence $\{y_k\}$ through the general operator $F(\{\bullet\})$:

$$m_k = F(\{y_k\}) \qquad (19)$$

Some simple cases may be considered:

(1) Direct measurement

$$m_k = y_k + n_k \qquad (20a)$$

where n_k is the white measurement noise with zero mean and variance σ^2.

(2) There is a processing of the output, leading to

$$\begin{aligned} z_{k+1} &= \boldsymbol{\Phi} z_k + \rho\, y_k \\ m_k &= \boldsymbol{c}\, z_k \end{aligned} \qquad (20b)$$

One especial case is the pure time-delay. In that case, the measurement states are the delayed outputs. For a delay of $d = 4$ instants, the model is

$$\boldsymbol{\Phi} = \begin{bmatrix} 0 & 1 & 0 & 0 \\ 0 & 0 & 1 & 0 \\ 0 & 0 & 0 & 1 \\ 0 & 0 & 0 & 0 \end{bmatrix} ; \qquad \rho = \begin{bmatrix} 0 \\ 0 \\ 0 \\ 1 \end{bmatrix}$$

$$\boldsymbol{c} = \begin{bmatrix} 1 & 0 & 0 & 0 \end{bmatrix} ; \qquad m_k = y_{k-d}$$

(3) There is a batch processing of the samples, considering the system process-sensor as an hybrid system. In this case, every available sample is considered as an event that starts the process of measurement. This batch process will give as output the measured value of the output process sampling. If the batch processing needs a constant computing time to compute the measurement, it can be written as:

$$m_k = g(y_{k-d}) \qquad (20c)$$

Again, the simplest case is for $g(\bullet)$ being a constant gain. A pure time-delay can be also considered in this way, as it has been considered in the development of this work. The most typical case is the pre-processing of the samples, either in measurement, transportation or general process.

In all cases, the general operator F can be considered to have a scarce sampling of the output, such that in addition to the measurement delay, a scarce measurement pattern is considered. Thus, the measured output will not be available in all the input updating instants.

The prediction error dynamic under a periodic availability of measurements corresponds to a linear time invariant system, as shown in (11). However, under random availability, that expression uses the random variable N. If N has a known set of possible values, the predictor error dynamic is given by structured linear time variant system, and the stability (election of l) must take into account stochastic process theory, as explained in (Sanchis, 1999).

6. REFERENCES

Albertos, P. and G.C. Goodwin (2002). Virtual sensors for control applications. *Annual Reviews in Control* **26**, 101–112.

Lozano, R., P. García, P. Castillo and A. Dzul (2003). Robust prediction-based control for unstable delay systems. *42nd IEEE Conference on Decision and Control* pp. 4014–4019.

Richard, J.P. (2003). Time-delay systems: an overview of some recent advances and open problems. *Automatica* **39**, 1667–1694.

Sanchis, R. (1999). *Control of Industrial Processes with Scarce Measurements*. Doctoral Thesis. Universidad Politécnica de Valencia, Spain.

Sanchis, R. and P. Albertos (2002). Recursive identification under scarce measurements. convergence analysis. *Automatica* **38**, 535–544.

ELSEVIER
IFAC
PUBLICATIONS
www.elsevier.com/locate/ifac

A MULTI-OBJECTIVE OUTPUT-FEEDBACK CONTROLLER FOR SYSTEMS WITH FRICTION

Karim Khayati* Pascal Bigras Louis-A. Dessaint

*Email: karim.khayati.1@ens.etsmtl.ca
École de Technologie Supérieure,
Department of Automated Manufacturing Engineering,
1100, Notre-Dame West, Montreal, Quebec, H3C 1K3,
Canada.

Abstract: In this paper, we develop a full-order dynamic output feedback positioning tracking for servo-systems with friction, based on LuGre friction observer dynamics with fixed model parameters. To this end, we use the position and velocity measurements. The compensator design is appropriately built to ensure a global asymptotic stabilization of the tracking error dynamics under a specific pole clustering. And, we take into account some H_2 cost performance to offer an optimal control structure. The subsequent stated controller matrices are found by using the Linear Matrix Inequality approach. Simulation results illustrate the effectiveness of the proposed compensator. Copyright ©2004 IFAC

Keywords: Friction compensation, LuGre model, Stabilization, Multi-objective control synthesis, LMI.

1. INTRODUCTION

The unavoidable friction phenomena cause in general the performance of servomechanisms to deteriorate due to non negligible tracking errors, limit cycles and undesired stick-slip motion (A.-Hélouvry, 1994). That being so, they pay out the most industrial application processes (for instance moving robots, pneumatic actuators, *etc.*). Faced with these economic difficulties, plant and controller designers have striven increasingly towards of more efficient and, thus, economically more competitive plants to compensate for friction aspects. And, several strategies have been investigated in the literature (see *e.g.* (Canudas *et al.*, 1995)) to address this issue, with a view to achieving a higher degree of servosystem accuracy without the need for excessive integral gains. In practice, friction involves very complicated phenomena which are difficult to describe analytically

and which rely on the physical properties of the contact surfaces. Early classical friction models are described by static mappings between velocity and friction forces which depend on the sign of the velocity, such as Coulomb friction and viscous friction (A.-Hélouvry, 1994). However, many of the interesting properties observed in systems involving friction cannot be explained by static models alone. This is because friction does not produce an instantaneous response in the form of a change of velocity, *i.e.* it has internal dynamics ; as stick-slip motion, pre-sliding displacement, Dahl effect and frictional lag (A.-Hélouvry, 1994). In addition, any friction compensation based on static maps defaults at very low velocities. For instance, as with the use of any Coulomb friction with stiction model, the control signal becomes discontinuous and chattering effects appear at velocities close to zero. So, it was argued in (A.-Hélouvry, 1994) that dynamic models are necessary to describe friction phenomena accurately, and there are many of them in the Dahl model, the Bliman-Sorine model and the LuGre scheme (Canudas *et al.*, 1995). For a review of some of the existing dynamic friction models, see (A.-Hélouvry, 1994) and references cited therein. In this paper, we assume that the friction phenomenon is of a dynamic nature, and is covered by the LuGre friction model. Such a model captures most friction characteristics and is generically modelled as the average deflection force of elastic bristles between two contact surfaces. The bristles represent the asperities and irregularities characterizing the surfaces at a microscopic level.

Several controller/observer designs for mechanical systems using the LuGre nonlinear friction model have been developed. In (Canudas *et al.*, 1995; Canudas and Lischinsky, 1997), we see that conventional linear controllers (*e.g.* PD, filtered PID, *etc.*) can be chosen such that the transfer function relating the position error to the friction estimation error is strictly positive real (SPR). The stability of the tracking and observation errors is then ensured. The response is regarded as reasonably accurate, but very slow due to the excess damping in the system. Later, Hirschorn and Miller proposed a continuous dynamic controller for a class of nonlinear systems (Hirschorn and Miller, 1999), which ensured sufficient conditions for a global stabilization. Nevertheless, the design parameters of this controller were, *a priori*, not arranged according to the appropriate specifications set out in estimation and tracking dynamics. In this paper, a full-order output feedback structure is proposed for both estimation and tracking strategies. This is followed by an extended feature on the static output feedback approach set out in (Canudas *et al.*, 1995). In fact, the higher degree of flexibility of such controllers is technically more conducive to improving the stability condition. Our controller can be considered as a generalization of the linear controller given in (Canudas *et al.*, 1995). We make use of an advanced solving method to attain an optimum observer based solution that satisfies simultaneously more than one objective. An SPR condition is firstly established, and then, enhanced with a stability region in order to enforce prescribed transient response specifications and to prevent limit fast controller dynamics. Finally, we elaborate a solution to the problem of any H_2 optimization of the feedback; also known as Linear-Quadratic-Gaussian (LQG) optimization (Zhou *et al.*, 1996). We assert to limit simultaneously the tracking errors and the input energy. Although, the LQG problems and related H_2 problems have usually transparent solutions in the time domain (Zhou *et al.*, 1996), the H_2 cost function offers a more natural way of representing certain aspects of the system performance. The present study examines the effectiveness of our controller in improving the transient positioning response. Particularly, the "substantial high closed-loop bandwidth", due to the deduced SPR conditions for the PD and filtered PID controllers as noticed in (Canudas and Lischinsky, 1997), can be remake "better" ; as the approach discussed herein acts directly on controller poles. Calculation of the stabilization control components is cast as a convex problem of multi-objective output feedback involving linear matrix inequalities (LMIs) (Scherer *et al.*, 1997). This paper is organized as follows. Section 2 introduces the model of a moving single mass with friction. We briefly recall the above-mentioned LuGre model. In section 3, we investigate the observer-based controller design, and, in section 4, the stability condition and additive clustering pole constraints inside one H_2 cost optimization. A multi-objective output feedback synthesis is then presented. A simulation and relevant discussions are presented in section 5. Concluding remarks appear in section 6.

2. MODEL DESCRIPTIONS

Here, we investigate a mass m at position x:

$$m\ddot{x} + f = u \qquad (1)$$

under the influence of a dynamic friction f and an input active force u. The control objective is to design a controller to track any prescribed trajectory in accordance with a best tradeoff between stability proof and transient performances. We model friction force variations using the LuGre friction model, as follows (Canudas *et al.*, 1995):

$$\dot{z} = \dot{x} - \frac{\sigma_0 |\dot{x}|}{g(\dot{x})} z \qquad (2)$$

and

$$f = \sigma_0 z + \sigma_1 \dot{z} + \sigma_2 \dot{x} \qquad (3)$$

where the internal friction state z describes the average relative deflection of the contact surfaces during the stiction phases. σ_0, σ_1 and σ_2 are the frictional stiffness, frictional damping and viscous friction coefficients respectively. The term $g(\dot{x})$ is a finite function which can be chosen to describe different friction effects. One parameterization of $g(\dot{x})$ to characterize the Stribeck effect is given in (Canudas *et al.*, 1995):

$$g(\dot{x}) = \mu_c + (\mu_s - \mu_c) \cdot e^{-\frac{\dot{x}^2}{\dot{x}_s^2}} \qquad (4)$$

where μ_c is the normalized Coulomb friction, μ_s the normalized static friction coefficient and $\dot{x}_s$ the constant Stribeck relative velocity.

The controller design presented in the next section is based on the following statements: **(A.1)** The position reference signal x_d is assumed to be differentiable twice. **(A.2)** The position x and the velocity $\dot{x}$ are assumed to be measurable. **(A.3)** The bristle deflection z is physically non measurable. **(A.4)** The complete friction model (2)-(4) is notably characterized by the 'static parameters' μ_c, μ_s and $\dot{x}_s$, and the 'dynamic parameters' σ_0, σ_1 and σ_2. These friction parameters are known *a priori*.

Remark 1: We note that the assumption (A.4) has to be only considered in order to simplify the problem formulation in this paper. Nevertheless, taking into account any variation of friction model parameters can easily be matched in our multi-objective context.

3. OBSERVER-BASED CONTROLLER DESIGN

In this section, we introduce a model-based friction compensation scheme designed to reduce the effect of friction. This compensation scheme, illustrated in Figure 1, is a generalization of others proposed in *e.g.* (Canudas *et al.*, 1995). In fact, we propose to replace the static output feedback designed for the controller and the observer in (Canudas *et al.*, 1995) by two combined full-order dynamic feedback mechanisms. As will be seen in the next section, this generalization allows us to design the controller in order not only to stabilize the tracking error, but also to solve the multi-objective optimization problem. Under assumptions (A.3) and (A.4), let the observer-based Lu-Gre friction compensation scheme (see Figure 1) be given by:

$$\dot{\hat{z}} = \dot{x} - \frac{\sigma_0 |\dot{x}|}{g(\dot{x})}\hat{z} - u_z \qquad (5)$$

and

$$\hat{f} = \sigma_0 \hat{z} + \sigma_1 \dot{\hat{z}} + \sigma_2 \dot{x} \qquad (6)$$

where $\hat{z}$ is the observed internal state, $\hat{f}$ is the friction estimate and u_z represents a deflection rate correction and constitutes any observer dynamic feedback term related to the measurable states of the system. Furthermore, we assume the control law to be given by:

$$u = m\ddot{x}_d + \hat{f} + u_e \qquad (7)$$

where u_e designates a direct active force and is any dynamic controller term related to the measurable states of the system.

Next, we introduce the trajectory tracking and observation errors respectively $e = x - x_d$; $\tilde{z} =$

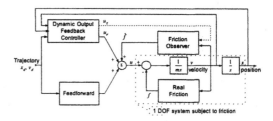

Fig. 1. Block diagram of the observer-based friction compensation scheme.

$z - \hat{z}$ and $\tilde{f} = f - \hat{f}$. It is straightforward to compute the error dynamics for the observer:

$$\dot{\tilde{z}} = -\frac{\sigma_0 |\dot{x}|}{g(\dot{x})}\tilde{z} + u_z \qquad (8)$$

and

$$\tilde{f} = \sigma_0 \tilde{z} + \sigma_1 \dot{\tilde{z}} \qquad (9)$$

The choice of our control results in the following tracking error dynamics:

$$m\ddot{e} = -\tilde{f} + u_e \qquad (10)$$

Consider the whole dynamics (8)-(10) and let $\mathbf{x} = \begin{pmatrix} e & \dot{e} \end{pmatrix}^{\mathrm{T}}$ be the measurable state vector and $\mathbf{u} = \begin{pmatrix} u_e & u_z \end{pmatrix}^{\mathrm{T}}$ the input vector. Therefore, the linear dynamic equation (10) is written in the state space representation as:

$$\dot{\mathbf{x}} = \mathbf{A}\mathbf{x} + \mathbf{B}\mathbf{u} + \mathbf{B}_{\tilde{f}}(-\tilde{f}) \qquad (11)$$

with

$$\mathbf{A} = \begin{pmatrix} 0 & 1 \\ 0 & 0 \end{pmatrix}; \ \mathbf{B} = \begin{pmatrix} 0 & 0 \\ \frac{1}{m} & 0 \end{pmatrix}; \ \mathbf{B}_{\tilde{f}} = \begin{pmatrix} 0 \\ \frac{1}{m} \end{pmatrix} \ (12)$$

Henceforth, the control vector $\mathbf{u}$ is considered the output of any LTI compensator, which can be represented in state-space form as follows:

$$\begin{cases} \dot{\xi} = \mathbf{A}_c\xi + \mathbf{B}_c\mathbf{x} \\ \mathbf{u} = \mathbf{C}_c\xi + \mathbf{D}_c\mathbf{x} \end{cases} \qquad (13)$$

where ξ represents a two-dimensional dynamic controller state vector.

Proposition 1: *Consider $\tilde{G}(s)$ to be the closed-loop transfer function from $-\tilde{z}$ to u_z. Then, $\tilde{G}(s)$ is a proper map and has a realization $(\tilde{\mathbf{A}}, \tilde{\mathbf{B}}, \tilde{\mathbf{C}}, \tilde{\mathbf{d}})$ given by (14) with $\mathbf{E} \triangleq \begin{pmatrix} 0 & 1 \end{pmatrix}$.*

$$\tilde{\mathbf{A}} = \begin{pmatrix} \mathbf{A} + \mathbf{B}\mathbf{D}_c & \mathbf{B}\mathbf{C}_c \\ \mathbf{B}_c & \mathbf{A}_c \end{pmatrix};$$

$$\tilde{\mathbf{B}} = \begin{pmatrix} \sigma_0\mathbf{B}_{\tilde{f}} + \sigma_1\mathbf{A}\mathbf{B}_{\tilde{f}} + \sigma_1\mathbf{B}\mathbf{D}_c\mathbf{B}_{\tilde{f}} \\ \sigma_1\mathbf{B}_c\mathbf{B}_{\tilde{f}} \end{pmatrix};$$

$$\tilde{\mathbf{C}} = \begin{pmatrix} \mathbf{E}\mathbf{D}_c & \mathbf{E}\mathbf{C}_c \end{pmatrix}; \ \tilde{\mathbf{d}} = \sigma_1\mathbf{E}\mathbf{D}_c\mathbf{B}_{\tilde{f}} \ (14)$$

The proof is omitted.

The full-order dynamic feedback controller presented here is a generalization of the one proposed in (Canudas et al., 1995). The controller in (Canudas et al., 1995) was limited to a static PD feedback inside the regulation objective, with a static control-action on the observer (see Figure 1) proportional to the position error ($\triangleq ke$), which facilitates the explicit stability feature. Indeed, in the literature, authors often use simple direct error signals between the two blocks cited within the control-scheme above (Canudas and Lischinsky, 1997; Canudas, 1998). Hence, partitioning the closed-loop system into linear and nonlinear parts and thanks to the passivity of the LuGre model, the passivity theorem can be used to derive conditions on the controller which guarantee the stability of the closed-loop system. The condition is that the resulting linear block is SPR. For the particular case of static output feedback, having the form of (13) with $\mathbf{A}_c = \mathbf{B}_c = \mathbf{C}_c \triangleq \emptyset$ (i.e. empty set) and $\mathbf{D}_c \triangleq \begin{pmatrix} -k_p & -k_d \\ -k & 0 \end{pmatrix}$. Canudas (Canudas et al., 1995) showed that $\tilde{G}(s)$ is SPR if $k > 0$, $k_p > 0$, $k_d > \frac{\sigma_0 m}{\sigma_1}$. But, in our general cases of full-order dynamic controllers, this passivity condition remains only barely established and/or analytically difficult to construct. In the next section, the general SPR condition will be cast as the feasibility of a convex objective expressed as an linear matrix inequality (LMI) constraint. Then, the whole observation-control context can include additive constraints (such as H_∞, H_2, etc.). Here, we will include an adequate pole clustering constraint inside an H_2 index optimization problem.

4. LMI FORMULATION

4.1 Stability condition

Consider system (1) and suppose that the friction force variations are captured by the friction model in (2)-(4). Let the friction observer be given by (5) and (6), and the control law be given by (7).

Proposition 2: For any $\epsilon > 0$, the observer and position errors are exponentially stable if there exists a matrix $\tilde{\mathbf{P}}_s = \tilde{\mathbf{P}}_s^T > 0$, such that

$$\begin{pmatrix} \tilde{\mathbf{A}}^T \tilde{\mathbf{P}}_s + \tilde{\mathbf{P}}_s \tilde{\mathbf{A}} & \tilde{\mathbf{P}}_s \tilde{\mathbf{B}} - \frac{1}{\epsilon^2} \tilde{\mathbf{C}}^T \\ \tilde{\mathbf{B}}^T \tilde{\mathbf{P}}_s - \frac{1}{\epsilon^2} \tilde{\mathbf{C}} & -\frac{2}{\epsilon^2} \tilde{\mathbf{d}} \end{pmatrix} < 0 \quad (15)$$

The proof is omitted.

Remark 3: The internal bristle deformation is manifestly rather smaller than the mass displacement. Thence, the coefficient ϵ is generically chosen too small to well make fit the matrix scaling in (15), and then to avoid an ill conditioning with the LMIs.

Remark 4: The inequality (15) is equivalent to the statement " $\tilde{G}(s)$ is SPR, i.e. $\tilde{G}(j\omega) + \tilde{G}^*(j\omega) > 0$, for all $\omega \in \mathbb{R}$" (Scherer et al., 1997).

Remark 5: In the particular case of static output feedback, we have $\tilde{\mathbf{d}} \triangleq 0$, and $\frac{dV}{dt}$ can only be negative semi-definite. Then, Barbalat's lemma and the SPR condition are sufficient to prove that $e(t)$ and $\tilde{z}(t)$ tend to zero. Details of this development can be found in (Canudas, 1998).

4.2 Multi-objective design

Furthermore, we can design a controller which ensures the SPR condition given by (15) while assigning the closed-loop transfer $\tilde{G}$ poles in any LMI region of the complex left-half plane, and achieving an optimal H_2 performance (Scherer et al., 1997). These augmented constraints on the eigenvalue clustering can be used to enforce some response specifications like stability margin, settling time and overshot, and the frequency response limits of $\tilde{G}$. For instance, to ensure a stability margin of λ and prevent fast controller dynamics (i.e. frequency of oscillatory modes) simultaneously, we propose to bunch the closed-loop poles in the intersection of the half-plane $\Re(z) < -\lambda$, with the disk of radius r centered at the origin. The LMI characterization for this stability region is now expressed as $\exists \tilde{\mathbf{P}}_r = \tilde{\mathbf{P}}_r^T > 0$, such that (Scherer et al., 1997)

$$\tilde{\mathbf{P}}_r \tilde{\mathbf{A}} + \tilde{\mathbf{A}}^T \tilde{\mathbf{P}}_r + 2\lambda \tilde{\mathbf{P}}_r < 0 \quad (16)$$

and

$$\begin{pmatrix} -r\tilde{\mathbf{P}}_r & \tilde{\mathbf{A}}\tilde{\mathbf{P}}_r \\ \tilde{\mathbf{P}}_r \tilde{\mathbf{A}}^T & -r\tilde{\mathbf{P}}_r \end{pmatrix} < 0 \quad (17)$$

In addition, some optimization design can be seen as an important issue to limit the cost over any branch inside the considered system dynamics. For instance, to find a control scheme that minimizes, in a compromise, the sensitivity of the error output and the control cost to the friction estimation error. This objective is introduced as a typical H_2/LQG optimization of the closed-loop transfer from $-\tilde{f}$ to $\begin{pmatrix} \mathbf{u} \\ \rho \mathbf{x} \end{pmatrix}$. In the following, we establish the LMI characterization for the proposed index optimization:

Proposition 3: Let $\tilde{G}_2(s)$ be the closed-loop transfer function from $-\tilde{f}$ to $\begin{pmatrix} \mathbf{u} \\ \rho \mathbf{x} \end{pmatrix}$. Then, $\tilde{G}_2(s)$ is a strictly proper map and has a realization $(\tilde{\mathbf{A}}, \tilde{\mathbf{B}}_2, \tilde{\mathbf{C}}_2)$ given by (14) and

$$\tilde{\mathbf{B}}_2 = \begin{pmatrix} \mathbf{B}_{\tilde{f}} \\ \mathbf{0}_{21} \end{pmatrix}; \ \tilde{\mathbf{C}}_2 = \begin{pmatrix} \mathbf{C}_2 + \mathbf{D}_2\mathbf{D}_c & \mathbf{D}_2\mathbf{C}_c \end{pmatrix} \quad (18)$$

Moreover, $\|\tilde{G}_2\|_2^2 < \nu$ if and only if $\exists \tilde{P}_2 = \tilde{P}_2^T > 0$ and $T = T^T$ such that:

$$\begin{pmatrix} \tilde{A}^T\tilde{P}_2 + \tilde{P}_2\tilde{A} & \tilde{P}_2\tilde{B}_2 \\ \tilde{B}_2^T\tilde{P}_2 & -I \end{pmatrix} < 0 \qquad (19)$$

$$\begin{pmatrix} \tilde{P}_2 & \tilde{C}_2^T \\ \tilde{C}_2 & T \end{pmatrix} > 0 \qquad (20)$$

and

$$\mathrm{Tr}(T) < \nu \qquad (21)$$

With the matrix inequality formulations (15)-(17) and (19)-(21) above, the controller design problem can be stated as follows: "Minimize ν subject to $\tilde{P}_s = \tilde{P}_s^T > 0$, $\tilde{P}_r = \tilde{P}_r^T > 0$, $\tilde{P}_2 = \tilde{P}_2^T > 0$, $T = T^T$, A_c, B_c, C_c, and D_c, such that the closed-loop systems (14) and (18) satisfy respectively (15)-(17) and (19)-(21)."

This problem is not tractable using the LMI formulation unless we require that the same Lyapunov matrix $\tilde{P}$ satisfies (14)-(21). We therefore restrict our attention to the following conservative formulation (Scherer et al., 1997): "Minimize ν subject to

$$\tilde{P} = \tilde{P}^T > 0 \qquad (22)$$

$T = T^T$, A_c, B_c, C_c and D_c satisfying (14)-(22), with $\tilde{P} = \tilde{P}_s = \tilde{P}_r = \tilde{P}_2$."

4.3 Output feedback synthesis

The conditions for design of the controller given in the previous section are bilinear in the variables $\tilde{P}$, A_c, B_c, C_c and D_c. By using the linearizing change of controller variables introduced in (Scherer et al., 1997), we convert these bilinear matrix inequalities (BMIs) into LMIs in a different set of variables. Partition $\tilde{P}$ and $\tilde{P}^{-1}$ as:

$$\tilde{P} = \begin{pmatrix} P & M \\ M^T & R \end{pmatrix} \text{ and } \tilde{P}^{-1} = \begin{pmatrix} Q & N \\ N^T & S \end{pmatrix} \qquad (23)$$

with $P = P^T$, $Q = Q^T$ such that the invertible matrices M and N satisfy

$$MN^T = I - PQ \qquad (24)$$

Let the new stated matrical variables $\hat{A}$, $\hat{B}$, $\hat{C}$ and $\hat{D}$ be given by:

$$\begin{cases} \hat{A} = MA_cN^T + MB_cQ + \\ \qquad PBC_cN^T + PAQ + PBD_cQ \\ \hat{B} = MB_c + PBD_c \\ \hat{C} = C_cN^T + D_cQ \\ \hat{D} = D_c \end{cases} \qquad (25)$$

Proposition 4: A full-order output-feedback controller (13) and a symmetric matrix $\tilde{P} > 0$ partitioned as in (23) exist such that (15)-(17) and

(19)-(22) with respectively (14) and (18) hold if and only if matrices P, Q, $\hat{A}$, $\hat{B}$, $\hat{C}$ and $\hat{D}$ exist such that the following LMIs are feasible:

$$\begin{pmatrix} Q & I \\ I & P \end{pmatrix} > 0 \qquad (26)$$

$$\begin{pmatrix} a_{11} & \star & \star \\ a_{21} & a_{22} & \star \\ a_{31} & a_{32} & -\dfrac{2\sigma_1}{\epsilon^2}E\hat{D}B_{\tilde{f}} \end{pmatrix} < 0 \qquad (27)$$

$$\begin{pmatrix} -rQ & \star & \star & \star \\ -rI & -rP & \star & \star \\ AQ+B\hat{C} & A+B\hat{D} & -rQ & \star \\ \hat{A} & PA+\hat{B} & -rI & -rP \end{pmatrix} < 0 \qquad (28)$$

$$\begin{pmatrix} a_{11} & \star & \star \\ a_{21} & a_{22} & \star \\ B_{\tilde{f}}^T & B_{\tilde{f}}^TP & -I \end{pmatrix} < 0 \qquad (29)$$

$$\begin{pmatrix} Q & \star & \star \\ I & P & \star \\ C_2Q+D_2\hat{C} & C_2+D_2\hat{D} & T \end{pmatrix} > 0 \qquad (30)$$

and

$$\mathrm{Tr}(T) < \nu \qquad (31)$$

with

$$a_{11} = AQ + QA^T + B\hat{C} + (B\hat{C})^T$$
$$a_{21} = \hat{A} + (A+B\hat{D})^T$$
$$a_{31} = \sigma_0 B_{\tilde{f}}^T + \sigma_1 B_{\tilde{f}}^TA^T + \sigma_1 B_{\tilde{f}}^T\hat{D}^TB^T - \frac{1}{\epsilon^2}E\hat{C}$$
$$a_{22} = PA + A^TP + \hat{B} + \hat{B}^T$$
$$a_{32} = \sigma_0 B_{\tilde{f}}^TP + \sigma_1 B_{\tilde{f}}^TA^TP + \sigma_1 B_{\tilde{f}}^T\hat{B}^T - \frac{1}{\epsilon^2}E\hat{D}$$

and $\star$ replaces blocks that are readily inferred by symmetry.

For the proof, see e.g. (Scherer et al., 1997). The matrices A_c, B_c, C_c and D_c can easily be derived from (25) by calculating invertible matrices M and N satisfying (24). LMIs (26)-(31) can easily be solved by using the interior point optimization method implemented in the ©MATLAB software using the LMI control toolbox (Gahinet et al., 1995).

5. SIMULATION RESULTS AND COMMENTS

Simulations were performed to illustrate the effectiveness of the proposed controller. The friction parameters were based on (Hirschorn and Miller, 1999): $m = 1.62\mathrm{kg}$, $\sigma_0 = 10^5\mathrm{N/m}$, $\sigma_1 = 495\mathrm{Ns/m}$, $\sigma_2 = 4.6\mathrm{Ns/m}$, $\mu_c = 2.0\mathrm{N}$, $\mu_s = 2.15\mathrm{N}$ and $v_s = 0.005\mathrm{m/s}$. The position reference tracking signal was chosen as follows: $x_d = 0.1\cos(0.32\pi t)$, by investigating the output-feedback to ensure the SPR condition on the linear part with a parameter

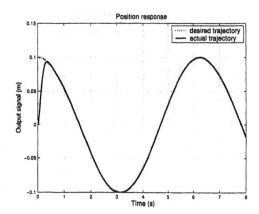

Fig. 2. Position response.

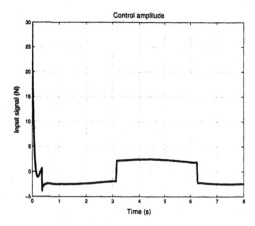

Fig. 3. Control effort.

scaling ϵ of 10^{-5}, with respecting to a stability margin λ of 10 and a disk of radius $r = 200$ centered at the origin. The parameter ρ of 0.2 is used to compromise between the control effort and the disturbance rejection performance. The actual position and the reference tracking signals are both shown in Figure 2. As discussed earlier, the tracking error converges asymptotically to zero. The Figure 3 shows the optimized control magnitude. We go on to show that the feasibility of the proposed method can be ensured under 'several' multi-objective constraints, thanks to the high degree of flexibility of such a controller. In other words, this output feedback structure induces susceptible progress in specific servo-system regulating tasks, compared to other controllers proposed in the literature. For instance, we note that the SPR conditions for the PD and filtered PID involve a substantially larger closed-loop bandwidth (Canudas and Lischinsky, 1997). Indeed, it is more convenient to overcome this discrepancy by using the submissive multi-objective synthesis and LMI frameworks. In particular, this dilemma should be investigated using the chosen stability region constraints.

6. CONCLUSION

Our paper is devoted to a common stabilization problem for a servo-system under the influence of friction. The goal of this paper is to extend the approach of (Canudas et al., 1995) to multi-objective output feedback synthesis, which offers more flexibility in terms of improving stability conditions. Besides, we have developed various performances of the closed-loop system encountering LMI-region stability constraints and an arranged H_2 cost optimization. Investigation of the sensitivity of the stability conditions to the two kinds of uncertainty (parameterized and/or non-parameterized) known a priori is expected to be a problem which it is feasible to solve using the multi-objective output feedback design proposed in this paper. This provides us with the motivation in our future work to look at these realistic hypotheses without relying on costly adaptive schemes.

REFERENCES

A.-Hélouvry, B. (1994). A survey of models, analysis tools and compensation methods for the control of machines with friction. *Automatica* **30**(7), 1083–1138.

Canudas, C. W. (1998). Comments on "a new model for control of systems with friction. *IEEE Trans. Automat. Contr.* **43**(8), 1189–1190.

Canudas, C. W. and P. Lischinsky (1997). Adaptive friction compensation with a partially known dynamic friction model. *Int. J. of Adaptive Contr. And Signal Processing* **11**, 65–80.

Canudas, C. W., H. Olsson, K. J. Aström and P. Lischinsky (1995). A new model for control of systems with friction. *IEEE Trans. Automat. Contr.* **40**(3), 419–425.

Gahinet, P., A. Nemirovskii, A. J. Laub and M. Chilali (1995). *MATLAB LMI Control Toolbox.* The MathWorks Inc.. Mass., USA.

Hirschorn, R. M. and G. Miller (1999). Control of nonlinear systems with friction. *IEEE Trans. Automat. Contr.* **7**(5), 588–595.

Scherer, C., P. Gahinet and M. Chilali (1997). Multi-objective output-feedback control via lmi optimization. *IEEE Trans. on Automat. Contr.* **42**(7), 896–911.

Zhou, K., J. C. Doyle and K. Glover (1996). *Robust and Optimal Control.* Prentice Hall.

ORTHOGONAL VS SKEWED INERTIAL SENSORS REDUNDANCY: A NEW PARADIGM FOR LOW-COST SYSTEMS

Richard Giroux [*,1]

* Ecole de technologie supérieure, LACIME Laboratory
1100 West Notre-Dame St., Montréal, Canada, H3C 1K3
rgiroux@ele.etsmtl.ca

Abstract: Systems integrity is of primary concern in numerous applications and inertial navigation systems is one of them. In the early days, orthogonal redundancy of sensors was used for its simplicity. From the 70's, optimal skewed redundant structure of sensors have been preferred because it reduces the number of sensors performing a given task. It is believe that the advent of low-cost sensors, including micro-electro-mechanical systems (MEMS), will make a change in the paradigm of skewed redundancy back to the implementation of orthogonal redundancy. Based on the information theory, it is shown that given a fixed number of sensors, the volume of information is the same for optimal (skewed or orthogonal) configurations. Also, the loss of information upon fault isolation in the orthogonal configuration is less damaging for the overall system performance than its skewed counterpart. Finally, cost figures for orthogonal configuration of sensors is also challenged, giving a relative importance to geometric complexity of skewed configurations, hence making the former structure more appealing for real system design. Copyright ©2004 IFAC

Keywords: Redundant sensor system ; Orthogonal redundancy ; Skewed-redundancy ; Cost analysis ; Inertial navigation

1. INTRODUCTION

Systems integrity is of great importance in many sectors of application. There are several levels of fusion management that leads to system robustness with respect to faults, from system high-level to the sensor low-level integrity verification. This paper addresses the sensors low-level integrity aspect.

Inertial sensor redundancy scheme used so far is mainly composed of high performance sensors, with the rationale of protecting the system against

sensor failure (Pejsa, 1974; Radix, 1993; Patton and Chen, 1994). Since those sensors are expensive, an optimal skewed structure of sensors is preferred to the orthogonal counterpart since it takes less sensor to perform the same task. So, the main benefit of the skewed configuration is more one of financial consideration. However, the mechanical arrangement of such a structure could be very complicated.

Low-cost sensors, and micro-electro-mechanical systems (MEMS) sensors particularly, are very good candidates for redundancy applications: they are small, they are cheap and they usually have low power consumption. Also, the accuracy of a single commercial sensor is weak so they

---

[1] At the time of submission, the author was a visiting postgraduate at the Australian Centre for Field Robotics (ACFR), The University of Sydney, Australia.

are usually classified as motion sensors, instead of inertial or tactical sensors. Hence, using many of them to perform the same task is highly intuitive. To the author knowledge, only few people have work in the area of low-cost redundancy, and all of them still assume skewed configuration of sensors (Sukkarieh et al., 2000; Ray and Phoha, 2002; Allerton and Jia, 2002).

MEMS sensors modify the skewed-redundancy paradigm. Given their small dimensions and low cost, it is believed there is no need for a skewed configuration of sensors anymore. Orthogonal redundancy of MEMS sensors should fulfill the task of detecting and isolating faults, and moreover increase the axis measurement accuracy compared to the one-sensor per axis orthogonal scheme.

First, the theory of optimal configuration for redundant sensors is revisited. Then, the performance of each configurations is assessed, the degradation modes performance are compared and the estimated cost is qualitatively derived for each geometric structure. Finally, concluding remarks highlight the paper's contributions and ongoing research on the integration of redundant low-cost sensors is discussed.

2. THEORY OF OPTIMAL REDUNDANT CONFIGURATION

Let the variable $\vec{x} \in \mathbf{R}^3$ to be measured, the N redundant measurement is given by:

$$\vec{m}_{(N \times 1)} = H_{(N \times 3)} \vec{x}_{(3 \times 1)} + \vec{v}_{(N \times 1)} \qquad (1)$$

where $\vec{m} \in \mathbf{R}^N$ is the vector of the N measurements, $H_{(N \times 3)}$ is the measurement matrix composed of line vectors $\vec{h}_i$ representing the directional unit vector of the sensor i with respect to the frame of the measured variable $\vec{x}$, and $\vec{v}_{(N \times 1)}$ is the measurement noise of the N sensors. The best estimate of the vector $\vec{x}$ can be determined by many ways, whether the discrete Kalman filter theory, the minimisation of the mean square error or by the maximum likelihood estimation theory, and is given by

$$\hat{\vec{x}} = \left(H^T H \right)^{-1} H^T \vec{m} \qquad (2)$$

On another side, the estimated error covariance matrix is given by Equation (3). To simplify the demonstration, and without limiting the result, it is assumed that each sensor has identical noise variance $\sigma_v^2 = \sigma_{v_i}^2 = \mathsf{E}\{v_i^2\}$.

$$\begin{aligned} P &= \mathsf{E}\{(\hat{\vec{x}} - \vec{x})(\hat{\vec{x}} - \vec{x})^T\} \\ &= \sigma_v^2 \left(H^T H \right)^{-1} \end{aligned} \qquad (3)$$

Hence, the general optimisation problem is to find a matrix H that minimise the estimated error covariance matrix, or better said that minimize a criterion involving the matrix P. As first shown by Sukkarieh (Sukkarieh et al., 2000), the information theory gives more insight in this optimisation process, and the information matrix will be also used through the rest of the paper. The inversion of the estimated error covariance matrix P is the information matrix, and will be expressed by I.

$$I = P^{-1} = \frac{1}{\sigma_v^2} H^T H \qquad (4)$$

2.1 Optimisation criteria

There are three known optimisation criteria (Vodicheva, 2001): the minimisation of the trace, the minimisation of the determinant or the minimisation of the condition number of the covariance matrix P. The two first optimisation criteria can be easily inverted (maximisation instead of minimisation) to be applied to the information matrix (maximisation of the trace and maximisation of the determinant of the information matrix I). The maximisation of the determinant is a better choice because it is related to the distribution of the information in the information space.

$$\max_H \mathcal{V} = \max_H \det(I) = \max_H \det(H^T H) \qquad (5)$$

The multiplicative factor σ_v^2 has been neglected because it does not influence the optimality. Geometrically speaking, the eigenvalues of the matrix I are equivalent to the three axis of an ellipsoid. By maximizing the determinant of the matrix I, the volume of information is maximized. The maximum volume for an ellipsoid is in fact a sphere, giving equal eigenvalues for the information matrix, which in turn is the expected result found in the litterature (Pejsa, 1974; Sukkarieh et al., 2000).

2.2 Optimal configuration of N one-axis sensors

To determine the optimal configuration, it is assumed that each sensor has only one axis of measurement. Figure 1 shows the geometrical orientation of the sensor i, with the elevation angle θ_i and the azimut angle ψ_i.

Given this geometrical configuration, the directional vector of the sensor i is given by:

$$\vec{h}_i^T = \begin{pmatrix} \mathsf{c}\theta_i \mathsf{c}\psi_i & \mathsf{c}\theta_i \mathsf{s}\psi_i & \mathsf{s}\theta_i \end{pmatrix} \qquad (6)$$

where $\mathsf{s}(\,\cdot\,)$ and $\mathsf{c}(\,\cdot\,)$ are the sine and cosine functions, respectively. The measurement matrix is constructed as follow:

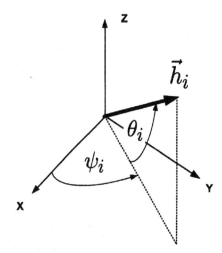

Fig. 1. Geometric configuration of sensor i

$$H = \begin{pmatrix} \vec{h}_1^T \\ \vec{h}_2^T \\ \ldots \\ \vec{h}_N^T \end{pmatrix} \qquad (7)$$

Hence, it is easy to compute the matrix of information for N sensors.

$$I = H^T H$$

$$= \begin{pmatrix} \vec{h}_1 & \vec{h}_2 & \ldots & \vec{h}_N \end{pmatrix} \begin{pmatrix} \vec{h}_1^T \\ \vec{h}_2^T \\ \vdots \\ \vec{h}_N^T \end{pmatrix} \qquad (8)$$

$$= \begin{pmatrix} \sum c^2\theta_i c^2\psi_i & \sum c^2\theta_i c\psi_i s\psi_i & \sum c\theta_i s\theta_i c\psi_i \\ \sum c^2\theta_i c\psi_i s\psi_i & \sum c^2\theta_i s^2\psi_i & \sum c\theta_i s\theta_i s\psi_i \\ \sum c\theta_i s\theta_i c\psi_i & \sum c\theta_i s\theta_i s\psi_i & \sum s^2\theta_i \end{pmatrix}$$

where the summation $\sum(\cdot)$ is for all the sensors, i.e. $\sum_{i=1}^{N}(\cdot)$.

The maximization of the volume of information with respect to the measurement matrix H can be replaced by its maximization with respect to the orientation angles θ_i and ψ_i. The volume of information of the information matrix for N sensors and its optimization can be written as follow:

$$\mathcal{V} = \det(I)$$

$$= \sum c^2\theta_i c^2\psi_i \sum c^2\theta_i s^2\psi_i \sum s^2\theta_i$$

$$\quad - \sum c^2\theta_i c^2\psi_i \left(\sum c\theta_i s\theta_i s\psi_i \right)^2$$

$$\quad - \left(\sum c^2\theta_i c\psi_i s\psi_i \right)^2 \sum s^2\theta_i \qquad (9)$$

$$\quad - \left(\sum c\theta_i s\theta_i c\psi_i \right)^2 \sum c^2\theta_i s^2\psi_i$$

$$\quad + 2 \sum c^2\theta_i c\psi_i s\psi_i \sum c\theta_i s\theta_i c\psi_i \sum c\theta_i s\theta_i s\psi_i$$

$$\{\theta_i^{\mathrm{opt}}, \psi_i^{\mathrm{opt}}\} = \arg \max_{\theta_i, \psi_i} \{ \mathcal{V} \} \qquad (10)$$

This optimization problem represented by Equation (10) is too difficult to solve numerically due to its infinity of solution and circular terms. Therefore, two hypothesis are introduced to simplify the formulation. First, it is assumed that the elevation of each sensors is identical[2]. The second assumption comes from the desire to maximize the volume of information. The negative terms in the Equation (9) will always be negative and therefore should be cancelled. The appropriate azimut angles which cancel the summations $\sum s\psi_i$, $\sum c\psi_i$ and $\sum c\psi_i s\psi_i$ are symmetric around a circle. Hence, the volume of information of a configuration of N one-axis sensors with equal elevation and symmetric around a circle is given by Equation (11). These two hypothesis also reflect the idea of equally space regions that minimize the variance of the systems measurement error given equally distributed spatial information (Pejsa, 1974).

$$\mathcal{V} = N s^2\theta c^4\theta \sum_{i=1}^{N} c^2\psi_i \sum_{i=1}^{N} s^2\psi_i \qquad (11)$$

Optimal elevation angle From Equation (11), it is easy to determine the optimal elevation angle.

$$\frac{\partial \mathcal{V}}{\partial \theta} = 2N \sum_{i=1}^{N} c^2\psi_i \sum_{i=1}^{N} s^2\psi_i \; s\theta c^3\theta \left(1 - 3s^2\theta\right)$$

$$= 0 \qquad (12)$$

Giving that the summations of the square of the sine and cosine of the azimut angles are non-zero, and that the elevation angle is between $0° < \theta < 90°$, the optimal elevation angle is given by Equation (13) and is in agreement with already known results.

$$s^2\theta^{\mathrm{opt}} = \frac{1}{3}$$
$$\Downarrow$$
$$\theta^{\mathrm{opt}} = 35,26° \qquad (13)$$

Optimal azimut angles It has already been assumed that the azimut angles were symmetric around a circle. This can also been inferred from the maximization of the information volume with respect to an azimut angle ψ_k.

---

[2] There is also another optimal configuration consisting of one axis at $\theta_1 = 90°$ and the remaining axis symmetrically distributed around a circle, and with elevation θ_i, $i = 2 \ldots N$ as a function of N. This particular configuration is not discussed in this paper.

$$\frac{\partial \mathcal{V}}{\partial \psi_k} = N s^2 \theta c^4 \theta \; s(2\psi_k) \Big(N - 2 \sum_{i=1}^{N} s^2 \psi_i \Big)$$
$$= 0 \qquad (14)$$

Taken into account that the elevation angle is comprised between $0° < \theta < 90°$, and by assuming that $s(2\psi_k) \neq 0$ (which can always be true by rotating the sensors configuration along the axis perpendicular to the horizontal plane), the summation of the square of the sine of the azimut angles equals the number of sensors divided by 2.

$$\sum_{i=1}^{N} s^2 \psi_i^{\mathsf{opt}} = \frac{N}{2} \qquad (15)$$

This relation holds only if the sensors are distributed around a circle. The usual admitted azimut sequence for an optimal configuration is given by Equation (16) and is represented at Figure 2 for the case of six one-axis sensors.

$$\psi_i^{\mathsf{opt}} = \frac{360}{N}(i-1) \quad , \quad i = 1 \dots N \qquad (16)$$

However, the previous configuration is not the only one satisfying Equation (15). The overlapping of the axis over three symmetric axes is also an optimal configuration with respect to the criterion of maximizing the information volume. The azimut angles are given at Equation (17) and represented at Figure 3 for six one-axis sensors.

$$\psi_i^{\mathsf{opt}} = \frac{360}{N} 2(i-1) \quad , \quad i = 1 \dots N \qquad (17)$$

It can be easily shown that the latest configuration is in fact rotated orthogonal triads. Lets have any two directional vectors $\vec{h}_k$ and $\vec{h}_j$ forming an angle β_{kj} between them,

$$\vec{h}_k = \begin{pmatrix} c\theta c\psi_k \\ c\theta s\psi_k \\ s\theta \end{pmatrix} \quad , \quad \vec{h}_j = \begin{pmatrix} c\theta c\psi_j \\ c\theta s\psi_j \\ s\theta \end{pmatrix} \qquad (18)$$

then, the scalar product of any two directional vector of this configuration is equal to zero

$$\vec{h}_k \cdot \vec{h}_j = c^2\theta \; c(\psi_k - \psi_j) + s^2\theta$$
$$= 0 \qquad (19)$$

because $\psi_k - \psi_j = \pm 120° \; \forall \; k \neq j - 3d$, $d = 0 \dots (\frac{N}{3} - 1)$. Since the scalar product of these two vectors is null, it implies that any non-collinear directional vectors are orthogonal.

$$\vec{h}_k \cdot \vec{h}_j = c\beta_{kj} = 0$$
$$\Downarrow$$
$$\beta_{kj} = 90° \qquad (20)$$

where $k, j = 1 \dots N$, $k \neq j - 3d$, $d = 0 \dots (\frac{N}{3} - 1)$. Then, the configuration shown at Figure 3 can be rearranged to be viewed as Figure 4.

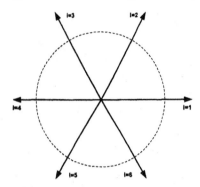

Fig. 2. Optimal skewed redundancy: in-plane projection of the sensor axes, 6 one-axis sensors

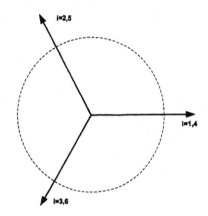

Fig. 3. Optimal orthogonal redundancy: in-plane projection of the sensor axes, 6 one-axis sensors

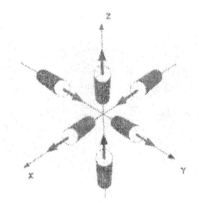

Fig. 4. Orthogonal redundancy, 6 one-axis sensors

It can be concluded that both optimal skewed and orthogonal configuration have the same volume of information. In fact, substituting Equations 13 and 15 in Equation 11, it is easy to come back to the well known relation where the amount of information in a optimal configuration is only a function of the number of sensors.

$$\mathcal{V} = N \mathsf{s}^2 \theta^{\mathsf{opt}} \mathsf{c}^4 \theta^{\mathsf{opt}} \sum_{i=1}^{N} \mathsf{c}^2 \psi_i^{\mathsf{opt}} \sum_{i=1}^{N} \mathsf{s}^2 \psi_i^{\mathsf{opt}}$$

$$= \frac{N^3}{27} \tag{21}$$

3. PERFORMANCE ASSESSMENT

Although redundancy of sensors brings more information and augment the accuracy of the global system, the strongest rationale for redundancy is fault detection and isolation. The number of fault isolation is a function of the number of triads formed with three non-collinear sensors. Hence, for the same isolation task, the orthogonal configuration needs more sensors than the skewed one, especially when it is designed on the worst case scenario. Figure 5 compares the two configurations with respect to the volume of information and the number of possible fault isolation. The number of sensors needed is indicated on the graph for each configuration.

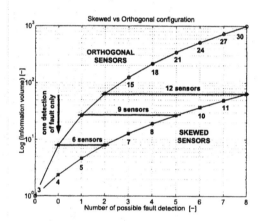

Fig. 5. Comparison of skewed and orthogonal configurations

Also, Figures 6 and 7 show the relative loss in information after successive fault isolations, for the worst and the best case. For the orthogonal configuration, the worst case happen when all the faults are on the same axis, whereas the best case distributes the faults along every axis. On the other side, the worst case for the skewed configuration is when all faults are adjacent, while symmetric fault occurrence is the best case.

From these three figures, the orthogonal configuration shows some advantages. First, the greater number of sensors enhances the quantity of information accessible to the systems. Furthermore, upon fault of sensors, the loss of information is less than the skewed configuration. It should be

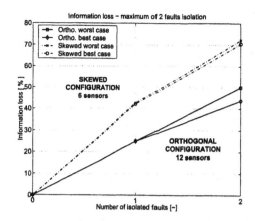

Fig. 6. Performance degradation upon fault isolation (maximum of 2 faults isolated)

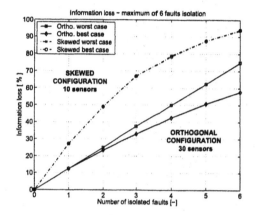

Fig. 7. Performance degradation upon fault isolation (maximum of 6 faults isolated)

noted that in the best case where each fault is distributed equally in the orthogonal configuration, the number of possible fault isolation is multiplied by 3. Although the skewed configuration shows nearly the same behavior for the best or worst case scenario, the global performance of the orthogonal configuration is better.

4. QUALITATIVE COST ASSESSMENT

Of course, the greater level of information present in the orthogonal structure is due to its number of sensors, and and the excessive number of sensors is always the argument against this configuration. Figure 8 shows the relative evolution of the cost of the orthogonal scheme with respect to the skewed one for a given task. The relative cost is given by Equation (22), where α is a parameter indicating the relative importance of mechanical complexity in the total system cost.

$$\begin{pmatrix} \text{Additional} \\ \text{cost} \end{pmatrix} = \begin{pmatrix} \text{Additional cost} \\ \text{from sensors} \end{pmatrix} - \alpha \begin{pmatrix} \text{Cost of geometry} \\ \text{complexity} \end{pmatrix} \tag{22}$$

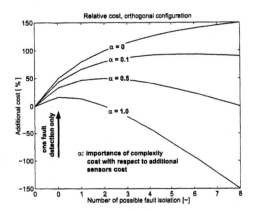

Fig. 8. Relative cost of the orthogonal configuration with respect to the skewed configuration

If $\alpha = 0$, only the cost for additional sensors is taken into account. However, it can be assessed that mechanical complexity has a cost and should be included in the cost analysis. Although purely qualitative, the influence of the geometrical complexity, as shown in Figure 8, is not negligible. Also, the α parameter can represent the lowering of individual sensor cost. Indeed, even if the individual cost of sensors is going down, as it is expected for MEMS inertial sensors, the mechanical complexity of skewed redundant scheme is still the same. Hence, the relative cost of orthogonal configuration will diminish, and even be cheaper at some point than skewed configurations.

5. CONCLUSION

This paper has introduced a change in paradigm over the configuration of redundant sensors for low-cost inertial navigation systems. It has first shown that the orthogonal configuration and the optimal skewed configuration have the same level of information, given they have the same number of sensors. So, for a specific isolation task, the volume of information in the orthogonal structure is more important, and the loss of information upon fault isolation is less damaging for the overall system performance.

The cost figures for orthogonal configuration of sensors have also been challenged, giving a relative importance to geometry complexity of skewed configuration, hence making the former structure more appealing for real system design.

This paper has not dealt with the actual implementation of redundant low cost sensors, i.e. the computational burden associated with the use of many sensors, the problem of redundant low-cost sensor error identification and the implication of this change of paradigm on the fault identification and isolation algorithms. All those issues are part of the author's ongoing research and concrete results are expected in the near future.

ACKNOWLEDGEMENTS

The financial support for this work was partly provided by the Natural Sciences and Engineering Research Council of Canada (NSERC), the "Fonds Nature et Technologie du Québec" (NATEQ) and the Ecole de technologie supérieure (ETS), Montreal, Canada. Also, the author would like to acknowledge the ARC Centre of Excellence programme, funded by the Australian Research Council (ARC) and the Australian New South Wales State Government, as part of his research work at the Australian Centre for Field Robotics (ACFR), The University of Sydney, Australia. Finally, the author would like to thank Dr. René Jr. Landry, from the ETS, Dr. Richard Gourdeau, from the Ecole Polytechnique of Montreal, and Dr. Salah Sukkarieh, from the ACRF, for their constant support.

REFERENCES

Allerton, D.J. and H. Jia (2002). An error compensation method for skewed redundant inertial configuration. In: *Institute of Navigation 58th Annual Meeting*. Albuquerque, USA.

Patton, R. J. and J. Chen (1994). Review of parity space approaches to fault diagnosis for aerospace systems. *Journal of Guidance, Control and Dynamics* **17**(2), 278 – 285.

Pejsa, Arthur J. (1974). Optimum skewed redundant inertial navigation. *AIAA Journal* **12**(7), 899–902.

Radix, J.C. (1993). *Systemes Inertiels a Composants Lies "Strap-Down"*. Supaero. Toulouse, France.

Ray, A. and S. Phoha (2002). Calibration and estimation of redundant signals. In: *Proceedings of the 2002 American Control Conference*. Vol. 2. pp. 1437–1442.

Sukkarieh, Salah, Peter Gibbens, Ben Grocholsky, Keith Willis and Hugh F. Durrant-Whyte (2000). A low-cost, redundant inertial measurement unit for unmanned air vehicles. *The International Journal of Robotics Research* **19**(11), 1089–1103.

Vodicheva, L. (2001). Fault-tolerant strapdown inertial measurement unit: Optimization approaches. In: *Proceedings of the 8th St-Perterburg Intl. Conf. on Integrated Navigation Systems*. pp. 108 – 110. St-Petersburg, Russia.

ELSEVIER
IFAC
PUBLICATIONS
www.elsevier.com/locate/ifac

INFORMATION FUSION IN A MULTI-BAND GPS RECEIVER

Jamila Kacemi[1] **Serge Reboul**
Mohammed Benjelloun

Laboratoire d'Analyse des Systèmes du Littoral (EA 2600)
Université du Littoral Côte d'Opale 50 Rue Ferdinand
Buisson, B.P. 699, 62228 Calais Cedex, France
jamila.kacemi@lasl.univ-littoral.fr
serge.reboul@lasl.univ-littoral.fr
mohammed.benjelloun@lasl.univ-littoral.fr

Abstract: The works presented in this paper deals with the design of a tracking fusion filter for the future multi-carrier G.P.S signal. The estimation of the distance and speed between the satellite and the G.P.S receiver is realized with a Kalman track fusion algorithm. In this system, inspired from the modified track-to-track algorithm, the observations are merged with a measurement fusion algorithm and the estimates are fused with a state vector fusion Kalman filter. This hybrid model can fuse the measurement obtained from the three frequencies of the future G.P.S. system. It can also fuse the state vectors that describe distances and speed which are supposed to come from different sensors. The simulation results obtained on synthetic data show the feasibility of the method and the contribution of the fusion over the localization accuracy. *Copyright © 2004 IFAC*

Keywords: Estimation, Data Fusion , Tracking

1. INTRODUCTION

The satellite navigation G.P.S. NAVSTAR is a three dimensional positioning system. It is based on the measurements of the distances between the receiver and a set of satellites. Each satellite is transmitting its own position in a navigation message and the receiver measures the transmission time from the satellites to the receiver. Actually in a civil G.P.S. receiver only one frequency L1 (1575.42 MHz), is used for broadcast of navigation data and ranging codes from the satellites, by the use of code division multiple access (CDMA). Each satellite uses a different ranging code. The navigation data provides the receiver the satellite location at the time of transmits. The ranging code may be used by the receiver to determine the propagation delay of the satellite signals. Then we can calculate the pseudo-distances between the receiver and a satellite with the speed of the satellite signals ($3 10^8 m/s$). The satellite and the receiver are not synchronized, then we must calculate the receiver position and synchronization (x, y, z ,Δt) with four different pseudo-distances.

Today the evolution of the GPS system passes by the increase of the number of carrier frequencies (evolution of NAVSTAR, GALILEO). For example in the future NAVSTAR GPS system a second civil signal will use the C/A code currently used at GPS L1 and will be located at GPS L2 (1227.6 MHz). The third civil signal, which is intended to meet the needs of critical safety-of-life applications, will be located at 1176.45 MHz

---

[1] supported by the french Region Nord pas de Calais

(L5) (Anon., 1999). The work presented in this article is about the designed of a tracking filter fusion for the data coming from the future system GPS multi-frequencies.

A common application of data fusion techniques is the estimation of target position or kinematique information from multiple measurements from a single or multiple sensors. The conventional state-vector fusion and measurement fusion are two kinds of methods for Kalman filter based data fusion. The conventional measurement fusion has lower estimation error but a higher computational cost(Liggins *et al.*, 1997). Several rules for measurements fusion have been proposed in the past decade. The most popular one is simple averaging, which is based on fixed combining schemes and the less popular one is order statistics combiners. In the case of measurement with different statistical parameters, weighted averaging gives better results (Fumera and Roli, 2002). One of the frequently used state vector fusion is the so-called track-to-track fusion algorithm which was first proposed by (Bar-Shalom, 1981). It is proposed in (Gao and Harris, 2002) a modified track-to-track fusion algorithm which gives better fusion results in the cases of different sensors. The tracking fusion filter proposed in this article is realized with a Kalman filter that fused the state vector and the measurement. It is inspired of the modified track-to-track algorithm. This hybrid model can fuse the measurement obtained from the three frequencies of the future G.P.S. system. It can also fuse the state vectors that describe distances and speed which are supposed to come from different sensors. The paper is organized as follow. Section 2 describes the GPS pseudo-distance model. We present the fusion algorithms in section 3 and in section 4 the fusion tracking filter of the GPS signal. In section 5 we present numerical experimentations.

2. THE GPS MODEL

We can express the pseudo-distances between the GPS receiver and the satellite k from the real distances and offsets and noise (we suppose here that the GPS and the receiver have been synchronized in the GPS signal acquisition process):

$$P_j^k = \rho^k + I_j^k + T^k + \epsilon_j^k \qquad (1)$$

P_j^k is the pseudo distances measure on the frequency j for the satellite k. We consider here the three frequencies of the future system G.P.S., j=1,2,3, and the 24 working satellite. ρ^k is the pseudo distance between the satellite k and the receiver (distance + synchronization). I_j^k , the ionospheric offset , is function of the satellite k and the frequency j. T^k , the tropospheric offset,

is function of the satellite k. ϵ_j^k , the thermal noise of the receiver, is function of the satellite k and the carrier frequency j. This error is assumed to be a white Gaussian noise which the variance is depending on the thermal noise power.

The correction of the tropospheric offset is realized with models that depend on the temperature, pressure, humidity, and elevation of the satellite (Leick, 1995). The correction of the ionospheric offset can be done with the pseudo-distances obtained on two different carrier frequencies(Kaplan, 1996). The corrected distance has the expression:

$$\rho^k = \frac{\gamma_{ij} . P_i^k - P_j^k}{\gamma_{ij} - 1} \qquad (2)$$

with:

$$\gamma_{ij} = (\frac{f_i}{f_j})^2$$

$i = 1, 2, 3$ et $j = 1, 2, 3$ $i \neq j$. We can then deduce three expressions, j=1,2,3, for the tropospheric offset I_{ij}^k associated to a carrier frequency i and a satellite k. We found for the offset of the carrier frequency L1 (i=1) and the satellite k, the following equation:

$$I_{11}^k = \frac{P_2^k - P_1^k}{\gamma_{12} - 1} \text{ with} : \gamma_{12} = (\frac{f_1}{f_2})^2 \qquad (3)$$

$$I_{12}^k = \frac{P_3^k - P_1^k}{\gamma_{13} - 1} \text{ with} : \gamma_{13} = (\frac{f_1}{f_3})^2 \qquad (4)$$

et,

$$I_{13}^k = \frac{P_1^k(\gamma_{23} - 1) - \gamma_{23}P_2^k + P_3^k}{\gamma_{23} - 1} \qquad (5)$$

$$\text{with} : \gamma_{23} = (\frac{f_2}{f_3})^2$$

We also found the same type of equation for the carrier frequency L2 and L5. The expressions 3,4 and 6 are correlated Gaussian random variables. In the proposed filter, the measurements fusion algorithm weight and sum the ionospheric offset and then track those fused measurements by a Kalman filter. We give in annex the expressions of the statistical parameters of the expressions, as a function of the power thermal noise. They will be used in the calculation of the weighted parameters values.

In practice the position of the satellite and the receiver changes, and the distance ρ^k between the satellite k and the receiver changes with time. Furthermore the satellite and receiver position evolutes with time. This movement provoked the increase or decrease of the carrier frequency at the receiver (Doppler effects). The difference Δf_j^k we observe on the carrier frequency f_j is function of the relative speed between satellite and receiver:

$$\Delta f_j^k = V^k.(\frac{f_j}{C}) + \epsilon_j \tag{6}$$

ϵ_j is the receiver error measurement on the difference Δf_j^k. This speed information, measured on the three carrier frequencies of the future G.P.S. system, will be fused. The fused measures will be used in a kinematic model of the pseudo-distance evolution.

3. THE FUSION MODEL

3.1 Measurement Fusion

Let us consider the weighted averaging of an ensemble of N correlated random variables. Let us consider $Y_i = [y_{1i}y_{2i}...y_{Ni}]^t$ the measurement vector coming from i, $i = 1...N$ sensors. The output of the weighted averaging combiner can be expressed as:

$$Y_i = \sum_{m=1}^{N} \alpha_m y_{mi} \tag{7}$$

With normalised weight:

$$\sum_{m=1}^{N} \alpha_m = 1 \tag{8}$$

The variance of the expression 7 is given by:

$$\sigma^2 = \sum_{m=1}^{N} \alpha_m^2 \sigma_m^2 + \sum_{m=1}^{N} \sum_{m \neq n} 2\sigma_{mn}\alpha_m\alpha_n \tag{9}$$

Where, σ_m is the variance of the noise measurement y_{mi} and σ_{mn} the covariance of the measures coming from the sensors m and n. The problem is to find the values of α_m that minimizes the variance of the expression 9. To solve the problem we used the Lagrange multiplier. Let:

$$F(\alpha_1, \alpha_2, ..., \alpha_N) = \sigma^2 + L(\alpha_1 + \alpha_2 + ... + \alpha_N - 1)$$

L is the Lagrange multiplier and we want to minimize F(..):

$$\frac{\delta F(\alpha_1, \alpha_2, ..., \alpha_N)}{\delta\alpha_m} = 0 \text{ with } \sum_{m=1}^{N} \alpha_m = 1$$

In this case we have N+1 equations with N+1 unknown and a solution can be found. It is easy to see that in the case of uncorrelated measurements the weights that minimize F(...) are:

$$\alpha_m = \frac{1}{\sigma_m^2 \sum_{i=1}^{N} \frac{1}{\sigma_i^2}} \tag{10}$$

And when we have the same statistical parameters for the measurement this expression becomes:

$$\alpha_m = \frac{1}{N} \tag{11}$$

, which is simple averaging. In the case of correlated measurement, it is not easy to derive a general analytical expression for the optimal weight. However, in the case of three sensors the expression of the α_m is given by:

$$\alpha_1 = \tag{12}$$

$$\frac{\sigma_2^2\sigma_3^2 - \sigma_2^2\sigma_{13} - \sigma_3^2\sigma_{12}}{\sigma_1^2\sigma_2^2 + \sigma_1^2\sigma_3^2 + \sigma_2^2\sigma_3^2 - 2\sigma_1^2\sigma_{23} - 2\sigma_2^2\sigma_{13} - 2\sigma_3^2\sigma_{12}}$$

$$\frac{-\sigma_{23}^2 + \sigma_{23}\sigma_{13} + \sigma_{12}\sigma_{23}}{+2\sigma_{12}\sigma_{13} + 2\sigma_{21}\sigma_{23} + 2\sigma_{31}\sigma_{32} - \sigma_{12}^2 - \sigma_{13}^2 - \sigma_{23}^2}$$

$$\alpha_2 = \alpha_1 \times \tag{13}$$

$$\frac{\sigma_1^2\sigma_3^2 - \sigma_1^2\sigma_{23} - \sigma_3^2\sigma_{12} - \sigma_{13}^2 + \sigma_{13}\sigma_{23} + \sigma_{12}\sigma_{13}}{\sigma_2^2\sigma_3^2 - \sigma_2^2\sigma_{13} - \sigma_3^2\sigma_{12} - \sigma_{23}^2 + \sigma_{23}\sigma_{13} + \sigma_{12}\sigma_{23}}$$

$$\alpha_3 = \alpha_1 \times \tag{14}$$

$$\frac{\sigma_1^2\sigma_2^2 - \sigma_1^2\sigma_{23} - \sigma_2^2\sigma_{13} - \sigma_{12}^2 + \sigma_{12}\sigma_{23} + \sigma_{13}\sigma_{12}}{\sigma_2^2\sigma_3^2 - \sigma_2^2\sigma_{13} - \sigma_3^2\sigma_{12} - \sigma_{23}^2 + \sigma_{23}\sigma_{13} + \sigma_{12}\sigma_{23}}$$

3.2 The modified track-to-track fusion model

It was shown in (Bar-Shalom, 1981) and (Saha and Chang, 1998) that the estimation errors of both sensors are correlated since the process noise of same target enters the equations of both sensors. The modified track-to-track fusion (Gao and Harris, 2002) then is performed by fusing the prefiltered state vector of a pair of associated tracks to generate a new estimate of the fused state vector. Let the state model for two sensors be:

$$x(t + 1) = Fx(t) + \nu(t) \tag{15}$$

The measurement equations for sensors n=1, 2 ($H^1 \neq H^2$) are given by:

$$z^m(t) = H^m x(t) + w^m(t) \tag{16}$$

In this algorithm the estimate state vectors associated to the two different measurements are fused. The unique prediction is obtained with the previous fused state. Let $\hat{x}(t/t)$ be the fused sate, and $\hat{x}(t + 1/t)$ the predicted state. We have:

$$\hat{x}(t + 1/t) = F\hat{x}(t/t) \tag{17}$$

The prediction $\hat{x}(t + 1/t)$ is combined with the measurements $z^1(t)$ and $z^2(t)$ in a step of correction and the two estimate $\hat{x}^1(t + 1/t + 1)$, $\hat{x}^2(t +$

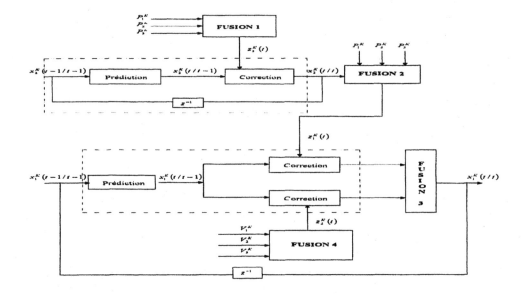

Fig. 1. Signal GPS tracking filter

$1/t+1$) are fused. The expression of the fused estimate of the state vector $\hat{x}(t+1/t+1)$ is generated from the static linear estimation equation.

$$\hat{x} = \hat{x}_1 + (P^1 - P^{12})$$
$$(P^1 + P^2 - P^{12} - P^{21})^{(-1)}(\hat{x}_2 - \hat{x}_1) \quad (18)$$

The covariance matrix of the fused estimate $\hat{x}(t+1/t+1)$ is then given by:

$$P = P^1 + (P^1 - P^{12})$$
$$(P^1 + P^2 - P^{12} - P^{21})^{(-1)}(P^{21} - P^1) \quad (19)$$

4. GPS SIGNAL TRACKING

We show figure 1 the tracking filter composed of several fusion blocks where the principle is described in section 3. The purpose of this filter is to estimate the distances and speed between the G.P.S. receiver and the satellite k. In practice these algorithms are simultaneously performed. Let x_1^k be the two dimensional state vector of speed and distances between the satellite k and the receiver. The filter associated to this state vector, is the well known $(\alpha - \beta)$ filter which the working mode condition and the parameters initialization are given in (Bar-Shalom and Li, 1993). Let x_2^k be the three dimensional state vector that represents the ionospheric offset for each frequency. The state equation of the filter is given by:

$$x_1^k(t+1) = \begin{pmatrix} 1 & T \\ 0 & 1 \end{pmatrix} x_1^k(t) + \begin{pmatrix} T^2/2 \\ T \end{pmatrix} \nu_1^k(t)$$

$$x_2^k(t+1) = \begin{pmatrix} 1 & 0 & 0 \\ 0 & 1 & 0 \\ 0 & 0 & 1 \end{pmatrix} x_2^k(t) + \begin{pmatrix} \nu_2^k(t) \\ \nu_3^k(t) \\ \nu_4^k(t) \end{pmatrix}$$

Let $z_1^k(t)$ be the distances measurement and $z_2^k(t)$ the speed measurement. $z_3^k(t)$ is a three dimensional vector measurement of the ionospheric offset. The measurement equations are given by:

$$z_1^k(t) = \begin{pmatrix} 1 & 0 \end{pmatrix} x_1^k(t) + w_1^k(t)$$

$$z_2^k(t) = \begin{pmatrix} 0 & 1 \end{pmatrix} x_1^k(t) + w_2^k(t)$$

$$z_3^k(t) = \begin{pmatrix} 1 & 0 & 0 \\ 0 & 1 & 0 \\ 0 & 0 & 1 \end{pmatrix} x_2^k(t) + \begin{pmatrix} w_3^k(t) \\ w_4^k(t) \\ w_5^k(t) \end{pmatrix}$$

We defined four fusion blocks on figure 1:

- In the block 1, the correlated measurements of the ionospheric offset defined in the expressions 3-6, are fused. We use the weighted averaging method and the statistical parameters of the offset defined in annex. The outputs of this bloc are three measures, used for the correction of $x_2^k(t/t-1)$ in a Kalman filter.

- In the block 2 and 4, the measurements supposed to be uncorrelated are fused. The output of these blocs are two measures, used for the correction of the predicted states $x_1^k(t/t-1)$ with the weight described in the expression

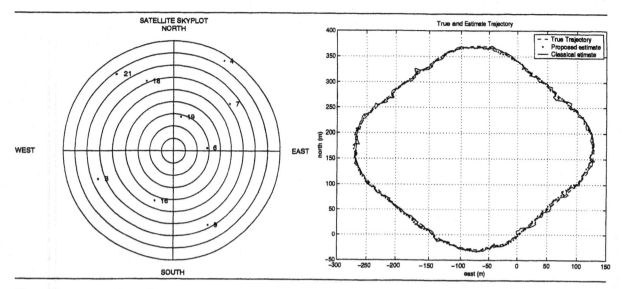

Fig. 2. Receiver and satellite trajectory.

10. In the block 2 we fused the satellite - receiver distances we obtained when we subtract the estimate ionospheric offsets $x_2^k(t/t-1)$ to the pseudo distances P_1^k, P_2^k, P_3^k. In the block 4 the speeds measurements deduced from the frequencies variations are fused.

- In the block 3 we fused the estimated state vectors with the modified track-to-track algorithm described in section 3.2. In this case we consider the measure of distances and speed coming from two different sensors.

The system is then composed of three Kalman filter which work simultaneously. The estimates obtained from these three filters are merged or fused to track the distance and speed. This data are then used in the calculations of the position. (Choi and Cicci, 2003).

5. EXPERIMENTATION

We show in this experimentation the contribution of information fusion in a multi-carrier-frequencies G.P.S. receiver. The proposed method has been applied to real GPS data obtained for a known static position (reference). This data are available as Rinex file at the National Geodetic Survey web site. For this data, a set of pseudo range measurements between the receiver and the satellites for the carrier L1 and L2 as well as the positions of the satellites were computed. Positions have been calculated with a classical and the proposed method. In the classical method the ionospheric offset are corrected with the expression 2. The error between the solutions and the reference are shown for the two methods in figure 3.

Let consider the trajectory defined in figure 2, in this case the G.P.S. data are obtained in simulation (Toolbox GPSoft). In this simulation the tropospheric and ionospheric errors are generated with the models described in (Leick, 1995). The value of the thermal power noise is defined as

in (Leick, 1995) for a signal to noise ration of 32 dBHz on each frequency. In this context of dynamical experimentation there is a change in the carrier frequency due to the Doppler Effect. From these noisy frequency variations we deduced the value of the relative speed. We show on figure 2 the evolution of the visible satellites during the experimentation and the trajectory calculated with the proposed filter and a classical mono frequency Kalman filter.

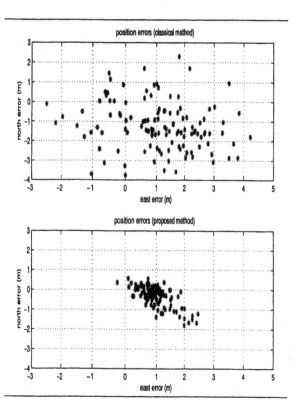

Fig. 3. Positions errors of the two methods.

We present figure 4 the position error obtained with the fusion filter we proposed and with the

107

classical Kalman filter performed on single frequency. We give in table 1 the variance error of different parameters estimated with the proposed and the classical filter. We report in this table the error variance on the pseudo distance and on the coordinates x and y. We also report the mean and the variance of the position error. The error value, obtained in this experimentation, show that the fusion filter we propose for a multi-frequency G.P.S. receiver permit to increase the precision of localization.

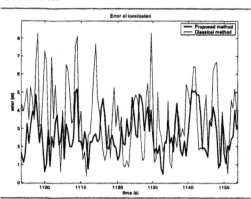

Fig. 4. Localization error.

Table 1. Error variance of different parameters.

	pseudo-distance	abcisse	ordinate	position (m/σ^2)
Classical receiver	100	8.9	8	3.7/3.8
Proposed receiver	34	2.8	3.4	2.7/1.3

6. CONCLUSION

In this article we present a fusion filter for the tracking of speed and distances between a satellite and a G.P.S. multi-carrier-frequencies receiver. The proposed filter has an hybrid structure for the fusion of measurements and state vectors. For the structure we proposed, inspired from the modified track-to-track Kalman filter, we give the expression of the weighted averaging combiner that fused the correlated measurement coming from the three frequencies of the future G.P.S. receiver. The experimentation shows the feasibility of the proposed filter and the improvement in the localization precision.

7. ANNEX

The offsets I_{i1}^k, I_{i2}^k and I_{i3}^k of the carrier frequency i, are correlated random variables with a Gaussian distribution. They are calculated with the pseudo-distances estimated on the three frequencies of the

G.P.S. We note σ_i^{k2} the power of the thermal noise for the pseudo-distances i and the satellite k. We note σ_{ij}^{k2} the variance of the ionospheric offset I_{ij}^k and σ_{ijl}^{k2} the covariance between the ionospheric parameters I_{ij}^k and I_{il}^k. For the ionospheric offset of the carrier frequency $L1$ and the satellite k, we have:

$$\sigma_{11}^{k2} = \frac{\sigma_1^{k2} + \sigma_2^{k2}}{(\gamma_{12} - 1)^2}$$

$$\sigma_{12}^{k2} = \frac{\sigma_1^{k2} + \sigma_3^{k2}}{(\gamma_{13} - 1)^2}$$

$$\sigma_{13}^{k2} = \frac{\sigma_1^{k2}(\gamma_{23} - 1)^2 + \sigma_2^{k2}\gamma_{23}^2 + \sigma_3^{k2}}{(\gamma_{23} - 1)^2}$$

$$\sigma_{112}^{k} = \frac{1}{(\gamma_{12} - 1)(\gamma_{13} - 1)}\sigma_1^{k2}$$

$$\sigma_{113}^{k} = \frac{1}{(\gamma_{12} - 1)}\left(\frac{\gamma_{23}}{(\gamma_{23} - 1)}\sigma_2^{k2} + \sigma_1^{k2}\right)$$

$$\sigma_{123}^{k} = \frac{1}{(\gamma_{13} - 1)}\left(\frac{1}{(\gamma_{23} - 1)}\sigma_3^{k2} + \sigma_1^{k2}\right)$$

$$\sigma_{112}^{k} = \sigma_{121}^{k}, \ \sigma_{113}^{k} = \sigma_{131}^{k},$$

$$\sigma_{123}^{k} = \sigma_{132}^{k}$$

REFERENCES

Anon. (1999). Vice president gore announces new global positioning system modernization initiative. Office of the Vice President.

Bar-Shalom, Y. (1981). On the track-to-track correlation problem. *IEEE Transactions on Automatic control* pp. 571–572.

Bar-Shalom, Y. and X. Li (1993). *Estimation and Tracking: principles, techniques and Softwares*. Artech House.

Choi, E. and D.A Cicci (2003). Analysis of gps static positioning problems. *Applied Mathematics and Computation* **140**, 37–51.

Fumera, G. and F. Roli (2002). Performance analysis and comparison of linear combiners for classifier fusion. *Proc of IAPR Int. Workshop on Statistical Pattern Recognition*.

Gao, J.B. and C.J. Harris (2002). Some remarks on kalman filters for the multisensor fusion. *Information Fusion* **3**, 191–201.

Kaplan, E.D. (1996). *Understanding GPS: Principles and Applications*. Artech House.

Leick, A. (1995). *GPS Satellite Surveying*. John Wiley and Sons publication.

Liggins, M., C. Chong, I. Kadar, M.Alford, V. Vannicolas and S. Thomopoulos (1997). Distributed fusion architectures and algorithms for target tracking. *Proceedings of IEEE* **85**, 95–107.

Saha, R. K. and K.C. Chang (1998). An efficient algorithm for multisensor track fusion. *IEEE Transactions on Aerospace and Electronic Systems* **34**, 200–210.

MONITORING OF THERMOFLUID SYSTEM USING LINEARIZED MULTIENERGY BOND GRAPH

W. El Osta*, B. Ould Bouamama⁺ and C. Sueur*

*LAIL, UMR CNRS 8021, Ecole Centrale de Lille, BP 48
F59651Villeneuve d'Ascq, France
El_Osta.Wassim@ec-lille.fr, Christophe.sueur@ec-lille.fr
+ LAIL, UMR CNRS 8021, Polytech'Lille
F59655 Villeneuve d'Ascq, France
belkacem.bouamama@univ-lille1.fr

Abstract: Process engineering systems are present in industry with risk. For the improvement of their safety the diagnosability (ability to detect and to isolate faults) study is needed. System modelling and Analytical Redundancy Relations (ARRs) generation are the main steps in the diagnosability analysis. The innovative interest of the present paper is the use of only one representation: the bond graph theory for modelling (as multidisciplinary tool and unified language) and structural analysis (for fault detection and isolation (FDI) design). The diagnosability are directly and graphically determined from the linearized multiport bond graph with out need of ARRs signature faults calculation. The developed approach is applied to a thermofluid installation. *Copyright © 2004 IFAC*

Keywords: Fault Detection and Isolation, Diagnosability, Bond graphs, Process Engineering, Structural analysis, Analytical Redundancy Relations.

1. INTRODUCTION

System modelling is the main step in the Diagnosabilitystudy. Dynamic models in process engineering industry are strongly non linear. The non linearities result mainly from coupling of different energies (thermal, chemical,...). Because of the multi-domain energies involved in such processes, the use of bond graph modelling as a multidisciplinary tool and an unified language is well-adapted [Karnopp and al., 1990].

These processes are present in industry with risk (nuclear, chemical, petrochemical). For the improvement of their safety the diagnosability (which actuator or sensor can be monitored with existing instrumentation architecture before any industrial implementation) study is needed. The fault detection and isolation (FDI) procedures consist on the comparison between the model and the real process. Depending on the kind of knowledge used to describe the plant operation, different approaches for the design of FDI procedures have been developed in the literature. The developed in the present paper approach belongs to model based FDI method.

The complexity of thermodynamic bond graph needs a linearization to use for monitoring. While the real variables in industry are controlled, the processes mainly run around the operating regime

and therefore the linearized model can be considered valid.

From the control point of view, structural properties (controllability, observability) of process engineering systems based on the linearized multienergy bond graph models have been developed in [El Osta and al., 2003].

For the FDI purposes, the proposed methods in the literature concerning the monitoring of physical processes using bond graph methodology are developed in the case of 1-port elements modelling single energy [Tagina, 1995].

In [Karnopp, 1977] and [Rimaux, 1995] is developed a methodology to derive from non linear monoenergy multiport bond graph a linearized model introducing modulated sources. This single port bond graph model has a linear structure, which means that the linearization procedure deals only with actuators and elements.

The innovative interest of the paper is to extend the developed linearization approach to a multienergy bond graph models. Furthermore on top of modelling, the diagnosability analysis of multienergy linearized bond graph model is developed. Moreover, the developed approach in the present paper can be applied to any thermofluid process using a generic classification of such processes. It will be shown also that the study of the actuator Diagnosabilitycan be limited on thermal system.

The paper is organised as follows: after a brief description of the multienergy non linear bond graph modelling in the second section, the constitutive elements and their linearized forms in thermodynamic bond graph are presented. A recall on the monitoring of physical processes using bond graph proposed for 1-port monoenergy bond graph followed by an extension to multienergy bond graph is given in the third section. Finally, this approach is illustrated by an application to a coupled thermofluid system.

2. MULTIENERGY BOND GRAPH MODELLING

2.1 Component Bond graph models

2.1.1. Introduction

Process engineering is characterized by the coupling of several phenomena such as chemical, thermal and hydraulic ones. As power variables, the temperature T or specific enthalpy h and pressure P are used as effort variables. The mass flow $\dot{m}$ and the thermal flow $\dot{Q}$ or enthalpy flow $\dot{H}$ are used as flow variables. The choice is not trivial and depends on the kind of considered physical phenomena [Ould Bouamama, 2003] [Thoma and al., 2000]. To distinguish the different types of thermal power, the enthalpy flow carried by convection and thermal flow carried by conduction are respectively denoted $\dot{H}$ and $\dot{Q}$.

Any thermofluid process can be considered as interconnected components. Three main types of functional component can be considered : the storage plant items (tank, boiler, condenser...), transportation plant items (pipe, pipe with valve, pipe with pump,...) and the energy sources. This functional clasiification based on bond graph theory is well developed in [Ould Bouamama, 2001].

In the following, non linear actuators (energy sources) are modelled by bond graph element. Then, the linearized bond graph is drawn directly from a graphical approach. Non linear passive elements are not modelled here, for more details see [El Osta and al., 2003]. While thermofluid sources and pumps (thermal flow by convection) act on both energies (thermal and hydraulic), heaters (thermal flow by conduction) $Sf_T : \dot{Q}$ participates only in the thermal balance.

2.1.2. Energy Sources

- *Thermofluid Source*

In thermofluid system, the coupling of both energies is quantified by equation 1.

$$\dot{H} = \dot{m}h = \dot{m}C_P T = F(\dot{m},T) \quad f_T = F(f_H, e_T) \quad (1)$$

We propose to represent the thermofluid source ST_f as defined in the figure 1.

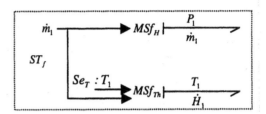

FIG. 1. Thermofluid source ST_f.

$Se_T : T_1$ can be used as thermal effort source or as a parameter if the temperature of the inlet fluid is constant.

- *Mechanical Elements (pumps)*

The pumps as mechanical elements are widely used in thermofluid systems. In the present paper are considered only centrifugal pumps, which are widely used in process engineering industry. The details of the models describing the technology of the pump depends on the purpose of the models. In [Thoma, 1990], the essential action of hydrostatic pump is given as a transformer between mechanical and hydrostatic power, or an MTF (modulated transformer) in case of adjustable or variable velocity. In some paper the pump is conceived of as a gyrator with some parasitic elements in terms of slips and leakages [Mukherjee and Karmakar, 2000]. In this consulted paper, the mechanical aspect is not modelled.

Depending on its position in the circuit, the mission of the pump can be transportation of the fluid (feeding a tank) or increasing the pressure. Two cases can be considered: the external pumps and the pumps installed in hydraulic circuits. For the first case the pump is feeding the system externally and the model is similar to the one given in figure 1. In the second case, the provided by the pump flow $\dot{m}$ depends on the difference between upstream and downstream pressure (ΔP) (figure 2).

Based on pump characteristic (figure 2), it is shown that at a constant rotational speed, the pump can be considered as a modulated flow source by the pressure difference: $MSf=f(\Delta P)$.

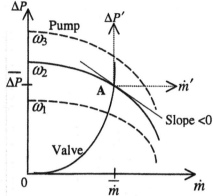

FIG. 2. Characteristic curve of a centrifugal pump with ($\omega_1 < \omega_2 < \omega_3$).

The corresponding bond graph model is given (figure 3). The pump carries also the thermal energy of the fluid which depends directly on the flow of the pump characterized by equation (1). The flow source $Sf^* : \dot{m}C_pT$ on a thermal 1-junction assures the flow $\dot{H}$ in thermal bonds. However, since we have $T_1 = T_2 = T$ a fictive flow detector ($Df^* : \dot{H}$) and a fictive modulated source ($MSf^* : \dot{H}$) are used to impose a non zero effort value on Sf^*.

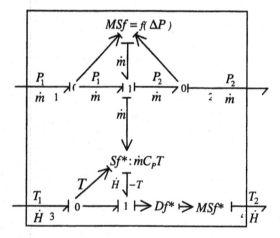

FIG. 3. Non linear model of an installed pump.

Fictive detectors and sources after linearization are represented adding "*" in order to distinguish them from real ones.

2.2 Linearization of Physical Systems using Bond Graph Approach

[Karnopp, 1977] describes the linearization procedure of the different components of bond graphs without the extension of the linearization on the whole bond graph models. The extension on the linearization of the whole bond graph is developed by [Rimaux, 1995] for the monoenergy systems. The operating point **A** is fixed by the process characteristic (it may be the intersection between pump and pipe, see figure 2). As illustration of the linearization methodology applied to a thermofluid processes, let us consider only the thermofluid source.

Linearizing the thermal flow source $MSf_{Th} : \dot{m}C_pT$ characterized by equation (1) around the operating point ($\overline{\dot{m}}, \overline{T}, \overline{H}$) gives:

$$\dot{H}' = \left(\frac{\partial F}{\partial \dot{m}}\right)_{(\overline{\dot{m}},\overline{T})} \dot{m}' + \left(\frac{\partial F}{\partial T}\right)_{(\overline{\dot{m}},\overline{T})} T'$$
$$= C_p\overline{T}\dot{m}' + \overline{\dot{m}}C_pT' = Sf_h^* + Sf_T^* \qquad (2)$$

where $\dot{m}'$, $\dot{H}'$ and T' are the incremented variables.

This is an algebraic constitutive bond graph R equation relating flow variables, $C_p\overline{T}\dot{m}'$ and $\overline{\dot{m}}C_pT'$ are represented in the linearized bond graph model (fig.4) by flow sources modulated by $\dot{m}'$ and T' respectively. The first term is modulated by a hydraulic flow variation; the informational link assures the modulation value $\dot{m}'$.

The linearized bond graph noted *LTS* for « Linearized Thermofluid Source » is represented by figure 4.

The fictive source Sf_h^* describes the influence of the variation of hydraulic flow on thermal energy.

The circled elements are the real sources, they are the active elements $\dot{m}'$ and T'.

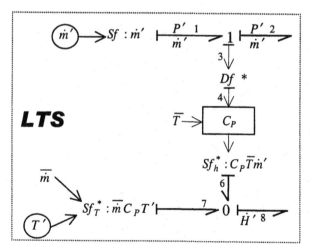

FIG. 4. Linearized BG Model of the Thermofluid Source *LTS*.

From figure 4, it is shown that only hydraulic energy can influence thermal energy (due to the presence of informational bond in only one direction). This coincides with the modelling hypothesis (under saturated case).

3. DIAGNOSABILITY BASED ON LINEARIZED BOND GRAPH MODEL

Definition 1. [Rimaux, 1995] (generalised causal path) A generalised causal path is a causal path that can follow power links or informational links, or both.

Definition 2. A variable is monitorable iff it is possible to detect and to isolate the faults that may affect it.

Using the bond graph approach, it is possible to deduce directly from a bond graph model if variables to be monitored are monitorable without any need to generate the ARR or the signature faults. Some propositions are given for simple bond graph with one port in [Tagina, 1995]. In this paper, the proposition is extended to thermodynamic multiport linearized bond graph models.

3.1 Diagnosability of Sensors

Proposition 1. If the number of sensors is higher than 1, then all sensors are monitorable. Indeed if r is the sensor number, then it comes that there are r independent ARRs [Tagina, 1995].

3.2 Diagnosability of Sources

Proposition 2. Two control sources S_i and S_j in thermodynamic linearized bond graph model are not monitorable if: (a) the covering of the generalised causal path linking the source S_i to any sensor D_l contains the covered causal path from the source S_j to the same sensor D_l, (b) the causal path between elements S_i-D_l that are not listed in S_j-D_l did not appear in any causal path or loop when the causal path is removed from the monitored bond graph.

Proposition 3. Using at least one hydraulic sensor, the hydraulic and thermal sources become monitorable.

Proof. The non linear equations in case of the under saturated fluid are written in equation (3).

$$\dot{x}_H = f(x_H, u_H)$$
$$\dot{x}_T = g(x_H, x_T, u_H, u_T) \tag{3}$$

and the measure equations :

$$De_H = p(x_H) \Rightarrow x_H = p^{-1}(De_H)$$
$$Df_H = q(x_H) \Rightarrow x_H = q^{-1}(Df_H) \tag{4}$$
$$De_T = r(x_H, x_T) \Rightarrow x_T = Fr^{-1}(De_T, De_H, Df_H)$$

where f, g, p, q, r, and Fr are non linear functions of x_H (hydraulic states), x_T (thermal states), u_H (hydraulic sources) , u_T (thermal sources) and sensor values. At the hydraulic level, we may have a pressure or level detector De_H or a flow detector Df_H. However, at thermal level only a temperature detector De_T may exist.

Substituting the known value of the state (from measured equations) in the state equation, we obtain the following ARRs :

$$ARR1 : \Phi_1(D_{eH}, Df_H, u_H) = 0$$
$$ARR2 : \Phi_2(D_{eT}, De_H, Df_H, u_H, u_H) = 0$$

The corresponding signature fault matrice is given below. "1" in the ith row and j^{th} colum means that the ith ARR is sensible the jth failure.

ARR	u_H	u_T	De_H	De_T	Df_H
ARR 1	1	0	1	0	1
ARR 2	1	1	1	1	1

The boolean failure signature vectors of the hydraulic source (u_H) (1 1) and thermal source (u_T) (0 1) are different, thus the failures which may affect them can be detected and isolated.

Proposition 4. In the linearized thermofluid systems, if thermal sources are monitorable and if we have at least one hydraulic detector then all the sources become monitorable.

Proof. In thermofluid systems, we have three kind of sources: a thermal source (heater), a thermofluid source and a pump. For the thermofluid source and the pump, the effect of $\dot{m}'$ on thermal system is represented by $Sf_h^* : C_P \overline{T} \dot{m}'$ (figure 4, 6) and a set of fictive flow and effort sources resulting from the linearization of MC and MR.

Recall the proposition given in [Tagina, 1995]: The fault affecting sources are isolable and detectable iff the column vectors of the transfer matrix are linearly independents. This approach is available in the generalized causal path [Rimaux, 1995]. The term m_{ij} of the transfer matrix $M(s)$ relating the output "i" to the input "j" is defined by equation 5, where $T_k(s)$ is the gain of the kth causal path which links the source (actuator) S_j to the sensor D_i, $D(s)$ and $D_k(s)$ are the bond graph determinant and reduced determinant respectively.

$$\frac{D_i(s)}{U_j(s)} = \frac{\sum_k T_k(s).D_k(s)}{D(s)} \tag{5}$$

Apply equation 5 to a term relating a hydraulic actuator $\dot{m}'$ (thermofluid source or pump) and a thermal detector De_T. One of the causal paths will contain the covered causal path from the corresponding fictive source $Sf_h^* : C_P \overline{T} \dot{m}'$ to the same thermal sensor De_T ($T_l(s) = \alpha \dfrac{D_i(s)}{Sf_h^*(s)}$; α is a constant). Consequently, the Diagnosabilitystudy of hydraulic sources can be limited on thermal system: indeed, if Sf_h^* are monitorable in thermal system then the hydraulic sources will be.

This result can also be deduced from the sources Sf_T^* : in the linearized thermofluid systems, thermofluid sources and pumps are linearly coupled by equation (2), the causal paths followed by the resulting hydraulic and thermal sources (Sf_h^* and Sf_T^*) to a thermal detector in the thermal system are the same. In addition to these sources, we have heaters (pure thermal sources Sf_T).

In addition, if we have at least one hydraulic detector then hydraulic and thermal sources will have different signatures in the transfer matrix (proposition 3).

The necessary conditions for all control sources to be monitorable in thermodynamic linearized bond graph model are: (a) at least one detector is mounted at hydraulic level, (b) thermal sources (Sf_T^* and Sf_T) are monitorable.

4. APPLICATION

4.1 Description of the installation

The developed approach is applied to a thermofluid system given by figure 5. It consists of a storage tank C_1 filled by the mass flow $\dot{m}_1$ at temperature T_1, the feed water supply system and a heated tank by a warming resistance supplying an electrical power $\dot{Q}$. The feeding circuit consists of a pump and a pipe R_1. The heated water flows from the tank C_2 through the valve R_2 to the outlet system imposing the P_S pressure. T_S is the oulet temperature. The valves R_1 and R_2 are on-off valve.

The system to be modelled has a purely non linear behaviour due to the coupling of different phenomena of diverse natures. We are in fact in presence of electrical energy (power generator), hydraulic (fluid circulation in pipes), thermal (heated tank) and mechanical (pump, valve).

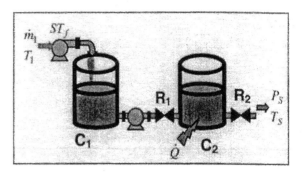

FIG. 5. Scheme of the installation.

We dispose of two thermal detectors allowing the measure of the temperature in the storage tank and in the heated tank, a hydraulic sensor measuring the pressure in C_1. They are supposed to be ideal, and are modelled by using signal bonds meaning that no power is transferred.

4.2 Linearization of Bond Graph models

The linearized bond graph model is given by figure 6. The linearized sources are the new active elements. While the hydraulical and thermal effort sources P_S' and T_S' are considered as disturbance. There are four active elements: the inlet mass flow and the inlet enthalpy flow (hydraulic source $\dot{m}_1'$ and thermal source T_1' of the thermofluid source ST_f), the pump MSf and the thermal heating flow Sf_T of the tank C_2. Real sensors measure the variation of the temperature in C_2 $De_T : T_{C_2}'$ (bond 86), the variation of the temperature in the storage tank $De_T : T_{C_1}'$ (bond 87) and the pressure variation in the storage tank $De_H : P_{C_1}'$ (bond 85). Active elements are indicated as circled vertices and sensors as squared vertices. The bonds 28, 53 and 73 (figure 6) are the special informational bond used to simplify the representation; each bond represents a fictive flow detector followed by a modulated fictive flow source (see figure 3).
We distinguish four dynamical elements, two hydraulical capacities and two thermal ones in integral causality and therefore the system order is equal to four.

4.3 Diagnosability

Since we have more than one sensor we deduce from proposition 1 that all sensors are monitorables. For actuators: based on propositions 3 and 4 our study is limited on T_1', Sf_2^*, Sf_T. Let us consider the covering of the bond graph causal paths from the considered control sources to each thermal sensor, which lead to:

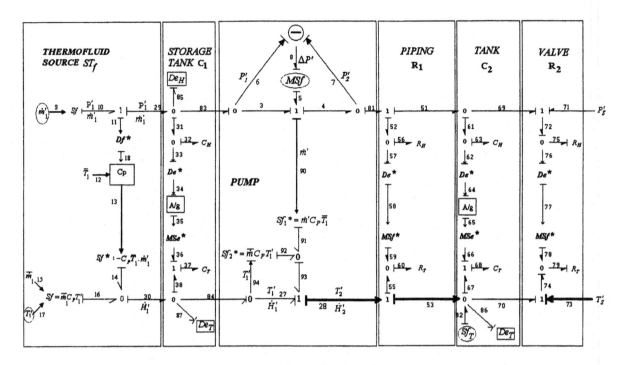

FIG. 6. Linearized Bond Graph model.

$T'_1 - 17 - 16 - 30 - 38 - 37 - C_T - 37 - 38 - 87 - De_T.$

$T'_1 - 17 - 16 - 30 - 38 - 37 - C_T - 37 - 38 - 84 - 94 - 92 - 93$
$- 28 - 53 - 67 - 68 - C_T - 68 - 67 - 86 - De_T.$

$Sf_2* - 92 - 93 - 27 - 84 - 38 - 37 - C_T - 37 - 38 - 87 - De_T.$

$Sf_2* - 92 - 93 - 28 - 53 - 67 - 68 - C_T - 68 - 67 - 86 - De_T.$

$Sf_T - 82 - 67 - 68 - C_T - 68 - 67 - 86 - De_T.$

Based on proposition 2, it is easy to deduce that these control sources are monitorable. And due to the hydraulic detector, all the actuators become monitorable.

5. CONCLUSION

Thermofluid systems are complex because of the energies coupling. They are present in industry with risk and need consequently a monitoring. The study of diagnosability of linear monoenergy systems was previously done using a bond graph approach. An extension on linearized multienergy thermofluid bond graph is made here. Conclusions on the monitoring of actuators and sensors for linearized thermofluid systems can now be derived directly from the topology of the bond graphs without deriving the state equations. The method is applied to a thermofluid system with pump. The further problem is to extend the Diagnosabilityanalysis for non linear thermodynamic bond graph.

6. REFERENCES

[El Osta and al., 2003] W. El Osta, B. Ould Bouamama, C. Sueur (2003). 'Analyse structurelle des bond graphs multienergies linearisés'. MAJECSTIC'2003 MAnifestation des JEunes Chercheurs STIC, 29-31 octobre, Marseille, France.

[Karnopp, 1977]. 'Power and Energy in Linearized physical Systems'. Journal of the Franklin Institute, Vol.303, No.1, pp.85-97, Janvier 1977.

[Karnopp and al., 1990] Karnopp, D. C., D. Margolis and R. Rosenberg (1990). Systems Dynamics: A unified approach, second ed..John Wiley. New York.

[Mukherjee and Karmakar, 2000] A. Mukherjee, R. Karmakar, Modelling and Simulation of Engineering Systems Through Bond Graphs, Alpha Sciences International, Pangbourne, UK, 2000.

[Ould Bouamama, 2001] Ould Bouamama B. (2001). Model builder for Thermo-fluid Systems Using a Bond Graph and Functional Modelling. 3th European Simulation Symposium ESS'01, pp. 822-826, Marseille France, October, 18-20, 2001.

[[Ould Bouamama, 2003] B. Ould Bouamama 'Bond graph approach as analysis tool in thermofluid model library conception'. Journal of the Franklin Institute, Vol.340, pp.1-23, 2003.

[Rimaux, 1995] Rimaux, S. (1995). Etude des propriétés structurelles de certaines classes de systèmes physiques non linéaires modèlisés par bond graph. Ph.D thesis. Université de Lille, France.

[Tagina, 1995]. Application de la Modélisation Bond Graph à la Surveillance Des Systèmes Complexes. PhD thesis. University of Lille, France.

[Thoma, 1990] Thoma, J. U. Simulation by bond graphs. Introduction to a Graphical Method. Springer Verlag, Berlin Heidelberg, 1990

[Thoma and al., 2000] Thoma, J. U. and B. Ould-Bouamama (2000). Modelling and Simulation in Thermal and Chemical Engineering. Bond Graph Approach. Springer Verlag.

Published by Elsevier Ltd on behalf of IFAC

GPS/INS SENSOR FUSION FOR ACCURATE POSITIONING AND NAVIGATION BASED ON KALMAN FILTERING

J.Z. Sasiadek and P. Hartana

Department of Mechanical & Aerospace Engineering
Carleton University
1125 Colonel By Drive
Ottawa, Ontario, K1S 5B6, Canada
e-mail: jsas@ccs.carleton.ca

Abstract: This paper presents the position and velocity determination by using INS and GPS. The measurement results from INS and GPS sensors are fused by using Kalman filter. Dilution of Precision (DOP) technique is used to select a combination of satellites to be used as measurement data. Two implementations of Kalman filter, feedforward and feedback are used. The experiment shows that the selection of the satellites affects the measurements. For biased and correlated data an adaptive technique, developed earlier was used. This paper is a demonstration of known techniques rather than an introduction of new ones.

Keywords—Sensor Fusion, GPS, INS, Kalman Filtering, Accurate Positioning

1 INTRODUCTION

The Inertial Navigation System (INS) is widely used as one component in guidance, navigation, and control systems. It includes accelerometers and gyroscopes to provide velocity and angular rate information. Integrating this information, the position and orientation of the vehicle can be calculated. Global positioning system (GPS) is commonly combined with the INS to bind its propagation error. These two types of signals are fused together to produce one accurate navigational information. Kalman filter is the common algorithm used to fuse the measurements (Rumeliotis *et al* 1999, and Santini *et al*, 1997).

Some authors (Jetto, *et al.*, 1999) and (Sasiadek and Wang, 1999 and Sasiadek and Wang, 2003) used advanced techniques of Kalman filter gain tuning based on fuzzy logic. There are many research works already conducted to improve position estimation when implementing INS and GPS, whether as a standalone estimation method, or as an integrated system. Sasiadek and Hartana (2004) made a more extensive introduction to these problems. Pozo-Ruz *et al* (1998), a satellite selection criterion was introduced in order to improve accuracy of the estimation when using GPS.

The distributed Kalman filter simulator (DKFSIM) was introduced Lawrence (1996). This filter is used to evaluate the performance of several different filter architectures and sensor model conditions for advanced, multi-sensor navigation systems. In (Sukkarieh *et al*, 1998), authors examined the integration of INS-GPS for autonomous land vehicle.

Currently, there are 28 global positioning satellites constellation available to be used for navigation. From any particular location, a receiver can view around 10 satellites at any given time. Although, the estimation is usually improved when using more satellites, the channels available to receive the measurement signals are usually restricted. Moreover, not all the visible satellites are good to be used as the sources of measurement. Therefore, the receiver has to choose the best combination of satellites in view to get the optimal estimation.

In this paper, INS and GPS measurements data are combined to find the optimal measurement results by using Kalman filter. The best combination of satellites is chosen from a number of visible satellites by using dilution of precision (DOP) technique. The effects of the change of satellites in combination to the estimation are examined. If the data are found to be biased and/or correlated, a special adaptive techniques described by Sasiadek and Wang (1999) was used.

2 POSITION DETERMINATION BY USING INERTIAL NAVIGATION SYSTEM

Position determination by using the INS can be briefly explained by the following steps:

- Measurement of the accelerations in the directions of the navigation axes. For vertical acceleration, the effect of the earth gravitation has to be included in calculation.
- Determination of distance and velocity by integrating the acceleration. This integration involves time interval, so this interval must be known accurately.
- Measurement of the rotation rates (from gimbals motion in a stabilized platform or from gyroscopes in a strapdown system) as a process to find the direction of the measured distance and velocity. This should include the compensation of the earth's rotation.
- Combining the distance and heading data gives an updated dead reckoned position to display.

3   POSITION DETERMINATION BY USING GPS PSEUDORANGE

To determine user position, GPS uses time-of-arrival of signals broadcast by satellites. The signals contain two types of ranging codes, which are pseudorandom noise (PRN), and navigation data.

In three dimensional spaces, user position can be determined from GPS satellites constellation by measuring the ranges from minimal four satellites positions. Since, the satellites and user receiver clock will generally have a bias error from GPS system time, the ranges determined from this process are called pseudorange (ρ), and contains the geometric satellite-to-user range, an offset caused by the difference between user clock and GPS system time, and an offset caused by the difference between satellites clock and GPS system time.

In three dimensions space, the range can be determined by using the following equations:

$$\rho_1 = \sqrt{(x_1 - x_u)^2 + (y_1 - y_u)^2 + (z_1 - z_u)^2} + ct_u \quad (1)$$

$$\rho_2 = \sqrt{(x_2 - x_u)^2 + (y_2 - y_u)^2 + (z_2 - z_u)^2} + ct_u \quad (2)$$

$$\rho_3 = \sqrt{(x_3 - x_u)^2 + (y_3 - y_u)^2 + (z_3 - z_u)^2} + ct_u \quad (3)$$

$$\rho_4 = \sqrt{(x_4 - x_u)^2 + (y_4 - y_u)^2 + (z_4 - z_u)^2} + ct_u, \quad (4)$$

where x_i, y_i, and z_i denote the satellites position in three dimensions, x_u, y_u, z_u are user position, c is the speed of light.

4.   GPS ERRORS

In determination of position by using GPS, the accuracy of the sensors fusion performance is also influenced by the geometry of satellites distribution, which is known as dilution of precision (DOP). It is defined as geometry factors that relate parameters of the user position and time bias errors to the pseudorange errors.

The more usable satellites that are available result in higher accuracy. In this case, the DOP parameter is calculated from all available satellites. However, the DOP criteria can also be used to find the four best satellites out of a geometric constellation consisting of more than four satellites. This gives advantages particularly if only a four-channel receiver is available.

The error of the pseudorange measurement from the approximate user position in this case can be formulated as:

$$\rho_i - \hat{\rho}_i = -\frac{x_i - \hat{x}_u}{\hat{r}_i}\Delta x_u - \frac{y_i - \hat{y}_u}{\hat{r}_i}\Delta y_u$$
$$-\frac{z_i - \hat{z}_u}{\hat{r}_i}\Delta z_u + c\Delta t_u, \quad (5)$$

or in matrix form as:

$$\begin{bmatrix} \Delta\rho_1 \\ \Delta\rho_2 \\ \Delta\rho_3 \\ \Delta\rho_4 \end{bmatrix} = \begin{bmatrix} a_{x1} & a_{y1} & a_{z1} & 1 \\ a_{x2} & a_{y2} & a_{z2} & 1 \\ a_{x3} & a_{y3} & a_{z3} & 1 \\ a_{x4} & a_{y4} & a_{z4} & 1 \end{bmatrix} \begin{bmatrix} \Delta x_u \\ \Delta y_u \\ \Delta z_u \\ c\Delta t_u \end{bmatrix} \quad (6)$$

Equation (6) can be written in compact form as:

$$\Delta\rho = H\Delta x. \quad (7)$$

From this equation, geometric dilution of precision (GDOP) is defined as:

$$GDOP = \sqrt{trace(H^T H)^{-1}}, \quad (8)$$

where trace[·] indicates the sum of diagonal elements of the matrix. Not all of the available satellites used in this work are visible all the time during the flight. Sasiadek and Hartana, (2004) presented the satellites visibility chart as a function of time when the INS and GPS measurement data are recorded.

The other possibility for localization using the GPS sensor is the carrier-phase method. Farell *et al* (2003) used this method to calculate lateral displacement of the vehicle.

5.   SENSOR FUSION PROCEDURE

When the GPS measurement result is used with INS, the characteristics of GPS measurement binds the error introduced by INS.

In this work, the GPS and INS sensors are used to find the user position. The problem can be summarized how to the optimal signal on the basis of two different signals available.

In this case, the Kalman filter is implemented to fuse the measurement output and feed the result to the control system.

When implementing Kalman filter in inertial navigation problems (or any other applications), two states can be used: total state space and error state space formulation. In this work, error state space formulation will be used, where the difference between INS and GPS measurement data is used as the input to the Kalman filter. The advantage of this configuration is the filter does not need to work at the same frequency as the INS, which is usually very high. The frequency at which the filter works is the same as the frequency of the availability of the GPS data, which is usually lower than the frequency of INS.

In general, there are two different implementations of the error state space configuration: feedback and feedforward. These implementations are based on how the estimated errors resulted from the filter are combined with Fig.1 and Fig. 2 display the appropriate configurations as block diagrams.

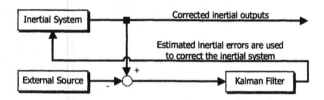

Fig. 1 Error Feedback Kalman Filter

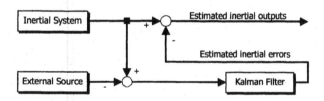

Fig. 2 Error Feedforward Kalman Filter

The measurements obtained from satellites should be unbiased and uncorrelated. Positions and velocities can be represented as a linear model (9) in state space form. In case of 8 states model, the particular states represent East position error, East velocity, North position error, North velocity, altitude error, altitude rate, clock bias, and clock drift respectively. For 8-state Kalman filter, model for the INS measurement can be written as:

$$
\begin{bmatrix} \dot{x}_1 \\ \dot{x}_2 \\ \dot{x}_3 \\ \dot{x}_4 \\ \dot{x}_5 \\ \dot{x}_6 \\ \dot{x}_7 \\ \dot{x}_8 \end{bmatrix} = \begin{bmatrix} 0 & 1 & 0 & 0 & 0 & 0 & 0 & 0 \\ 0 & 0 & 0 & 0 & 0 & 0 & 0 & 0 \\ 0 & 0 & 0 & 1 & 0 & 0 & 0 & 0 \\ 0 & 0 & 0 & 0 & 0 & 0 & 0 & 0 \\ 0 & 0 & 0 & 0 & 0 & 1 & 0 & 0 \\ 0 & 0 & 0 & 0 & 0 & 0 & 0 & 0 \\ 0 & 0 & 0 & 0 & 0 & 0 & 0 & 1 \\ 0 & 0 & 0 & 0 & 0 & 0 & 0 & 0 \end{bmatrix} \begin{bmatrix} x_1 \\ x_2 \\ x_3 \\ x_4 \\ x_5 \\ x_6 \\ x_7 \\ x_8 \end{bmatrix} + \begin{bmatrix} 0 \\ w_2 \\ 0 \\ w_4 \\ 0 \\ w_6 \\ w_7 \\ w_8 \end{bmatrix} \quad (9)
$$

or in matrix form as:

$$ \dot{\mathbf{x}}(t) = \mathbf{F}(t)\mathbf{x}(t) + \mathbf{w}(t), \quad (10) $$

where $\mathbf{w}(t)$ represents the accelerations and clock noises of the INS sensor. It is assumed that the noise is purely white noise, zero mean, and Gaussian with covariance $\mathbf{Q}$. Transforming (10) into discrete form results:

$$ \mathbf{x}_{k+1} = \mathbf{\Phi}_k \mathbf{x}_k + \mathbf{w}_k, \quad (11) $$

where:

$$ \mathbf{\Phi}_k = e^{\mathbf{F}T} $$

$$ \mathbf{w}_k = \int_{kT}^{(k+1)T} e^{\mathbf{F}[(k+1)T - \tau]} \mathbf{w}(\tau) d\tau $$

The covariance of $\mathbf{w}_k$ in discrete form ($\mathbf{Q}_k$) can be found as:

$$ \mathbf{Q}_k = \int_0^T e^{F\tau} \mathbf{Q}\, e^{F^T \tau}\, d\tau. \quad (12) $$

The observation for the Kalman filter is the difference between pseudorange as measured by GPS and the predicted pseudorange, which can be derived from (6). This equation is defined in Earth-Centered, Earth-Fixed (ECEF) coordinate system. Transforming it into local coordinate frame results:

$$
\begin{bmatrix} z_1 \\ z_2 \\ z_3 \\ z_4 \end{bmatrix} = \begin{bmatrix} a_{x1} & 0 & a_{y1} & 0 & a_{z1} & 0 & 1 & 0 \\ a_{x2} & 0 & a_{y2} & 0 & a_{z2} & 0 & 1 & 0 \\ a_{x3} & 0 & a_{y3} & 0 & a_{z3} & 0 & 1 & 0 \\ a_{x4} & 0 & a_{y4} & 0 & a_{z4} & 0 & 1 & 0 \end{bmatrix} \begin{bmatrix} x_1 \\ x_2 \\ x_3 \\ x_4 \\ x_5 \\ x_6 \\ x_7 \\ x_8 \end{bmatrix}
$$

$$
+ \begin{bmatrix} v_1 \\ v_2 \\ v_3 \\ v_4 \end{bmatrix} \quad (13)
$$

The last term in (13) is the measurement noise. It is also assumed as zero mean Gaussian white noise with covariance $\mathbf{R}$.

In practice, this assumption is not always valid. In that case the Kalman filter gain would diverge and filter would not perform its role. In that case the adaptive Kalman filter has to be employed. The details of such filter could be found in Sasiadek and Wang (1999 and 2003).

Measurement (13) is already in discrete form, so it can be implemented directly into discrete Kalman filter. Writing this equation in matrix form result:

$$z_k = H_k x_k + v_k \qquad (14)$$

Therefore, the set of equations for the GPS aided inertial navigation system as shown in (11) and (14).

6. EXPERIMENTS AND RESULTS

The real experiments with aircraft flying and measuring simultaneously position with GPS and INS were conducted. Subsequently, the measurements were fed into the sensor fusion algorithm including the implementation of the Kalman filter in the GPS-INS sensors fusion. Two implementations of the filter were considered: error state feedforward and error state feedback Kalman filter.

The data were recorded on board of an aircraft flying along a designated path. The data in form of the INS measurements, GPS satellites' pseudorange measurements, and the satellites' locations from all visible satellites were recorded.

Number of visible satellite has to be at least four. The selection of the satellites used in the experiment was done by using the DOP method as explained before. The simulation experiment has been conducted for 2.5 hours long. The results of the experiment are displayed as the position and velocity errors. Figures 3 until 10 display these results. The dash line represents the result of the experiment when using error state feedforward Kalman filter and the solid line represents the result when using error state feedback Kalman filter.

From the figures, it can be seen that the error state feedback Kalman filter implementation results in smaller error than the feedforward one. It can also be seen that there are some spikes of the error. These spikes are resulted when the filter switches into different satellites combination. The switches are needed because one or more satellites used in previous combination are not visible anymore, or because the combination will produce greater DOP value than the other possible combinations.

For filter covariance, shown only for velocity error in East direction, in general the feedback Kalman filter gives smaller uncertainty. It can also be seen that the uncertainty increases before the change of satellite combination occurs. This is reasonable, because when the satellites become not visible, the measurements produced from GPS receiver also become unreliable. From the plot of covariance can be seen that the third combination gives smaller uncertainty than the first two. The third combination also gives good results until the end of simulation. This pattern is valid for position error covariance as well as velocity error covariance.

For Kalman gain, in the first combination, the feedback configuration gives more weight to the INS measurement

than the feedforward configuration. The opposite is resulted in the third combination, where the feedback gives more weight to the GPS measurements.

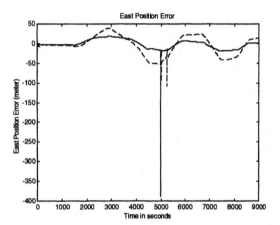

Fig. 3   East position error.

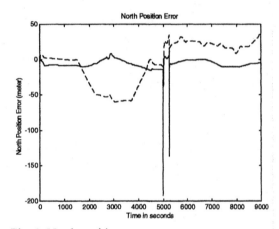

Fig. 4   North position error.

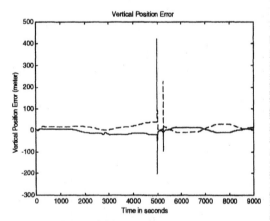

Fig. 5   Vertical position error.

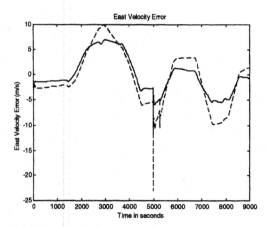

Fig. 6 East velocity error.

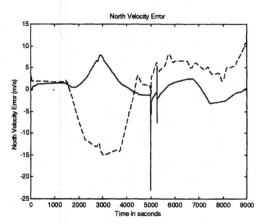

Fig. 7 North velocity error.

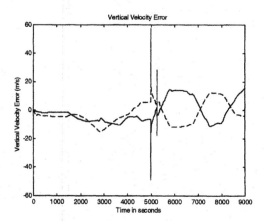

Fig. 8 Vertical velocity error.

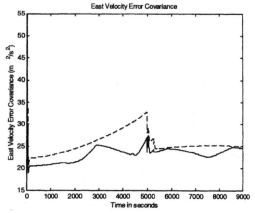

Fig. 9 Velocity error covariance in East direction

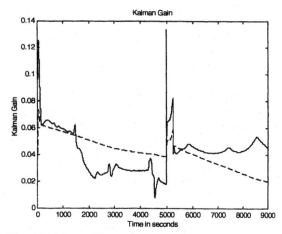

Fig. 10 Kalman filter gain.

The results shown in Figs. 3 to 10 are obtained from the experimental data involving the aircraft flying over a test range during several hours.

Sometimes, the data from GPS as well as INS cannot be considered to be a unbiased and uncorrelated, and therefore the basic assumption of the Kalman filter – white noise would not hold. The colored noise will cause the filter gain to diverge. The adaptive gain has to be used. Earlier, the authors suggested an efficient method of adaptive Kalman filtering based on fuzzy logic method. Sasiadek and Wang (1999) used fuzzy logic adapted controller (FLAC) to prevent the filter from divergence when fusing signals coming from INS and GPS on flying vehicle. Nine rules were used. There were two inputs, which are the mean value and covariance of innovation, and the output is a constant that is used to weight exponentially the model and measurement noise covariance.

Figure 11 explains the idea of fuzzy adaptive Kalman filtering.

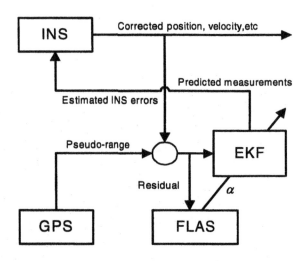

Fig. 11 Adaptive sensor fusion based on fuzzy logic

The adaptation of Kalman gain is done on the base of covariance and mean value. This adaptation prevents filter from divergence and improves the performance of the algorithm. The details of this method were reported by Sasiadek and Wang (1999, 2003).

7. CONCLUSIONS

This paper presents methods of sensor fusion based on Kalman Filter and analyzes various design approaches. The method was applied to the GPS/INS integration. The experimental results were presented and discussed. The objective of this paper was to demonstrate a known engineering technique and verify procedure developed by authors earlier. The performance of this method was checked and it was possible to verify the applicability in various conditions.

8. REFERENCES

Farrell, J.A., Tan, H.S., and Yang, Y., CarrierPhase GPS-Aided INS-Based Vehicle Lateral Control, *ASME Journal of Dynamic Systems, Measurement, and Control*, September 2003, Vol. 125, pp. 339-353

Jetto, L., Longhi, S., and Vitali, D., Localization of a wheeled mobile robot by sensor data fusion based on a fuzzy logic adapted Kalman filter. *Control Engineering Practice 7*, 1999, pp. 763-771.

Jetto, L., Longhi, S., and Venturini, G.,Development and experimental validation of an adaptive extended Kalman filter for the localization of mobile robots. *IEEE Transactions on Robotics and Automation 15*(2), 1999, pp. 219-229.

Lawrence, P. J., Navigation Sensor, Filter, and Failure Mode Simulation Results Using the Distributed Kalman Filter Simulator (DKFSIM), *IEEE 1996 Position, Location and Navigation Symposium*, Atlanta, Georgia, April 22-26, 1996, pp. 697-710.

Pozo-Ruz, A., Martinez, J. L., and Garcia-Cerezo, A., A New Satellite Selection Criterion for DGPS Using Two Low-Cost Receivers. *Proceeding of the 1998 IEEE International Conference on Robotics & Automation*, Leuven, Belgium, May, 1998, pp. 1883-1888.

Roumeliotis, S. I., Sukhatme, G.S., and Bekey, G. A., Circumventing Dynamic Modeling: Evaluation of the Error-State Kalman Filter applied to Mobile Robot Localization, *Proceeding of the 1999 IEEE International Conference on Robotic & Automation*, pp. 1656-1663.

Santini, A., Nicosia, S.and Nanni, V.,Trajectory estimation and correction for a wheeled mobile robot using heterogeneous sensors and Kalman filter. *Preprints of the Fifth IFAC Symposium on Robot Control ,SYROCO 1997*, pp. 11-16.

Sasiadek, J. Z., and Wang, Q, Sensor fusion based on fuzzy Kalman filtering for autonomous robot vehicle. *Proceeding of the 1999 IEEE International Conference on Robotics & Automation*, pp. 2970-2975.

Sasiadek, J. Z., and Wang , Q., Low cost automation using INS/GPS data fusion for accurate postioning, *Robotica*, Vol. 21, 2003, pp. 255-261

Sasiadek, J.Z. and Hartana, P. "Sensor Fudion for an Unmanned Aerial Vehicle" *Proceedings of the 2004 IEEE Intl. Conf. On Robotics and Automation*, New Orleans, April 2004

Sasiadek, J. Z. and Hartana, P., Odometry and Sonar Data Fusion for Mobile Robot Navigation. *Proceeding of the 6th IFAC Symposium on Robot Control – SYROCO 2000*, 531-536.

Sukkarieh, S., Nebot, E. M., & Durrant-Whyte, H. F., Achieving Integrity in an INS/GPS Navigation Loop for Autonomous Land Vehicle Applications. *Proceeding of the 1998 IEEE International Conference on Robotics & Automation*, Leuven, Belgium, May, 1998, 3437-3442.

ELSEVIER

IFAC
PUBLICATIONS
www.elsevier.com/locate/ifac

MANAGING IT/IS PROJECTS FOR ENTERPRISE INTEGRATION

Renate Sprice and Janis Maknia

Riga Technical University
Latvia

Abstract: Enterprise integration itself causes changes in enterprise infrastructure as well as in enterprise information technology and information systems. This paper discusses the depth of the problem of handling changes during enterprise integration and points to the fact that the need for different types of changes is caused not only by enterprise integration activities but also by other factors. Due to variety of change causes, choice of enterprise integration projects to be initiated and estimation of their value for future enterprise becomes a task of a very high complexity. To accomplish this task, the paper suggests to combine three techniques, namely, (1) enterprise integration requirements analysis, (2) information systems changes predictions framework, and (3) enterprise integration oriented project evaluation framework. *Copyright © 2004 IFAC*

Keywords: Enterprise Integration (EI), Business Integration, Change Management, IT project feasibility.

1. INTRODUCTION

Enterprise Integration (EI) is seen as one of the means for delivering paramount business performance in today's uncertain economy (Sheratt, 2003). The uncertainty of economy is caused by rapid changes in technologies, peculiarities of global marketing, and unpredictable political and natural events. EI itself causes changes in enterprise infrastructure as well as in enterprise information technology and information systems (IT/IS). This paper discusses the depth of the problem of handling changes during enterprise integration and points to the fact that the need for different types of changes is caused not only by enterprise integration activities but also by other factors. Due to variety of change causes, choice of IT/IS enterprise integration projects to be initiated and estimation of their value for future enterprise becomes a task of a very high complexity. The paper suggests to combine three techniques for identifying IT/IS project scope for enterprise integration. These techniques are (1) enterprise integration requirements analysis, (2) information systems changes predictions framework, and (3) project

evaluation framework. The paper is structured as follows: Section 2 introduces the concept of Enterprise Integration and analyses IT/IS role in enterprise integration, Section 3 discusses two main approaches for enterprise integration from the point of view of IT/IS. Section 4 focuses on derivation of IT/IS changes requirements from analysis of enterprise integration incentives at several levels of abstraction. Section 5 introduces general framework for IT/IS changes prediction, Section 6 proposes enterprise integration oriented project evaluation framework, which integrates business level requirements and IT/IS changes predictions in analysis of projects relevant for enterprise integration. Section 7 presents brief conclusions and directions of further research.

2. ROLE OF IT/IS IN ENTERPRISE INTEGRATION

Interest in EI is associated with changes at the business level, Markus (2000), Mc Keen (2002), and progression at technology level (Markus, 2000). These two streamlines point to *business integration* and *systems integration* (Markus, 2000), and possibilities to integrate these two concepts.

Business integration may be interpreted as creation of tighter coordination among the discrete business activities conducted by different individuals, work groups, or organizations, so that unified business process (BP) is formed (Markus, 2000). Business integration enables the enterprise to operate in a more effective way, e.g. by providing opportunities to employees to work more efficiently, minimizing overtime and allowing departments to accomplish more without hiring additional personnel (EI, 2002), reducing unnecessary communication, improving performance of BP (inside and outside the enterprise), etc., e.g. Enterprise Information Portal used by regional law firm in USA in order to understand their clients' needs and envision where their businesses will be tomorrow (Wojtkowski, 2004).

Business integration is the most important concept from the viewpoint of an enterprise because of opportunities associated with this idea. We can divide all these opportunities in two groups: *internal* and *external opportunities*. Internal opportunities are those inside the organization, e.g., reduction of unnecessary work, and cost-effective performance. External opportunities root outside the organisation and display direct relationship between effectiveness and competitiveness of an enterprise (Thompson, (1997), Wang, (2002)). They concern relationships between customers and internal BP (Markus, (2000), Hohpe, (2002)), supply chains, and participation in virtual organizations. The exploitation of these opportunities is a prerequisite for survival of an enterprise in existing globalization age (Thompson, 1997). Therefore well-balanced, in terms of internal and external opportunities, business integration scenario is essential for enriching competitiveness of the enterprise.

System integration is an instrument for achieving the goal, which in this case is business integration in the enterprise.

System integration is the creation of tighter linkages between different computer-based information systems and databases (Markus, 2000). *Enterprise Application Integration* is almost a synonym for system integration (EI, (2002), McKeen, (2002)). We can sum up these opinions and say that *System integration* is merging of information systems (computer-based and manual) using particular rules and standards in order to achieve business integration.

The fact that many of core technologies have now reached their level of maturity and they are viable and becoming standardized, is one of the reasons why we can talk about EI (RIA, 2002). On the one hand headway of IT/IS facilitates EI by providing different resources like WEB services, Internet protocols, software packages (Markus, 2000), on the other hand, there are still many problems related to integration of different technical platforms, efficient, reliable, and secure data exchange between multiple enterprise applications (RIA, 2002), etc., that hinder EI.

3. TYPES OF ENTERPRISE ARCHITECTURES

As it was mentioned before there are two kinds of EI: business integration and system integration. There are different integration architectures that can be applied in every case.

Enterprise Modeling Approach (EMA) and *Workflow approaches* may be used for business integration (Juesfeld, 2001). According to Juesfeld (2001) in the EMA models represent the key entities of an enterprise, i.e. it's organizational structure, resources, business processes, goals, etc. EMA has a descriptive nature. The enterprise model is mainly used to understand the enterprise and to support development of information systems. In workflow approaches the enterprise is seen as a network of agents whose collaboration is defined by process or workflow model (Juesfeld, 2001). The purpose of these approaches is to prescribe behaviour of actors in the enterprise or dynamic processes of enterprise performance.

There are many methodologies, which are suitable for these two enterprise integration approaches (Markus, 2000). Organizational and workflow models used in those methodologies have to enable a clear awareness of organizational goals, values, and pride on the organizational behalf (Wang, 2002).

There are four categories of system integration architectures that can be employed in system integration (McKeen, 2002): (1) *data integration*; (2) *application integration*; (3) *process integration* (4) *inter-organizational*. It depends on the purposes of an enterprise which integration architecture is more appropriate for it.

Data integration. At this level of the EI not only required data transformation and normalization from different (often hard coded) application has been performed, but as pointed out by Juesfeld (2001) data dynamically based on a set of pre-configured rules can be routed and distributed. There are many methods available for accomplishing data integration: batch data exchange, shared data base, raw data exchange, remote procedure calls, messaging, data warehouses, etc. (RIA, 2002).

Application integration. This is the next level of EI. At this level diverse applications have been linked to accomplish specific business processes. According to McKeen (McKeen, 2002) the most popular method is messaging-oriented middleware (MOM). These messages brokers transport information ("messages") between application by identifying, transforming, and routing messages to the appropriate applications on an event –driven, asynchronous basis. Application-specific adapters enable the conversion between different types of applications based on different technologies. One more integration method (McKeen, 2002) is screen scraper that allows input data captured in one application share with other applications.

Process integration. The next level of integration is to coordinate the logical flow among the integrated applications. In other words, it is "event-oriented" or "transaction-oriented" integration where transactions / events provide the linkage among various applications. Tools that are suitable for process integration are business process automation products and application servers and they are based on workflow technology.

Inter-organizational integration. This type of integration links processes beyond the enterprise to include trading partners – both suppliers and customers in a whole chain, e.g. supply chain, Business to Business (B2B), Business to Customer (B2C), and others.

An enterprise has to decide which is the most appropriate type of integration architecture in order to support requirements arisen from business integration. In order to do this the detailed analysis of possible changes and impact of this changes on the enterprise has to be done.

4. DERIVATION OF IT/IS CHANGES REQUIREMENTS FROM ANALYSIS OF ENTERPRISE INTEGRATION

Having analysed the changes of enterprise IT/IS we noticed that these changes depend on the types of enterprise integration architecture. It is not sufficient to use a single architecture as an integration basis (McKeen, 2002). Observations show that, in order to reach the business goals during integration successfully, several types of integration architecture are to be combined.

Data level integration is based on shared data and access to them. The types of communication are to be taken into consideration (Woolf, 2002). Due to the fact that the data within applications are maintained first of all for the needs of the application itself, quite often the data are not designed for sharing (Smith, 2002), data are placed on different levels of detailisation and are selected from different sources (Dignum, 1999). However, data level integration is most widely used in practice due to the following reasons:
- standardised data access technologies (SAS, (2003), Artim, (2000));
- low costs of making changes of enterprise IS;
- a possibility of adding functions to applications, which provide data exchange or sharing;
- changes are based on the existing structures.

However, it is not always sufficient to integrate only data by data sharing. Different levels of data detailisation and the different sources may determine the quality of the data-completeness, unambiguousness, meaningfulness and correctness (Huang, 1999), and require exchange with logically organised data. In such cases application to application integration architecture could be used.

As mentioned above, application integration is based on MOM. This software provides data exchange across several applications (McKeen, 2002). MOM integration patterns are dealt with in (Hophe, (2002), Woolf, (2002)). As applications function asynchronously, there arise difficulties in communication and data exchange across applications. Usage of the following solutions can be observed in practice:

1) matching data formats across applications:
- data formats within applications are changed according to the standard;
- MOM maintains the different data formats.

2) a mechanism is created to inform about the moment that data are available within applications and can be delivered to other applications.
- Defining time intervals when the data should be ready for delivery.
- Inserting an indication that the data are ready.

Process-level integration concerns logical integration and restructuring of several enterprises BP (Harrington, (1997), Cassidy, (2001)). Changes in BP are widely dealt with in literature (Grover, 2000A). The paper (Kettinger, 2000) gives analysis of 25 leading approaches of BP change. It is based on transition from "as is " state to "to be" state. Changes within an enterprise IT/IS are directed with the following purposes:
- forming new and more intensive co-ordinating structures across the processes (Teng, 2000);
- to develop metrics of business and of IT/IS process performance (Cassidy, 2001);
- to monitor changes of BP;
- to involve the customers and suppliers into BP;
- to organise continuous improvement.

Inter-organisation integration determines changes in compliance with the business goals and vision of the enterprise. This determines the following changes of IT/IS:
- Improving of data exchange among enterprises. In this case data formats are highly significant.
- Part of IS functions are transported from one enterprise to another.

- The existing structures are perfected or new ones are designed for co-ordination of applications.

When analysing IT/IS changes in compliance with integration architectures we found that changes occur in the following IT/IS elements: data, communication, IS data users, and organisational changes in the enterprise, e.g. changes in responsibilities after implementation of Enterprise Information Portal (Wojtkowski, 2004).

Figure 1 gives a graphic representation of basic issues for gathering requirements for IT/IS changes and illustrates the fact that processes, applications and inter-organisation enterprise integration use common data. IT/IS changes requirements are depicted as rectangles in Figure 1. The arrows pointing to the rectangles indicate interdependence with other IT/IS changes requirements. The newly obtained data from the integration should be available for the future work of the enterprise and they should be accumulated. Figure 1 illustrates also that during enterprise integration the requirements corresponding to enterprise goals should be

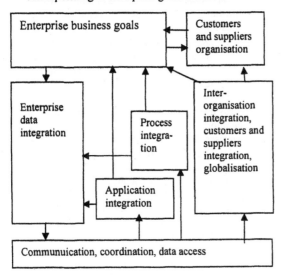

Fig. 1. Graphical presentation of requirements for IS changes.

understood and that within the enterprise IT/IS the following requirements should be met:

- enterprise data should be available for all relevant processes in the enterprise; data should accumulate; all data characterising the new integrated processes should be stored;
- flexible communication and coordination structures ensuring data exchange between the processes and

applications within the enterprise should be supported.

5. GENERAL FRAMEWORK FOR IT/IS CHANGES PREDICTION

Certainly, it is difficult to predict feasible IS changes during enterprise integration. This is determined by several reasons and factors. There are several trends in IT/IS changes (Grover, 2000B). They are as follows: re-informating, de-massifying, re-mediating, and interfacing. There are also the following organisational changes stemming from reengineering (Teng, 2000): (1) forming a new and better coordination between the processes, applications, and IT/IS users; (2) defining the metrics for the BP and IT/IS processes; (3) involving suppliers and customers in production processes; (4) maintaining continuous improvement.

Analysing changes which have occurred within enterprises in recent years the following dimensions of changes might be mentioned: communication and coordination, organisational structure, the data and the knowledge accumulated in the enterprise, and the production processes within the enterprise. Let us deal with each dimension separately.

Communication and coordination develops within the enterprise with the increase of the rate of data exchange and formation of new coordination structures. Development of technologies offers possibilities of fast access to and search of data in large databases. This opens up an opportunity for global and mobile users to have access to the enterprise data. According to Grover (Grover, 2000B) coordination expenses tend to be 0.

Organisational structure becomes more flexible. A transition from enterprise integration based on its structure to integration, which is based on information, is taking place. New solutions for the enterprise structure are applied; they are based on faster and easier communication, coordination, and shared data application that allows to shorten the time of decision making. The structure of the enterprise is designed in a way to simplify the inner control. This is reach through involvement of suppliers and users in the work of the enterprise.

The data and the knowledge accumulated in the enterprise are available for the users both within the enterprise and outside of it. With the increase of the number of IT/IS users also the data about the users, their groups, requirements, and motivation keep growing. Segmentation of the products or services of the enterprise occurs. The volume of knowledge increases as well, and data accumulation characterising the production processes takes place. This gives a better possibility to identify those parts of the processes, which add value.

The performance of the enterprise is divided into processes. Value adding activities are identified. The metric of the processes is defined and applied. With the purpose of improving the quality the expenses the suppliers are involved in the enterprise processes. The control over the processes is simplified. When developing simpler methods of control, the enterprise products or services change as well.

Trends of IS changes are presented in Figure 2. Development of IT/IS is shown by two intersecting lines, the movement of the lines is similar to that of scissors. Two lines enclose the four areas of enterprise IT/IS changes - on top - the organisational, at the bottom - performance, on the right - communication, coordination and the number of IS users; on the left - the knowledge of the enterprise, data, information, and metric of the processes.

The trend of IS changes is indicated by arrows 1, 2, 3, and 4. The first arrow indicates increase of the volume of data of the enterprise, the second one indicates the increase of communication and coordination of the enterprise, the third arrow indicates the growth of the volume of the data characterising enterprise performance, the fourth - indicates the increase of the number of enterprise IT/IS users.

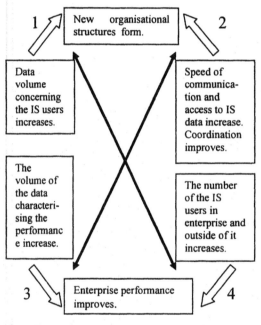

Fig. 2. Presentation of the trends of IT/IS changes

The framework presented in Figure 2 shows the following trends of the enterprise IT/IS changes:

- coordination within the enterprise and outside it improves;
- the number of IS users increases;
- the volume of the data in the enterprise increases, repositories of knowledge of enterprises are formed;
- new organisational structures are developed, which are based on better coordination and wider involvement of suppliers and customers;

- the volume of data characterising the production processes in the enterprise increases, the processes in the enterprise are improved through a wider use of data characterising the processes;
- the processes in the enterprise are improved by determining the value adding components. The suppliers and customers are involved in the production processes.

All these trends can be supported by EI. In order to get necessary result the most appropriate framework has to be chosen. The EI oriented project evaluation framework can be used for this purpose.

6. ENTERPRISE INTEGRATION ORIENTED PROJECT EVALUATION FRAMEWORK

IT/IS project evaluation is always important because successfully integrated applications are likely to show a higher return on investments (EI, 2002) and reduce risks (Wang, 2002). Appropriate evaluation framework may help to make right decision regarding the changes in the enterprise. SARAF is one solution that enables evaluation of project alternatives in terms of organizational and interorganizational feasibility (Stapleton, 2003). SARAF is an IT/IS evaluation framework applicable to all sorts of IT/IS projects. In order to get results of higher quality we expand this framework on evaluation factors specific to EI projects (see Section 4 and Section 5).

SARAF has three main dimension *organizational*, *task environmental*, *societal environmental*. We expand SARAF with one more dimension *technological* that includes evaluation factors related to system integration. In case of internal integration greater attention has to be paid to the evaluation factors of organizational dimension, in case of external integration – to task environmental and societal environmental dimensions. But in both cases there are some types of evaluation factors that have to be assessed regardless of type of integration: (1) ensuring of data reliability, comparativeness and unique interpretation (Balakrishnan, 1999); (2) necessary standards for communication and data exchange (Hohpe, (2002), McKeen, (2002), Wang, (2002)); (3) security (inside and outside an enterprise) McKeen, (2002); (4) easy of use McKeen, (2002). These evaluation factors were observed also in case study report describing the implementation of Enterprise Information Portal (Wojtkowski, 2004).

Organizational level
1) EI architecture has to be compatible with the following elements of the organization (Stapleton, 2003) that have been defined in the enterprise modelling process: *goals, structure,*

resources (*human, technical, finance*), *existing IS/IT, processes.*

2) EI architecture has to be compatible with following elements of the organization (Stapleton, 2003) that have been defined in the enterprise modelling process: *problem solving capacity, availability of the information.*

3) EI architecture has to ensure the following quality aspects of IT/IS solution: *quality of information in terms of data reliability, comparativeness and unique interpretation, security of the solution, easiness of use.*

Task environmental level

4) The chosen EI architecture have to be compatible with following elements of the task environment of organization (Stapleton, 2003) that have been defined in the analysis of the environment of organization: *customer IT/IS, supplier IT/IS, government IT/IS, requirements for information from the task environment.*

5) The following aspects have to be evaluated: *customers satisfaction, communication between the members of the task environment and organization, available standards, data formats.*

Societal environmental level

6) The chosen EI architecture have to be compatible with the following elements of the societal environment of organization (Stapleton, 2003) that have been defined in the analysis of the environment of organization: *technological progression, economical conditions, legal requirements, socio-cultural conditions.*

Technological level

7) The following aspects have to be evaluated in order to get information about technological suitability of the chosen EI architecture: *capabilities of the chosen technologies, vendors of chosen technologies.*

7. CONCLUSION

In this paper the results of investigation of the relationships between Enterprise Integration and changes in IT/IS of the enterprise have been presented. The results are exploited to create IT/IS change management framework and EI oriented project evaluation framework (expanded SARAF).

More research is needed to test these two frameworks in context of EI, in order to create better tools for managing IT/IS changes in EI context and probably discover additional evaluation factors to be included in the SARAF.

REFERENCES

Artim, J. M. (2000). Enterprise Frameworks and Enterprise Integration: A Usability Perspective. available at http://www.cse.unl.edu/~fayad/workshops/oopsla2000/ws/ws00-Artim.doc

Balakrishnan, A., Kumara, S.R.T., Sundaresan, S. (1999). *Manufacturing in the Digital Age: Exploiting Information Technologies for Product Realization.* Kluwer Academic Publishers, Boston.

Cassidy, A., Guggenberger, Keith. (2001). *A Practical Guide to Information Systems Process Improvement.* CRC Press LLC

Dignum, F., Houben, G.J. (1999). Integrating Information Systems: Linking Global Business Goals To Local Database Applications. *Journal of Integrated Design and Process Science.* March 1999, Vol.3, No.1,44 available at http://www.sdpsnet.org/journals/vol3-1/hoube.pdf

EI, (2002). Enterprise Integration: A Key To Successful Public Sector Customer Service (2003)– White Paper, Available at:http://motorolla.com/csr.

Grover, V., Kettinger, W. (2000A) *Process think: winning perspectives for business change in the information age.* Idea Group Publishing.

Grover, V., Kettinger, W. (2000B) *IT: The next 110010_2 Years. Process think: winning perspectives for business change in the information age.* Idea Group Publishing.

Harrington, H.J., Esseling, E. K.C., Van Nimwegen, H. (1997). *Business Process improvement.* McGraw-Hill

Hohpe, G. (2002). Enterprise Integration Patterns. Available at http://www.enterpriseintegrationpatterns.com/

Huang, K., Lee, Y. W., Wang, R. Y. (1999). *Quality Information and Knowledge* Prentice-Hall, Inc.

Juesfeld, M.A., deMoor, A. (2001). Concept Integration Precedes Enterprise Integration.In: *Proceedings of the Hawai'i International Conference On System Sciences,* Maui, Hawaii.

Kettinger, W.J., Teng, James T.C. (2000) Conducting Business Process Change: Recommendations from a Study of 25 Leading Approaches.

Markus, M.L. (2000). Paradigm Shifts-E-Business And Business/System Integration.In: *Communications of AIS,* Volume 4 Number 10. Available at http://www.ais.com

McKeen, J.D., Smith, H.A. (2002). New developments in practice II: Enterprise Application Integration . *Communications of the Association for Information Systems,* Volume 8, 2002, pp. 451-466.

RIA, (2002). *The Role Of Integration Architectures: Roll With The Changes* Deloite Consulting

SAS (2003). SAS Integration Technologies: A Roadmap. available at: http://support.sas.com/rnd/itech/doc9/overview/index.html

Sheratt, M. (2003). Aligning costs with revenues, *Financial Executive*, October, 2003, pp. 59-62

Smith, D., O'Brien, L., Kontogiannis, K., Barbacci, M. (2002). Enterprise Integration. available at: http://interactive.sei.cmu.edu/news@sei/columns/the_architect/2002/4q02/architect-4q02.pdf

Stapleton, L., Kirikova, M.;Sprice, R. (2003). Knowledge Models for Adopting Advanced Technologies in SMEs. In: *Preprints of the 8th IFAC Symposium on Automated Systems Based on Human Skill and Knowledge*, Gothenburg, Sweden.

Teng, James T.C., Grover, V., Friedler, K.D., Jeong, S.R. (2000) Initiating and Implementing Business Process Change: Lessons Learned from Ten Years of Inquiry. In: Grover, V., Kettinger, W., *Process think: winning perspectives for business change in the information age*. Idea Group Publishing.

Thompson, J.L. (1997). Strategic Management Awareness and Change - International Thomson Business Press.

Wang, C. (editor). (2002). An Integration Architecture For Process Manufacturing Systems. In: *International Journal of Computer Integrated Manufacturing*, Volume 15, Number 5., pp.413-426. Available at: http://www.tandf.co.uk/journals.

Wojtkowski, W., Major M. (2004). Enterprise Information Portals:Efficacy in the Information Intensive Small to Medium Sized Business. In: *Annals of Cases in Information Technology*, Volume 6, by Khosrow-Pour M.(ed). Idea Group Publishing.

Woolf, B., Brown, K. (2002). Patterns of System Integrtation with Enterprise Messaging. Available at http://www.enterpriseintegrationpatterns.com/

ELSEVIER
IFAC
PUBLICATIONS
www.elsevier.com/locate/ifac

A REFERENCE MODEL FOR ENTERPRISE-CONTROL SYSTEM INTEGRATION – OVERVIEW ON A STANDARDISATION INITIATIVE

David Chen

LAP/GRAI, University Bordeaux 1
351 Cours de la libération, 33405 Talence Cedex, France
Tel: (0)5 4000 65 30, Fax: (0)5 4000 66 44
Email: chen@lap.u-bordeaux1.fr

Abstract: This paper aims at presenting an overview of an on-going international standardisation initiative carried out by the joint working group JWG15 of IEC and ISO. It is concerned with the development of a multipart standard named IEC 62264: Enterprise-Control system integration. After a brief introduction on the background, goal and the approach, some hierarchy models are reviewed, which are used as a basis to define the scope of the standard. Then functional data flow model, object models and associated data attributes are outlined. The activity models of manufacturing operations management are discussed. Future works and conclusions are given at the end of the paper. *Copyright © 2004 IFAC*

Keywords: Enterprise integration, Enterprise control, Production Planning, Production Management

1. INTRODUCTION

1.1 Background of the work

The IEC 62264 multipart standard "Enterprise-Control system integration" is based on the work previously performed by ISA (Instrumentation, Systems and Automation Society) and published as a ISA standard. In 2001, a joint working group JWG15 of IEC/SC 65A and ISO/TC 184/SC5 was established to further develop the ISA standard into a joint IEC and ISO international standard.

Some enterprise models were used as a basis to develop the standard: The Purdue Reference Model for CIM; the MESA International Functional model; and the equipment hierarchy model from the IEC 61512-1 standard.

1.2 Problematic

It has been considered that shop floor coordination is a weak point in production management and control. There are many information/data flows between production control functions and other enterprise functions. Factory people have a lot of difficulties to manage these information flows because in most of the cases their origins, destinations and the content are not always clearly defined. Data reconciliation is therefore considered as a serious issue for enterprise-control system integration (IEC 62264-1, 2002). The data have to be valid to be useful for the enterprise systems. Any error in the data sent to production and from production may alter the performance of the manufacturing system.

1.3 The goal of the standard

The primary goal of this standard is therefore to define the interface content between the production control system and the rest of the enterprise in order to reduce the risk, cost, and errors associated with implementing these interfaces. In other words, the standard aims at specifying the data flows between enterprise's business systems and its manufacturing control systems. This allows to separate business processes from manufacturing processes and provides a clear demarcation of responsibilities and functions as well as a clear description of exchanged information (Brandl, 2001).

More specifically, IEC 62264 provides a standard terminology and a consistent set of concepts and models for integrating production control systems with enterprise systems that will improve communications between all parties involved (IEC 62264-1, 2002). Furthermore, a shared ontology in manufacturing domain can be derived from the definitions in this standard (DelaHostria, 2003) to improve interoperability of enterprise applications.

1.4 The integration focus

The IEC 62264 standard fits within the area of enterprise integration. Enterprise integration can be approached in various manners (Chen *et al.* 2004). CEN TC310/WG1 recognises three levels of integration: (1) Physical Integration (interconnection of devices, NC machines,... via computer networks), (2) Application Integration (dealing with interoperability of software applications and database systems in heterogeneous computing environments) and (3) Business Integration (co-ordination of functions that manage, control and monitor business processes). Michel (1997) considers that integration can be obtained in terms of: (1) data (data modelling), (2) organisation (modelling of systems and processes) and (3) communication (modelling of computer networks, for example the 7-layer OSI model). Integration can also be developed through consistent enterprise-wide decision-making. The IEC 62264 standard is concerned with integration by data i.e. data flows and data semantics.

1.5 The approach

The approach followed to develop the standard is illustrated in figure 1. At first, the domain of production control and the domain representing the rest of the enterprise are defined. Functions within each of the domains are then identified and functions relevant to this standard targeted (functions of interests). Information flows between functions of interests are then represented as functional data flow model. A categorisation of the information is proposed and related (information) object models defined.

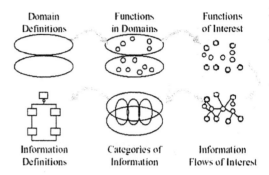

Figure 1. The approach (IEC 62264-1, 2002)

2. HIERARCHY MODELS

IEC 62264 starts by presenting some hierarchy models to define the scope of the standard.

2.1 Functional hierarchy

The functional hierarchy is shown in figure 2. It is based on the Purdue hierarchy model. The standard covers the level 3 and a part of level 4. Levels 2,1,0 are cell or line supervision functions, operations functions, and process control functions which are not considered by this standard.

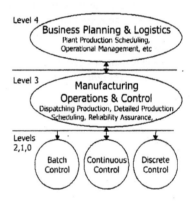

Figure 2. Functional hierarchy (IEC 62264-1, 2002)

At the current stage of development, the IEC 62264 standard contains three parts.

- Part-1 addresses the interface between levels 4 and 3 by identifying relevant functions in both levels and objects that are normally exchanged between level 3 and level 4.
- Part-2 is concerned with describing the attributes associated with the object models defined in Part 1.
- Part-3 defines activity models of manufacturing operations management at the level 3.

A possible work on part 4 is planned to deal with attributes of object models of manufacturing operations management defined in part 3.

2.2 Equipment hierarchy model

The equipment hierarchy model describes how production resources are usually organised (see figure 3).

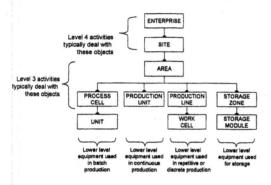

Figure 3. Equipment hierarchy (IEC 62264-1, 2002)

This standard is concerned with the interface between enterprise/site and area. Nevertheless some data flows between activities at level 3 are also described in Part-3 of the standard.

Remarks:
- Enterprise-control system integration can also be studied from the point of view of resource organisation in terms of organisation structure, responsibility and authority. This aspect is out of the scope and not covered by this standard.
- This equipment hierarchy is slightly different from the hierarchy defined in ISO TR 10314 (1991) which deals with a Reference Model for Shop Floor Production. In this ISO approach, resources are structured into six levels: enterprise, facility, section/area, cell, station, and equipment. The last four levels constitute the shop floor level.

2.3 Decision hierarchy model

There is also a hierarchy of decision-making activities involved in enterprise-control systems. The decision hierarchy model based on the GRAI decisional approach is illustrated in figure 4.

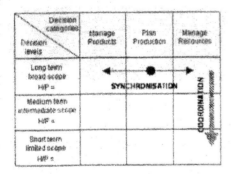

Figure 4. Decisional model (for illustration only)

The decision-making activities deal with products, resources and time. The different combinations of these lead to a categorization into three basic types of decisions: Manage products, Manage resources and Plan production. Decisions are also classified into three basic time horizons: Long term and broad scope that are concerned with the definition of the objectives consistent with the global objectives of the enterprise; Medium term and intermediate scope that deal with the allocation of resources to meet the objectives defined in the long-term time horizon; Short term and limited scope associated with the planning and execution of actions, using the means implemented at the medium-term time horizon, to reach the objectives of the long-term time horizon (CEN 14818, 2003).

Remarks:
- Long term decisions are mainly concerned with level 4 activities while short term decisions are at level 3. Medium term decisions cover part of levels 4 and 3 and can be considered as interface between the two levels.
- Enterprise-control system integration through consistent decision-making is not the focus of this standard and this aspect is not developed in detail.

3. FUNCTIONAL DATA FLOW MODEL

The objective of the functional data flow model is to identify the boundary between the production control domain and the rest of the enterprise (see figure 5).

The wide dotted line illustrates the boundary of the enterprise-control interface. This is equivalent to the level 3 – level 4 interface defined in figure 2. The labelled lines indicate information flows of importance to production control. The wide dotted line intersects functions that have sub-functions that may fall into control domain, or fall into the enterprise domain depending on organisational policies of a particular company. Indeed different companies may place the functions in different organisation groups (IEC 62264-1, 2002).

Part-1 of the standard describes in detail sub-functions of each function (bubbles) identified and information flows labelled in figure 3 as well as object models (see detail in IEC 62264-1, 2002).

4. OBJECT MODELS

4.1 Categories of information

Information exchanged between functions as identified in figure 5 can be categorised into three main classes (IEC 62264-1, 2002):

- Information required producing a product: this is 'Product Definition Information' dealing with 'how to make a product'
- Information on the capability to produce a product: this is also called 'Production Capability Information'. It answers 'what is available'

- Information about actual production of the product: also called 'Production Information'. It is concerned with 'what to make and results'.
Some information in each of these three areas is shared between the production control systems and the other business systems (see figure 6).

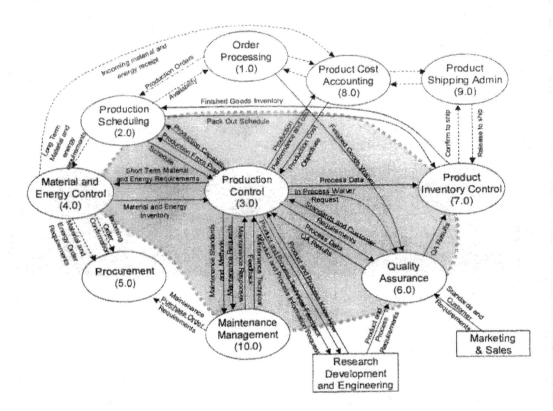

Figure 5. Functional enterprise-control model (Functional data flow model) (IEC 62264-1, 2002)

These three information categories are used to define formal information models which are detail enough for actual enterprise integration projects.

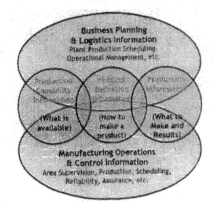

Figure 6. Areas of information exchange (IEC 62264-1, 2002)

4.2 Object model structure

The part-1 standard also presents an overview of the information contained in the object models. Object models aim at better defining the semantics (meanings) of the terms used in the standard.

The approach is as follows: Objects of interest (exchanged between identified functions) are identified. Relations between these objects are defined using UML notation. Attribute types associated to each object are described using templates.

For example, we will consider in this paper one of the three information categories: Production capability information. For this category there are also three main areas of information about the production capability that have significant overlap as shown in figure 7.

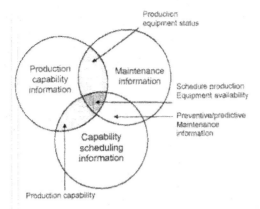

Figure 7. Production capability information (IEC 62264-1, 2002)

Four categories of resources are defined: (1) Personnel, (2) Equipment, (3) Material (and energy), (4) Process Segments.

A production capability is a collection of capabilities of personnel, equipment, material and process segment as illustrated in figure 8.

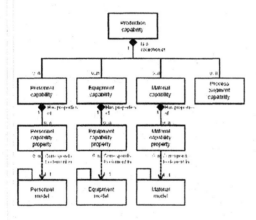

Figure 8. Production capability model (IEC 62264-1, 2002)

The Personnel model, Equipment model, Material model and Process segment capability model are then described in detail using UML notation. For example, the Personnel model is shown in figure 9.

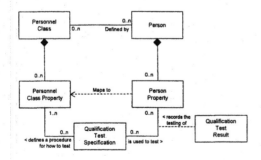

Figure 9. Personnel model (IEC 62264-1, 2002)

4.3 Object model attributes

Object model attributes are described in the part-2 of the standard. It adds details to the object models through the definition of attributes, so that interfaces that can be implemented may be constructed (IEC 62264-2, 2002). The objective is to define a minimum set of industry-independent information as attributes types. The attributes are represented in forms of tables of attributes associated to the object models elaborated in part-1 of the standard.

As an example, the personnel class defined in the personnel model shown in figure 9 is described in detail with attributes (see table 1).

Table 1 - Attributes of personnel class property (IEC 62264-2, 2002)

Attribute Name	Description	Examples
ID	An identification of the specific property, unique under the scope of the parent personnel class object. For example, the property "Has Class 1 Safety Training" (with values of Yes or No) may be defined under several different Personnel Class definitions, such as Fork Lift Operator and Pipe Fitter classes, but has a different meaning for each class.	Class 1 Certified
		Night Shift Available
		Monthly Exposure Hours Maximum
Description	Additional information and description about the personnel class property.	"Indicates the certification level of the operator"
		"Indicates if operator is available for night shift."
		"Indicates the maximum monthly exposure hours that can be used."
Value	The value, set of values, or range of the property. This presents a range of possible numeric values, a list of possible values, or it may be empty if any value is valid.	(True, False)
		(True, False)
		[0..20]
Value Unit of Measure	The unit of measure of the associated property values, if applicable	Boolean
		Boolean
		hours

Remark: Only attribute types are defined. Values for all attributes must be instantiated when using the models to a specific company.

5. ACTIVITY MODELS OF MANUFACTURING OPERATIONS MANAGEMENT

Activity models of manufacturing operations management are described in part-3 of the standard. It is concerned with the activities associated with manufacturing operations and control of the Level 3 functions. Some of the data that flows among these activities are also identified (ISA 95, 2003).

Functions concerned are: Production Scheduling (2.0), Production Control (3.0), Material and Energy Control (4.0) and Product Inventory Control (7.0) as shown in figure 5. Functions 2.0, 4.0 and 7.0 are only partially described. Quality Assurance (7.0) and Maintenance management (10.0) are also partly described.

5.1 Categories of operations

Four categories of operations are defined: (1) Production operations, (2) Inventory operations, (3) Quality operations, and (4) Maintenance operations. Inventory operations are concerned with both Material and Energy control (4.0) and Product Inventory control (7.0).

For each of the category of operations, an activity model is elaborated. Each activity is then described in terms of tasks. A generic activity model is defined to use for all four categories of operations.

5.2 Generic activity model of operations management

The generic activity model is shown in figure 10. The ovals indicate collections of tasks, identified as the main activities. Lines with arrowheads indicate an important information flow between the activities (ISA 95, 2003).

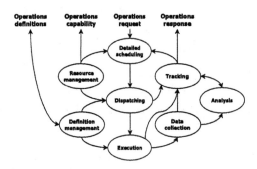

Figure 10. Generic activity model of operations management (ISA 95, 2003)

This generic model is used as a template to define the models of management for production operations, maintenance operations, quality operations, and inventory operations.

Remark:
- The term 'Operations' means categories of operations which correspond to the four categories defined previously. Each category specifies a set of activities. Within an activity, there is a set of tasks. A task can have sub-tasks.
- It was considered that it is a good level to standardize on the activity level, not on the task level since tasks are not completely the same in different industries.

6. CONCLUSION

This paper presented an overview of the IEC 62264 standard. Relating to ISO 15704 (1998), the IEC 62264 can be considered as a reference model (or partial model) of data flows between business domain and production control domain. It mainly focuses on function and information views defined in ISO 15704. The various models defined are detail enough so that interfaces may be implemented in industries. However the success of this standard largely depends on industry acceptance and use. Consequently, a strong dissemination and training actions are necessary.
Future work planned by JWG15 is to define data attributes of objects identified in the activity models of manufacturing operations management. Furthermore some preliminary discussions started within ISA SP95 to deal with business to manufacturing transactions that may be used on the objects defined in the object models.

ACKNOWLEDGEMENT

The author wishes to acknowledge the members of JWG15, Joint Working Group of ISO TC 184/SC 5 and IEC SC 65A, who have prepared the IEC 62264 multipart standard and ISA SP95 for their original contribution. Special acknowledgements are extended to Denis Brandl for his determinant position as JWG 15 convener and SP 95 active member.
This paper is not intended to represent the view of ISO/IEC JWG15 and ISA SP95. The author is to be taken responsible for the content of his paper.

REFERENCES

Brandl, D. (2001), Enterprise/Control system integration using the ISA95 standards, slides presentation, IEC/ISO JW15 meeting, Paris.

CEN prCEN/TS 14818:2003(E), Enterprise Integration – Decisional Reference Model, Technical Specification, ICS 03.100.01.

Chen, D. and Vernadat, F. (2004), "Standards on Enterprise Integration and Engineering: A state-of-Art", in International Journal of Computer Integrated Manufacturing (IJCIM), Volume 17, n°3, March-April, 2004, pp.235-253.

DelaHostria, Em (2003), 'Interoperability of Standards to support Application Integration', in *Enterprise Inter-and-Intra Organisational Integration*, Eds. K. Kosanke *et al.*, Kluwer Academic Publishers, ISBN 1-4020-7277-5.

IEC/FDIS 62264-1:2002(E), Enterprise-control system integration - Part 1: Models and terminology, 2002.

IEC/FDIS 62264-2:2002(E), Enterprise-control system integration - Part 2: Object Model Attributes, 2002.

ISA 95.00.03-2003, Enterprise/Control System Integration – Part 3: Activity Models of Manufacturing Operations Management, Draft 15, November 2003.

ISO TR 10314, 1991, Reference Model for Shop Floor Production Standards, Part 1 - Reference model for standardisation, methodology for identification of requirements, ISO TC 184 SC5/WG1.

ISO 15704 (1998), Requirements for Enterprise Reference Architecture and Methodologies, ISO TC184/SC5/WG1, N423.

Michel, J.J. (1997), Manufacturing, Modelling and Integration, Presentation at a meeting of the computer department at CETIM (slides).

ELSEVIER

IFAC

PUBLICATIONS
www.elsevier.com/locate/ifac

MANUFACTURING EXECUTION SYSTEMS (MES) BASED ON WEB SERVICES TECHNOLOGY

L. Canché[1], M. de J. Ramírez[2], G. Jiménez[3], A. Molina[1]

Integrated Manufacturing Systems Center[1]
Mechatronics and Automation Department[2]
Informatics Research Center[3]
Instituto Tecnológico y de Estudios Superiores de Monterrey
Av. E. Garza Sada 2501, 64849, Monterrey, Nuevo León, México
al785356@mail.mty.itesm.mx, miguel.ramirez@itesm.mx, guillermo.jimenez@itesm.mx,
armolina@itesm.mx

Abstract: The manufacturing plant environment is a mishmash of different systems that require a lot of integration and interoperability. This paper discusses how the application of Web services, self-describing chunks of software, leveraging XML and other Internet standards and protocols, are used to facilitate data sharing between dissimilar systems in the shop-floor. This solution could represent a lower cost alternative in comparison with other distributed technologies. *Copyright © 2004 IFAC*

Keywords: Web Services, Manufacturing Execution System, Low Cost Integration, Interoperability.

1. INTRODUCTION

Today, many manufacturers use a Manufacturing Execution System (MES) that replaces many of the manual shop-floor management functions. The continued strong growth of the MES market indicates that manufacturers indeed recognize and increasingly value these solutions (Fulcher, 2000). Currently, much of the production is still directed by the plant-floor control system, but higher-level decisions --such as revising production flow to meet changing shop-floor conditions-- are quickly calculated and passed down from the MES.

Hooking up a MES system with the rest of the enterprise is not an easy task (Vijayan, 2000) but it is relevant in enterprises with a mix of vendor platforms and legacy systems. The scenario involves multiple communication protocols and application programming interfaces (APIs) to connect to other applications such as ERP, Supply Chain, Product and Process Engineering, Sales Management, and Service Management. Integrating a MES into an enterprise wide system requires very tight connectivity between systems that are as varied as Order Entry Systems, Product Configurators, Planning and Control Systems, as well as Sales Force and Delivery Systems.

Each software component in an integrated enterprise should interface with other components sharing information and communicating in a common format. It represents a major issue and challenge to achieve integration and interoperability across whole shop-floor systems.

Industries need a mechanism that provides a quick and simple way of implementing interoperable and integrated infrastructures for manufacturing software systems. Recently, new architectures based on proven computing standards, called Web services, are being championed by Microsoft, Sun, IBM and other major technology providers, which promise to allow users to bridge these problem domains more easily (Sussman, 2002).

This paper presents a low cost approach to integrate MES systems in the shop-floor, and discusses how this technology makes possible to share real-time data from the plant floor with disparate manufacturing systems relatively easily. Its structure is as follows: in the second section, the concept of MES is presented. In the third section, an analysis of approaches and technology trends regarding MES is presented. The fourth section describes how Web services technology can solve the integration and interoperability problem in MES. Section five presents the conclusions.

2. MES CONCEPTS

Manufacturing Execution Systems (MES) can be defined as "tools for manufacturing management". They consist of one or more computer software applications that provide a real-time view of manufacturing operations while providing seamless integration of real-time data with information systems such as Production Planning, Inventory and Shop Floor Control Systems. Their real value results in data that is automatically converted into "actionable information" that can then be used for improved decision making at all levels of the manufacturing enterprise (operation, manufacturing management, engineering, IT, planning, materials management, etc.).

The Manufacturing Execution Systems Association International (MESA, 1997a), defines MES as a:

"Manufacturing Execution Systems (MES) deliver information that enables the optimization of production activities from order launch to finished goods. Using current and accurate data, MES guides, initiates, responds to, and reports on plant activities as they occur. The resulting rapid response to changing conditions, coupled with a focus on reducing non value-added activities, drives effective plant operations and processes. MES improves the return on operational assets as well as on-time delivery, inventory turns, gross margin, and cash flow performance. MES provides mission-critical information about production activities across the enterprise and supply chain via bi-directional communications."

Benefits of an MES implementation include automating the flow of information between MRP and accounting systems and providing a foundation for integration with future finite scheduling or capacity planning systems. These systems are notoriously "information hungry", and run on diverse software presenting an opportunity to reduce the cost associated with obtaining this information, while at the same time increasing its timeliness and accuracy, ultimately enabling better production decisions to be made and more accurate customer commitments to be made by the sales and marketing organization.

MESA recently has conducted research concerning actual benefits that manufacturers have obtained by implementing MES. This study titled "Report From The Field" (MESA, 1997b) reports the following results:

Reduced manufacturing cycle time by an average of 45%; reduced data entry time, usually by 75% or more reduced Work in Progress (WIP) an average of 24%; reduced paperwork between shifts an average of 61% reduced lead time by an average of 27%; reduced paperwork and blueprint losses an average of 56%; reduced product defects an average of 18%.

Finally, an MES may play a critical role in a company's ability to achieve and maintain efficient operation due to the functions that many MES support. This often includes core functions such as planning system interface, work order management, workstation management, inventory tracking and management, material movement management, data collection and exception management. Additionally, MES supports other functions such as: maintenance management, time and attendance, statistical process control, quality assurance, process data/performance analysis, document/product management, product trace ability and supplier management (McClellan, 2000).

3. APPROACHES AND TECHNOLOGY TRENDS

MES is a relatively new concept, introduced in the 1990's and the MES software systems (MESA 1997, Huch 2000, Miklovic 1999) are either mainly based on traditional technology. Nowadays most of the MES software systems on the market are characterized by inflexible, rigid and monolithic architectures (Boyle 1999). Although most systems come with a user friendly, graphical specification and configuration interface to configure the software for a particular manufacturing system through the application programming interfaces (APIs), they usually fail to address special features which are out of their limited scope.

MES is essentially an integrating set of functions providing links between Planning Systems used in Strategic Production Management (such as ERP) and Manufacturing Floor Control. So the APIs and data transport or communications mechanisms are, in some ways, a core piece of MES functionality, not just an incidental technical detail (MESA, 1997a). Due to this, new approaches have been proposed like integrated MES framework (Choi et al., 2002), a component-based application framework for MES (Füricht et al., 2002) and distributed and object-oriented MES framework (Cheng et al. 1999).

These proposals make emphasis on the use of two different distributed object technologies: CORBA and COM/DCOM (Windows-DNA). Common Object Request Broker Architecture specified and developed by the Object Management Group. Microsoft's solution is Windows Distributed internet Applications. Windows-DNA is an application infrastructure for creating distributed multi-tier computing solutions. These approaches follow the direction suggested by MESA (MESA, 1997) and OMG (OMG, 1999) in its Future Technology Model, depicted in Fig 1.

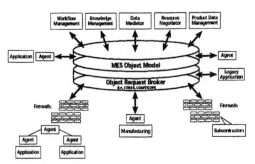

Fig.1 Future Technology Model (MESA, 1997b).

According to Fig.1, MES is incrementally evolving towards a consistent object model, along with the rest of the software industry. In this future information systems model, MES uses an object request broker to pass manufacturing events to workflows, agents, and external systems (SCM, ERP, Legacy, SSM, P/PE, Controls, Data warehouse) through data mediation.

Unique plant business policy is represented as sets of rules within knowledge management which can initiate manufacturing events.

Distributed Component Object Model (DCOM) is a set of Microsoft concepts and program interfaces in which client object can request services from server object on other computers in a network (Cauldwell et al., 2001). DCOM is based on the COM framework. Using DCOM interfaces, the client object can forward a Remote Procedure Call (RPC) to the application server object which provides the necessary processing and returns the result. This can then be sent out in the response object. However COM/DCOM did not solve the problem of the industry as it needed an architecture for creating interoperable software. DCOM failed because it cannot work through firewalls and because it works only in Windows platforms (Skonnard, 2001).

Common Object Request Broker Architecture (CORBA) is an architecture and specification for creating, distributing, and managing distributed program objects in a network. It allows programs at different locations and developed by different vendors to communicate in a network through an interface broker (Ayala et al., 2002). CORBA was developed by a consortium of vendors through the Object Management Group (OMG).

The essential concept in CORBA is the Object Request Broker (ORB). It means that a client program can request services from a server program or object without having to understand where the server is located in a distributed network or what the interface to the server program looks like. To make request or return replies between the ORBs, client programs use the General Inter-ORB Protocol (GIOP) and, for the Internet its Internet Inter-ORB Protocol (IIOP). IIOP maps GIOP request and replies to the Internet's Transmission Control Protocol (TCP) layer in each computer.

Notably, CORBA and Microsoft support a gateway approach so that a client object developed with the Component Object Model (COM) will be able to communicate with CORBA server and vice versa.

However, CORBA is not a mature technology (Asaravala, 2002), because after so many decades of computing, the industry has not yet achieved the complete maturity and is still evolving. Though CORBA is an excellent architecture for the development of distributed applications, it does not solve all the problems in this area. It has several limitations such as: Uses IIOP (Internet InterORB Protocol) for communication between client and server. This requires the use of custom marshalling and security code and restricts the interoperability in the long run. IIOP may require puncturing firewalls in the organization. CORBA implementation is usually complex, expensive and requires special skills to design, install and maintain.

The problems involved in implementing business systems based on these technologies have often been difficult, expensive, and time consuming and this ultimately led to emergence of Web services.

4. APPROACHING MES WITH WEB SERVICES TECHNOLOGIES

The Web services set of standards and technologies represent the evolution of past distributed component technologies like remote procedure calls (RPC), ORPC (DCOM, CORBA, Java RMI), messaging (MSMQ, MQSeries), and even modern Web applications (like Google.com). Because RPC was so difficult, developers layered object facades over the RPC mechanism to hide complexity. This led to the many flavors of object-oriented remote procedure calls (ORPC). Of course, not all application communication can be modeled with RPC, so other messaging paradigms (like those provided by MSMQ) were necessary. Then, over the years, as developers struggled with interoperability between the various ORPCs and messaging systems, they turned to the evolving Web as a potential solution for these challenges (Skonnard, 2001).

Web services are a form of service oriented architecture (SOA), intended to enable developers to create components that can be assembled and deployed in a distributed and heterogeneous environment. As their name implies, Web Services consists of a set of core standards that are based on existing Internet and Web standards. The three primary Web Services standards (Roy, 2001) are the Web Services Description Language (WSDL), Simple Object Access Protocol (SOAP), and Universal Description, Discovery and Integration (UDDI). WSDL is an XML-based language that is used to describe both the business and technical characteristics of Web services and how to interact with them. SOAP is a messaging framework that utilizes the HTTP as its default transport mechanism to enable communication between client and server application programs. SOAP messages are created using XML which provides a flexible method for creating structured, self-describing messages (Albornoz, 2002). In the original architectural model for Web services, UDDI was intended to serve as a public directory service where WSDL documents could be stored for public access by external parties. In practice, many organizations are currently deciding to limit public access to information about the services they offer and have created private UDDI directories or use alternative mechanisms for publishing WSDL documents and service binding information. Both SOAP and WSDL have been submitted to the World Wide Web Consortium (W3C) for standardization. W3C is supporting their continued development and has published their specifications. The UDDI directory was originally developed under the auspices of a consortium called uddi.org. In July of 2002, development work for UDDI was transitioned to the OASIS standards consortium. Each of these standards relies on existing standards-based protocols and methods such as HTTP and XML.

Like their predecessors, Web services are intended to enable application programs to communicate and share functionality across networks. However, unlike their predecessors, Web services are designed to leverage the standards-based architecture of the Internet and World Wide Web to promote interoperation across a wide variety of computing platforms. Web services provide an improved environment for distributed computing because they enable a loose coupling between client and server applications that support more dynamic and flexible integration of application modules.

The Web services architecture is based upon the interaction between three roles: service provider, service registry, and service requestor (IBM, 2001). The interactions involve publish, find and bind operation. Figure 2 shows, the components providing these operations and their interactions. A services provider hosts a network accessible Web service (a software module with specific functionality), which is an implementation of a Web service. The service provider defines a service description for the Web services and publishes it to a service registry. The service provider can also publish the description to other places, including directly to a service requestor, if desired. The service requestor issues a find operation to retrieve the service description locally or from the service registry, and uses the service description to bind with the service provider. The requestor then binds with the service provider, and invokes or interacts with the Web service.

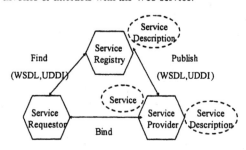

Fig.2 Web services Architecture (IBM, 2001)

Various software vendors are introducing Web service application environments such as IBM's Websphere, Microsoft's .NET, or Sun Microsystem's SunONE. The open source community also has Web services initiatives such as DotGNU and OpenSOAP under development. All of these environments promise the following benefits (Lubinsky, 2002):

Better interoperability: By minimizing the requirements for shared understanding. XML-based Web Services Description Language (WSDL) and a collaboration and negotiation protocol (UDDI) are the only requirements for shared understanding between a service provider and a service requester. By limiting what is absolutely required for interoperability, collaborating Web services can be truly platform- and language-independent.

Just-in-time integration: Collaborations in Web services are bound dynamically at run time. A service requester describes the capabilities of the service required and uses the service broker infrastructure to find an appropriate service. Once a

service with the required capabilities is found, the information from the service's WSDL document is used to bind to it. Dynamic service discovery, invocation (publish, find, bind) and message-oriented collaboration yield applications with looser coupling, enabling just-in-time integration of new applications and services. This in turn yields systems that are self-configuring, adaptive, and robust with fewer single points of failure.

Reduced complexity by encapsulation: All components in Web services are services. What is important is the type of behavior a service provides, not how it is implemented. A WSDL document is the mechanism to describe the behavior encapsulated by a service. Encapsulation is key to:

- Coping with complexity. System complexity is reduced when application designers are free from worrying about the implementation details of services they are invoking.

- Flexibility and scalability. Substitution of different implementations of the same type of service, or multiple equivalent services, is possible at runtime.

- Extensibility. Behavior is encapsulated and extended by providing new services with similar service descriptions.

Better interoperability of legacy applications: By allowing legacy applications to be wrapped in WSDL and exposed as services, the Web services architecture easily enables new interoperability between these applications. In addition, security, middleware, and communications technologies can be wrapped to participate in a Web service as environmental prerequisites. Directory technologies, such as LDAP, can be wrapped to act as a service broker. By wrapping the underlying plumbing (communications layer, for example), services insulate the application programmer from the lower layers of the programming stack. This allows services to enable virtual enterprises to link their heterogeneous systems as required (through http-based communications) and/or to participate in single, administrative domain situations, where other communication mechanisms can provide a richer level of functionality.

All these make Web services an easy, quick and cost effective integration medium. Web Services can provide the infrastructure to support a whole manufacturing system and allow hooking an MES into an enterprise wide system. MES has functions that support, guide, and track each of the primary production activities, core functionalities are (Meneses, 2002, MESA, 1997a):

Process Routing Management. As inputs, it receives a State of Resources report (this report, when displays critical numbers, acts as a trigger for production stop). Besides that, it receives the basic information related with Work Orders, including Order's Priority Hierarchy, and some other features like Size, Product ID, etc.

ERP's Info Receiving or Planning System Interface. This functionality acts as an information gate, without significant info processing operations. Strategic information received includes Tentative Due Dates, BOP's (Bills of Processes, according to the received Work Order), BOM (attached Bill of Materials for the Work Order) and, besides that, it receives the approvals for the MES Scheduling Simulations.

Scheduling Simulation. It acquires all the information related with the updated state of resources (such resources are labor force, materials and equipment availability) and, through a simulation procedure, it delivers forecasted state of resources after work order completion and finishing time.

Exception Management. This functionality is considered as a subsection of Process Routing Management. It manages all pre-programmed routines and algorithms required when there is a non-normal condition in the system, such as material scarcity, a down machine, status or date change in a work order, among others.

Performance Analysis. It receives information from Time and Attendance during working sessions, from Material Control as a report of finished goods and reworks, from Quality Management as dimensional or featuring non conformities and, from Process Routing Management as the forecasted Work Order time and the real processing time.

Quality Management. Receives information from Quality Control Variables from CTQ (Critical to Quality Variables) -in-Product and in-process Measurement Devices. Its outputs are sent, mainly, to the Performance Analysis Functionality and to the user's interface through a Global Quality Report.

Time and Attendance. Acquires data related to Human Resources Availability, in order to generate the adequate updated information related with Operative Labor Force and its performance.

Control of Materials. It plays a vital role. This functionality doesn't only counts and tracks inventory, but it is in charge of all equipment related with Material Handling and Movement.

Operation Interface or Data Collect and Supply Interface. As the ERP's Info Receiving or Planning System Interface, this functionality acts as an information gate, without significant information processing operations.

Under a web services approach, these functionalities become a collection of Web services that work together to control the information flow between different manufacturing systems. When information is required, the application client invokes remote functions on the server to send and receive data using SOAP. Also a single Web service could serve two o more distinct applications. For example, scheduling system could use a Web service to check the status of

a work order before planning a new job and the same service is used to show a costumer the state of the job through a Customer Relationship Management system.

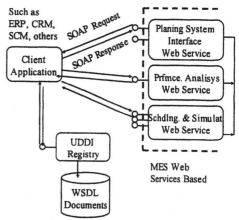

Fig.3 Conceptual model Web services Based

Figure 3 shows the conceptual model of a MES based on Web services. The architecture of Web services is an innovative concept that is intended to reduce several existing constraints that have reduced the utility of previous distributed computing technologies. Intrinsic capabilities of Web services are loose coupling of application components, separation of concerns, and interoperability across multi-vendor environments. These capabilities can provide significant advantages and benefits for integrating MES in the manufacturing environment.

Gaxiola, et al (2003) explains a first approach to this focus in a MES Holon, addressed to integrate systems in small and medium manufacturing enterprises where the use of commercial management software is reduced, due to high cost and poor adaptability to SMEs from developing countries.

When considering Web services with application integration, there are some missing pieces. For example, Web services don't provide the mechanism to leverage user interfaces. The Web Services User Interface (WSUI) initiative, announced in June 2001, continues to move towards a solution for this problem, but the technical obstacles are significant. In addition, current Web services do not address security well, lacking support for authentication, encryption, and access control. Indeed, Web services do not have the ability to authenticate publishers or consumers of the Web services.

The XML-Based Security Services Technical Committee from the Organization for the Advancement of Structured Information Standards (OASIS) is looking to shore up security within Web services with the Security Assertion Markup Language (SAML). This security standard allows organizations to share authentication information among those they wish to share Web services with as partner organizations. Other emerging security standards include the XML Key Management

Specification (XKMS), based on Public Key Infrastructure (PKI).

5. CONCLUSIONS

This paper introduced the concept of Web services technology and shows why Web services are a simply and practical extension of distributed computing technologies that have been under development for several years.

Also, discusses how Web services could solve the integration and interoperability problems faced when implementing MES systems. Web services would constitute an infrastructure for the quick and simple implementation and deployment of manufacturing software systems.

The conceptual model of MES based on Web services is shown and explains advantages of implementing with Web-based standards and why they constitute a quick and cost effective way of integration for data exchange.

For future works, this concept, the potential deployment of Web Services and related service-oriented architectures could be extended to meet the strategic goals of the increasingly virtualized manufacturing enterprise.

ACKNOWLEDGEMENTS

The research reported in this paper is part of a Research Chair in Mechatronics from ITESM titled "Design, Manufacturing and Integration of Reconfigurable and Intelligent Machines". The authors wish to acknowledge the support of this grant in the preparation of the manuscript.

REFERENCES

Albornoz J. (2002), *Understanding the intricacies of SOAP*, Software Engineer, Advanced Internet Technology, IBM, October 2002

Asaravala A., (2002) *Can public Web services work?*, New Architect, Vol. 7, No. 11, Page Start 34. Nov 2002.

Ayala D., Browne C., and Chopra V, Other (2002): *Professional Open Source Web Service*. Wrox Press, United State.

Boyle W., (1999): *Distributed manufacturing execution framework*. In Domain-specific Application

Cauldwell P. Chawla R. and Chopra V., Other (2001). *Professional XML Web services*, Wrox Press, Spain.

Cheng F., Shen E., Deng J., Nguyen K (1999): *Development of a system framework for the compute r-integrated manufacturing execution system: a distributed object-oriented approach In Domain-specific Application*. Int. J. Computer Integrated Manufacturing, 1999, Vol. 12, No. 5, 384 -402.

Choi B.K. and Kim B.H. (2002) *MES (manufacturing execution system) architecture for FMS compatible to ERP (enterprise planning system)*, Computer Integrated Manufacturing, 2002, VOL. 15, NO. 3, 274–284

Fulcher J. (2000). *The right tool for the job*, Manufacturing Systems, pages A6-A8, Apr 2000.

Fürieht R.,Prähofer H., Hofinger T. and Altmann J.(2002) *A Component-Based Application Framework for Manufacturing Execution Systems in C# and .NET* ,This paper appeared at the 40th International Conference on Technology of Object-Oriented Languages and Systems (TOOLS Pacific 2002), Sidney, Australia.

Gaxiola L., Ramírez M. de J., Jiménez G. and Molina A. (2003). *Proposal of Holonic Manufacturing Execution Systems Based on Web Service Technologies for Mexican SMEs*. Procedings of Holomas 2003 in LNAI 2744, pages 156-166, Prague, Czech Republic, Sep 2003.

Huch, B. (2000) *Analysis and Optimization of processes in manufacturing trough Manufacturing execution systems*. Diploma thesis (in German). Technical University Graz, Austria.

IBM (2001) *Web Services Conceptual Architecture (WSCA 1.0)*, IBM Software Group, May 2001

Lubinsky B. and Farrell M. (2002).*Web Services and Distributed Component Platforms, Part 2*, Web Services Journal, Vol. 1, No.3

McClellan (2000): Introduction to Manufacturing Execution Systems. MES Conference & Exposition, June 12-14; Phoenix, AZ, USA (2000) 3 – 11

Meneses. (2002). *Diseño e implementación de un sistema MES Flexible y de bajo costo para la mediana empresa*, MSc. Thesis in ITESM Monterrey, May 2002

MESA. (1997a). *MES Explained: A High Level Vision*, MESA, Sep 1997.

MESA. (1997b). *The Benefits of MES: A Report from the Field*, MESA, May. 1997.

Miclovic, H. (1999): *MES Market: Spectacular but short lived*. Monthly Research Review, Gartner Group.

OMG (1999) *NIST Response to MES Request for Information*. OMG Group

Roy J. and Ramanujan A. (2001), *Understanding Web Services*, IT Pro IEEE Computer Society, Nov 2001

Skonnard A. (2002) The Birth of Web Services, MSDN Magazine, Oct 2003

Sussman D. (2002). New glue for age-old gaps, MSI, pages A6-A8, Sep 2002.

Vijayan J. (2000). *Manufacturing execution systems*, ComputerWorld, page 38, Jul 2000.

DESIGN METHODOLOGY FOR CNC APPLICATIONS BASED ON OPEN SYSTEMS

*Miguel de J. Ramírez C., **Arturo Molina G., ***Guillermo Jiménez P., ****María A. Noriler

*Mechatronic Department, ITESM Campus Monterrey,México. miguel.ramirez@itesm.mx
**Manufacturing Integrated Systems Center, ITESM Campus Monterrey, México. armolina@itesm.mx
***Informatic Research Center, ITESM Campus Monterrey, México. guillermo.jimenez@itesm.mx
****Student of Control and Automation Engineering, Federal University of Santa Catarina, UFSC, Brazil.

Abstract: This document presents a design methodology for building applications based on the Universal Numerical control (UNC) reference architecture based on open systems developed on previous research works. UNC architecture map the openness concept to the needs of countries in development which justification emerges from the need of acquiring low cost CNC technology and making retrofit of manual machine tool. The methodology discusses guidelines for the design documentation and mainly addresses the kind of information needed using design notations such as UML. The project was developed under a collaboration agreement between ITESM in Mexico and the Santa Catarina University in Brazil. *Copyright © 2004 IFAC.*

Keywords: CNC, Control Design Methodology, Open System Controller.

1. INTRODUCTION

The Research Chair on Mechatronic of the Monterrey Institute of Technology (ITESM) has developed the concept of Universal Numerical Control (UNC) based on open system concept to develop a control architecture. The main aim of an open architecture control is the easy implementation and integration of customer-specific controls by means of open interfaces and configuration methods in a vendor-neutral, standardized environment (Jovane, 1998) (Binder, 1996). With an open architecture is possible design and implement systems formed with different combinations of software modules that can plug and play on the architecture to reach the needed functionality (Ramírez, 2002).

The rationale of UNC project is related with the fact that in countries like Mexico, small and medium metal-working industry rarely has access to computer numerical control (CNC) machine tools and then they work with conventional machine tools (Ramírez, 2001). On the other hand, only well established companies have access to automated machine tools due to its cost. Also the research group has pointed out that Mexican small and medium metal-working industry has been facing major problems in international markets due to the lack of CNC technology which is the basis for competition in this sector (Ramírez, 2002).

Therefore an area of opportunity has been identified to develop national technology in the area of CNC using low cost PCs, object oriented programming technologies and software based open system architecture implemented with real-time operating systems.

Pritschow et al (2001) mention that to achieve a re-configurable and adaptable control the internal architecture of the control system is based on a platform concept. The main aims are to hide the hardware-specific details from the software components and to establish a defined but flexible way of communication between the software components. An application programming interface (API) ensures these requirements. The whole functionality of a control system is subdivided into several encapsulated, modular software components interacting by the defined API. (Pritschow et al, 2001).

The figure 1 show the concept reached by UNC in order to do retrofit on diferents machine-tools, this concept get a library of software modules which can integrate to UNC reference architecture that is see like a "Black Box". The customer can choose the specific sofware modules to do his own CNC controller (Ramírez, 2001).

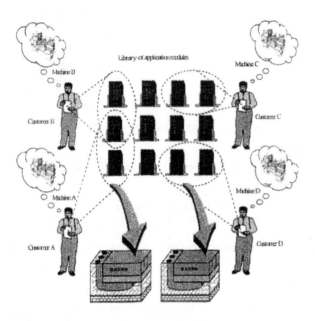

Figure 1. The UNC application concept.

The UNC concept will offer a system software layers which will serve as a link between applications software and hardware. It will offer an exactly specified application interface which will provide standard services in the fields of communication, data storage, graphics, dialogue management, configuration and operating system.

This document shows a design methodology for developing software modules that fulfill the UNC reference architecture requirements.

2. PROJECT DEVELOPMENT

2.1 UNC Reference Architecture

Reference architecture is a conceptual model that specifies rules and methods for the integration of the components of a system by means of standardized interfaces to achieve a structure that permits different developers to build systems with characteristics specified by the architecture (Ramírez, 1998). The UNC architecture must present a modular software structure to admit a division of the CNC functions and to give flexibility to the controller platform as there are a lot of types of machine tools like: lathes, milling machines, drilling machines, etc. Moreover, these machines use different types of components as: step motors, DC motors, AC motors, a large variety of sensors, screw, etc.

Figure 2 shows the UNC architecture which belongs to the classification of PC-based open CNC control systems that have Man Machine Interface openness as Kernel openness. UNC reference architecture is a multi layer system grouped in a hardware and software platform that provides services to the application software functional units (FU's). The hardware system consists of the hardware layer which includes electronic components that make part of the controller (motor drivers, sensors, etc.) as the circuits of the computer and its peripherals, such as I/O and network boards. The software system

contains the operational system layer, communication layer and the application program interface (API).

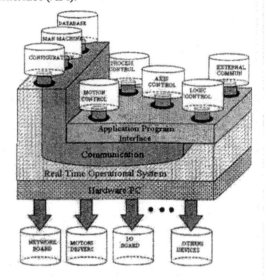

Figure 2. UNC Reference Architecture.

The basic application FU's defined are: Motion Control (MC), Axis Control (AC), Logic Control (LC), Process Control (PC), External Communication (EXC) and Simulation (SIM). The system FU's (that provide services for the application FU's) are: Man-Machine Control (MMC), Database (DB) and Configuration (CONF). Each FU is shortly explained below (Ramírez, 1998):

a) Man-Machine Control (MMC): represents the machine, or part of the machine, to the external entities, such as the operator, the supervising system over network, and CAD/CAM systems for the machine, etc., and allows these entities to control the operation of the machine. It could further provide utilities for preparation of jobs for the machine as well as general utilization and management of the machine and the computing resources.

b) Motion Control (MC): enables the machine to produce relative motion of a given degree of freedom. It does so by issuing appropriate commands to the axis controls. It is responsible for calculating the trajectories that must be executed.

c) Axis Control (AC): includes all the means necessary for activating the axis to execute movement. It is responsible for generating the signal tables and sends it to each axis of the machine tool.

d) Process Control (PC): represent, when needed, the auxiliary systems of the machine, such as spindle, coolant, lubrication system, etc. They are also responsible for the data management and processing of the auxiliary systems they represent.

e) Logic Control (LC): responsible for the operation of the actuators and data taking of the sensors built in the machine. They should also ensure the consistence of the operations and the data taken from the process.

142

f) External Communication (EXC): provides the communication of the system to the exterior by an industrial net of communication or load programs by a computer (Direct Numerical Control, DNC).

g) Simulation (SIM): permits the machining simulation.

2.2 Design Methodology

Based on the concept of the Universal Numerical Control reference architecture and its Functional Units, this paper discusses the necessity of developing a simple and applicable methodology and proposes it for building CNC applications over the UNC platform. The design methodology must reflect the guidelines of the reference architecture, such as:

a) Must permit the representation and the implementation of the software by a modular way or in parts in order to fulfill the modularity requirement;

b) Must achieve the concept of software reusability in order to facilitate building applications and shorten its development time;

c) Must have an abstraction of the real world in order to represent all the elements of the environment where the control systems interact;

d) Must permit the right establishment of the relationship and interactive information among different functional units in order to represent the architecture topology and the application program interface of the control system;

e) And finally must permit a high degree of detail documentation in order to facilitate for the application programmer understand the system and develop the applications.

The proposed methodology presents the three main phases of a software development that represent different aspects of the system:

a) Analysis: discusses what the system does and provides a clearly understanding about the system requirements;

b) Design: discusses the architecture and structure of the system;

c) Implementation: codify the design in a programming language.

d) Software Integration.

The UNC design methodology is shown in figure 3.

2.2.1 Analysis. The analysis phase is an iterative and co-operative process of analyzing a problem, documenting the resulting observations, and checking the accuracy of the understanding gained. It involves technical concerns of how to represent the requirements.

This phase serves as a starting point to the next phase, the design, and consists of the following steps: System Requirements Specification, System Design and FU's realization.

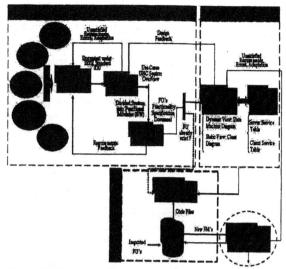

Figure 3. Design methodology for building CNC applications over UNC platform.

2.2.1.1 System requirements specification. The requirement specification is the starting point of the analysis phase. It reflects the mutual understanding of the system requirements to be developed between the people involved. Eventually, the system delivered will be assessed by testing its compliance with the requirement specification.

For the requirement specification some factors must be analyzed:

a) User's Needs: It refers to the requirements that the user formulate about the use and the application environment of the machine tool, such as quality of the required production, product machining time, user interface functions, etc.

b) Process Characteristics: It involves the specific aspects of the machinating process (lathing, milling, drilling, etc.) that the machine tool will do, such as tool and piece geometry, cut parameters, tolerances, surface finish, etc.

c) Machine Tool Characteristics: It refers to the mechanical aspects of the machine tool that must be analyzed for its control, such as machinating area, actual state of the mechanical structure of its axis motion system, rigidy, screw pitch gage, etc.

d) Electrical and electronics devices: It includes the characteristics of the electrical components, like alternated current motors, direct current servomotors, step motors, etc. And includes the electronics components such as sensors, drivers, personal computer, etc.

e) CNC Basic Functionality: It refers to the basic functionalities of a CNC system, such as: motion control, axis control, etc.

The result of this step is documented in the requirement specification. As a guideline for the contents of a requirement specification the methodology follows IEEE Standard 830. This standard does not give a rigid form for the requirements specification but has a structure which adheres to the constraints needed.

2.2.1.2 Use cases overview. Analysis information can be cleared or additional information can be represented using the use case diagram language defined in UML (Unified Modelling Language). This diagram allows the designer to complete the Requirements Specification with information about the actors and their interaction with the system. Use case diagrams capture a broad view of the primary functionality of a system in a manner that is easily understood by a non-technical audience.

The use case view captures the behavior of a system, subsystem, or class as it appears to an outside user. It partitions the system functionality into transactions meaningful to actors-idealized user of a system. The pieces of interactive functionality are called use cases. A use case describes an interaction with actors as a sequence of messages between the system and on or more actors. The term actor includes humans, as well as other computer systems and processes. Each use case is shown as an ellipse with the name of the use case. The use cases are enclosed by a rectangle denoting the system boundary. An actor that initiates or participates in a scenario is shown as a stick figure with the name of the actor below. The output of this step is a use case diagram view of the system.

2.2.1.3 FU's realization. Considering the amount of functionality which has to be included in each FU field, it becomes clear, that a further subdivision is necessary. This is the only way to substitute or complete part functionalities.

This step of the methodology just specify clearly the functionality of each FU needed, as in order to design, define their architecture and their interface parameters properly, the process and functionality of each FU must be analyzed.

It describes the decomposition of each FU into modules: (attributes) identification, type, purpose, function and subcomponents. Provides the information about the function, use of each FU and its modules and superficially its interfaces to others functional units. The output of this step is a document describing the modules, subcomponents of each FU and its functionality.

2.2.2. Design. In the design phase the architecture of the system is devised in terms of its components and interfaces between those components. The design phase results in a specification as well: a precise description of the design architecture, its components and its interfaces. This phase serves as a starting point to the next phase, the implementation, and consists of the following steps: FU's Architecture Specification and Design and API Establishment.

2.2.2.1 FU's architecture and design. This step is responsible for the specification and design of each functional unit. In order to fulfill the requirements of the reference architecture and facilitate the application programmer understanding two views of

different levels and areas were chosen to represent each FU: static and dynamic view.

Static view models concepts in the application domain, as well as internal concepts invented as part of the implementation of an application. This view is static because it does not describe the time-dependent behavior of the system, which is described in the other view. The diagram constituent of this static view chosen to be used in this methodology is the class diagram.

Dynamic view describes the behavior of a system over time. Behavior can be described as a series of changes to snapshots of the system drawn from the static view. About this view, for the design documentation of this methodology will be used the state machine diagram.

This view and diagram were specially chosen because FU's are described by the functionality that it provides and a state machine that determines its behavior. Some states are mapped to the logic channels of the communication interface servers (see API Establishment). Those states will be discharged by the clients FU's. The specification of each UF define the states that make it up, however, the figure 4 are shown three states that will always be presented inside each FU (Ramírez, 2002):

a) Active State: This state initializes the state machine and is activated by the configuration FU at boot-up. The variables are initialized and other processes for the normal operation are run.
b) Normal Operation: Inside this state are involved all the states defined to fulfill the functionality of the FU (that must be followed by the application programmer).
c) Error: Any state can lead the flow to the error state. This happens when the FU had a processing or a logic error in its operation.

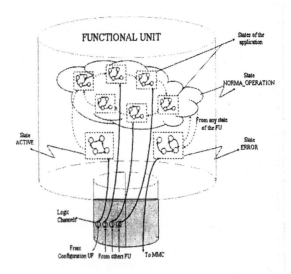

Figure 4. FU's State Machine.

2.2.2.3 API establishment. Besides the class diagram and the state diagram that specify the structure of each FU, must be defined a service table (see Table 1) that the servers provide. In this table must be indicated the Server Channel, FU Client, Communication Class (CC), Message Name, State and the Service Description.

Table 1. Server Table.

There are three types of communication class: variable for reading and writing data; process that triggers actions in server state machine; and events that are used for unsolicited sending of events and reports (from servers).

The state refers to the server (and client) state depending on the type of the communication class. If the communication class is a variable, the state will be the server state which this variable is mapped. Each FU internal variable that must be accessed by the environment is mapped in some state and at a logic channel. If the communication class is a process then the state is the server state which is activated by the message to execute a process (service) to its client. And finally, if the communication class is an event, for a clearly understanding, the state is specified as 'Server Mapped State' and 'Client Activated State'. Server mapped because usually the events are variables sending (for example a client does not need to send a message to its server every time to know the new value of a actualized variable; the server sends it when the variable is actualized without a solicitation) and this is the state which the variable is mapped in the server. However it works to others cases also, it is the state which the event comes from. And client activated because generally when the client receives an event (message) a new state in its machine state is activated. Also, when the communication class is as event, the channel defined is the channel of the client, which the server (when the event happens) has to send it. It is necessary to give a unique identifier to each FU server, such SERV_AC for the axis control server.

Also exist a Client table. Table 2 show a client requesting table on which must be indicated the Connection Identifier (Id_conn), the destinated Server and Channel, Communication Class, Message Name, Requesting State and the Requesting Description. The requesting state refers to the state of the FU which requests the service.

Table 2. Client Table.

In this design phase were proposed the steps and the documentation necessary so the application programmer will not need to be worried about 'what' the FU does (functionality) since he will make the 'how' (implementation).

The next steps to be followed by the methodology are the implementation (each FU) and the integration. As written before, the implementation is codifying the design in a programming language and it is responsibility of the application programmer. Each FU implemented makes part of a FU's library that later will be used to build another UNC system.

3. CASE STUDY

The methodology was applied for the UNC development of the EMCO Compact 5 Lathe. The EMCO Compact 5 is a lathe with Computer-Numerically Controlled functions and was aimed at the basic CNC training market and manufactures who wanted the capacity to mass produce limited numbers of small parts with the minimum of trouble. The swing over the bed was 100 mm, over the cross slide 60 mm and the capacity between centres 310 mm. The cross-slide travel was 50 mm and the headstock spindle bore a generous 16 mm. It was driven by a 440W motor and had a speed range from 50-3200 rpm with a rapid-traverse rate for the carriage of 700 mm/min. The machine is programmed in Standard G and M Codes and is just 32 in long, 20 in high and sitting on a base 20 in deep.

The figure 5 illustrates the implementation architecture of this work under the design methodology. In this picture can be seen the division into the four layers and your respective hierarchic disposition:

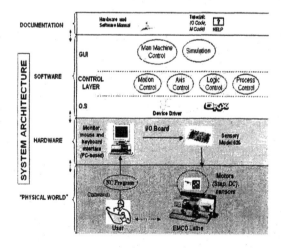

Figure 5. Design Methodology and Implementation.

• The first layer, on a lower level, represents the physical world, where are located all the external elements that interact with the UNC. Stepper motors and proximity sensors was mounted on lathe EMCO. The second layer, the intermediate layer, represents all hardware devices of the UNC that define interfaces to the physical world and to the execution platform (O.S./device driver) of the software components. Sensoray model 626 I/O board was used, this data acquisition board has 48 digital I/O channels, 16 differential analog inputs and 4 analog outputs. In a Pentium II PC desktop was installed the control software. The third layer, on a higher level, represents the software components. It's divided into O.S. (Operating System), Control Layer and GUI (Graphical User Interface). QNX was the real-time operating system that was used for implement the control application. The GUI was developed with Photon Application Builder. The fourth and last layer represents the documentation that must support the UNC system.

Figure 6 show the GUI of UNC. The functionality consist mainly of management file menu, edition menu, compilator menu, machining menu, manual menu, and MDI menu.

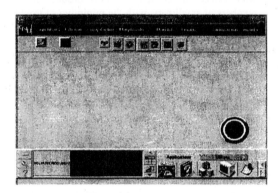

Figure 6.   GUI of the UNC.

4. CONCLUSION

The project has demonstrated the applicability of the design methodology proposed. It gets as a result a powerful design tool in order to build new CNC system with features such as openness, PC-based, low cost and software-based. This design tool belongs to the design tools group that formed the general design methodology developed by Research Chair on Mechatronic.

One of the major facilities of the methodology is the standardization of the rules for the development of applications for machines. The next step of application will be for integrated intelligent characteristics to UNC such chatter monitor, tool wear, tool break, backslash monitor, as so on.

This work will direct the future works on Universal Numerical Control developed by research and graduation students of ITESM. The next step of the project is the developed of CNC controllers for machines focus on modeling process with the same concept.

5. ACKNOWLEDGEMENTS

The research reported in this paper is part of a Research Chair in Mechatronics of ITESM titled "Design, Manufacturing and Integration of Reconfigurable and Intelligent Machines". The authors wish to acknowledge the support of this grant in the development of this research.

6. REFERENCES

Binder, D. (1996). Wie offen hätten Sie's denn gern?. Offene Systeme in der Fertigung, Aachener Werkzeugmaschinenkolloquium.

Jovane, F. (1998). Open Architecture Control Systems. Summary of Global Activity", ITIA Series.

Pritschow G., Altintas Y., Jovane, F., Koren Y., Mitsuishi, M., Takata, S., Van Brussel, H., Weck, M., and Yamazaki, K. (2001). Open Controller Architecture – Past, Present and Future-. *Annals of CIRP.* pp

Ramirez, M. de J. (1998). Desarrollo de un Control Numérico Universal de Bajo Costo Basado en Software y Sistemas Abiertos. *Master Tesis,* ITESM Campus Monterrey.

Ramírez, M. de J., (2001). Sofware based Computer Numerical Controller for Low Cost Automation in Small and Medium sized Metal-Processing Enterprises in Developing Countries. *Memories of the VI IFAC Symposium on Cost Oriented Automation,* Berlin, German. October 8-9.

Ramírez, M. de J. (2002). Tecnología Mexicana de CNC para la Modernizacion de la Planta Productiva de la Micro, Pequeña y Mediana Industria Metal-Mecánica de Países en Vías de Desarrollo. *Extensión Congress,* Sistema ITESM.

ELSEVIER

IFAC

PUBLICATIONS
www.elsevier.com/locate/ifac

THE USE OF GERAM IN THE FORMALIZATION
OF THE GLOBAL VIRTUAL BUSINESS FRAMEWORK

Raymundo Carrasco, Nathalie Galeano, Arturo Molina

Centro de Sistemas Integrados de Manufactura (CSIM),
Instituto Tecnológico y de Estudios Superiores de Monterrey (ITESM),
Avenida E. Garza Sada 2501 Sur, C.P. 64849, Monterrey, NL, México
raymundo.carrasco.aguirre@indsys.ge.com, ngaleano@solutionmes.com, armolina@itesm.mx

Abstract: A formalization of the Framework for Global Virtual Business (COSME Model) to create a reference model for Virtual Enterprise design, creation and operation using GERAM is presented. This reference model has been used to replicate the project in other regions of Latin American, specifically El Salvador. The most important lesson learned is that the use of formal frameworks and methodologies reduces time effort and cost in the implementation of similar projects. GERAM has been proved to be useful in determining the necessary elements for the creation of Enterprise Integration Architectures. *Copyright © 2004 IFAC*

Keywords: Reference Architecture, Enterprise Integration, Enterprise Modelling, Modelling, Virtual Enterprise.

1. INTRODUCTION

Today customers demand high quality products manufactured at a lower cost with a short delivery time. To succeed in this scenario enterprises are adopting manufacturing philosophies such as CIM, JIT, lean manufacturing, concurrent engineering, the extended enterprise and the virtual enterprise. The trend in manufacturing industry is that in the future products will not be manufactured by a single enterprise; instead the enterprise will be a node in a network of enterprises. This is the concept of Virtual Enterprise. A Virtual Enterprise (VE) is a network of enterprises that share its core competencies to gain access to new markets and fulfil a customer need. The creation and operation of a VE is a complicated task, necessary conditions for this are co-operation and co-ordination of all members of the team. The activities required to achieve this co-ordination have to be carefully engineered in a structured manner. To achieve this co-ordination a guide or roadmap must be develop that outlines the activities needed for the creation and operation of the VE. Reference Architectures (RA) can be used as guidelines. Several architectures have been proposed for the integration of single enterprises. PERA, CIMOSA, and GRAI-GIM are recognized among the most complete ones (Williams 1994). These architectures

were the foundations for the creation of GERAM. GERAM is the result of a standardization effort to create a RA with the best characteristics of the three architectures mentioned earlier. GERAM specifies the requirements needed in a Enterprise Reference Architecture. GERAM is so general that can be applied to any enterprise (Bernus and Nemes 1996). This paper describes how GERAM concepts were employed to analyze and formalize the Framework for Global Virtual Business, developed by the COSME Network to make possible the creation and operation of VE's in a global environment (Molina et al 2001).

2. ARCHITECTURES FOR VIRTUAL ENTERPRISES

2.1 GERAM

In 1990 the IFAC/IFIP Task Force on Architectures on Enterprise Integration was formed. The mission of the Task Force was to select the best RA from the existing ones to be used as a single universally accepted architecture in the field of Enterprise Integration. The principal architectures analyzed by the Task Force were CIMOSA, GRAI-GIM and PERA. The result of this study was GERAM (General Enterprise Reference Architecture and

Methodology). GERAM includes the most important contributions of the three mentioned architectures (Bernus). GERAM is a generic architecture, so it can be used in many kinds of enterprises including Virtual Enterprises

2.1 Reference Architectures for Virtual Enterprises

Several reference architectures for Virtual Enterprise have been defined:
- The National Industrial Information Infrastructure Protocol, (NIIIP Consortium, 1998).
- Framework for Global Virtual Business (Molina et. al, 1998)
- VEGA (Zarli and Poyet 1999).
- Virtual Enterprise Reference Architecture – VERA (Tolle et al. 2002).
- PRODNET (Camarinha-Matos et al. 1998; Camarinha-Matos and Afsarmanes 2003)

The Framework for Global Virtual was proposed by the ALFA project COSME-GVE (Molina et al 2001). The purpose of the model was to provide a sound conceptual reference model to understand and analyze how VE can be configured, created and dissolved. COSME is a network created by four European and two Latin American Universities. The framework has three main objectives (Molina et. al. 1998, Bremer et al. 1999):

- Explain how Global Virtual Business can be conceived using three business entities: Virtual Industry Clusters (VIC), Virtual Enterprise Broker (VEB), and Virtual Enterprises (VE).
- Provide a comprehensive set of business entities defined in terms of core products, core processes, and core competencies to understand and analyze Global Virtual Business.
- Establish a common framework for future research work on Global Virtual Business.

The results reported in this paper are part of a research project focused in defining each of the three components of the framework (VEB, VIC, and VE) based on GERAM. The results are described for the Virtual Industry Cluster (VIC) in figure 1. The different components of GERAM are used to create a partial model of the VIC using the life cycle, In the next section one of the components, the VIC is described in terms of GERAM, as an example of how the complete COSME model has been formalized.

3. VIRTUAL INDUSTRY CLUSTER REFERENCE MODEL WITH GERAM

The Virtual Industry Cluster (VIC) is an aggregations of small and mediums companies from diverse industries, with well defined and focused competencies, with the purpose of gaining access to new markets and business opportunities by leveraging their resources and therefore their competencies (Molina and Flores 1999). The present work

proposes a framework and a methodology based on the seven GERAM components for the creation of a Virtual Industry Cluster (VIC), see figure 1. The principal component is the architecture (GERA-VIC). The architecture is a model or representation of the VIC through its life and different modelling views. GERA-VIC uses three main concepts: Processes (Life Cycle and Modelling views), Human Roles and Technological Tools). The life cycle phases define pertinent activities executed during the life of the entity, in this case the entity is the VIC. There are eight different life cycle phases defined in GERAM. Each phase describes the modelling views (functions, information, resources and organization) used in the design, creation and operation of a VIC. These phases are shown in detail in Table 1. The Generic Enterprise Modelling Concepts are defined in a VIC ontology which describes the main elements of the VIC, i.e. Clusters and Core Competencies. The Enterprise Engineering Methodologies are the process required to design, create and manage a VIC. The Enterprise Modelling Language used is eEPC (Event driven processes chain) diagram. Therefore Enterprise Engineering Tools was ARIS Toolset software. The Partial Enterprise Model created is the VIC Reference Model (FRAMEVIR). The results of the instantiation of FRAMEVIR are the operating clusters in Mexico and El Salvador (mexican-industry.com, elsalvador-industry.com)

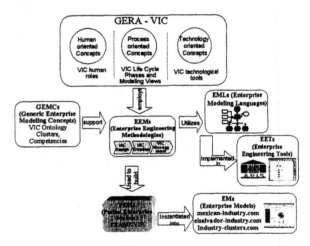

Figure 1. Virtual Industry Cluster Reference Model using GERAM

3.1 GERA – VIC Life Cycle

Table 1 describes all the life cycle for the design, creation, and operation of a Virtual Industry Cluster. In the first two phases the need for the creation of a VIC is detected and the mission, vision, and business plan for a VIC are elaborated. In the next phase the requirements for the VIC are defined, these include human, technological, and organizational resources. The first three phases of the VIC life cycle are executed by organizational entities like governmental agencies, universities, and industrial associations interested in the development of enterprise's core competencies. At the preliminary design a new organizational entity is formed. This entity will be called VIC business entity, and it is in charged of the creation and administration of the VIC.

The VIC business entity can be a non-profit non-Governmental Organization (NGO) or a business that commercializes the core competencies of the enterprise members of the VIC, such as a broker company. In the preliminary design, an overall specification of the VIC is made, to do this the VIC design Methodology is employed. This methodology includes the next steps:

1. Identify industrial sectors. In this step the industrial sectors are identify within the economic region under study.
2. Definition of Cluster. The industrial sector with greatest potential of growth is then selected, and the enterprises with best core competencies in the sector are the possible members of the VIC.
3. Cluster Organization. This step consists in organizing the selected enterprises into one or more clusters classified by product, process, or technology.

The next life cycle phase is Detailed Design, in this phase the VIC organizational structure, human roles, and Information and Communication Technologies (ICTs) are defined. After design phases the implementation process begins. This process consists in the application of the VIC creation Methodology:

1. Contact the enterprises, and invite them to form part of the VIC. Elaborate list of enterprise that accepted the invitation.
2. Enterprise evaluation. In this step the enterprise is evaluated with the aid of IMMPAC (Integration and Modernization of the Small and Medium Enterprise to reach competitiveness) methodology. IMMPAC first evaluates the competitive position of the enterprise, this by analyzing the enterprise products, markets, customers and suppliers. Next IMMPAC analyzes the productivity indicators of the enterprise. The third step of the methodology is to analyze the customers' critical successful factors (Caballero et al 2000).
3. Qualify enterprises. This step assigns a qualification to the enterprises according to the IMMPAC results.
4. VIC formalization. Finally the VIC organization is formalized in a general meeting.

After the implementation phase the operation of the VIC starts. In this stage the Core Competencies (CCs) of the enterprises are developed by best practices sharing, combining individual functions with group functions, and improvement projects. The activities of the operation phase are included in the VIC management Methodology (Galeano and Molina 2003):

1. CCs selection. Abilities that differentiate the members from competitors are selected.

2. CCs development. Implementation of improvement projects to strengthen the CCs.
3. CCs deployment. Commercialization of individual CCs of the members or CCs of the cluster.
4. CCs protection. Development of new clusters with complementary CCs, legal protection of CCs processes and technology.

Table 1 - VIC life cycle phases

Life Cycle Phase	Function View	Information View
IDENTIFICATION	• Detect the need for a VIC • Define the entity, the VIC (Use COSME model)	• Economic data (government) • Chambers of commerce studies • COSME model
CONCEPT	• Outline Mission, Vision, strategies and policies (for a generic VIC) • Define Business Plan	• Documentation of mission, vision, strategies, policies, and Business Plan
REQUIREMENTS	• Select economic region to analyze • Identify internal and external customers (OEM's, MSME's) • Identify customer needs • Identify Human Resources and ICT requirements	• Economic, political, and demographic data • Organization chart
PRELIMINARY DESIGN	• Apply VIC design Methodology • Identify Industrial Sector • Identify and select possible clusters within the sector with best competencies • Organize Cluster (Products, Technologies, and Processes)	• Economic census • Chambers of commerce studies
DETAILED DESIGN	• Design of organizational structure • Define human roles. • Design ICTs • Define business processes	• Organizational chart • Enterprise database • IMMPAC model
IMPLEMENTATION	• ICT installation • VIC creation Methodology • Contact and invite possible enterprises to be VIC members • Evaluate VIC members • Qualify members • Cluster formalization	• Methodology for cluster formation (Formation meeting, commitment to cooperate, agreement, contract) • Evaluation and Qualification data • CC data
OPERATION	• Commercialization VIC management Methodology • Select CC • Develop CC • Deploy CC • Protect CC	• Enterprise development Plan • New products database • Customer database • Individual record for each VIC member
DECOMMISSION	• VIC disintegration • VIC members re-missioning • Assets distribution	• Disintegration Plan

Table 1 - VIC life cycle phases (Continue)

Life Cycle Phase	Organization View	Resource View
IDENTIFICATION	Government, Educational Institutions, Private organizations	Describe the human resources needed to perform identification process: researchers, industry managers, government officials.
CONCEPT	Organizational entities that will elaborate mission, vision, and business plan: Business entrepreneurs, non-governmental associations (NGOs), chambers of commerce.	Human resources: Stakeholders of the business.
REQUIREMENTS	Organizational entities that will identify the VIC requirements: (Governmental agencies, chambers of commerce, NGOs)	Human resources: Stakeholders of the business. Technological resources: Government and private economic databases.
PRELIMINARY DESIGN	Description of the organizational entities that define an overall specification of the VIC: the VIC Business Entity.	VIC staff. ICT's: • IMMPAC • Web Page • Electronic Database of VIC members • Business portal • Online training and "Best Practices" sharing
DETAILED DESIGN	The VIC Business Entity is the organization responsible for this phase.	VIC staff. ARIS Toolset Modeling software.
IMPLEMENTATION	VIC Business Entity and Enterprises members of the VIC	VIC staff, software programmers, and staff from the different enterprises. VIC server, and network,
OPERATION	VIC Business Entity and Enterprises members of the VIC.	VIC staff. Electronic IMMPAC, Business Portal, Enterprise database.
DECOMISSION	VIC Business Entity and Enterprises members of the VIC	VIC and enterprise staff.

The final phase of the VIC life Cycle is decommissioning. The members of the VIC could decide to separate, therefore there is a need for a plan to finish this association, or redefine its mission.

3.2 GERA – VIC Modelling Views

The modelling views serve to represent the activities, roles, organization and resources in the VIC. The modelling views are: function, information, organization and resource. These views are presented in the life cycle activities of Table 1.

a) Function View. Characterizes all the activities to be executed throughout the VIC life cycle.

b) Information View. Represents the information/knowledge of the objects in the VIC (economical data, industry performance measures, evaluation data)

c) Resource View. Models the human, physical, or technological resources employed by the VIC such as: employees, databases, information systems.

d) Organization View. Describes the responsibilities of all the entities identified in the other views.

To exemplify the life cycle phases employed in the reference model, four modelling views of the "Preliminary Design" phase will be described in more detail. This research uses ARIS Toolset software to model the different business processes through the life cycle of the VIC. The four modelling views can be represented by ARIS Toolset.

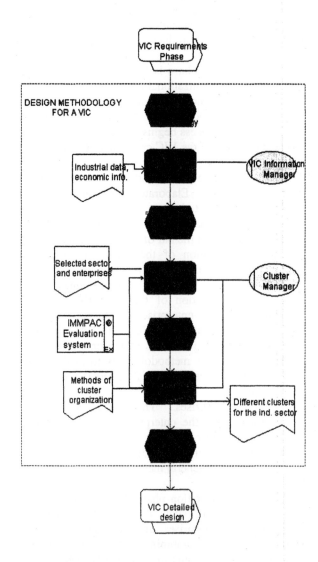

Figure 2. eEPC model of the design methodology for the VIC

Design Function View (Figure 2):
- Functions: Identify industry sectors, Define clusters and Organize clusters.

150

- Information: Industrial data, economical information of the region, and IMMPAC Evaluation System
- Human Resources: Information Manager and Cluster Manager
- Technological resources: IMMPAC Information System
- Outputs: Different clusters for specific industrial sectors.

Design Organization View: This organization view defines the entities responsible for the design of the VIC, in this case the VIC Business Entity which includes the high level managers defined in Figure 3.

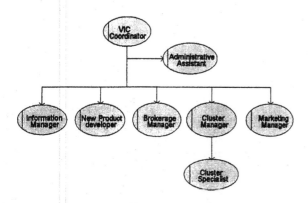

Figure 3. VIC Organizational structure and human resources

Design Information View: This is the data/information regarding the companies to be included in the clusters. This information is kept in the IMMPAC database system (Figure 4).

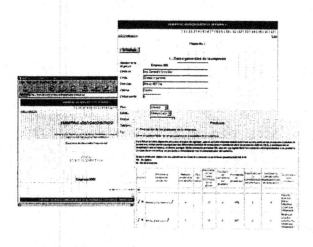

Figure 4. Information View: IMMPAC Internet System (http://csim.mty.itesm.mx/immpac/)

Detailed Design Resource View: These are the necessary resources to complete the design task, in the human part they include the VIC staff and in the technological part they include business modelling software (Figure 5).

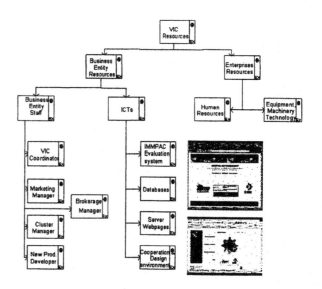

Figure 5. Resource View of the VIC

3.3 GERA – VIC Instantiation process

The instantiation process is described in figure 6. A Partial Model FRAMEVIR was defined based on the GERAM components. This partial model included a life cycle definition with their four modelling views. Three key processes were defined, namely: design, creation and manage. FRAMEVIR uses ARIS to model the different business processes through the life cycle of the VIC. The reference model was used to create two VICs, one in Mexico (Mexican-industry.com) and one in El Salvador (elsalvador-industry.com).

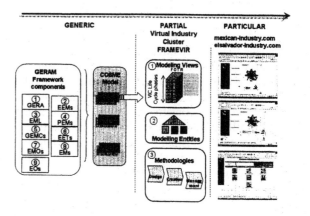

Figure 6. Instantiation process to design, create and manage VICs in Mexico and El Salvador.

4. DISCUSION OF EXPERIENCIES

GERAM proved to be a useful tool to map the necessary elements for Enterprise Engineering for the design, creation and manage of the Virtual Industry Cluster concept. However, at the beginning of the formalization process, to understand all the GERAM concepts was a difficult task in order to achieve a mental model that could be useful to describe all the activities that were required to create VICs.

Today, Generic Enterprise Modelling concepts and operational systems are the less developed elements of VIC methodology; glossaries and ontological theories have been developed to achieve a better understanding among other VIC implementations.

The creation of a reference model to replicate the first experiences of creating a Virtual Industry Cluster in Mexico was imposed by the need of experimenting the conceptualization of Virtual Enterprises in other Latin America Regions, for example Brazil (Bremer et al. 2001). Based on these experiences the framework has been used to create a new VIC in El Salvador. The application of the reference model in El Salvador project allowed the implementation team to reduce considerable the time and effort spent in relation to the Mexican experience. The reason to work with other region in Latin America is that the cultural, legal and technology aspects are similar; therefore the reference model can easily be adjusted to comply with a specific LA country or region.

5. CONCLUSION

A Reference Model for Virtual Industry Cluster (FRAMEVIR) has been formalized based on GERAM to replicate the experiences of designing, creating and managing Virtual Industry Clusters (VICs). FRAMEVIR complies with most of the elements stated in GERAM, especially with the definition of a Generic Architecture which includes: a life cycle process, modelling views and underlying methodology to instantiate VICs. Further research needs to be undertaken to prove the validity of the VIC methodologies in more industrial regions. A project to replicate the model other regions of Mexico is underway. The eEPC modelling language has been used to represent VIC processes; however it is the author's opinion that UML will be a better modelling language to use. FRAMEVIR allows stakeholders of industrial development programs to engineer the replication of the VIC concepts, in the future the other two entities of the COSME model (Virtual Enterprise Broker and Virtual Enterprise) will be formalized too. FRAMEVIR needs to be refined to be more user friendly to industrial managers or business entrepreneurs that decide to create a new Virtual Industry Clusters.

6. ACKNOWLEGDMENTS

The research reported in this paper is part of a Research Chair in Manufacturing of ITESM titled "Enterprise Integration Engineering to develop Extended Enterprises for Mass Customization". The authors wish to acknowledge the support of this grant in the preparation of the manuscript.

7. REFERENCES

Bernus P, Nemes L. (1996). A framework to define a generic enterprise reference architecture and methodology, *Computer Integrated Manufacturing Systems*, **9 (3)**, 179-191.

Tolle, M., Bernus P., and Vesterager, J. (2002). Reference Models for Virtual Enterprises. *in Collaborative Business Ecosystems and Virtual Enterprises*. L.M. Camarinha-Matos, (Ed.), Kluwer Academic Publishers, Boston, pp. 3-10.

Bremer CF., Eversheim W., Walz M., Molina Gutiérrez A. (1999). Global Virtual Business: A Systematic approach fo Exploiting Business Opportunities in Dynamic Markets. *International Journal of Agile Manufacturing*. 2, Issue 1. 1-11.

Bremer CF., Michilini FVS., Siqueira JEM., Ortega L.M. (2001). VIRTEC: An example of a Brazilian virtual organization. *Journal of Intelligent Manufacturing*. **12 (2)**, 213-221.

Caballero D., Molina A., Bauernhansl T. (2000). A Methodology to Evaluate Enterprises to become members of Virtual Industry Clusters. , *in E-Business and Virtual Enterprises: Managing Business-to-Business Cooperation*, L.M. Camarinha-Matos, H. Afsarmanesh, Rabelo (Eds.), Kluwer Academic Publishers, pp. 443-454 .

Camarinha-Matos LM, Afsarmanesh H, Garita C, Lima C. (1998). Towards an architecture for virtual enterprises. *Journal of Intelligent Manufacturing*, **9 (2)**, 189-199.

Camarinha-Matos LM, Afsarmanesh H. (2003). Elements of a base VE infrastructure. *Computers in Industry*. **51 (2)**, 139-163.

Galeano N, Arturo Molina, Modelo para el desarrollo de competencias clave en clusters virtuales industriales. (2003). *XXXIII Congreso de Investigación y Extensión del Sistema ITESM*, Enero 23 y 24, 2003, in Spanish.

Molina A., Ponguta S., Bremen C., Eversheim W. (1998). A Framework for Global Virtual Business. *Agility & Global Competition*, Vol. 2, No. 3, 56-69.

Molina A., and Flores M. (1999). A Virtual Enterprise in Mexico: From Concepts to Practice. *Journal of Intelligent and Robotics Systems*, **26**, 289-302.

Molina A., Bremer C.F., Eversheim W. (2001). Achieving Critical Mass: a global research network in systems engineering. *Foresight*, Vol. 3, No. 1, February 2001, 59 - 64.

NIIIP Consortium, (1998). http://www.NIIIP.com

Vernadat, F. B. (1996). Enterprise modelling and Integration. London; New York Chapman & Hall.

Williams, T.J et al. (1994). Architectures for integrating manufacturing activities and enterprises. *Computers in Industry*. Volume 24, pages 111-139.

IFIP–IFAC Task Force 1999, GERAM Version 1.6.3.

Zarli, A., and Poyet P., (1999). A framework for distributed information management in the virtual enterprise: the VEGA project, in: *Infrastructures for Virtual Enterprises—Networking Industrial Enterprises*, L.M. Camarinha-Matos, H. Afsarmanesh (Eds.). Kluwer Academic Publishers, Boston, 1999.

ELSEVIER

IFAC

PUBLICATIONS
www.elsevier.com/locate/ifac

REALTIME COLLABORATIVE MIXED REALITY ENVIRONMENT WITH FORCE FEEDBACK

Yong-Ho Yoo, Wilhelm Bruns

Bremen University, Germany
Research Center for Work, Environment and Technology (artec)
bruns@artec.uni-bremen.de
yoo@artec.uni-bremen.de

7th IFAC Symposium on Cost Oriented Automation, 2004, Ottawa

Abstract: A low cost *Mixed Reality* implementation with force feedback as a base for further empirical studies of collaboration in distributed real and virtual environments will be presented. It will be shown, how the system can be used to get more insight in tangible cooperation between humans, avatars or in general real and virtual systems.

This application is related to *Hyper-Bonds*, a unified concept to describe complex effort/flow driven automation systems distributed over real and virtual worlds. It allows selected materialization of parts of the system into reality and their functional connection to a simulation model. *Copyright © 2004 IFAC*

Keywords: mechatronics, shared virtual environments, force feedback devices, simulation, mixed reality, co-presence, object manipulation

1. INTRODUCTION

A concept of mixed reality, the blending of real and virtual realities (Ohta & Tamura, 1999), from toys at an early school level to tools at a work level is introduced to support experience and understanding of principles of physics, human-computer interaction and automation. Our aim is to use a unified concept of physics and control theory, implement them in low cost hard- and software, to support a continuous expansion of experience and knowledge about automated systems towards a cost oriented experimentation with alternatives (van Amerongen, 2001).

In some previous work, the concept of complex objects was introduced (Bruns, 2000) being objects with a real concrete part coupled to various virtual representations (simulation, animation, symbolic) by means of grasp- or image-recognition. This coupling introduces the possibility to build and change real systems and synchronously generate their functional representatives. Simulation may be carried out with the virtual model and compared with the desired or actual behavior of the real system. This concept has been extended by bi-directional links between the virtual and the real model, being able to sense and

generate various relevant physical continuous effort and flow phenomena via universal connections: **Hyper-Bonds** (Bruns, 2003). A unifying concept supporting this approach is the Theory of Bond-Graphs (Paynter, 1961; Karnopp, 1995). Bond-Graph theory considers a continuity of energy (Effort x Flow) flow in abstract networks. Effort can be electric voltage, air pressure, force, momentum, temperature etc. Flow can be electric currency, air volume flow, velocity, heat-flow etc. An implementation of Hyper-Bonds for simple pneumatics and electrical interfaces between computer internal and external system-components has been demonstrated. Its extension towards force and momentum mechanics will be shown. This extension is motivated by a need to have a better feeling for remote process control and for collaboration in virtual environments. The former is a well studied problem known from remote robot control in astronautic or surgery applications. The latter has only recently found consideration with the widespread use of multi-user environments in games, entertainment, learning and tele-work. The co-operation of several dislocated humans in a shared virtual space, communicating with and sensing each other in a tangible way is a challenging task.

Several authors investigate the role of touch in shared virtual environments (SVE). Basdogan et al., 2000, studied the influence of haptic feedback on task performance and the sense of being together. An interesting vision is their haptic version of the "Turing Test": one real experimental person has to press a stick against a virtual brick and move it to a position, supported by an imitated or real person . The experimental person is asked to recognize who is currently co-operating. Their actual experiment however was the "Ring on a Wire" scenario: two persons had to move a ring along a curved wire without collision. They developed multithreading techniques for integration of vision and touch and found the graphics and haptic update rates to be of at least around 30Hz and 1000Hz, respectively to have a satisfying experience. Their system enabled haptic interactions of two users on one computer. Because of these severe time restrictions, a collaboration via Internet, even Internet2, with force feedback proved to be unpractical at the moment.

Ruddle et al., 2002 therefore addressed the *piano movers' problem* (manoeuvre a large object through a restricted virtual space) only with visual feed-back. They systemized this rather uncovered field of human-machine interaction in three levels of cooperation: 1) users can perceive each other, 2) individually change the scene, 3) simultaneously act on and manipulate the same object (independent or co-dependent). The co-dependent simultaneous action has been further distinguished as being symmetric (both actors perform the same role, like wearing and fine manipulation) or asymmetric (one actor is wearing, the other one is fine manipulating).

Haptic communication and cooperation may play an important role in future preparation and training of humans in hybrid production systems. We therefore introduce a low cost solution for the study of force feedback phenomena based on toys, embedded in a concept suitable for extension to real automation problems and distributed applications.

2. BOUNDARIES AND INTERFACES

A virtual system behaviour may be studied and controlled through well defined boundaries, figure 1. If these boundaries are dislocated and connected to the virtual world via Internet, we face severe problems of time delay and synchronisation. This is a limitation for the distribution of hard real time processes. Nevertheless, for slow real-time or state oriented, event driven processes a distribution is possible and offers interesting perspectives.

In order to provide arbitrary boundary conditions, a mechanism is necessary to generate and sense various physical phenomena. An implemented coarse prototype for electrics (voltage and current) and

pneumatics (pressure and volume-flow) demonstrated its successful integration into a virtual construction and simulation for learning applications (Bruns, 2003).

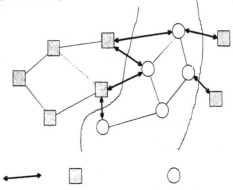

Hyper-Bonds Real Components Virtual Components

Fig. 1. Boundaries cutting a system

The concept of Hyper-Bonds is being applied in a learning environment for electro-pneumatics, where students can work on complex systems, freely switching between virtuality and reality[1], figure 2. The modeling desk can be used in a distributed way according to Ruddle's category 2: multi-user may place objects on the virtual table and modify the circuit synchronously but only related to different objects. However, the system is able to support several cooperation tasks close to category 3 in a distributed way. Figure 3 shows a situation where a real pressure valve can be pressed to drive a virtual cylinder and a virtual valve can be pressed at the same time to drive a real cylinder. This cross-circuit may be changed to a safety circuit, allowing to activate a pneumatic-cylinder only if two pressure valves are pressed simultaneously. These two pressure valves may be distributed in a virtual environment and a real place. This concept is now being extended to mechanical phenomena (force, momentum).

Fig. 2. DERIVE Learning Environment

---

[1] EU-IST Project DERIVE (Distributed Real and Virtual Learning Environment for Mechatronics and Tele-Service)

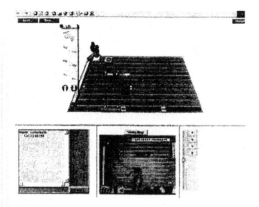

Fig. 3. Distributed Cross-Mixed Reality Circuit

3. FORCE FEEDBACK SOLUTION BASED ON LEGO BRICKS

A Low Cost Momentum Handle (LoMo), based on Lego Bricks, which can be used to feel the artificial force feedback generated from a DC motor and sensed by a pressure sensor, was developed (figures 4-7). The LoMo can be used in various distributed mixed reality applications, implementing a flexible force feedback.

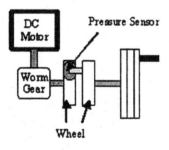

Fig. 4 Low Cost Momentum Handle (LoMo)

Fig. 5. Upper view of LoMo

If a force is applied to the handle, trying to rotate the wheel, a signal is generated by a wheatstone bridge equipped with the pressure sensor and forwarded to a micro processor through an A/D converter; a PLC based on Infineon's C164CI (20 MHz) micro-processor was used as a micro station. The voltage of

the wheatstone bridge is directly proportional to the torque applied by the user. To generate the relevant pulses by PWM (Puls Width Modulation) to drive the DC motor, a Dual Full-Bridge Driver L298 was used.

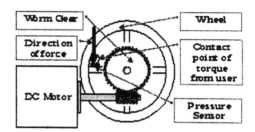

Fig. 6. Side view of LoMo

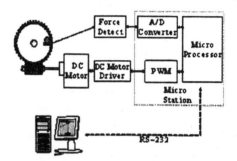

Fig. 7. Hardware implementation of LoMo

The duty ratio of pulses were calculated at the microprocessor level. A worm gear connects the rotational axis of the motor with the axis of the handle in such a way, that no rotation can be directly transmitted from handle to motor but only from motor to handle. The motor only rotates in a software-controlled way in the direction of the force given by the user. As the resistor of the pressure sensor is inverse proportional to the imposed force F, the torque τ_s applied to the sensor-wheel is (1.2).

$$V \propto F \ , \quad F = \frac{r_u}{r} \cdot F_u \qquad (1.1)$$

$$\tau_s = r \cdot kV \qquad (1.2)$$

with

V , voltage detected from the wheatstone bridge;

k , constant,

F_u, force imposed on the handle by the user,

r_u, distance from the axis to the grip of the user ,

r , distance from the axis to the sensor.

A LoMo is able to communicate with another LoMo via internet using UDP sockets (Fig. 8). When a signal comes from the serial port (RS-232) module, it is processed by Hyper-Bonds Operation module, and then it is forwarded to the GUI (Graphic User Interface) module and TCP/IP module.

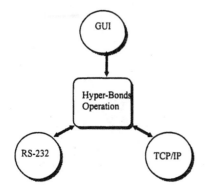

Fig. 8. Software implementation of LoMo in a PC

4. IMPLEMENTATION OF HYPER-BONDS

A distributed mixed reality cooperation model (Fig. 9) is used to investigate the process of Hyper-Bond implementation, its time critical behavior and its ergonomics. Two networked users can carry a mass from one side to the other via internet using handles connected via virtual rope. They can feel the artificial force feedback from the handle and see a virtual mass in the 3D virtual environment. The law of gravity is simulated. Drawing the bond-graph model for a completely real environment (Fig. 10) may give some insight into how to draw physical borders between real parts, virtual parts and the hyper-bond interface and to find key equations (hyper-bond equations). These equations apply to core algorithms of the hyper-bond interface. Assuming that it will be cut and implemented as shown in Fig. 10, *Se1, Se2, TF1, TF2* are real elements, *Se3, I* and *node 1* are virtual elements.

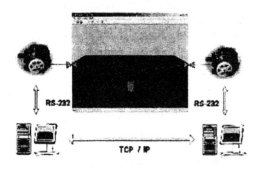

Fig. 9. Distributed Cooperation Mixed Reality model

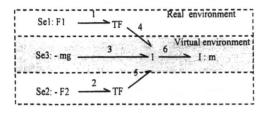

Fig. 10. Bond-graph model of figure 9 (g: acceleration by gravity, m: mass of the object.)

Implementing Hyper-Bonds allows the control of a virtual mass through signals from real parts and vice versa. Both handles are controlled by signals from the virtual part, using (1.3) and (1.4) derived from bond-graph (Fig. 10). The energy source Se3 and storage *I* is translated into effects to both handles and the display by using these equations as core algorithms of a Hyper-Bond Interface System.

$$e_6 = e_4 + e_3 + e_5 \qquad (1.3)$$

$$f_6 = \frac{1}{m} \int e_6 dt \qquad (1.4)$$

$$e_4 = ne_1 = nF_1, \; e_5 = ne_2 = -nF_2, \; e_3 = mg$$
(*e* : effort, *f* : flow, *n* : constant, *F* : force)

In the GUI (Graphic User Interface) the mass is moved in two dimensions, consequently equations (1.3) and (1.4) have to be calculated as vector sum (Fig. 11). The force feedback value used to rotate the right handle is a scalar sum of *F2* and *p*, the projection of vector sum *F1* and *mg* to the *F2* direction (Fig. 12).

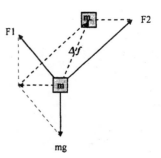

Fig.11. A graphic display algorithm using vector sum of the equation (1.3)

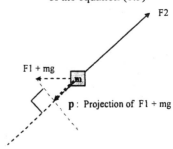

Fig. 12. Algorithm deriving force feedback value

From (1.3) and Fig. 11 the resulting torque of the motor for feedback is

$$\tau_2 = r_2 \cdot (F_2 + p) \qquad (1.4)$$

This torque-value is used to drive the right motor by pulse-width modulation. The left motor is driven accordingly.

5. CONCLUSION

We tested the implementation with the above scenario (Fig. 9) using a 100 MBit dedicated LAN, 2 PCs with AMD Athlon 600 processors and 256 MB RAM, Windows 2000 (no realtime!), 2 Conrad PLC with 20 MHz, resulting in a sensor-PLC-motor-cycle of 1 msec, a PLC-PC-cycle time of 8 msec and a distributed sensor-PLC-PC-LAN-PC-PLC-motor-cycle of 18 msec + LAN-time (order of msec depending on load). Although these 18+ msec for the whole cycle are well below a desirable 1000Hz found by Basdogan et al., 2000, the "feeling" of the weight was already appreciated by several test persons. To compare this distributed solution with a local real-virtual connection, we use a scenario of Fig. 13, which may also be used to teach basic physics in a tangible way. Further evaluation experiments to compare various modes of control (Melchiorri, 2003) and network-influences (Hirche & Buss, 2003) will be undertaken (Yoo & Bruns, 2004).

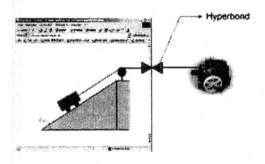

Fig. 13. Local Hyper-Bond Solution

Together with a general interface concept of Hyper-Bonds, this solution is not only a stand-alone experiment, but can be integrated in a general consideration of systems from a point of view of continuity of energy flow. This opens up a broad area of empirical investigations on how physical phenomena and logics can be experienced in a tangible and transferable way, thus contributing to new concepts of distributed real and virtual laboratories[2].

REFERENCES

Amerongen, J. van (2001): Modelling, Simulation and Controller Design for Mechatronic Systems with 20-sim. Proc. 1st IFAC conference on Mechatronic Systems, pp. 831-836, Darmstadt, Germany

Basdogan, C., C.-H. Ho, M. A. Srinivasan. (2000): An Experimental Study on the Role of Touch in Shared Virtual Environments. ACM Transactions on Computer-Human Interaction, Vol. 7, No. 4, pp 443-460

Bruns, F. W. (2000). Complex Objects and Anthropocentric Systems Design. In: Advances in Networked Enterprises (L. M. Camarinha-Matos, H. Afsarmanesh, H.-H. Erbe (Ed.)), Boston, pp.249-258

Bruns, F. W. (2003): Hyper-Bonds. Human Skills Oriented Systems Design with Mixed Reality. Proc. of 6th IFAC Symp. on Automated Systems based on Human Skill and Knowledge. Göteborg

Hirche, S., Buss, M. (2003). Study of Teleoperation Using Realtime Communication Network Emulation. Proc. IEEE/ASME Int Conf. Advanced Intelligent Mechatronics, Kobe, Japan

Karnopp, D. C., D.L. Margolis, R.C. Rosenberg (1990). System Dynamics – A unified Approach. John Wiley, New York

Melchiorri, C. (2003). Robotic Telemanipulation Systems: An Overview on Control Aspects. 7th IFAC Symposium on Robot Control, Wroclaw, Poland

Ohta, Y., H. Tamura (1999). Mixed Reality – Merging Real and Virtual Worlds. Tokyo

Paynter, H. M. (1961). Analysis and Design of Engineering Systems, MIT Press, Cambridge, MA

Ruddle, R., Savage, J. C. D., Jones, D. M. (2002). Symmetric and Asymmetric Action Integration During Cooperative Object Manipulation in Virtual Environments. ACM Transactions on Computer-Human Interaction, Vol. 9, No. 4, pp 285-308

Yoo, Y.-H., Bruns, W. (2004). Motor Skill Learning with Force Feedback in mixed Reality. 9th IFAC Symp. on Analysis, Design, and Evaluation of Human-Machine Systems, Atlanta

---

[2] EU-IST Project Lab@Future
(http://www.labfuture.net/)

ELSEVIER

IFAC

PUBLICATIONS
www.elsevier.com/locate/ifac

HUMAN CENTERED DESIGN IN ASSEMBLY PLANTS FOR A COST EFFICIENT IMPLEMENTATION OF INNOVATIONS - HAND GUIDED ROBOTS

Petra Kohler, Florian Sarodnick, Tim Lum, Hartmut Schulze, Siegmar Haasis, Ralf Giessler

*Daimler Chrysler AG, Wilhelm-Runge-Str. 11, D-89081 Ulm & Bela-Barenyi-Str. 1
D-71059 Sindelfingen, petra.p.kohler@daimlerchrysler.com*

Abstract: a study was conducted to analyze the application of a "hand-guided robot" for an assembly plant. Main focus was the direct co-operation between the employee and robot. We were interested in the usability of the robot, its effects on work processes and consequently the acceptance by using a hand-guided robot instead of a handling device in the assembly line. An interdisciplinary team (process engineers, mechanical engineers, and working psychologists) designed a prototypical work place after working at the assembly line and a detailed activity analysis. Template for the prototypical work place was a workstation moving suspension struts between two assembly lines and the insertion in the corresponding brackets. Results indicated that the process can be optimized by using the hand-guided robot. Furthermore physical work load for the employees could become better in particular to prevent back problems. An issue could be the required space for the robot as well as the work processes and consequently the mental work load of the employees. *Copyright © 2004 IFAC*

Keywords: Hand-guided robots, man-machine-interaction, participatory design, ergonomics, cost-oriented production

1. INTRODUCTION

Technological advances in robotics loosen the strict detachment from human and robotic labor. This is the focus of the project "hand-guided robotics" at DaimlerChrysler AG. We tested the potential introduction of an "Assist Mode" robot in the assembly plants. With part automating in assembly lines expenses can be saved and work can be divided. Considering the age distribution of the population in the next years (Hradil, 2000) it is important to take care about working places for older employees. By using hand-guided robotics physical work load can be taken over by the robot and actions requiring highly precise motorical skills as well as flexible work can be done by the employees. This new mode of close co-operation between employees and robots raises questions about the interaction employee-robot (ergonomic), the working structure, and consequently the acceptance of using a hand-guided robot in the assembly line. In this paper we describe an investigation about how employees can work with a hand-guided robot in the assembly line. What has to be considered that employees accept the close cooperation with the robot? Can overall costs be reduced?

2. BACKGROUND

2.1 Hand-Guided Robots

Nowadays, most human-robot interaction capabilities are developed by robotics experts for robotics experts. Because robots will soon be deployed in a great number, it is necessary to adapt technology to human skills. (Scholtz, 2003). That means, designers must use robot technology to support human skills instead of substituting them. To handle the robot successfully, the operator must be able to co-ordinate and to interact with it. In other words, he ought to have a high level of situational awareness. This means that the operator knows what the robot is doing and what he is likely to do in the near future. Situational awareness appears at 3 levels (Scholtz, 2003): The first and lowest level is the perception of cues in the environment. In a second step, the person can integrate multiple pieces of information and evaluate its importance to reach his goals. Full situational awareness means that the operator is able to forecast future events based on his perception and interpretation of the situation (Scholtz, 2003). Although a relation between situational awareness and performance exists, a person can have achieved level 3 situational awareness and still not perform well. Reasons for this are poorly designed systems or cognitive failures of the operator. Situational awareness plays an important role in making decisions because it is closely tied to a person's mental model which provides the basis to recognise unexpected system behaviour and to cope with it (Dudenhoeffer, 2001). Another important feature to be considered by the engineer is the level of automation. Operators of fully automated systems often fail in responding to emergency situations (Dudenhoeffer, 2001). This decrease in situational awareness is a consequence of vigilance and a shift to the role of a passive recipient. Which task is to be automated depends on its complexity as well as on the perspective of efficacy. The worker is superior to technology in his sensing and processing abilities. On the other hand, he tends to tire and is vulnerable to injuries resulting from cognitive and motor effort. That is why it is important to find the right level of automation.

Tools, intended for the direct interaction with a human worker, handling a shared payload have been developed. They are called collaborative robots (cobots) and are a subset of the material handling equipment named Intelligent Assist Devices (IADs) (Pershkin & Colgate, 1999). Cobots produce software-defined virtual surfaces which guide and constrain the motion of the shared payload. IADs meliorate ergonomic working conditions, product quality and productivity by combining human and robotic work in an appropriate manner. This can be reached by the combination of the robot's strength with the sensing and dexterity of the human worker. Institutions that have made a large contribution to the development of IADs are e.g. General Motors and the Northwestern University. Some cobots have been designed for the automobile assembly line. They help reducing ergonomic concerns in terms of physical and cognitive loading by providing some amplification of human power (Akella, 1999). Furthermore, they improve safety, quality, and productivity.

Although research on cobots is still in its infancy, IADs are likely to play an important role in near future technology.

2.2 Ergonomic

In occupational life sometimes it is necessary to change work systems and operational procedure shortly after introducing them into the company in order to adapt them to consolidated findings in ergonomics. These corrections refer to ergonomic, physiological, psychological, legal, and safety features of work organization. If working conditions are not designed ergonomically, employees' illness rises and causes high costs for the enterprise (Lawaczeck, 2003). That means that the status of employees' illness must be kept low for being able to compete with other companies. Diseases of muscles, cellular connective tissue and skeleton have a leading position among the causes of disability in assembly plants. Main reason for this is the bondage of a particular posture while assembling (Lawaczeck, 2003). By making work more ergonomic, it would be possible to save expenses of millions of Euro (Lawaczeck, 2003). Introducing a hand-guided robot to an assembly plant would not only facilitate work because of reduced work load, especially for the older worker, but also make assembling more rational. Such a reduced work load would lead to a declining absence from work and save money. Besides, the insertion of a robot would change work organization. This would urge the company to guarantee the necessary space for the work and to avoid loop ways.

If changes in work organization are taken up subsequently, they often require enormous costs, but to dispense with them would lead to psycho-physiological and physical harms or to psychosocial impairment of the workers.

In comparison with this retrospective correction it is obviously the better method to consider concepts and rules of ergonomics already when designing systems and procedure. In preventive work organization people anticipate possible threads for health and well-being and try to prevent them right from the start. Furthermore, the so called prospective work organization seems to be the best solution. Here workers are given opportunities for actions and designing and by this they get the chance to develop their personality at work.

In particular, it is recommendable to design human labor and automatization simultaneously because work systems have to be regarded as socio-technical systems which consist of a social and a technical part (Ulich, 1998). The social subsystem is made up of the members of the organization with their claims, knowledge and skills. The technical subsystem, on the other hand, contains the technological and spatial working conditions. These 2 parts of the socio-technical system are connected by occupational roles, which mark different arrangements of task sharing between human being and machine. Ulich (1998) advises the "joint optimization" of technology, organization and inserted human resources in this context. He claims that the retrospective adaptation of one system to the other leads to suboptimal solutions. Human being and technical subsystem are connected by the task. According to Ulich (1998), the division of labor between man and machine determines the development and construction of production systems to a great extent.

2.3 Activity Analyses

Psychological activity analyses is supposed to provide an insight into work procedure as well as into frequency of particular activities, its sequence, and the time needed for every activity (Ulich, 1998). Especially, it shall provide information about how much several activities contribute to the completion of the assignment. This can be used to identify which activities determine productivity. Psychological activity analyses includes 3 steps: the registration of activities by observational interviews as the first stage. The configuration and procedure of the job is being examined by drawing a sample of observations and interviewing workers afterwards. In the second step, the work process is devided into observable categories. Those categories are necessary for the differentiated and precise registration of every activity. Finally, the categories are used for an activity observation recording time and succession of appearance for every relevant work station (Ulich, 1998). To get a representative sample of activities, it is of great importance to spend enough time observing. The workers are competent experts in their field, which makes it necessary to watch them in their working environment. As the result a diagram can be generated. Thereby it is possible to display the structure of the working process and to compare different working processes. Besides, several features of the working process can be compared, e.g. how often certain activities are repeated, how predictable faults are, the variety of assignments, the frequency of communication and cooperation with colleagues, or the time needed for special tasks Ulich, 1998).

3. METHODOLOGY

The cooperation between worker and robot was focus of our investigation. Acceptance is mainly determined by the task and task context of an employee. In a first step we did an activity analyses by working in the assembly line for one day and participatory observations of the employees for two days. Parallel analyses of document was conducted. This data was used to design a prototypical work place for the hand-guided robot and to develop an structured interview guide for the employees working at the prototypical work place.

3.1 Co-operation

Three employees from the research department worked for a day at the assembly line. It was important for us to grasp as much as possible the work with the handling device and the working context. In particular our own co-operation helped us to have a better understanding about the work. For the following observation we deduced observation categories. Beyond we got to know the workers at the assembly line and build up mutual trust. That was important for the following observations.

3.2 Document analyses

Documents analyzed were the standard work sheets and an overview about all work stations. A standard work sheet is an instruction about the work steps at a work station. The overview shows all workstations in the assembly line.
For our category system we used both the co-operation and the document analyses.

3.3 Activity Analyses

We did the activity analyses in the early shift and late shift at a particular assembly line of DaimlerChrysler in Sindelfingen. The early shift was from 6:00 a.m. to 2:07 p.m., the late shift was from 2:15 p.m. to 10:22 p.m. The employees belonged to a work group. Group members rotated between different work stations in within the shift. Every employee passed five stations in this time. During their shift they had four breaks (two breaks with ten minutes, one break with fifteen minutes and one break with thirty-five minutes). After a break the workers moved to another work station. PDA's were used for the observations. A special software, installed on the PDA's, displaying all activity categories needed for the observations was used. The observer could press the category if he noted a behavior for this category. Succession and duration of the behavior are recorded. Three researchers observed every work station directly before or after the handling device in the assembling process and at least two the work with the handling device. Altogether we observed six work stations independently. This procedure should prevent observation mistakes or individual working procedures of an employee. This covers an objective view of the work.
The handling device was a work station with several work steps. Table 1 shows the work steps an employee had to do at this work station.

Table 1. Work steps of an operator

| Move the aggregate card |
| Take up the suspension struts with the handling device |

Close the pair of tongs
Give the line free
Move the suspension struts with the handling device to the L-line
let the suspension struts off and insert them in the brackets
move the handling device off
move back and twist

Figure 1 gives an impression about the work stations. There are two assembly lines, an oval line and an L-line. Between the L-line and the oval line the employees used the handling device for moving the suspension struts from the oval line to the L-line, inserting the suspension struts in the brackets on the L-line. On the left side (according to the movement of the assembly line) there were three work stations mainly for bolting screws. On the right side there were three work stations one of them was the handling device work station. After the suspension struts were inserted in the brackets they had to be screwed up and controlled (second work station on the right).

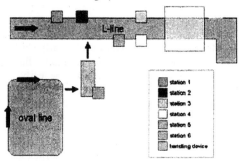

Fig. 1. Structure of the work stations

3.4 Prototypical Work Place

From the co-operation, the document analyses, and the activity analyses we deduced the prototypical work place. The prototypical work place contained three possible work stations:
- to bolt eight screws with a compressed air screwdriver
- to joint the suspension struts with the robot
- to bolt the traction- and rod strut of the suspension struts

The first and the last step were from the every day work. The second step replaced the work at the handling device.

In particular we were interested in the handling with the robot. After the subjects bolted the eight screws they had to steer the robot. The robot was placed in security cell providing access to the worker through an automatic rolling window. The workers had to steer the robot through the opened window. The robot was steered through two levers attached to the robot that were activated by pressing a button on the back of each lever for two seconds.

One lever was rigid the other could be moved like a joystick. While steering the robot the buttons on the back of the levers had to be pushed at all times. Then they had to steer the robot with the suspension struts to the brackets by using the steering lever attached to the robot. When the suspension struts were in the right position over the brackets a visual feedback from two lamps in eye level and a squealing sound from the robot appeared. Then the suspension struts could be released by pressing a button on top of the lever. To activate the automatic part of the robot without human interaction a button had to be pressed to close the rolling window. Finally they pressed a further button to activate the robot for taking new suspension struts. In our study the suspension struts were fixed at the robot. The robot only moved back and forth and did the same steps in the next cycle again. Figure 2 shows the prototypical work place.

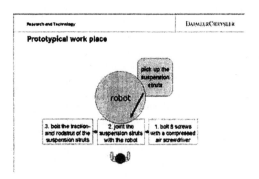

Fig. 2. Prototype of a workplace

The test subjects cycled. They did the three work steps successively and then began with the first step again. We observed the steps with a video camera and activity analyses, to compare the prototypical work place with the every day work places. In particular we were interested in the needed time, the usability of the robot, and the acceptance of the robot. The work context was very important, because persons perceive the problems of a new device when they have to use it like in every day work. The prototypical work place had one restriction, the missing assembly line. One researcher turned the work steps one and three back after the subject completed them to simulate the moving assembly line. It was a static work place. The subjects worked about twenty minutes at the prototypical work place.

3.5 Structured Interview

After the subjects worked with the robot they were interviewed. One researcher interviewed one, two, or sometimes three subjects. The interview lasted about one and a half hour. Contents of the structured interview were the following categories:

- general interest (assessment of our prototypical work place, emergency strategy, etc.)
- ergonomics (working with the robot) safety
- perception and acceptance of the robot (attitude to robot work)

Because of the interviews and the observations we and analyzed the subjective impressions as well as the objective conditions.

3.6 Time Period

The investigation took place in January and February 2003. The co-operation and activity analyzes were in January, the prototypical working with the robot was in February.

3.7 Sample

Because of the work rotation at the assembly line we observed different workers at the same station. Altogether we observed twelve persons (one woman), six of them in the early shift and six employees in the late shift.

At the prototypical work place all subjects worked at the assembly line C-class. Nineteen subjects attended at our investigation, all of them were male.

4. RESULTS

4.1 Activity Analyses

The test subjects needed 94 seconds for one cycle at our prototypical work place. The rhythm for one cycle at the assembly line was 72,5 seconds. An analyses with a t-test for independent samples (α = 5%) at the first fictive working step resulted in a significant difference. For this step the subjects needed significant more time than they needed at the assembly line. For the insertion and the screw up of the traction- and rod strut a t-test (α = 5%) didn't show any significant differences. Comparisons of the cycles of the fictive work station a subject did in the beginning with cycles the same person did in the end of the 20 minutes didn't show any significant differences.

At the assembly line the employees needed 53 seconds with a standard deviation of 7 seconds for one cycle at the handling station. The part of hard physical workload (take the suspension struts with the handling device and move them to the L-line and let them out) was 42 seconds with a standard deviation of 9 seconds. For the direct interaction with the robot the subjects needed 33 seconds.

The activity analyses (figure 3) shows a homogenous working process human-robot.

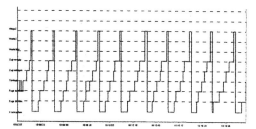

Fig.3. Human – Robot working process

4.2 Structured Interview

General Interest:
Five employees had experiences with robots. The attitudes towards robots were diverse. The employees were surprised about the easy handling with the robot. They estimated 5-10% of the physical effort by working with the robot compared with the handling device. All subjects were aware of rationalization.

Most employees had doubt about the space the robot would need. The handling device was perceived more flexible. In case of failure it can be replaced quickly. At the moment employees can move freely cross the handling work place when they support each other. With a robot they are not anymore so flexible.

Ergonomic:
The delay of the automatic start by the start was a disturbance of the working process. The two handed action was seen as disturbance as well. The employees assumed that within a short time one joystick would be taped with adhesive tape, so that one handed action would be possible. Some employees recommended to use visual or haptic sensors.

Furthermore the levers were to high for smaller persons, that could be a problem, because shorter persons had to stretch their arms for steering the robot.

Lamps were put up at eye level and should give the subjects feedback when the suspension struts were fixed in the brackets. Nearly all subjects didn't notice or care about the lamps.

Most of the subjects criticized the mechanism to shut the rolling gate. Both buttons should be put together and the reaction of pressing the buttons should be immediately.

The width of the window was rated too small, because with a moving assembly line employees need more space to steer the robot.

Safety:
Work process design and security were the center of our investigation. Aspects concerning security will be discussed in a further forthcoming publication.

Acceptance and Perception of the Robot:

Altogether the employees assessed the robot positive, because of the easy handling and the less physical work load. Most of them saw the robot as a machine or intelligent system (Figure 4). Problems were seen in the space the robot would need and an emergency strategy when the robot would not work. The current spatial layout of the assembly line would not allow the installation of the robot due to missing space. Furthermore they saw problems in an emergency strategy when the robot turn out. At the moment an alternative handling device is provided that could not be used with the robot installed.

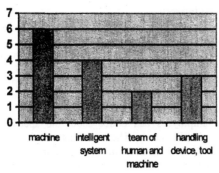

Fig. 4. Operators feeling of an assistant robot

5. DISCUSSION

5.1 Activity Analyses

The prototypical work station contained to many work steps in view of the assembly line rhythm. The significant differences at step one can be caused by damaged thread of the screws at the prototypical work place. At the beginning we used the same screws for 2-3 participants. After they were broken we used them only for one participant. Comparisons of the cycles a subject did in the beginning with cycles the same person did in the end didn't show any significant differences. That means employees can quickly learn the handling of the robot.

Some employees complained about back ache when working to long with the handling device. Employees are working about 80 minutes in one shift with the handling device. The introduction of the robot would relieve the hard physical work load. Besides, the employees needed less time by working with the robot compared to the handling device.

Figure 3 shows an homogenous working process at the prototypical working place. In particular for vacation workers and disabled persons it would be a good work place. But for every day workers it could be a problem to have less change (monotony) in their work. Especially the underestimation of the robot's power and the attribution of intelligence in

combination with and the monotonous work could bear a safety problem.

5.2 Structured Interview

General Interest:
The space was limited at the current assembly line. Installing the robot would limit or even prevent the flexibility of the workers to move between the workstations to help out colleagues. When designing future assembly lines installing a hand guided robot spatial flexibility has to be provided at the work stations that employees can help each other out.

Ergonomic:
The levers should be adjusted individually. Taller and shorter persons can consequently determine the best height for themselves.
The lamps at eye level have to be bigger and the lights have to give a clearer feedback, e.g. like a traffic light.
The buttons for closing the rolling gate should be put together.
The window has to become bigger.
But all in all the handling with the robot was assessed as very good.

Acceptance and Perception of the Robot:
In view of safety it could be difficult that the robot was seen as an intelligent system. Most of the employees underestimated the danger of a robot in comparison with the handling device (figure 5). But employees with experience with robots had more respect.
Furthermore it has to be thought about an emergency strategy when the robot falls out.

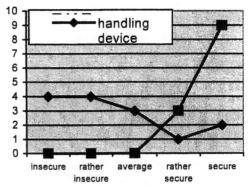

Fig. 5. Operators evaluation

6. CONCLUSION

All in all hand guided robots can be applied in assembly lines. The hard physical work load with the handling device can be reduced. Our investigation showed that it is not only a problem for older employees. In particular back problems can be avoided and therefore costs can be reduced. But the robot has to be tested in a real working surrounding. Furthermore we estimate that half of a work station can be rationalized if the robot were installed in the current assembly line. Training should be considered, because employees did not see the power of a robot, in particular with the combination of monotonous work.

When introducing hand guided robots at the assembly line employees should participate in the design and layout. Persons working every day at the assembly line know their work place very well and can support by designing the work process. Employees' work should include more work steps besides handling with the robot to avoid monotony; for example, little programming work or maintenance of the robot. This would support a good assessment of the robot.

REFERENCES

Akella, P. et al. (1999). Cobots for the automobile assembly line. *Proceedings of the IEEE 1999 International Conference on Robotics and Automation.* Detroit, MI.

Dudenhoeffer, D. et al. (2001). Modeling and Simulation for Exploring Human-Robot Team Interaction Requirements. *Winter Simulation Conference 2001, pp. 730-739.*

Hradil, Stefan (2000). Sozialer Wandel. Gesellschaftliche Entwicklungstrends. *In: Schäfers, Bernhard Zapf, Wolfgang (Hrsg.): Handwörterbuch zur Gesellschaft Deutschlands, 2. Auflage, Opladen: Leske und Budrich, S.645f.*

Lawaczeck, M. et al. (2003). Ergonomische Beurteilung von Montagetätigkeiten in der Automobilindustrie. *Zeitschrift für Arbeitswissenschaft 2003/1, p. 35.*

Pershkin, M. and Colgate, J. (1999). *Cobots. Industrial Robot: An International Journal Volume 26, 5 , pp. 335-341.* MCB University Press.

Scholtz, J. (2003). Theory and Evaluation of Human Robot Interactions. *HICSS, 36.*

Ulich, E. (1998). *Arbeitspsychologie, 89-99; 161-179.* Zürich: vdf, Hochschulverl. AG an der ETH Zürich; Stuttgart: Schäffer-Poeschel.

www.elsevier.com/locate/ifac

ON HUMAN ROBOT COLLABORATION - A SURVEY ON COST ASPECTS

Heinz-H. Erbe

Center of Human-Machine Systems, Technische Universität Berlin

Abstract: Collaboration of human operators and automation systems like robots are under development for achieving more flexibility in production and for saving cost when avoiding repeated reconfiguration of the systems. Tele-operation is considered with the aspect of tactile interaction over remote workplaces supporting maintenance. Cobots are presented to relieve workers from physical load together with path guiding assistance. Recent research results of robot assistants of workers at shared workplaces are discussed with respect to their implementation in industry. *Copyright © 2004 IFAC*

Keywords: intelligent assisting devices, continuous variable transmission, force feedback devices, mixed reality, human-robot communication

1. INTRODUCTION

A recent IFAC-Technical Board workshop (Rotterdam, August 2003) identified Collaborative Control as an emerging area for automatic control among others. It has been stated that "distributed systems are typically composed of numerous lower-level sub-systems with their individual control tasks and responsibilities. Collaboration among such interrelated systems is clearly essential in order to benefit from the respective strengths of the several "partners"".

Machine-Machine- and *Human-Human* - collaboration are subjects to consider, but this contribution concentrates on *Human-Machine or Human-Robot*-collaboration.

Collaboration demands a deep involvement and commitment in a common design, production-process or service; i.e. to work jointly with others on a project, on parts or systems of parts.

Cooperation on the other hand demands only to share the same intentions with others; i.e. the formation of partnerships and commitments among enterprises. Therefore collaboration is more strong than cooperation.

Three different developments regarding collaboration will be considered:
- hand guided robot (Cobot)
- tele manipulation, tele-operation
- robot guided by communication (gestic, speech recognition).

Hand guided robots are considered as Intelligent Assisting Devices (IAD), tele-manipulators/tele-operators as semi-intelligent devices and the third category as human co-workers.

2. HAND GUIDED ROBOT (COBOT)

The development of collaborative robots (Cobots) or intelligent assisting devices (IAD) was motivated by ergonomic problems in assembly of parts, where its weight endangered long-term the human body. A complete automation of assembly processes is complicate if not impossible at all. The reconfiguration of Robots doing assembly-tasks could be very costly. Humans on the other hand have capabilities that are difficult to automate, such as parts-picking from unstructured environments, identifying defective parts, fitting parts together despite minor shape variations, etc.

Physical teaming of workers and robots in a shared workspace can help to solve assembly problems and protect workers from long-term health endangering. There exist different concepts to do this. One is called extenders (Kazerooni, 1996), where the robot amplifies the human effort (servo motor concept, when steering an automobile or amplifying the brake-force). In another concept the human force is primarily a source of information rather than power

(Deeter et al, 1997). In contrast the Cobot-concept supposes that shared control, rather than amplification of human power, is the key enabler (Peshkin et al, 2001). The main task of the cobot is to generate a virtual environment, defined in software, into physical effect on the motion of a real payload, and thus also on the motion of the worker. An overhead rail system in gantry-style as used in many shops can be considered a cobot but without a virtual surface.

Virtual surfaces separate the region where the worker can freely move the payload from the region that cannot be penetrated. These surfaces or walls have the effect to the payload like a ruler guiding the pencil. To draw a straight line free hand is not easy just as the unguided movement of a payload is for a crane driver. Virtual surfaces are generated by software, and therefore the payload are moved by a shared control of computer and worker.

Fig. 1. Unicycle, a single wheel in contact with a planar surface.

In a two-dimensional task-space a virtual surface is a curve in a plane. In a three-dimensional one, it is a two-dimensional surface. The payload can slide along the curve or surface thereby relieving the worker from muscular force. Of course, he is anyway relieved from the payload.

The virtual surface can be hard like a ruler for a pencil or compliant. That depends on the task to be done and the environment, for example at an assembly line.

Figure 1 shows the simplest cobot, a single free rolling wheel (a Rollerblade) as the steerable transmission in contact with a flat surface, hold upright by a rail system. A steering motor reorients the rolling direction of the wheel. This mechanism is equipped with sensors for position, velocity and applied x-, y-forces of the worker. A payload would be decoupled from the force sensor. See figures 1 to 7 of Peshkin et al (2001).

The planar virtual curve in the task space has an effect like a rail, guiding the cobot when sliding

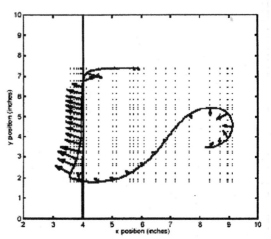

Fig. 2. Unicycle trajectory with the applied forces.

along and prevents the penetration of the "forbidden" region from the side of free movements. The steering angular velocity is $\omega = u/r$, (u measured tangential speed along the virtual curve, r the instantaneous radius of curvature). In the experimental case of figure 2 with a straight line as the virtual curve the angular velocity is zero. The free region is on the right side of this line. The arrows map the force of the worker. As sliding perpendicular to the rolling direction of the wheel is not allowed, one can see that the virtual curve is not hard at the beginning of the contact. This is caused through the open loop control, accumulating errors with deadreckoning. The movements in the free region are controlled by the

Fig. 3. "Scooter", three steered wheels with separate devices for measuring the rolling speed.

measured applied force of the worker to the handle.

The one-wheeled cobot has a two dimensional task space but only a single mechanical degree of freedom. Two mechanical degrees of freedom are possible with the three-wheeled cobot in figure 3. This enhances the task space dimensionality to three (x, y, θ). Whereas in the first case the worker can only control the speed of the movement and the direction is computer controlled by steering based on the measurement of the applied force to the handle, in

Fig. 4. Cobot at a General Motors assembly line

the second case the worker has also control over the direction.

Peshkin et al (2001) report on an application of the three-wheeled cobot at an assembly line of General Motors (Fig. 4).

The rolling wheels of the cobots described above can be considered as translational transmission elements. The one-wheeled cobot holds the velocities v_x and v_y in a proportion to the steering angle of the wheel.

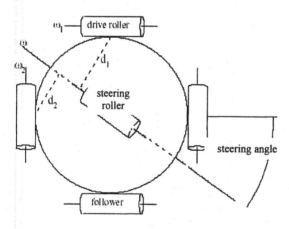

Fig. 5. Scetch of a spherical CVT

Fig. 6. Planar task space cobot with revolute joints

These steerable passive rolling wheels fall in the class of continuously variable transmissions (CVT).

To enlarge the dimension of the task space of cobots an articulated design with revolute joints is desirable. That means the construction of mechanical elements analogue to the rolling wheels, that relates a pair of angular velocities instead of translational velocities in a planar task space.

Figure 5 shows the principle of a solution where a

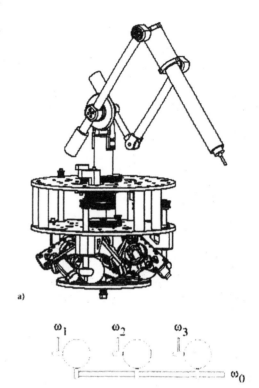

Fig. 7. Powered cobot with three revolute axes,

sphere caged by rollers is the basic element, the rolling surface. The angular velocities ω_1 and ω_2 are related to the steering angle by $\omega_2/\omega_1 = \tan \phi$.

A more compact solution needs only four rollers hold in frames drawn together by a spring thereby caging the sphere.

Figure 6 demonstrates the application of two CVTs in a planar task space. The endpoint with a handle is on the far right. Two steering motors with their rollers generate a transmission ratio of ω_3/ω_1 like the steering angles of the wheels of the planar cobots discussed above. The first two links without the

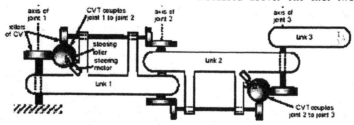

second CVT work like the cobot of figure 1, and all three links like the cobot of figure 3.

The CVTs in figure 6 communicate with each other in a serial chain. The transmission ratio constraining non-adjacent angular velocities (ω_1 and ω_3) is a product of all intervening CVTs. When the intervening angular velocity ω_2 is required to be zero, any value of the triple $\omega_1 : 0 : \omega_3$ is mechanically permitted. Peshkin et al (2001) propose a parallel structure of the CVTs to avoid this problem. Each joint's angular velocity is individually coupled to a common one ω_0. This causes an internal motion not coupled to a task space motion (for details see Peshkin et al, 2001). The internal motion can be powered to assist or impede the endpoint motion. Figure 7 shows a powered cobot for a x-y-z workspace with parallel arranged CVTs and a power-wheel at the bottom. This wheel contacts the sphere of each CVT introducing the common angular velocity ω_0.

Surdilovic et al (2003) proposed another concept of CVTs for Intelligent Power Assisting Devices (IPAD). The continuous variable transmission is a

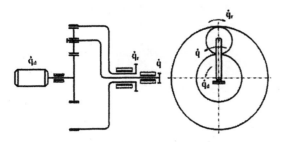

Fig. 8. Elemental hypocyclic differential gear

differential gear as explained in figure 8 (Fig. 8-11 of Surdilovic (2003)). It is cheaper to manufacture than the spherical CVT discussed more above. The stationary transmission ratio is $k = (\omega - \omega_r)/(\omega_d - \omega_r)$, which represents the angular velocities of the different parts of the gear. The drive velocity ω_d can be controlled based on the measured ring velocity ω_r.

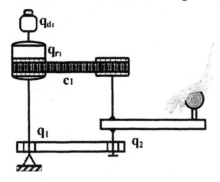

Fig. 9. Serial configuration

Figure 9 shows the application for a cobot of a serial structure. If the ring gear is fixed then the manipulator-endpoint can only rotate around joint 1.

If the drive is inactive, then the motion of the endpoint on joint 2 results via the tooth-belt, the ring gear and the differential gear in a motion along a fixed path defined by gear ratios and kinematic parameters. By activating the drive (the control system) a programmed trajectory will be followed. Measured is only the position of the end point as input for the drive control. A force sensor is not required. Figure 10 shows a cobot mechanism where the basic axes are coupled through one differential CVT.

Surdilovic et al (2003) enlarged this cobot mechanism with a parallel structure allowing path-

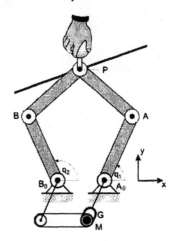

Fig. 10. Five-bar serial configuration

guidance and power assistance (Fig. 11). In cobotic mode with both drives controlled the motion depends on drive torques and the forces applied by the worker at the end point. The path within the free mode region is defined by the movements of the worker until the virtual wall is reached. Positions are measured by sensors. The accuracy to follow the virtual path

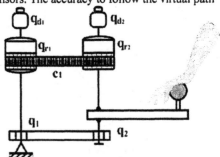

Fig. 11. Parallel configuration

depends on the law for closed loop control. Surdolovic et al (2003) report good results with a simple proportional controller. More details are presented in their original paper.

Kohler et al (2004) report on investigations with hand guided devices at assembly lines of an automobile factory. The analysis is restricted to robots assisting the workers when reducing the physical work load.

But the robots power can not be controlled through the workers as cobots can. Nevertheless the investigation found that workers at assembly lines accept the assistance through robots with a suitable work organization. However, cobots could be more appropriate. It seems that some more research and development is necessary to convince managers of the economic advantage of cobots by improving assembly work when used the skills of workers assisted by intelligent devices.

3. TELE-OPERATION

A multi-modal tele-presence technology, i.e. the extension of human sensory and manipulative capabilities to remote environments, attracted research and development efforts in recent years. The application is focussed to tele-operation in hazardous environments (space/nuclear), long-distance tele-maintenance and tele-service. Realistic tele-presence requires feedback of information in multiple modalities of human perception: visual, auditory and haptic. Several solutions for tele-operation are reported in the literature. See for example the overview of Melchiorri (2003). The reflection of the applied force to the environment on the remote site back to the operator is a requirement for an effective operation. Experiences with developed force feed back devices have been reported ((Buss, and Wollherr, 2003)

Yoo, and Bruns (2004) propose a low-cost momentum handle for force feedback (Fig. 12). The handle, actively driven by a motor, is always in a momentum-equilibrium through the wheel with a virtual force/momentum. A pressure-sensor attached to the handle senses a force if the user applies a momentum. This analog signal is fed via an A/D-converter to a microprocessor. The microprocessor controls via a D/A-converter the motor driving the wheel. This micro controller is connected to a serial port of a PC, where it gets the virtual force/momentum from. As an example Yoo, and

Fig. 12. Force feedback handle

Bruns (2004) demonstrate the lift of a virtual mass represented by an algorithm of a program running at a PC. The respecting signals are transferred to the micro controller generating via the motor moment a force at the handle. The momentum handle could be connected to the lever of a bench drill at a remote site. The handle (or lever) at the remote site senses a force if the drill touches the work piece. The micro controller transfers the signals to the local PC. The PC is connected via the internet to the PC at the master site, generating via the momentum handle on this site a force related to the force at the remote site.

The transfer of signals via the internet causes time delays. This affects the choice of the local controllers. Melchiorri (2003) discusses different strategies with respect to stability, inertia and damping, tracking, stiffness and drift.

The intentions of tele-presence or tele-operation are different from remote collaboration. No person is necessary at the remote site. But this could turn out as unrealistic, at least with respect to tele-maintenance. Then a bi-directional tele-presence is desirable. The concept of mixed reality, developed by Bruns (2003) using Hyper-Bonds as a universal interface type goes in this direction. This indication refers to the theory of bond graphs, promoted by Karnopp et al (1990). Mixed reality was first implemented for supporting e-learning, connecting virtuality to reality (Bruns and Erbe, 2003).

A first realization of a distributed collaboration has been presented by Yoo, and Bruns (2004). These developments could be enlarged for realistic human-human collaboration over distant sites.

4. HUMAN-ROBOT COMMUNICATION

Recent research results presented in the literature regarding a robot as a human co-worker are focused mainly on humanoid robots. See for example the conference documentation of Humanoids 2003 (Knoll, and Dillmann, 2003). These developments would have their application fields in offices and households as service robots and for assisting elderly and disabled persons.

An application in the manufacturing industry is very interesting. While industrial robots at production lines usually are working in a well known environment, the challenge is now to work together with human operators in an unstructured, uncertain environment.

One of the first developments for getting information of the human behavior when working was called "behavioral cloning" (Bratko, 1995). It is a process of reconstructing a skill from an operator's behavioral traces by means of machine learning techniques. However, the research was aimed to replace humans with robots but not for collaboration. This has changed now. Kimura et al (1999) recognized the object and the human grasp by two vision systems for analyzing the tasks done by the operator. With this information the robot analyses the operator demonstration and generates a task model for assisting the operator. Sato et al (2002) improved this development. Laengle, and Woern (2001) consider a robot as an intelligent assistant for the worker. Instead of researching for solutions to achieve a complete autonomous execution of complex tasks in an uncertain environment the authors enter into a compromise wherein the operator helps the robot to finish the tasks correctly when uncertainties occur. The robot can switch from automatically to semi-automatically executed tasks with help of the human operator. For observing the tasks force and vision sensors are used.

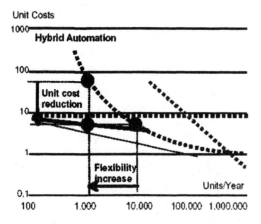

Fig. 13. Reduction of robot system costs compared to labor cost and assumed cost potentials of hybrid automation (Hägele et al, 2002).

5. CONCLUSIONS

Three different developments and their results regarding collaboration of human operators and robots have been presented and discussed. Cobots are promising for relieving workers from physical strain and can guide the path when moving parts for assembling. The implementation at workplaces in industry are still rare. Collaborating tele-operation with force feedback seems to be very interesting for supporting tele-maintenance. More effort has to be done to make it applicable for maintenance-service providers. Robot assistants for task sharing with human operators are still in its infancies but promising. All these developments have to be considered for their cost effectiveness. Not only the cost of the components, but the cost aspects within their working environment as well have to be assessed.

REFERENCES

Kazerooni, H. (1996). The human power amplifier technology at the University of California, Berkeley. *J. Robotics and Auton. Sys.*, Vol. 19, pp. 179-187.

Deeter, T.E., G.J. Koury, K.M. Rabideau, M.B. Leahy, Jr., and T.P. Turner (1997). The next generation munitions handler advanced technology demonstrator program. *IEEE Intnl. Conf. Robotics and Auto.*, pp. 341-345.

Peshkin, M.A., E. Colgate, W. Wannasuphoprasit, C.A. Moore, R.B. Gillespie, P. Akella (2001). Cobot Architecture. *IEEE Transactions on Robotics and Automation.*

Surdilovic, D., R. Bernhardt, L. Zhang (2003). New intelligent power-assist systems based on differential transmission. *Robotica*, Vol. 21, pp. 295-302.

Kohler, P., F. Sarodnik, T. Lum, H. Schulze, S. Haasis, R. Giesler (2004). Hand guided robots-Human centered design in assembly plants. *Preprints of this symposium.*

Melchiorri, C. (2003). Robotic Telemanipulation Systems: an overview on control aspects. *Preprints of the 7th IFAc Symp. on Robot Control*, Wroclaw, Poland.

Yoo, J-H., F.W. Bruns (2004). Realtime collaborative mixed reality environment with force feedback. *Preprints of this symposium.*

Wollherr, D., M. Buss (2003). Cost-oriented virtual reality and real-time control system architecture. *Robotica*, Vol. 21, pp. 289-294.

Karnopp, D. C., Margolis, D. L., Rosenberg, R. C.(1990). *System Dynamics – A unified Approach*. John Wiley, New York.

Bruns, F.W., Erbe, H.-H. (2003). Didactical Aspects of Mechatronics Education. In: *Proc. SICICA 2003 - Intelligent Components and Instruments for Control Applications*. L. Almeida, S. Boverie (eds.), Elsevier Ltd., Oxford

Knoll, A.C., R. Dillmann (eds.) (2003). Int. Conf. On Humanoid Robots, VDI/VDE-GMA, Düsseldorf, Germany, ISBN 3-00-012047-5.

Bratko, I., T. Urbancic, C. Sammut (1995). Behavioral cloning: phenomina, results and problems. *Proc. IFAC Symp. Automated Systems based on Human Skills*, Elsevier Ltd. pp. 143-149.

Laengle,T., H. Woern (2001). Human-robot-cooperation using multi-agent-systems. *J. Intelligent and Robotic Systems*, pp. 143-159.

Kimura, H., T. Horiuchi, K. Ikeuchi (1999). Human robot cooperation for mechanical assembly using cooperative vision systems. *Proc. IEEE & RSJ Intelligent Robotics and Systems 99*, pp. 701-706.

Sato, Y.K.B., H. Kimura, K. Ikeuchi (2002). Task analysis based on observing hands and objects by vision. *Proc. IEEE/RSJ Intl. Conference on Intelligent Robots and Systems*, pp. 1208-1213.

M. Hägele, W. Schaaf, E. Helms (2002). Robot assistants at manual workplaces: Effective cooperation and safety aspects. Proc. 33rd Intnl. Symp. on Robotics, Stockholm.

ELSEVIER
IFAC
PUBLICATIONS
www.elsevier.com/locate/ifac

ON WEB-BASED ARCHITECTURES FOR SIMULATION SUPPORTED TRAINING IN TECHNICAL NETWORKS

Alexei Lisounkin, Alexander Sabov, Gerhard Schreck

Fraunhofer Institut für Produktionsanlagen und Konstruktionstechnik
(Fraunhofer Institute for Production Systems and Design Technology)
Pascalstraße 8-9, 10587 Berlin, Germany
Phone: +49 (30) 39006 142,   fax: +49 (30) 3911037
e-mail: {alexei.lisounkin, alexander.sabov, gerhard.schreck}@ipk.fhg.de

Abstract: The paper presents an approach for the cost effective implementation of simulation supported training for plant operators. Simulation functions are seen as practical support for human cooperation with complex automation systems. A web-based system architecture is proposed and a generic approach dedicated to modeling and simulation of technical networks is applied. The concept has been developed within a project of the Fraunhofer Association e-Industrial Services. Aspects of system implementation and application are outlined. *Copyright © 2004 IFAC*

Keywords: Water industry, Web-based training, Software architectures, Process simulators, XML.

1. INTRODUCTION

Increasing requirements on the optimality and safety of water, gas, and oil supplying networks demand the use of advanced simulation based training systems, as well as their close interaction with the supervision and control modules of the networks. High skilled operators promote the concept of cost oriented automation, leading to less system failures and shorter down times and set up times.

Design of simulation systems and integration into daily business of network operators need remarkable human efforts. The attempts to make design and use of model-based simulators more practicable and more efficient are extremely significant. Observing the current situation with respect to staff qualification tasks in relation to limited resources for development of training applications, we recognize a dramatic need of cost effective training systems over the entire life cycle of automation solutions.

In the scope of this paper we focus aspects of web-based simulation systems for such networks like water, gas, and oil distribution, as mentioned above.

The operation of such system is rather challenging business. Process and system knowledge is distributed between network developers and installers as well as network operators. Moreover, the know-how may also be accumulated in different organizational units of the companies as well as at different locations. Naturally, use and maintenance of related simulation-based tasks must be provided on a distributed software platform by means of generic system configuration procedures.

There are some basic properties that facilitate the modeling of the mentioned networks and that will be exploited by our approach. Although the networks possess a relatively complex topology, the local functionality of the network elements can be represented by a small set of "standard" process units: pumping stations, filters, storages, valves, and connecting pipes. The main function of the facilities is the processing and transport of materials as homogeneous media. Therefore, the modeling of such network is based mainly on the energy and mass balance laws. The topology together with the settings of the processing units substantially determines the energy and mass flows in the facility. As a result, the topology also determines the structure of the equation system to be interpreted when performing the process control.

The modeling of simulation tasks by means of a proper modeling language together with a supporting interpreter increases the efficiency of application development. Recently, internet infrastructure with available tools offers a sufficient communication medium for development of distributed services. The remote access to training services reduces investments on site of the technology user and allows an effective development, operation, and maintenance of the server hardware and software.

2. GENERAL SYSTEM REQUIREMENTS

At the beginning, let us consider requirements on the system as well as server-client interaction model. Here, we take into account such application field like operation of process facilities. The requirements have been specified by associations of users in related businesses (e.g. NAMUR - association of users of process control technology, 1995).

In general, computer-based training, distance learning, and web-based education are being developed and applied more and more extensible. Especially internet plays an increasingly role as communication medium (see Page, *et al.*, 2000), and amount of tools available for development of web-based training services is huge. Such web-based training solutions allow a cost effective implementation over the complete life cycle of the system (Schreck, 2002). Within a short time an offer with respect to e-Learning solutions has been escalated. Also manufactures of technological and automation components are beginning to organize educational support and training for their products via internet (Langmann and Hengsbach, 2003).

Our intention is to provide technology and tools for training with respect to system view and dynamic behavior of the facilities. Here, system instructor and operator are in the focus of use case studies.

2.1 Functional Requirements

The functional requirements will be considered with respect to server and to client sides. Obviously, the server must provide functionality for an adequate - in terms of training tasks - simulation of a relevant facility. Moreover, performance of training tasks can be fulfilled if simulator supports the following functionality (compare to NAMUR, 1995):

- Simulator initialization with respect to training session objectives;
- Simulation time management (time slower/ faster);
- Simulator pause, continue, stop;
- Snapshot, save and load of simulation variables;
- Management of access to simulation variables (input/output access);
- Data access management and data monitoring depending on user role (instructor/ operator);

We emphasize that the simulation means reproduction of a facility dynamic behavior with respect to given values of input variables. The result of simulation is represented in form of an output vector with new values for model state variables. With respect to simulation complexity and its correspondence to the facility functionality, the possibility of operation with partial models should be available.

The client site must provide facilities for:
- Session management (create, edit, start, interrupt, pause, continue, stop) and
- User role management (instructor/operator).

From the user point of view, user interface should differ the user roles as well as abstraction levels and complexity of simulation tasks:
- Full scope native interface;
- Reduced scope task specific interface;
- Generic interface.

Obviously, the use of generic methods for model construction, interface configuration, and communication language increases the efficiency of the development. Here, we make use of internet specific tools like XML language for facility and training scenario data modeling as well as XML tools for model verification, for interaction with training system on client site, and for communication between client and server modules.

2.2 General System Architecture and System Behavior Modeling

Use of the client-server architectures is common for web-based applications. We focus on specific aspects of client-server architecture for training systems with dynamic behavior:
- Synchronization of client-server actions;
- Management of client and server states.

Naturally, the interaction of client and server via web is possible by means of messaging. The guiding role is played by the client. The server processes the requests of the client and provides them with data response. We use a synchronous mode of interaction. When a request has been sent by the client, the client is waiting for the response. Respectively, when a

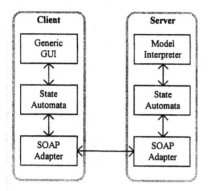

Fig. 1. General System Architecture

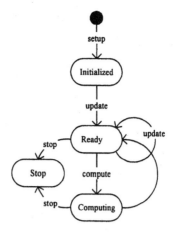

Fig. 3. Server State Automata

response has been sent by the server, the server is waiting for a new request.

The implementation of such interaction mechanism can be done using SOAP protocol (see http://www.w3.org/TR/SOAP/). The implementation base for this protocol is provided by a function-call-interface between the client and server modules. Here, the client has a calling role, the server is functionally subordinated to the client. The general architecture is depicted in Fig. 1. Implementations of SOAP protocol can be downloaded in internet.

Modeling of client and server states evolution is provided by means of finite automata. The state automata of the client is given in Fig. 2. The client features two main states - "Ready" and "Wait". The first state can be used for updating input information for the simulator. State "Wait" is achieved after transmission of new input vector to the server and will be finished by the corresponding server response. The user interface can be maintained in a separate thread in order to support a "smooth" communication with the user.

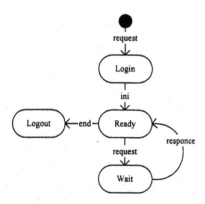

Fig. 2. Client State Automata

A simplified state automata of the server is depicted in Fig. 3. State "Ready" is used for the communication phase and state "Computing" for model processing.

State "Computing" may have internal complexity depending on evaluation scenarios of the simulator. In our application, we define two internal states - state "Listening" and state "Evaluation". "Listening" means numerical initialization of simulation attributes. For example, some process history which is used for measurement data filtering can be built during this phase. "Evaluation" means solving the simulation task. The interaction of these two phases is managed by the simulation module itself. Due to such structure of client and server automata, the mapping of client states and server states and synchronization of processing are very intuitive.

3. USER ROLES AND USE CASES

As already mentioned, web-based simulators promise to combine a localized development and maintenance of the simulator with distributed use of it. Naturally, generic aspects of simulator parameterization with respect to user specific facility, user role, and training tasks together with generic building of user interface are very important.

Related to simulator use cases, we consider the following user roles:
- "System access and administration" corresponds to user roles "Administrator" and "User";
- "Application design and maintenance" corresponds to user roles "Supervisor" and "Engineer";
- "Training" is performed by users in roles "Instructor" and "Operator".

In the framework of client use cases, the following use cases are considered:
- Session management represented by such functions like "Login", "Logout", etc.;
- Task parameterization with such functions like "Load/ Edit/ Save model", "Add/ Edit/ Delete element", "Change attribute", etc.;
- Application control including such function "Evaluate task", "Change input", etc..

With respect to client module use cases, all user roles may possess relations to all listed use cases. The existing trend by implementation of web-applications is to avoid saving session and user context information on the client side due to problems of access security. From this point of view, the user interface is functionally indifferent to user role.

4. SYSTEM COMPONENTS

This chapter is devoted to description of main technological components used or developed for the implementation of the system. These main components are:

- Simulation model;
- Modeling language for configuration of simulation model and user interface;
- Model interpreter;
- Graphical user interface;
- Communication.

In the following sub-chapters, main characteristics of the components are emphasized.

4.1 Simulation Model and Modeling Language

The modeling technology must allow an efficient structural and parametrical adaptation of the model to particular implementation variants of the supplying networks as well as to the individualities of control rules.

Assuming the network-like structure of the facilities, this goal can be achieved by framing "elementary" sub-models into more complex modeling structures, which represent the entire system. The sub-models are organized in application field specific libraries of task oriented functional modules. The functional modules represent physical and logical behavior of partial processes.

Thus, the modeling procedure is structured in two tasks. The first one is the design of a library of modeling units representing a relevant physical behavior of partial process elements. Hereby, due to the complexity of the technological processes, phenomenological modeling plays a great role.

The second task is modeling the network (or facility) topology and its interpretation. This includes specification of model unit interfaces (input, output, parameterization possibilities) as well as definition of connecting rules. The elements of the library together with the connecting rules form a modeling algebra allow construction of facility models in a formal (abstract) way using predefined formal operations – connecting rules (compare Lisounkin, et al., 2003).

Application of this approach to systems with a regular structure – networks for material flow – promises a high efficiency of the modeling. As already mentioned, we consider systems under following assumptions:

- Systems possess network structure with technological nodes and technological links;
- Nodes are elements which change properties of material flow;
- Links are elements for technical transportation of the processed material;
- System design and the corresponding system modeling are the superposition and association of technological elements by means of technological links.

The set of processing elements included in the system constructs the core of the system data model. The topological (or functional) relations between the elements serve as references between the nodes. It is important to emphasize that advanced simulation-based tasks may evaluate the node relations of different nature – technology-caused connections, supervision/control reasoned data flow, resource logic relations etc. An allied processing of relations, that may have different types and nature, is increasingly important for the advanced modeling technologies.

The design of advanced modeling and configuration languages is being considered with respect to the approach of generic modeling and reusable software. Development of an XML-based language for describing the network topology in conjunction with application line specific XML languages for describing the node characteristics, we determined as an appropriate technology. Herewith, the network modeling and verification language has an integrating role. The fact that the scanning, parsing, tools for XML-based languages are available justifies the use of XML grammar.

In general, XML-based grammars are appointed to represent data models with a tree structure. Obviously, the depicted class model can be also depicted by means of an XML-based language.
Using the natural structure of the XML grammar, the model of a network structure easily transformed in a form with four main semantic parts:

- a set of functional elements of a network ("node"-elements) – nodes in a graphical representation of the model,
- a set of networks ("net"-elements) with their connections ("pin"-elements) – edges in graphical representation of the model,
- a set of verification rules ("rule"-elements) which may define legal and illegal network patterns for the relevant network, and
- a set of attributes ("attr"-elements) related to node elements, network definitions, and connection descriptions.

The modeling language provides a rather conservative set of conventions for the definition of the set of "node"-elements and the topology of the related networks.

4.2 Model Interpreting

Speaking about model interpreting, we focus not only model simulation, but rather all information processing steps exploiting network structure of the technical system. Thus, the tasks listed below use substantially the network structure of the model for the processing:

- Model and process visualization;
- User input of model and process controls;
- Model-based simulation;
- Model data communication;
- Model data storage/editing/maintenance.

In the framework of the training simulator, the processing of the listed functions was implemented in a generic way. Thus, the functionality of technological units – pumping stations, filters, storages, valves, and connecting pipes – with respect to all mentioned tasks was implemented in form of object-oriented libraries. Based on these libraries, the processing of the tasks is based on formal methods of model building with respect to relations between the processing elements given by the topological relations.

4.3 Graphical User Interface

Considering the requirements of cost efficiency and web based technologies, we use Java together with Java3D by Sun Microsystems (http://java.sun.com) for the model viewing and user interaction. While Java3D serves as a free 3D viewer for the model, the direct relation to Java provides a powerful and free language for developing internet tools. Using these technologies, the graphical user interface can be integrated into a standard web browser (see Figure 5 for used technologies).

Our GUI in Figure 4 consists of:

- a 3D-View Panel showing the model;
- a navigation panel for the 3D world;
- a table showing the attributes and its values for a selected technological unit;
- a start/stop panel for simulation;
- a file panel for opening and saving models.

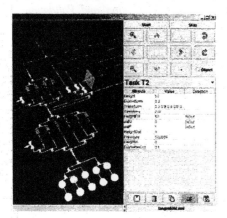

Fig. 4. GUI for real water supplying system

4.4 Communication

Communication between user and simulator is done by using the SOAP protocol. SOAP is a standard protocol intended for exchanging structured information in a decentralized, distributed environment (http://www.w3.org/TR/SOAP/). For the two programming languages (Java and C++) we used, there are plenty of free SOAP libraries available, providing the complete functionality for connection and error handling.

Clients can be developed rapidly for any language or software tool by using standards for communication description like the Web Services Description Language WSDL (http://www.w3.org/TR/wsdl).

A typical communication between the client and the server on runtime consists of:

- Client sends required process values to server;
- Client sends a compute command;
- Client requests process values from simulator;
- Server sends the requested process values.

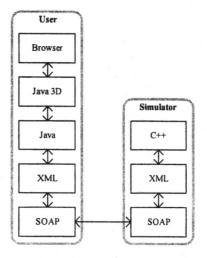

Fig. 5. Technologies for GUI and communication

5. TRAINING SCENARIOS

Within the previous chapters the infrastructure for simulation based training is outlined. Its configuration and application to specific training tasks strongly depend on the qualification goals for plant operators, which may be focused to:

- Training of the start-up and shut-down procedure,
- Training of the normal system operation,
- Training of exceptional situations and how to handle disturbances,
- Development of execution alternatives and optimization.

Accordingly training scenarios are to be elaborated and have to be provided to the network operator by training sessions and according exercises. Training sessions can be executed self-contained by the operator or by support of an human instructor.

Training scenarios are represented by specific models or sub-models of the plant which are loaded to the simulator and initialized by the according starting state. The dynamic behavior of the system is provided by the execution of the network simulator. Hereby, individual process units (e.g. pumps, pipes) may be setup specifically to the required behavior within training exercises. The resulting system operation is a consequence of the user input against the dynamic simulation.

In order to control and document the dynamic course of events within the training session, the relevant variables of state are recorded in data archives. This information represents a profile of the system's running and will be used for analysis and assessment. It is possible to evaluate the profile of the training exercise as well as the profile of the actual running of the system. From this, training elements can be constructed that coordinate with specific system situations with the training phase of the operator.

The intention is to provide the training scenarios in form of scripts with input and output values for the training tasks. The elaborated modeling language defines a semantic-syntactic framework for the scripts. The scripts can be produces involving facility supervision records originated from experienced operators (Lisounkin and Schreck, 2002).

6. CONCLUSION

Taking into consideration that the installation of training simulators is a rather expensive endeavor, to offer a simulation-based training as a service via internet is a promising approach. The developed approach was used for implementation of a training simulator for a real water supplying system. The actual work is devoted to elaboration of training scenarios for the operators. Simultaneously, the library of functional processing elements is being permanently extended. Further research work is also devoted to further development of modeling tools, especially model verification methods and extension of modeling language with respect to specification of model validation rules (Yagüe, et al., 2004).

7. ACKNOWLEDGMENTS

The initiative "e-Industrial Services" sponsored by Fraunhofer Gesellschaft zur Förderung der angewandten Forschung e.V., Munich, Germany, has declared the objective to provide a service platform, which allows both to order and to execute value-added services through the internet (http://www.e-industrial-services.de/), (Berger and Hohwieler, 2003). A main focus of development at Fraunhofer IPK is the planning and design of qualification services for operator personnel, especially the implementation of simulation-based training systems for system operators.

REFERENCES

Berger, R. and E. Hohwieler, (2003). Service Platform for Web-based Services for Production Systems. In: *Proceedings of the 36th CIRP International Seminar on Manufacturing Systems*, University of Saarbrücken, Germany.

Fraunhofer e-IS, http://www.e-industrial-services.de/.

Langmann, R. and K. Hengsbach, (2003) E-Learning & Doing Automation. In: *atp* 45, Heft 2, p. 58-68.

Lisounkin, A. and G. Schreck, (2002). Water Consumption Profile Analysis for Facility Control and Operator Training. In: *Proceedings of IASTED International Conference APPLIED SIMULATION AND MODELLING (ASM 2002)*, Crete, Greece, p. 234 - 239.

Lisounkin, A., G. Schreck, P.P. Alacórn, J. Garbajosa and A. Yagüe, (2003) XML-Based Modelling Language for Technical Networks. In: *Proceedings of the Industrial Simulation Conference ISC 2003*, Valencia, Spain, p. 541-545.

NAMUR (1995). Management von Trainings-simulatorprojekten. *NAMUR-Arbeitsblatt NA 60*, Version 22.11.1995.

Page, E.H. et al. (2000). Web-based Simulation: Revolution or Evolution?. In: *ACM Transactions on Modeling and Computer Simulation*, **Vol. 10**, No. 1, p. 3-17.

Schreck, G. (2002). Simulation Services for Training of Plant operators. In: *Knowledge and Technology Integration in Production and Services. Balancing Knowledge and Technology in Product and Life Cycle* (V. Marik et al. (Ed)) IFIP, Kluwer Academic Publishers, p. 79-86.

SOAP, http://www.w3.org/TR/SOAP/.

Sun Microsystems, http://java.sun.com.

Yagüe, A., P.P. Alarcón, J. Garbajosa, A. Lisounkin and G. Schreck, (2004). Construction of Verified Models for Systems Represented as Networks. In: *Proceedings of the 2nd International Workshop on Verification and Validation of Enterprise Information Systems VVEIS 2004*, Porto, Portugal, p. 44-49.

Web Services Description Language WSDL, http://www.w3.org/TR/wsdl.

ELSEVIER

IFAC
PUBLICATIONS
www.elsevier.com/locate/ifac

AN INTERACTION SYSTEM FOR EASY ROBOT PATH CORRECTION

Raffaello LEPRATTI
Ulrich BERGER

Brandenburg Technical University at Cottbus
Chair of Automation Technology
P.O. Box 10 13 44, 03013 Cottbus – Germany
{lepratti;berger.u}@aut.tu-cottbus.de
Tel.: 0049 - (0)355 – 69 2457
Fax.: 0049 - (0)355 – 69 2387

As logical consequence of a drastic growing automation degree the relevance of improved Human-Machine Interaction Systems is stressed as one of the essential key issues towards the future production scenario. In particular in the automotive sector market competitive reasons lead to increasing product diversification and, therefore, to the need of innovative concepts for flexible manufacturing systems. Consequently, also developments like standard and user-friendly interaction systems assume a paramount importance as contribution in stabilizing innovation leadership. An advanced factory automation system for easy corrections of deburring and dressing results, based on industrial robots, has been developed and successfully employed in industrial context. Since it has shown high efficiency, it has been further developed towards its use in further application fields. This paper points out related development aspects. *Copyright © 2004 IFAC*

Keywords: Man-Machine-System, Manufacturing System, Programming Systems, Robot Programming, Automobile industry.

1. INTRODUCTION

The innovative core production element of the future enterprise is represented by networked, knowledge-driven and agile manufacturing systems. These should ensure enterprise adaptability and innovation for any market demand connecting industrial competences, intelligent infrastructures and human centred operation.

According to this premise, it can be affirmed that one of the prime objectives of future research activities would focus on the development of knowledge-based and user-friendly manufacturing systems. These should aid a new production paradigm concept, whose results must lead to an innovative manufacturing infrastructure, economically and eco-efficiently scalable as far as down to the lot-size-one.

This aspect represents the most important pre-requirement in order to promote the transition process of industry from a mass production towards a full knowledge-based customer- and service-oriented one. Connected solutions should enable traditional industries to profitably produce semi-automated at scalable lot-sizes even down to one, and thus to

customize, to react faster to market changes, to launch rapidly new innovative or value added products as well as risk and, thus, to allow industry long-term competitiveness improvements.

A breakthrough in new applicable knowledge should be achieved by combining a framework of proven standardized technologies (production equipment as well as robots) with new co-operation and control approaches (automatic/interactive online planning and programming, cognitive sensing and control, reconfigurable systems). In addition, novel operational knowledge-gaining and -feedback mechanisms, machine intelligence and human machine interaction should be brought to a level of efficiency and robustness far beyond state-of-the-art.

The main radical innovation in manufacturing should focus on the realization of a new paradigm of knowledge-based human-centred production systems for the lot-size-one paradigm in a wide range of manufacturing scenarios.

Inherent to this innovation are:

- Concepts and technologies to acquire and transfer knowledge to next product variations and even generations, based on automated knowledge-gaining and feedback mechanism. Advanced supervision concepts cope with complexity and required rapid convergence.

- Scalable, flexible and (re-)utilisable production technologies covering the full spectrum: prototype production, mass-customization, production on demand, rapid time to market of products at uncertain and dynamically scalable volumes.

- Tool chain for interactive PPR, automatic on-line planning, programming and execution of workcells, including robots and other equipment (fixtures, tools, conveyors) with full support to automatic detection of reconfiguration situations.

- Realisation of lot-size-one robot based reconfigurable production cells combining real autonomy and operational safety, based on advanced automatic planning and programming, evolving machine intelligence and new interaction methods.

2. HUMAN-MACHINE INTERACTION ISSUES

Following the vision described in Section 1, also standard and user-friendly platforms for easy correction of process procedure and parameters emerge as necessary key elements towards a better embedding of human skills in future production scenarios.

Therefore, research efforts must focus also on the development of new strategies providing standard solutions for the human-centred manufacturing architecture. Its structure must feature a systemic dynamic learning behaviour, in which innovation arises from new complex interaction forms between integrated technologies, human resources, management and organizations in all phases of the valued-added chain, i. e. (i) production preparation, (ii) planning and programming as well as (iii) process execution.

Under these circumstances a standardized and smooth global information flow between all actors (humans and machines) involved in manufacturing processes is the most important pre-requirement for ensuring the correct process behaviour. However, too many complications still evolve when trying to find standard criteria for interoperability across the entire heterogeneous human qualifications and machine programming languages setting (Lepratti et. al., 2004)

As already stressed in (Lepratti and Berger, 2004), interoperation barriers could be identified in each of the main communication forms, which characterized today's manufacturing environment: beginning from the Human-Human Communication and the Machine-Machine Data Exchange up to the Human-Machine interaction. In this paper it will be paid attention exclusively to the recent communication issue, i. e. the Human-Machine Interaction, stressing

the importance of a novel solution for a unified Graphical User Interface (GUI), which allows machine operators user-friendly process corrections at robot cells for multi-task purposes.

3. STATE-OF-THE-ART

As seen in Section 2, Human-centred technological improvements in today's manufacturing scenarios require radical evolutions, in order to keep step with the internationalisation process of markets. Thus, due to competitive reasons, OEM manufacturers tend to (re-) integrate intelligent and user-friendly interaction technologies back in manufacturing plants.

As classical example of a possible application, state-of-the-art fuel and diesel engine cylinder heads are usually made of aluminium alloys in a mold casting process. After the casting, solidification and downcooling phases the cylinder head shows air inclusions, microcracks and pipes on its surfaces. Adjoin to the investment casting process, risers, gates and burr formations appear. Additionally, more or less strongly minted burrs form in dependency of the wear state of forming tools and in correspondence with tolerances between all boundaries of tool components (e. g. steel mould sections and sand core). As result: The form of the cylinder heads changes consistently up to some millimetres during the entire production time.

Thus, during the pouring process of different engine parts form deviations arise inevitably. These can endanger, obstruct and eventually prevent their further handling and treatment. Due to molds wear and additional different manufacturing tolerances the location of burrs, microcracks, pipes and so on differs each time. The unforeseeable location of manufacturing form deviations is the reason of huge problems in reaching a continuous quality of deburring results (see Fig. 1).

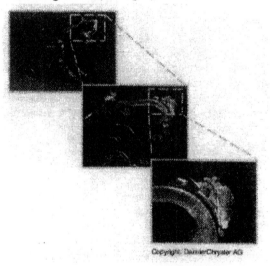

Fig. 1: Fig. 1: Example of arising burrs during the mold casting process. A form deviation on the work-piece upper right corner can be evidently recognized.

According to economical and safety requirements and as consequence of an ever more increasing automation degree (World Robotics, 2003) deburring and dressing processes are mostly performed by industrial robot systems. However, when deburring and dressing with industrial robots a time- and cost-intensive continuous machine re-programming is required. Therefore, it is necessary to develop an intelligent interaction system for on-line correction of industrial robot programs, in order to be able to react rapidly, surely and straightforward to these form intolerances enabling lasting functional and aesthetical quality results.

By means of such an interaction system simple intervention possibilities have to be easily carried out and correctly implemented in the main programs on the Robot Controller (RC). However, this must be made in accordance with the system user qualifications.

Thus, it becomes crucial in terms of the future manufacturing vision (Lepratti and Berger, 2004) to conceive intuitive multi-modal Human-Machine Interfaces, which represent everlasting challenges due to professional and cultural backgrounds of the involved personnel.

This paper describes a graphical interface, which has been developed for user-friendly robot path corrections at shop-floor level. Also further steps towards the development of a standard modular solution will be presented.

4. A FLEXIBLE HUMAN-ROBOT INTERACTION SYSTEM FOR ROBOT PATH CORRECTIONS

In this section some requirements for the development of a Human-Robot Interaction System are summarized before starting with the description of the performed solution.

Some related requirements are:

- In order to solve the problem described, correction values are to be given at a central standard terminal for all workpiece outlines, which must be processed.

- If the central terminal is a standard PC, the correction value input has to take place by means of an appropriate software tool in a simple and self-describing way, for example by a visually descriptive, graphically interactive control surface.

- A simple possibility must be given to the system user to input the correction values for removing the form deviations of all cylinder head types and surface areas in order to arrange a global correction of the production system.

- Furthermore, a constant high deburring and dressing quality has to be guaranteed during the entire production period without additional investments or downtimes for robots re-programming. This has to be achieved by fast reaction to all different component tolerances and wear statuses with direct intervention into the robot control.

Above all the main goal of the project described in this paper is a quality and productivity increase by lowering of reprogramming costs.

4.1 A first development: RobKorr V1.

A first development of a graphical interactive programming system (*RobKorr V1*) has been developed in co-operation with a big German car manufacturer, with which it is possible at present to correct in simple and user-friendly manner deburring and dressing results at cylinder heads for a large number of workpiece types.

Since a machine operator does not possess necessary pre-knowledge in programming industrial robots, a simple possibility has been given to him for entering determined deviation values into the system, in order to arrange a correction of the process and, thus, a removal of these errors.

The input mode takes place by means of a graphical interactive picture screen surface on a PC-Terminal at shop-floor level (see Figure 2). It could not be expected by it to determine and independently enter the necessary correction values manually. Therefore, this became possible by visualizing each workpiece form and its corresponding sides with the option to choose individual outlines on the PC screen, which should be corrected.

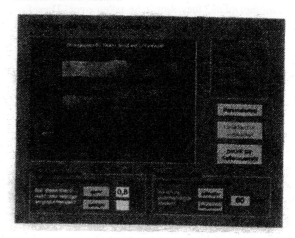

Fig. 2: *Robkorr V1* example of a program frame. In the shown situation a workpiece side as well as four areas have been selected. With the entered correction value of *0,8 mm* (see green field in the lower left part of figure) a corresponding material layer is going to be removed.

In order to develop the described interaction system and to allow its proper integration in production plants, following subtasks have been successfully carried out:

- Draft and production of a graphical interactive PC picture screen surface (visualization program): The correction value input should happen on a sample picture of the workpiece, at which the incorrect deburring areas are selected and corrected by indication of *more* or *less* accordance. Likewise a change of feed speed, necessary for deburring, should be possible by means of the selection of *faster* or *more slowly*.

- Production of a robot subroutine for the conversion of the shift of the robot course process for all different correction types on the basis, the correction values entered by the plant operator at the PC and for the shift of the work object.

- Development of a production-suited data security concept for the avoidance of errors at the PC during the correction value input, of overrun during the data communication between PC and robotic control, as well as of data back-up and misinterpretation in robotic control.

Further information regarding the step-by-step development mothodology of the described *RobKorr* version can be found in (Berger and Lepratti, 2002).

4.2 Necessity for a further development.

The main problem of *RobKorr*'s first version consists of the fact that the program structure does not allow parameter changes, which have a global effect on the program. This was conceived for a single application, so that parameters are freely selectable only in a pre-defined framework. Likewise, functions and procedures are shape-cut into an arrangement and in case of employment into another plant, they won't be taken over. Therefore, fundamental changes arise and *RobKorr V1* becomes useless. In order to make possible its further employment towards a variety of similar applications, the program code would have to be modified, whereby inclusively-calibrating errors accumulate up with increasing interference into the program.

Due to the latter, the program code has been further developped towards a next version, the *V2*. The layout of the user surface has been maintained, in order to offer the users an already familiar handling interface. With the production of this second version, attention has been paid from the beginning on to an open program structure, which allows a simple extension of the program source code at any time.

The philosophy consists in storing parameters of different applications into an external file and loading this into the *RobKorr V2* environment. Due to this data the program configures and, thus, adapts itself to different conditions. If one provides such a configuration file for each application, then *RobKorr V2* is executable without further expenditure of each of these plants.

With the employment in further plants new tasks result for *RobKorr* V2 from the conditions prevailing there. By adding the appropriate procedures the function range of the program increases. It is to be made constantly certain that a general validity must prevail during the integration of new functions, so that their use becomes possible, also in other plants. Likewise, special functions have to be brought into a general form and their special characteristics are to

be generated over configuration data. Thus, the requirements substantially rise for the programmer. More and more complex configuration files originate while the time goes by, which are not anymore realisable by simple input of parameters by hand. The task for the production of such configuration files is given therefore to a further program, which can read, represent and edit the configuration data in a suitable way.

5. DEVELOPMENT OF THE SECOND ROBKORR VERSION

The program packet *V2* represents a two-component solution, consisting of a main program and an editor. Figure 3 illustrates the *Robkorr* principle structur.

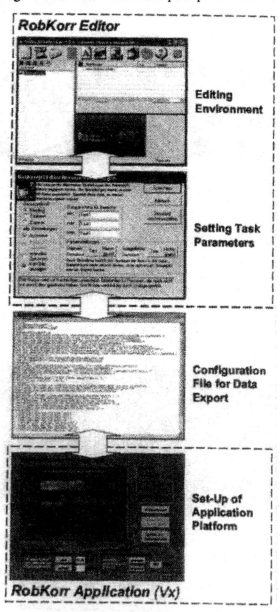

Fig. 3: The *Robkorr* principle structur: By means of the *RobKorr Editor* picture as well as further features of workpieces can be first loaded and specific application parameters set. On this basis a configuration data file is generated, which is used for structuring the *RobKorr* application interface plattform.

182

By the openly held program structure the package is adaptable and expandable according to the pursued goals. Object-oriented programming provides a simple handling. A short briefing into the program is sufficient, in order to make the operation possible also for an untrained user. Hereby, one sould be able to make changes at the correction values and to co-ordinate the program with the help of the so called *RobKorr* editor to new applications.

For the *RobKorr* editor and the *RobKorr V2* different requirements apply. While for *RobKorr V2*, which works on-line, i. e. directly in production plants, the process security plays the paramount role, the *RobKorr* editor, that is used off-line, is appropriate for a comfortable editing of new application features and parameters. Extensive editing functions offer various possibilities of the manipulation, automatic support during processing, as well as error recognition in case of contradictory inputs.

RobKorr V2 offers process-safe functions permitting little clearance. Furthermore, while in the first *RobKorr* version no unauthorized changes were possible, which could entail disturbances or even a breackdown of the entire manufacturing cell, these are now avoided due to a separation of setting data (*RobKorr Editor*) and process data (*RobKorr V2*) (see Figure 3).

In the next sections a short overview of this editor and the program organisation structure will be given.

5.1 The RobKorr Editor.

By the development of the *RobKorr Editor* the expenditure for the adjustment of the second version of *RobKorr* to new plants has been substantially simplified. The interface of the program has been generated on the source code basis of *RobKorr V2*. The fundamental difference consists of the fact, that in the editor objects concerned e.g. robots, workpiece lists, layouts of correction outlines and areas, etc. are changeable. Custom made menus for objects give access to the object characteristics and methods holding processing functions ready. The operation takes place in a simple manner only by means of the mouse and *Pop UP* menus (see Figure 4).

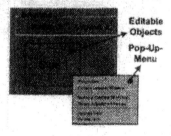

Fig. 4: Adding *RobKorr V2* features with custom-based menus. Various alternative features can be added a posteriori according with custom's specific requirements.

Since the editor shows the same layout as *RobKorr V1* and *V2*, during setting process the user can immediately get a picture of how his provided configurations will look like later on in *RobKorr* V2.

After the parameter setting process a configuration file is produced. It contains all necessary information, with which provided objects and their features are re-constructed as well as linkages re-built. It has two different functions: (i) As link between the editor and *RobKorr V2* (see Fig. 3) and (ii) as backup copy for later processing. By continuous advancements the format of the configuration file constantly changes. The *RobKorr* editor recognizes independently during the loading procedure the configuration version it is accounted for, inserts automatically missing data, specifies default values, converts to the new format or requests for the additional manual correction and examination by the user. For keeping of process security and compatibility within the program packet, *RobKorr V2* accepts only configuration files written with the same editor version. The program packet always remains thereby on the newest conditions.

5.2 V2 Program Structure.

Not all objects in *RobKorr* V2 are editable. Likewise, the forms for the input of the robot and workpiece data, of workpiece sides and outlines/areas are constant. These rigid objects form the program core, on which the objects capable for editing are put on. Also the functions of the object administration belong to the core. They organize the allocation of objects belonging together. New objects can thereby be provided and their characteristics such as position, color, quantity etc. be changed. Between them a hierarchy prevails. For example workpieces are assigned to certain robots and refer at the same time to their own workpiece sides and corresponding outlines. Figure 5 shows the hierarchical structure of the editable objects.

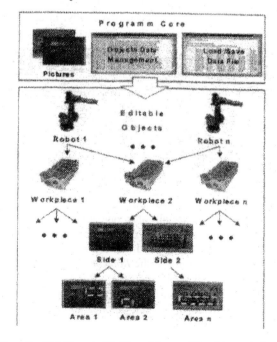

Fig. 5: *RobKorr* editable objects. The hierarchal structure of selectable objects within *RobKorr* foresees that to each industrial robot a predefine number and type of workpieces have been assigned and so on.

The program core forms the basis for the entire package and was very exactly examined for accuracy and internal conclusiveness. In addition, the compatibility between the *RobKorr* editor and *RobKorr V2* is guaranteed by it. All remaining functions are separate, closed in itself and contain the desired manipulation and input modes. They are integrated and called by the linkage with an object. Additionally, as many desired functions as needed can be added and linkages with the objects can be made. On this basis the program packet is arbitrarily expandable, since the program structure remains flat and open (Figure 6).

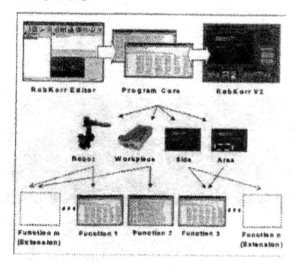

Fig. 6: Integration of Functions within *RobKorr V2*.

5. CONCLUSIONS AND OUTLOOK.

In order to support the enterprise transition process from muss production to the customized one, a set of new strategies and tools for enhencing interoperability is needed. Especially, in connection with the Human-Machine Interaction at shop-floor level new intelligent solution must ensure a better embedding of human skills in manufacturing plants.

This paper presented a development of a Graphical User Interface (GUI), which enables through easy interventions at a central PC correction at industrial robot programms.

By the successfully integration of a first version (*RobKorr V1*) in production field plant's availability has been clearly increased as well as additional costs for re-programming tasks have been drastically economised. Since this first version has been developed especially for one application, i. e. the deburring and dressing of cylinder heads, it has been extended and improved towards a second and more flexible version, i. e. the *RobKorr V2*. By means of a further development, the *RobKorr Editor*, an adaptation of *RobKorr* to other production requirements or to new application fields became possible.

However, in view of coming versions many possibilities still remain open (see (Lepratti et al., 2003)).

Advancement of *RobKorr V2* takes place already towards a third version *RobKorr V3*. Particularly, the function range of the program is increased and further auxiliary functions are integrated (such as embedding *Plug and Play.* functionalities or the use of dynamic code libraries (*DLL*'s), in order to permit, that several programmers work at the same time on the verion, without causing data collisions.

Most important improvements from the process-technical and program-technical points of view are to ensure downwards compatibility guarantying compatibility between all *RobKorr* versions and, consequently, also to permit their smooth update and replacement. It must become possible at any time to open a configuration in the editor *V2*, which was previously generated with the editor *V3* and, likewise, to open the same one in *RobKorr V2*.

ACKNOWLEDGMENT

The project paper is a result of a co-operation project with the *DaimlerChrysler AG (PWT-VEO)* of Stuttgart (Germany).

REFERENCES

Berger, U. und Lepratti, R., (2002): "Intelligent PC-based User Control Interface for On-line Correction of Robot Programs". In: Proceedings of the 7th International Conference on Control, Automation, Robotics and Vision (ICRACV 2002), Singapore, December 2 - 5, p. 276-281.

Lepratti, R.; Berger, U. (2004): "Enhancing Inter-operability through the Ontological Filtering System". In "Processes and Foundations for Virtual Organisations", Proceeding of the 5th IFIP Working Conference in Virtual Enterprises (PRO-VE 2004), Toulose, France, August 23-26, L. M. Camarinha-Matos and H. Afsarmanesh (Ed.), Kluwert, Boston-London, 2004, (in print).

Lepratti, R.; Jing, C.; Berger, U.; Weyrich, M. (2004): „Towards the knowledge-based Enterprise". In: Proceedings of ICEIMT 2004, Toronto, Canada, 2004, (in print).

Lepratti, R.; Luderer, A.; Berger, U. (2003): RobKorrV2: Eine flexible Lösung zur Mensch-Roboter-Interaktion. Im: Wissenschaftsmagazin Forum der Forschung der BTU Cottbus, 7. Jahrgang, Heft 16, 2003, p. 129-134.

World Robotics (2003): "Statistics, Market Analysis, Forecasts, Case Studies and Profitability of Robot Investment", USA; 2003.

www.elsevier.com/locate/ifac

IMPROVED GENETIC ALGORITHM AND ITS APPLICATION IN OPTIMAL COMBINATION STACKS OF STEEL ROLL*

Xiao-Dong WANG [1]   Ji-Ling JIN [2]   Quan-Li LIU [2]   Da-bo ZHANG [2]   Wei WANG [2]

1. Information and Engineering College, Northeastern University, Shenyang, 110006, China

2. Research Center of Information and Control, Dalian University of Technology, Dalian, 116024, China

Abstract For overcoming the disadvantage of local optimum and slower convergence speed of general genetic algorithms, an improved genetic algorithm with dynamic regulation parameters was proposed in this paper by introducing the variance and expectation of individual adaptive value to describe concentration-dissipation degree of population. The validity of this improved genetic algorithm is verified by simulation examples. The improved algorithm is also been applied to optimal combination stacking of steel roll in a batch annealing shop and a satisfactory result is obtained in production. *Copyright ©2004 IFAC*

Key words Genetic algorithm, optimal combination, low cost automation

1. INTRODUCTION

Since the genetic algorithm was proposed in 1970 's (Holland, 1975), many scholars have done a lot of research work on theory and application of genetic algorithm. But, one of the important problems still is how to improve its search ability and convergence speed and make it to solve practice problems better. Actually, crossover probability and mutation probability determine the convergence of genetic algorithm (Srinivas, 1994). In simple genetic algorithm, both crossover probability and mutation probability are invariable, so the convergence speed

is slow and can't get away from the area of the local optimal solution, therefore some algorithms of changing crossover probability and mutation probability have been suggested in recent years in order to have dynamical regulation function. The most representative work is the adaptive genetic algorithm (Srinivas and Patnaik, 1994). It regulates crossover probability and mutation probability adaptively with the population evolution. It suits better at the later period of population evolution. But it makes the evolution slowly at the early period because the better individuals among early population are in an unchangeable state, and the evolution may stays at a local optimal point.

Concentrating on how to change the crossover probability and mutation probability dynamically,

* The work was supported by the National "863 Project" of China (2002AA412010),
Corresponding author: Wei Wang,
wangwei@dlut.edu.cn

this paper suggests an improved genetic algorithm with adjustable parameters dynamically by using the variance and average of individual fitness to describe the variety and the convergence of population more exactly. This improved genetic algorithm can dynamically adjust the crossover probability and mutation probability according to the variety of individual in population. It not only increases the whole search ability of the algorithm, but also increases mountain climbing ability. The simulation examples indicate that the genetic algorithm given in this paper improves the quality of solution and the search efficiency and saves the optimization time. It is very suitable to solve the complex problem of simulate optimization. The improve genetic algorithm has been applied to optimal combination stacking of steel roll in a batch annealing shop, the application indicates that the improved genetic algorithm has obtained a good effect.

2. GENETIC ALGORITHM WITH ADAPTIVE PARAMETERS

Among the genetic algorithms with dynamic regulation function, the most representative is the algorithm given in (Srinivas and Patnaik, 1994). It regulates the crossover probability and mutation probability of individual according to the fitness of chromosome. The regulation equations are given below.

$$p_c = \begin{cases} k_1(f_{max} - f')/(f_{max} - f_{avg}), & f' > f_{avg} \\ k_3, & f' \leq f_{avg} \end{cases} \quad (1)$$

$$p_m = \begin{cases} k_2(f_{max} - f)/(f_{max} - f_{avg}), & f > f_{avg} \\ k_4, & f \leq f_{avg} \end{cases} \quad (2)$$

where p_c represents crossover probability; p_m represents mutation probability; f_{max} represents max fitness of population; f_{avg} represents the average fitness of population; f' represents the larger fitness of two crossed individuals; f represents the fitness of aberrance individual. k_1、 k_2、 k_3 and k_4 are constants between 0 and 1. The algorithm can protect the efficiency pattern and accelerate the worse solution to be washed out. Because the parameters

are conformed by the individual's fitness, it reduces the relation of individuals. At the same time, it can be seen from (1), the crossover probability of excellent individual is low, thus the good gene can't be promulgated effectively. The regulation policy makes the individual in population lacking the cooperation spirit, and causes the evolution speed being slower. Especially at the initial stages of evolution, it can't get into the neighborhood of optimal solution. In addition, the regulation policy will calculate the control parameters for every individual. When the population size is larger, it will take so much time that the optimization efficiency is reduced.

The process of genetic algorithm evolution includes two stages. The first stage (called rough search) is to close to the optimal solution nearby quickly in the greater search space. The second stage (called refined search) is searching the optimal solution around the neighborhood of the optimal solution. How to deal with the balance between rough search and refined search is critical of ensuring the quality of solution and increasing the convergence efficiency. This section improves on the equations (1) and (2), uses variance and average of fitness to describe the population state, regulates p_c and p_m adaptively according to the concentrated degree of fitness to ensure the global convergence and the convergence efficiency of the algorithm. The specific formula with adaptive regulation parameters is suggested below.

$$p_c = \begin{cases} p_{c1} + k_1 \exp(-\beta)\alpha & k_1 \exp(-\beta)\alpha \leq k_2 - p_{c1} \\ k_2 & k_1 \exp(-\beta)\alpha > k_2 - p_{c1} \end{cases}$$

$$(3)$$

$$p_m = \begin{cases} p_{m1} + k_3 \exp(-\beta)\alpha & k_3 \exp(-\beta)\alpha \leq k_4 - p_{m1} \\ k_4 & k_3 \exp(-\beta)\alpha > k_4 - p_{m1} \end{cases}$$

$$(4)$$

where, $\alpha = f_{ave}/f_{max}$, $\beta = f_{dev}/f_{dev-max}$, f_{ave} represents the average fitness of individual, f_{max} represents the maximum fitness of individual, f_{dev} represents the variance of fitness,

$f_{dev-max}$ represents the maximum variance of fitness of individual at present. p_{c1} and p_{m1} are constants, according to different problem, they usually have the value from 0.2 to 0.65 and 0.001 to 0.2. The constants k_1 and k_3 are ranged between 0 and 1.0. k_2 is a constants approaching to 1.0, it usually takes 1.0, k_4 is also a constants, ranging between 0.2 and 0.4.

The equations (1) and (2) regulate parameters for individual in population. This kind of regulation policy makes the individual lacking cooperation, leading to the slow evolution speed, and also increasing the computation quantity. Different form the equations (1) and (2), the equations (3) and (4) regulate parameters for the whole population by introducing variance and average of fitness to describe the congregating degree of individual, it is in favor of cooperating between individual and increasing of evolution speed.

In addition, during the search of genetic algorithm, when the convergence degree of population is large and the population variety is lost, we should increase p_c and p_m properly for jumping form the local solution; otherwise, we should decrease p_c and p_m properly for avoiding destroying the optimal solution. The main idea of the equations (3) and (4) is to regulate p_c and p_m dynamically to remain the variety of population according to the congregation degree of individual and make the algorithm get to global convergence quickly. In the above equations, β describes congregation degree of individual. β is smaller means that the fitness is more concentrated; The variety of population is lost more sharply. So p_c and p_m should decrease when β increases. a describes the evolution degree and congregation degree of population. a is bigger means that the evolution nears to the later stage, congregation degree of individual is bigger, and more probably gets into the local optimal solution. So p_c and p_m should increase when a increases. At the beginning of evolution, β is very big, a is very small, so p_c and p_m will be very small. For accelerating search of the optimal solution, p_c and p_m are given smaller values, called p_{c1} and p_{m1}. Along with the process

of evolution, a and β can fully denote the stage of evolution and the congregation degree of individual at present and regulates dynamically p_c and p_m based on the value of a and β. We can decide the operation of next step according to the stage of population, thus the algorithm can make the whole search process approach to the optimal solution quickly.

3. THE SIMULATION EXAMPLES

For validating the performance of searching solution of the genetic algorithm with the dynamic regulation parameters suggested in this paper, this section compares the improved genetic algorithm with the adaptive genetic algorithm (Srinivas and Patnaik, 1994). The choice method and scale method of the two kinds of genetic algorithm are same.

Function 1:

$$f_1(x) = \left|\sin(30x)\right| * (1 - 0.5x), \quad x \in [0,1]$$

This function has multi-peak values, the maximum value point is $x = 0.0517900$, and the corresponding maximum value is $f = 0.9739626$.

Function 2:

$$f_2 = 0.002 + \sum_{j=1}^{25} \frac{1}{j + \sum_{i=1}^{2} (x_i - a_{ij})^6}$$
$$(-65.536 < x_i < 65.536)$$

This function is multi-mode function with twenty-five peaks; the aim is to find the maximum value of this function.

The adaptive genetic algorithm (Srinivas and Patnaik, 1994) is abbreviated as AGA, and the improved genetic algorithm suggested in this paper is abbreviated as MAGA. For the randomicity of the original population, each algorithm has been tested thirty times independently. The AGA parameters can be found in (Srinivas and Patnaik, 1994). The MAGA parameters are choose as follows. For the function f_1,

$p_{c1} = 0.65$, $k_1 = 1.0$, $k_2 = 1.0$, $p_{m1} = 0.02$, $k_3 = 1.0$, $k_4 = 0.3$. For the function f_2, $p_{c1} = 0.25$, $k_1 = 1.0$, $k_2 = 1.0$, $p_{m1} = 0.1$, $k_3 = 0.1$, $k_4 = 0.4$. Table 1 gives a compared result between two algorithms. The valve value in the table represents the condition of stopping optimizing. The maximum generation number represents the most generation number. It is thought getting into the local solution when the optimal solution still doesn't approach with maximum generation number. The average generation number represents the average of evolution generation when the algorithm gets to valve value. It can be seen from the table 1, the MAGA takes precedence obviously compared with the AGA on jumping out local optimal solution and developing new search space.

3) Heating and cooling. The time of heating is about 20 hours. After heating, it needs to cool, and then, the traveling crane takes the steel rolls out of the stove. Thus an anneal production plan of steel rolls is over. The stove can proceed for the next production plan.

The heating time of steel rolls in every stove is determined by the heating curve. This curve is plotted according to the total weight of steel rolls in single stove. When the total weight of steel rolls increases 5 tons, it will need to change a heating curve, the corresponding heating time will increase a half of hour, the cooling time also need to increase. Practically, the weight and specification are different between every steel roll, so the total weight in stove

Table 1 the contrast between AGA and MAGA

Function	Valve value	Max generation number	Average generation number		Times of getting into local best solution	
			AGA	MAGA	AGA	MAGA
f_1	0.9739626	200	60	24	4	0
f_2	1.000	100	26	8	8	2

4 APPLICATION OF THE IMPROVED GENETIC ALGORITHM IN OPTIMAL STACKING OF STEEL ROLL

4.1 The problem description of steel roll tacking
Anneal is an important procedure of the production of cool sheet metal in metallurgical industry, it usually adopts the manner of batch annealing shop, a batch annealing shop commonly has 30-50 annealing stoves. The procedure of anneal production is given below.

1) Building pool. Choosing different type steel rolls (about 100 rolls) everyday needed from the frontal storeroom according to the compact order form.

2) Stacking. Assigning the production plan of stacking the steel rolls into stove, that is, 3 or 4 steel rolls are combined (forming an anneal production plan), and ready being put into a stove to anneal. Each stove usually can load 4 steel rolls.

changes from 30.0 to 60.0 tons. The idiographic production criterion is as follows. If the total weight is no more than 35 tons, the heating time is 19 hours; if it is more than 35 tons and less than 40 tons, the heating time is 19.5 hours; and so on. Obviously, for increasing the produce efficiency and reducing energy cost, it need to make the total weight approach the upper limit of boundary of production criterion when the steel roll is loaded. That is to say that the total weight of steel rolls after loading approaches 35 tons, 40 tons, 45 tons, 50 tons, 55 tons, 60 tons, not 45.1 tons, 50.2 tons, and so on. But it is not easy to get to the aim when combination stacking of steel roll, especially, when the number of steel roll is very large, it is more difficult for manpower stacking. In the local practical produce, it usually depends on the experienced worker to stack, so it is difficulty to find the best combination solution.

4.2 using the improved genetic algorithm to solve the optimal stacking of steel roll
In theory, the problem of the combination stacking of steel roll can be abstracted a kind of problem of

188

discrete combination optimization. The aim is to make the difference between the weight of steel rolls of an anneal production plan and the upper limit of heating curve is least. The problem can be abstracted the following optimization combination model.

$$f = \frac{1}{N}\sum_{i=1}^{N}\left(\overline{y}_i - y_i\right) \qquad (5)$$

where f represents the average error between the weight of steel rolls after combination and the setting weight in each stove; N represents the total counts of combination stove; y_i represents the total weight of steel rolls (3 or 4 rolls) in each stove; $\overline{y}_i$ represents the setting weight in each stove corresponding y_i. Because using the method of imitated optimization, the decision variable does not exposed in the equation (5). The steps of applying the improved genetic algorithm to solve the combination stacks of steel roll are given below.

Step 1. Representing each steel roll in the pool by j, recording the total counts of steel rolls by M and the weight of each steels roll by w_j.

Step 2. Initiating the genetic algorithm to bring the original population, here, adopting the natural number as code manner.

Step 3. If getting to the maximum generation number, taking the combination manner corresponding to the minimal valve of f as the best combination result, the optimization is over. If not getting to the maximum generation number, going to the step 4.

Step 4. Using the corresponding production rule, stacking the individual in population, calculating y_i for each individual (y_i represents the total weight of several steel rolls in the ith anneal plan of the combination result) and calculating f.

Step 5. Selecting operation depending on f of individual by using 'wheel select method'. The new generation population is yielded according to the adaptive crossover probability and mutation probability suggested in this paper.

Step 6. Repeating to execute the step 3 to step 5 until the optimization is over.

The improved algorithm mentioned above has been applied in optimal combination stacks of steel roll in anneal shop of Shanghai Baosteel Yichang Steel Rolls Co. Ltd. Taking a poll having 28 steel rolls as an example, its optimal combination result is shown in the table 2 and table 3. The table 2 is the 28 steel rolls being chosen and the table 3 is the optimal combination result.

It can be seen from the table 3 that this optimization yields 7 production plans, the steel roll number of each plan is given in the table 3, and the error between the total weight of steel rolls in each production plan and the setting weight in each stove are all less than 1 ton, except the 7th plan (the error is 1.07 tons), the error is also very small. In this case, the total weight of 7 plans is 374.08 tons, the value of f is 0.632.

Table 2: Weight of steel roll

Serial number	Weight (ton)	Serial number	Weight (ton)	Serial number	Weight (ton)	Serial number	Weight (ton)
1	14.44	8	14.26	15	12.37	22	14.76
2	14.74	9	15.13	16	8.03	23	14.57
3	14.74	10	12.83	17	14.83	24	14.78
4	14.74	11	14.73	18	14.60	25	14.80
5	14.75	12	14.76	19	14.30	26	15.21
6	14.79	13	14.88	20	7.36	27	7.89
7	14.50	14	14.59	21	7.85	28	8.85

Table 3: Optimal combination results

Number of anneal plan	Total weight (ton)	The combination number of steel rolls				The error with the upper limit (ton)
1	49.10	1	8	15	16	0.90
2	59.20	2	9	22	23	0.80
3	44.27	3	10	21	28	0.73
4	59.09	5	11	17	24	0.91
5	44.34	7	14	20	27	0.66
6	59.15	12	13	19	26	0.85
7	58.93	4	6	18	25	1.07

The real time simulation system with the improved genetic algorithm as core was implemented by using a personal computer. It has been run well in the anneal shop of Shanghai Baosteel Yichang Steel Rolls Co. Ltd for more than one year. The system has some useful functions, including setting the rules of optimal combination stacking according to production plan, automatic combination optimization stacking, ensuring the steady stacking, special stacking for special order, stacking manually and so on. The using of this system has improved the produce efficiency and has decreased the energy cost. The optimal combination stacking has obviously optimization effect. We take the combination stack in Dec. 6, 2002 as example. There were 100 MR steel rolls in the pool. It was required to set the stack rule according to full thick, full middle, full thin. After stacking, 25 production plans were yielded. The time of optimal combination stack is 35 second. The average of the error for every plan's weight difference to its upper limit is 0.297 tons

5 CONCLUSIONS

The main contribution of this paper is to use the variance and average of individual fitness to describe the convergence degree of population and propose an improved genetic algorithm with dynamic regulation parameters. The improved genetic algorithm can regulate the crossover probability and mutation probability adaptively according to the convergence degree of individual in population. It has the ability of global convergence and can converge quickly.

Since the algorithm can run in a personal computer, the cost is low. The improved genetic algorithm has been applied in steel roll optimal combination in annealing shop and the satisfied result has been obtained in practice.

REFERENCES

Holland J.H. (1975), Adaptation in natural and artificial systems. Ann Arbor: The University of Michigan Press.

Srinivas M (1994), Genetic algorithms: A survey, Computer, 26(6):17-26.

Srinivas M. and Patnaik L. M (1994), Adaptive probabilities of crossover and mutation in genetic algorithms. IEEE Transactions on System, Man and Cybernetics, 24 (4): 656–667

www.elsevier.com/locate/ifac

INTELLIGENT SHUTTLES FOR FMS USING ANT-BASED APPROACH

Y. Sallez, T. Berger, C. Tahon

Equipe Systèmes de Production
Laboratoire d'Automatique, de Mécanique et d'Informatique Industrielles et Humaines
UMR/CNRS 8530, Le mont Houy, F-59313 Valenciennes cedex 9
Email : yves.sallez@univ-valenciennes.fr

Abstract : This paper describes the interest of "virtual" pheromones to build intelligent routing systems in FMS (Flexible Manufacturing Systems). The first part describes the bases of the stigmergy concept using pheromones. In the second part, these concepts are applied to build our Product-Based Heterarchical (PBH) system. The third part is dedicated to a real implementation on a flexible assembly cell (in the Valenciennes's Aip-Primeca pole). *Copyright © 2004 IFAC*

Keywords : flexible manufacturing systems, routing, scheduling, stigmergy, pheromone.

1. INTRODUCTION

The manufacturing industry must deal with continually changing and increasingly complex product requirements as well as mounting pressure to decrease costs. To meet this challenge, Flexible Manufacturing Systems (FMS) must become more robust, scalable, reconfigurable, dynamic, adaptable and even more flexible. In this context, either a production program modification or resource failures implies partial or total reconfiguration of the system. This reconfiguration is difficult to accomplish because the problem is huge (combinatory explosion) and also because a partial solution must absolutely be anticipated for each failing resource situation.

Nowadays, communication technologies allow a new trend in FMS control, towards distributed systems composed of numerous autonomous and cooperative entities/agents.

Heterarchical/non-hierarchical architectures play a prominent role in FMS research field (Pujo and Kieffer, 2002). Heterarchical scheduling, a non-centralized approach, aims at choosing the best resource, performing the service, and taking into account service quality and traffic fluidity towards this resource. The result is more autonomous, robust and adaptive scheduling systems as compared to centralized scheduling systems, even highly reactive ones (Duffie and Prabhu, 1996).

The first part describes the bases of the stigmergy concept using pheromones. The different pheromone's characteristics are introduced. Some interesting insect-based methods are described for different manufacturing applications. In the second part, these concepts are applied to build our Product-Based Heterarchical (PBH) system. Basics principles and system architecture are detailed. The third part is dedicated to a real implementation on a flexible assembly cell (in the Valenciennes's Aip-Primeca pole) based on an innovative conveying system.

2. AN APPROACH BASED ON INSECT SOCIETIES

2.1. Stigmergy concept.

In nature, chemicals called pheromones are widely used by insects to build organized group activity. Ants are known to use pheromones for various communication and coordination tasks. Foraging ants lay down chemical trails, and all ants following a useful path add their odor to the trail, reinforcing it for future use.

This phenomenon, called stigmergy, allows indirect communication between creatures through the sensing and modifying of the local environment. This communication influences the creature's behavior. The history of stigmergy in the context of social insects is described in (Theraulaz and Bonabeau, 1999).

In the real world, three basic operations are associated with chemical pheromones : information fusion, information removal and local information distribution. In the first category, deposits from individual entities are aggregated to allow the easy fusion of information. In the second, pheromone evaporation over time removes obsolete or inconsistent information. In the last, local information is provided according to the diffusion of chemicals only in the immediate neighborhood.

In all of these operations, the pheromone field displays several characteristics :
- independence : The sender of a pheromone message does not know the identity of the potential receiver and does not wait for any acknowledgement. This characteristic makes pheromone use very useful for communication within large populations of simple entities.
- local management : Because diffusion falls off rapidly with distance, interactions through pheromones are only local, thus, avoiding the need for centralized interaction management.
- dynamism : Continuous reinforcement and evaporation respectively integrate new information and delete obsolete information.

In applications like dynamic products routing, these characteristics provide two main advantages : robustness and adaptability to the environment.

2.2. Applications in FMS.

The first experiments on the industrial use of stigmergy were lead by Deneubourg (1983) in the early 1980's using simulated "ant-like robots". Many researchers (Ferber, 1995 ; Arkin, 1998 ; Dorigo and Colombetti, 1998) have applied this concept to their studies of robot collectives and the resolution of optimization problems (Traveling Salesman Problems, Network Routing for telecommunications and the Internet).

Based on the ant foraging analogy, Dorigo developed the Ant Colony Optimization (ACO) metaheuristic, a population-based approach to the solution of combinatorial optimization problems (Dorigo et al., 1999). The basic ACO idea is that a large number of simple artificial entities are able to build good solutions to hard combinatorial optimization problems via low-level communications. ACO-based approach can be applied to almost any scheduling problem as job shop scheduling, vehicle routing...
Researchers have also applied the stigmergy concept to specific situation of manufacturing control systems :

- Parunak et al. (2001) emphasizes the importance of the environment in agent systems. Information flows through the environment are complementary to classical message-based communications between agents. Parunak focuses on the dynamics that emerge from interactions between multi-agent systems. In

this approach, the environment is computational and he uses agents moving over a graph to study supply network among manufacturing companies. Interactions among agents produce emergent dynamics that are analyzed using methods inspired of statistical mechanics.

- Brückner (2000) applies stigmergy concept to manufacturing control. He presents an extensive set of guidelines to enable design of synthetic ecosystems. This application is supported by an agent-system approach. Different types of agents are used to model resources, parts flow and control units. The application context is the routing of cars bodies in a Mercedes Benz paint shop.

- In (Peeters et al., 1999), the authors propose a pheromone based control algorithm designed in a bottom-up way. Peeters's as well as Brückner's works are based on the PROSA reference architecture (Wyns, 1999). A description of the agents and of the pheromone's life-cycle is available in (Mascada-WP4, 1999). As in the Brückner's work, a paint shop (Daimler-Chrysler) is used as application context.

- In (Cicirello and Smith, 2001) a method used by wasps to assign tasks within the nest is applied to dynamic shop floor routing. Each wasp-based agent is affected to a job in the queue and enters in concurrence with other scheduling wasps. The results of this contest is used to allocate tasks taking into account priorities and social rank. Experiments on different factory configurations have demonstrate that this approach is robust and capable of efficient adaptations in dynamically changing environments.

3. A PRODUCT-BASED HETERARCHICAL APPROACH FOR DYNAMIC PRODUCTS ROUTING

3.1. Basics principles.

Generally in FMS control, part flow is passive (Pujo and Kieffer, 2002). In the PBH approach, part flow is linked to the behavior of autonomous product entities, which move on a resource network. The approach is based on behavioral biological systems like ant colonies (Di Caro and Dorigo, 1998 ; Deneubourg et al., 1983).

In this proposed paradigm, each product entity has a list of services to be obtained successively from the resources (service stations) located on the nodes. These nodes are interconnected by paths used by moving entities to go from one node to another until they reach their destination node where the desired service can be obtained. Autonomous entities must choose both the destination node where it can obtain the desired service and the successive paths that will bring it from one neighbor node to another until it reaches that destination node. To make these choices, the entities use the data stored on each node (see figure 1 for a schematic portrayal of typical entity behavior).

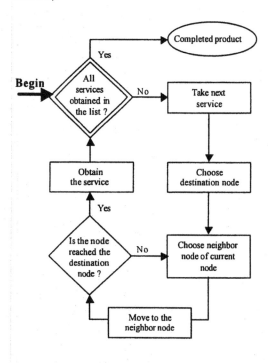

Fig. 1. PBH system model.

3.2. Description of the PBH approach.

- Choice of neighbor node

The entity chooses the shortest route from its current location to the destination node, according to the information provided by the neighbor nodes, in which coefficients associated with each destination node are stored.

Let P_{dn} represent the preference for the neighbor node n_n that comes with a short traversal time to move from the current node n_k to destination n_d via node n_n, which must belong to the n_k neighbor node set. In the following notation :

$P_{dn} \in [0;1]$

$P_{dj} > P_{dp} \Rightarrow n_j$ allows the entity to reach n_d more rapidly than n_p. n_j and n_p are neighbors of n_k.

$V_{n_k} \Leftrightarrow \{ n_k \text{ neighbor} \}; n_n \in V_{n_k}$

After standardization : $\sum_i P_{di} = 1 ; i \text{ with } n_i \in V_{n_k}$

The best neighbor n_c is such that $P_{dc} = \max(P_{dn})$.

If the entity receives insufficient information ($\forall i, j; P_{di} = P_{dj}$), it chooses n_c randomly.

- Choice of destination node

The service S_j belongs to the ordered list of services needed by the entity. S_{jl} is the S_j service obtained at

node n_l. Coefficient Q_{jl} describes the S_j service quality at node n_l. In the following notation :

$Q_{jl} \in [0;1]$

$Q_{j1} > Q_{j2} \Rightarrow S_j$ service quality is best at node n_1

Both Q_{jl} and P_{ln} are used by the entity to choose n_d in n_l nodes where service S_j can be obtained, the first neighbor $n_c \in V_{n_k}$ ($n_n \in V_{n_k}$; n_k being the node where the entity currently is). The following function : $\alpha.Q_{jl} + (1-\alpha).P_{ln}$ must be maximized, with $\alpha \in [0;1]$. In the end of the procedure, the destination node n_d, with best neighbor n_c, is such that

$\alpha.Q_{jd} + (1-\alpha).P_{dc} = \max(\alpha.Q_{jl} + (1-\alpha).P_{ln})$

Updating P and Q coefficients : The experience of each entity that has obtained a service is used to update the P and Q coefficients. During moving phases, each entity stores data in an embedded memory. This data describes the entity's path through the nodes n_k, including crossing time and, in the case of servicing, S_j and its quality Q_{jl}. Every time an entity reaches a destination node, a fictitious entity retraces its path virtually, updating coefficients at each crossing node.

Updating P : A node n_k has got :
- a P_{dn} matrix whose rows denote all possible destinations n_d and whose columns, all existing V_{n_k} neighbors,
- a matrix including the mean μ_{dk} and the standard deviation σ_{dk} of the time needed to go from n_k to n_d for all possible destinations.

T_{dk} is the time span needed to go from n_k to n_d. When the fictitious entity is located at node n_k, a comparison is made between T_{dk} and the mean μ_{dk} of previous T_{dk}. This comparison is valid only if μ_{dk} is stable enough in terms of the σ_{dk} of the previous T_{dk}.

In the end, three cases can be characterized :

- when μ_{dk} is stable and $T_{dk} < \mu_{dk}$, the use of n_c is reinforced by increasing P_{dc},
- when μ_{dk} is stable and $T_{dk} \geq \mu_{dk}$, the use of n_c is forgotten by decreasing P_{dc},
- when μ_{dk} is unstable, some adjustments are made to allow the stability of μ_{dk}.

Q coefficients are updated using a very similar approach.

4. REAL IMPLEMENTATION

4.1. AIP FMS cell description.

The AIP (Atelier Inter-établissements de Productique) FMS cell (Université de Valenciennes) reproduces, on a smaller scale, a real assembly line with most of the operational characteristics that can be found in an industrial environment.

The FMS is composed of seven workstations Wi placed around a flexible conveying system (see figure 2). This conveying system ensures a flexible flow of pallets to each workstation.

At the beginning, plates are loaded on shuttles by workstation W7. Shuttles reach the various assembly stations according to a predefined assembly process plan. For each station, a robot is able to assemble three kinds of components accessible in different kinds of storage (structured pallets, random position storage or flexible feeder). Every component may be picked from two different places. This redundancy introduces flexibility in the system, this being particularly useful in case of malfunction of a robot.

The conveying system is based on the Montrac concept of the Montech society (Montech, 2003). Montrac is a monorail transport system that uses self propelled shuttles to transport materials on tracks (see figure 3.a). Each shuttle is individually controlled and equipped with collision free optical sensor.
Transfer gates are used to connect the different tracks. Shuttles can change direction at each transfer gate (see figure 3.b).

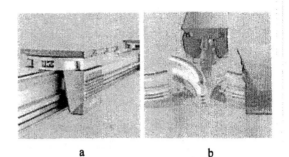

a   b

Fig. 3. Montrac system.

4.2. Implementation principle.

Following the simulation phase, which illustrated the feasibility of the PBH approach (Sallez *et al.*, 2001), an actual implementation is currently under way.

As previously explained, a node is assigned to each location where a decision must be made. At the node, a shuttle queries the server to read coefficients Q and P, which are stored in one of the server databases.

According to the PBH philosophy, each shuttle must know its location in the cell. To ensure this, two types of equipment must be installed :

- <u>Node instrumentation</u> : a label (tag) is physically associated to each node, identifying it by number.

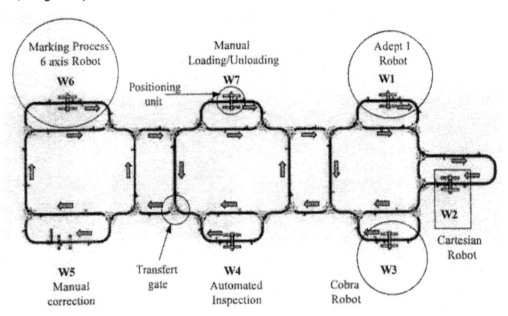

Fig. 2. Schematic view of the flexible cell.

- Shuttle instrumentation (See figure 4 for a schematic presentation) :

    - At each node, a shuttle reads the label to obtain the node's number. An identification system embedded in the shuttle performs this task using electro-magnetic reading technology (1).

    - According to the PBH principle, the shuttle asks the server for the coefficients. The shuttle sends the node's number and its identification number via hertzian transmission, and coefficients P (of the neighbors) and Q (of services needed) are sent by the server in return (2).

    - The shuttle begins the processing that will allow it to choose the best path. This processing is supported by a micro-controller (3).

The shuttle then sends a request to the Transport server (4) to be moved along its chosen path. The Transport server positions transfer gates according the chosen track (5).

Implementation details are available in (Berger et al., 2004). The shuttle instrumentation architecture is designed to be robust and open. A CAN bus is chosen to communicate between the different units (identification, processing and communication units). Each unit has a specific usage but all units are build around a microcontroller RISC compliant with CAN.

5. CONCLUSION

Applied in many research fields (robotics, network routing...), the stigmergy concept is an original answer to FMS scheduling for large flow management. As seen in the first part, stigmergy offers robustness and adaptability to environment variations. These properties are direct outcomes of virtual pheromone. PBH inherits of these proprieties, and in addition integrates service quality resource system.

Research on the theoretical framework of PBH is on going. Our works need to focus on reinforcement functions and the choice of the best one to converge as quickly as possible near an optimal balance in the flow parts, taking into account fluidity and service quality. The implementation actually in course in the Aip-Priméca Pole focuses on the realization of "intelligent" shuttles. This implementation is a first step towards the development of a real more self-organized flow part.

Our perspectives are to study the emergence of behaviors in large flexible manufacturing cells where entities (parts, resources, operators...) can communicate directly or indirectly (by stigmergy). We are more particularly interested in the concept of bionic manufacturing introduced by Vaario and Ueda (1998) or bio-inspired systems. In this work, the authors use local attraction fields to direct transporters carrying jobs to particular resources. A dynamic scheduling emerges of interactions between the different entities.

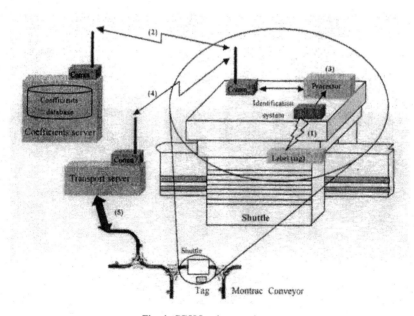

Fig. 4. PBH Implementation

REFERENCES

Arkin, R. C. (1998). *Behaviour-based robotics*. The MIT Press.

Berger, T., Y. Sallez and C. Tahon (2004). Intelligent routing in FMS using virtual pheromones, IFAC INCOM'2004, Bahia, April 2004

Brückner, S. (2000). Return from the Ant synthetic ecosystems for manufacturing control, Thesis Humboldt-University of Berlin, June 2000.

Cicirello, V. A. and S. F. Smith (2001). Wasp nests for self-configurable factories. In : *Agents 2001*, Proceedings of the Fifth International Conference on Autonomous Agents. ACM Press, May-June 2001.

Deneubourg, J. L., J. M. Pasteels and J.C. Verhaeghe (1983). Probabilistic Behaviour in Ants : a Strategy of Errors ? *Journal of Theoretical Biology*, **105**, 259-271.

Di Caro, G. and M. Dorigo (1998). AntNet : Distributed Stigmergic Control for Communications Networks. *Journal of Intelligence Research*, **9**, 317-365.

Dorigo, M. and M. Colombetti (1998). *Robot shaping : An experiment in Behaviour Engineering*. The MIT Press.

Dorigo, M., G. Di Caro and L. M. Gambardella (1999). Ant algorithms for discrete optimization. *Artificial Life*, **5(2)**, 137-172.

Duffie, N. A. and V. Prabhu (1996). Heterarchical control of highly distributed manufacturing systems, *international Journal of Computer Integrated Manufacturing*, **9(4)**, 270-281.

Ferber, J. (1995). *Les systèmes multi-agents*. InterEditions.

Montech : Montrac conveying system (2003) : http://www.montech.ch/montrac/content/

Mascada-WP4 (1999). WP4 Report : ACA (Autonomous Co-operating Agents) Framework for Manufacturing Control Systems.

Parunak, H. V. D., S. Brueckner and J. Sauter (2001). ERIM's Approach to Fine-Grained Agents. Proceedings of the NASA/JPL Workshop on Radical Agent Concepts (WRAC'2001), Greenbelt, MD, Sept. 19-21, 2001.

Peeters, P., H. Van Brussel, P. Valckenaers, J. Wyns, L. Bongaerts, T. Heikkilä and M. Kollingbaum (1999). Pheromone based emergent shop floor control system for flexible flow shops, International Workshop on Emergent Synthesis (IWES'99), Kobe, Japan, Dec. 6-7, 1999.

Pujo, P. and J. P. Kieffer (2002). Méthodes du pilotage des systèmes de production (Traité IC2, série Productique), Hermes Lavoisier.

Sallez, Y., T. Berger and C. Tahon (2001). Heterarchical product-based system for FMS dynamic scheduling, 29th ICCIE, Montréal, Nov. 1-3, 2001.

Theraulaz, G. and E. Bonabeau (1999). A brief history of stigmergy, *journal of Artificial Life*, **5(2)**, 97-116.

Vaario, J. and K. Ueda (1998). An emergent modeling method for dynamic scheduling, *Journal of Intelligence Manufacturing*, **9**, 129-140.

Wyns, J. (1999). Reference architecture for Holonic Manufacturing Systems – the key to support evolution and reconfiguration, Ph.D. Thesis K.U. Leuven.

ELSEVIER

IFAC
PUBLICATIONS
www.elsevier.com/locate/ifac

LOW COST SIMULATION OF PLC PROGRAMS

Richard Šusta[*]

*Department of Control Engineering,
Faculty of Electrical Engineering, Prague
http://dce.felk.cvut.cz/susta/,
susta@control.felk.cvut.cz

Abstract: The paper discusses the conversion of PLC (programmable logical controller) programs into forms suitable for their emulation by any software tool that includes some programming language capable of evaluating mathematical formulas, if-then instructions, and time tests, in the case of timers. The method offers advantage of low cost portability into a whole range of environments because it does not require the embodiment of large additional programming support for the simulation of various PLC instructions. The conversion, either of whole PLC program or its part, can also be utilized for auxiliary tests when a finished PLC program is transformed into a new hardware. *Copyright © 2004 IFAC*

Keywords: PLC, emulation, conversion, transfer sets

1. INTRODUCTION

Some PLC emulators are offered by their manufactures, [1] the others were written by third parties for special purposes, e.g. COSIMIR Freund *et al.* (2001). Such standalone tools usually perform very accurate simulations of PLCs including their time characteristics, but ordinarily, they can be embedded into another program only by limited or by complicated non-standard ways.

Before discussing the emulation itself, we first consider source codes of PLCs. The part 3 of IEC1131 standard, IEC (1993), defines a suite of programming languages recommended to be used with PLCs:

(1) Structured text — textual language with PASCAL like syntax and functionality;

(2) Instruction list — language resembling a typical assembler;
(3) (a) Ladder diagram — graphical language that appear as a schematics of the relay diagram; and
 (b) Function block diagram — graphical language resembling a logical schematics.

The structured text could the best source for emulations, unfortunately, few PLCs offer it. In contrast, the both graphical languages, which are implemented in a great number of PLC types almost exactly according to IEC 1131 specifications, only visualize internal PLC codes. Thus an instruction list ordinarily remains only one source for any PLC emulation.

Although many PLC types exist with various configurations, the number of their different instruction lists is much smaller due to backward compatibility of their software. For instance, Allen-Bradley's PLCs from families PLC-5, SLC-500, MicroLogix, and ControlLogix utilize similar in-

---

[1] For instance, Allen-Bradley sells emulators for its PLCs, but they only simulate PLC processors, so appropriate RSLogix development environments are also necessary for monitoring programs downloaded into emulated PLCs.

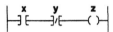

Fig. 1. Assignment $z = x \wedge \neg y$ in ladder diagram

struction lists, characterized in a lot of common features, which can be considered as belonging to one 'language group'. The rung of PLC ladder diagram depicted in Figure 1 corresponds to the source code "*XIC x XIO y OTE z*" in all mentioned PLC types.

If we define x, y, z variables and write proper emulation functions for all used PLC instructions, the rung above can be emulated by calling three functions, for instance: "*FnXIC(x); FnXIO(y); FnOTE(z);*". Such approach also allows modeling execution times of PLC instructions (if required) but it leads to long programs with many excessive calls, although the rung above can be emulated by one assignment $z = x \wedge \neg y$.

Assignments can emulate only pure control algorithm, but it is all we really need in many cases, and besides they are portable across whole range of programmable tools as a low cost emulation.

We based the conversion of PLC programs on the theory of the transfer sets that was originally designed for APLCTRANS algorithm Šusta (2003).

The emulation does not require the composition of whole PLC program into one automaton — every PLC rung (or PLC program block, respectively) can be converted into a separated statement. Therefore, it is possible to process a wider range of operations over APLCTRANS. [2]

2. OVERVIEW OF TRANSFER SETS

Here, we outline the transfer set theory adapted for a PLC emulation. In short, the theory formalizes concurrent evaluations of several expressions all at once. Suppose having variables x, y and C language assignments "*x = 2*x; y = x+1;*". Their classical consequent evaluation yields "*y = 2 * x + 1;*" for y variable, but their concurrent evaluation (utilized by the transfer sets) gives "*y = x+1;*" because x, y variables were assigned after evaluating the both expressions, thus the result corresponds to the program: "*$temp_x = 2 * x$; $temp_y = x + 1$;*

---

[2] APLCTRANS (Abstract PLC Transformation) was created for the verification of PLC programs. It performs an associative composition of some subset of PLC instructions into mathematical formulas and converts PLC program without loops in linear time in the size of its source codes at the most cases, though the program has an exponential complexity of its execution time. It was also proved that a PLC program can be modeled by a automaton of Mealy's family if and only if its operations are expressible with the aid of the instructions that are allowed by APLCTRANS.

$x = temp_x$; $y = temp_y$;" where $temp_x, temp_y$ are some temporary variables.

Transfer sets transparently specify complex transfer operations with program states. For instance, "push x" on an evaluation stack e_1, e_2, e_3 corresponds to three concurrent assignments "*$e_1 = x$; $e_2 = e_1$; $e_3 = e_2$;*" — notice that they may be listed in any order.

To reduce the size of this paper, we only present a rough definition of PLC storage and we replace the exact syntax of t-assignment expressions by the assumption of their similarity to C language, including ? : construction for the conditional assignments, but omitting pointers, ++ and - - operators. We hope that these short cuts do not confuse readers.

The conversion requires unambiguous relation between variables and memory. Let R be a set of PLC variables. If a boolean variable $b \in R$ is mapped to $I : 1/0$ bit address (the least significant bit of $I : 0$ input word) and an integer variable $w \in R$ is mapped to $I : 0$ word address then the both variables share $I : 1/0$ bit. We exclude this case by defining PLC storage.

Definition 1. (PLC storage). Let S be a set variables of a PLC. We will suppose that some given mapping of S into PLC memory is always firmly associated with S. Let $x \in S$ be any variable. The value of x evaluated with respect to S and to its mapping will be denoted by $[\![x]\!] S$. S is called a *PLC storage* if $[\![x]\!] S$ does not depend on $[\![y]\!] S$ for all possible contents of S and for arbitrary variables $x, y \in S$ such that $x \neq y$.

Let us write $EXP(S)^+$ for the set of all meaningful expressions over a given PLC storage S, i.e., their evaluation is known. $EXP(S)^+$ is nonempty because it contains at least numerical constants. If two (possibly different) expressions satisfy $[\![exp_1]\!] S = [\![exp_2]\!] S$ for all contents of S, we will write $exp_1 \equiv exp_2$, otherwise $exp_1 \not\equiv exp_2$.

Definition 2. Let $exp \in EXP(S)^+$ be any expression. The domain of exp is defined as:

$$\text{dom}(exp) \stackrel{df}{=} \{v_i \in S \mid v_i \text{ is used in } exp\}$$

Definition 3. (T-assignment). Let S be a finite non-empty PLC storage and $v = [\![exp]\!] S$ be the assignment of $exp \in EXP(S)^+$ value to $v \in S$ variable. We define:

$$\hat{v}[\![exp]\!] \stackrel{df}{=} v = [\![exp]\!] S$$
$$\text{dom}(\hat{v}[\![exp]\!]) \stackrel{df}{=} \text{dom}(exp)$$
$$\text{co}(\hat{v}[\![exp]\!]) \stackrel{df}{=} v$$

where $\hat{v}[\![exp]\!]$ is called a *t-assignment*, and notations $\mathrm{dom}(\hat{v}[\![exp]\!])$ and $\mathrm{co}(\hat{v}[\![exp]\!])$ stand for its *domain* and *codomain*. If $exp \equiv v$, then $\hat{v}[\![exp]\!]$ is called a *canonical t-assignment*. The set of all t-assignments for S variables is denoted by $\widehat{T}(S)$.

We have labeled t-assignments according to variables, but hat-accented, i.e., x variable has $\hat{x}$ t-assignment. We will also hat-accent all further objects related to t-assignments and our momentary assumptions about t-assignments will be expressed by their following forms:

$\hat{x}[\![exp]\!]$ represents a fully defined t-assignment for x variable with exp expression,

$\hat{x}[\![x]\!]$ stands for the canonical t-assignment for x variable, and

$\hat{x}$ denotes any t-assignment for x variable with an arbitrary $exp \in EXP(S)^+$.

T-assignments can be primed or subscribed, so $\hat{x}_i$, $\hat{x}_j$, and $\hat{y}$ represent t-assignments for three (possibly different) variables x_i, x_j, and y. If we need to distinguish among several t-assignments for one identical variable, we will always write them in their full forms — symbols $\hat{x}[\![exp_1]\!]$ and $\hat{x}[\![exp_2]\!]$ stand for two (possibly different) t-assignments for one x variable. The equality of t-assignments is determined by belonging to the same variable and their equivalent expressions.

Definition 4. Let $\hat{x}[\![exp_x]\!], \hat{y}[\![exp_y]\!] \in \widehat{T}(S)$ be two t-assignments. Binary relation $\hat{x} \cong \hat{y}$ is defined as the concurrent satisfaction of two following conditions: $\mathrm{co}(\hat{x}) = \mathrm{co}(\hat{y})$ and $exp_x \equiv exp_y$. If $\cong$ relation is not satisfied, we will emphasize this fact by $\hat{x} \not\cong \hat{y}$.

Lemma 5. Binary relation $\cong$ on set $\widehat{T}(S)$ is an equivalence relation.

Definition 6. (Transfer Set). *A transfer set on S storage* of a PLC is any subset $\widehat{X} \subseteq \widehat{T}(S)$ that satisfies for all $\hat{x}_i, \hat{x}_j \in \widehat{X}$ that $\mathrm{co}(\hat{x}_i) = \mathrm{co}(\hat{x}_j)$ implies $i = j$. We denote the set of all transfer sets for S variables by $\widehat{S}(S)$, i.e., $\widehat{X} \in \widehat{S}(S)$.

In other words, any transfer set contains at most one transfer function for each variable in S. The composition of transfer sets is based on the *concurrent substitution* defined here as a mapping from variables in S to terms of $EXP(S)^+$.

Definition 7. Let $\widehat{X} \in \widehat{S}(S)$ be a transfer set and $exp_{dest} \in EXP(S)^+$ be any expression. *Concurrent substitution* $\widehat{X} \rightsquigarrow exp_{dest}$ is defined as such operation whose result is logically equivalent to the expression obtained by these consecutive steps:

(1) For all $\hat{x}_i[\![exp_i]\!] \in \widehat{X}$:
while $x_i \in \mathrm{dom}(exp_{dest})$, this x_i occurrence in exp_{dest} is replaced by some not interchangeable reference to x_i, where x_i represents $\mathrm{co}(\hat{x}_i[\![exp_i]\!])$.

(2) For all $\hat{x}_i[\![exp_i]\!] \in \widehat{X}$:
while the result of the previous step (modified expression exp_{dest}) contains a reference to $x_i = \mathrm{co}(\hat{x}_i[\![exp_i]\!])$ then x_i reference is replaced by "(exp_i)" i.e., the expression of $\hat{x}_i[\![exp_i]\!]$ enclosed inside parentheses.

Example 8. Given concurrent substitution:
$$\{\hat{x}[\![x \wedge y]\!], \hat{y}[\![\neg x \wedge \neg y]\!]\} \rightsquigarrow \hat{c}[\![(x \vee y) \wedge x]\!]$$
Direct application of the first step described in the definition above yields
$$\hat{c}[\![(\underline{\hat{x}} \vee \underline{\hat{y}}) \wedge \underline{\hat{x}}]\!]$$
where underlining emphasizes that we have replaced variables by some unique references to the t-assignments that are not be their identifiers. The second step yields
$$\hat{c}[\![((x \wedge y) \vee (\neg x \wedge \neg y)) \wedge (x \wedge y)]\!]$$
but another acceptable results are also
$$\hat{c}[\![(x \equiv y) \wedge (x \wedge y)]\!] \quad \text{or} \quad \hat{c}[\![x \wedge y]\!]$$
because all expressions in three last t-assignments are equivalent.

Definition 9. (Weak composition). A weak composition $\widehat{Z} = \widehat{X} \circ \widehat{Y}$ of two given transfer sets $\widehat{X}, \widehat{Y} \in \widehat{S}(S)$ is the transfer set $\widehat{Z} \in \widehat{S}(S), |\widehat{Z}| = |\widehat{X}|$, with t-assignments $\hat{x}_i[\![exp_{z,i}]\!] \in \widehat{Z}$ that are constructed for all $\hat{x}_i[\![exp_{x,i}]\!] \in \widehat{X}$ as:
$$\hat{x}_i[\![exp_{z,i}]\!] = \hat{x}_i\left[\!\!\left[\widehat{Y} \rightsquigarrow exp_{x,i}\right]\!\!\right]$$
where $i \in I, |I| = |\widehat{X}|, I = \{1, 2, \ldots, |I|\}$.

Example 10. Let $S = \{b, x, y, z\}$ be a PLC storage then the week composition of two given transfer sets is:

$$\begin{aligned}
\widehat{C}_\circ &= \widehat{C}_2 \circ \widehat{C}_1 \\
&= \{\hat{x}[\![x * -y]\!]\} \circ \{\hat{x}[\![z/y]\!], \hat{y}[\![b \; ? \; y : z]\!]\} \\
&= \{\hat{x}[\![(z/y) * -(b \; ? \; y : z)]\!]\}
\end{aligned} \tag{1}$$

Lemma 11. The weak composition $\circ$ is not associative on $\widehat{S}(S)$.

Weak composition $\widehat{C}_2 \circ \widehat{C}_1$ can be modified to associative one, which we denote by $\circledcirc$, if we extend the leftmost transfer set $\widehat{C}_2$ before any composition by adding canonical t-assignments for all S variables, whose t-assignments are missing in the transfer sets. Let us write $\uparrow S$ for the extension operator described above, then $\widehat{C}_2' = \widehat{C}_2 \uparrow S$ always satisfies $|\widehat{C}_2'| = |S|$.

After the composition, all canonical t-assignments are removed by the compression operator $\downarrow$, so the

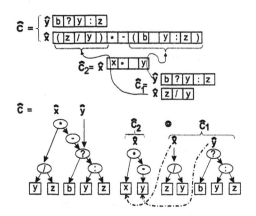

Fig. 2. Array versus graph implementation

associative compositions is defined as:
$$\widehat{C}_2 \circledcirc \widehat{C}_1 = ((\widehat{C}_2 \uparrow S) \circ \widehat{C}_1) \downarrow$$

The exact definitions are not too complex, but they need many auxiliary specifications. Thus we replace them by the following example. Readers may find the definitions including proofs and further details in Šusta (2003).

Example 12. We take PLC storage $S = \{b, x, y, z\}$ and transfer sets from Example 10.

$$
\begin{aligned}
\widehat{C} = \widehat{C}_2 \circledcirc \widehat{C}_1 &= \left((\widehat{C}_2 \uparrow S) \circ \widehat{C}_1 \right) \downarrow \\
&= \left((\widehat{C}_2 \cup \{\hat{b}[\![b]\!], \hat{y}[\![y]\!], \hat{z}[\![z]\!]\}) \circ \widehat{C}_1 \right) \downarrow \\
&= \left(\left\{ \hat{b}[\![b]\!], \hat{x}[\![x * -y]\!], \hat{y}[\![y]\!], \hat{z}[\![z]\!] \right\} \circ \widehat{C}_1) \right) \downarrow \\
&= \left(\widehat{C}_\circ \cup \left\{ \hat{b}[\![b]\!], \hat{y}[\![b\ ?\ y : z]\!], \hat{z}[\![z]\!] \right\} \right) \downarrow \\
&= \left\{ \hat{x}[\![(z/y) * -(b\ ?\ y : z)]\!], \hat{y}[\![b\ ?\ y : z]\!] \right\} \quad (2)
\end{aligned}
$$

Notice that $\widehat{C}$ in Equation 2 differs from $\widehat{C}_\circ$ from Equation 1 only by the presence of $\hat{y}$ t-assignment.

2.1 Implementation of transfer sets

When transfer sets are stored as numeric arrays, the top part of Figure 2, each variable is coded as a unique number with the size according to requisite number of variables. The arrays have very simple implementation, but drawback in the multiplication of expressions. Each composition replaces every occurrences of one variable in the expressions by new string, which gradually increases the sizes of arrays.

The second proposed implementation utilizes the structure similar to Binary expression diagrams (BEDs), e.g. Andersen and Hulgaard (1997); Hulgaard *et al.* (1999), where logical subexpressions are not repeated, but shared, bottom of Figure 2.

Nowadays transfer sets are implemented only with the aid of byte arrays. New graph based version is under development and is expected to be available by the end of 2004.

3. CONVERSION OF PLC PROGRAM

This sections describes the steps necessary for converting a PLC program. First, we outline the conversion of PLC instructions to transfer sets, then we will illustrate the method by an example.

The simplest way of the conversion represents composing each rung, or a program block respectively, to one transfer set $\widehat{C}_i$, for $i = 1$ to *count_of_rungs*, and converting $\widehat{C}_i$ into programming statements of required language. Because one transfer set describes the concurrent assignments, we must convert t-assignments to programming assignments with the aid of temporary variables outlined at the beginning of Section 2 on page 2.

The composition is very fast operation, so we can also try composing two or more following $\widehat{C}_i$ one transfer set and then we select combination with lesser size of final source code.

The transfer sets for basic PLC instructions are listed in Šusta (2003) for some PLCs of Rockwell Automation (Allen-Bradley) and Siemens. These PLCs have instructions lists, where conditions are stored in f_{reg} boolean register and rung is evaluated with the aid of a boolean evaluation stack with limited depth.

Some transfer sets for PLC instructions are also presented in the example in Section 4.

3.1 Conversion of arithmetic instructions

Arithmetic instructions of PLC correspond to a conditional assignment $\hat{x}[\![f_{reg}\ ?\ exp : x]\!]$, where *exp* describes the operation and $x \in S$ specifies a destination address. The assignment is evaluated as "if(f_{reg}) $x = exp$; else $x = x$;" imaginary program. In the case above, we may also remove else condition. For instance, ADD $y\ 1\ x$ instruction of PLC, which assigns $y + 1$ to x variable, is represented by $\{ \hat{x}[\![f_{reg}\ ?\ y + 1\ :\ x]\!] \}$ transfer set, which corresponds to the statement: "if(f_{reg}) $x = y + 1$;". If f_{reg} was always true, i.e., the instruction had no condition then it would be represented by $\{ \hat{x}[\![y + 1]\!] \}$ transfer set.

3.2 Side-effects of instructions

Side-effects mean that a PLC instruction alters its actual arguments or changes other variables.

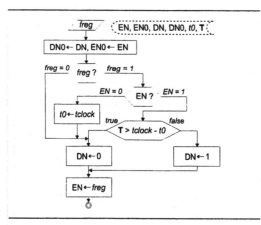

Fig. 3. Diagram of on-delay timer subroutine

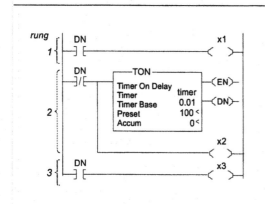

Fig. 4. Side-effects of on-delay timer (TON)

Instruction with side-effects cannot be normally converted into expressions where any of the other operands of the expression would be affected. They should be processed by special way.

Figure 4 shows PLC ladder diagram with on-delay timer TON. Its behavior can be emulated by the timer subroutine in Figure 3. The subroutine takes an evaluated input condition f_{reg} as its argument. It sets DN (done) and EN (enabled) output bits, which are tested by PLC instructions. $DN0$ and $EN0$ are their memories utilized for dealing with side-effects, $t0$ remembers the beginning of timing and it is loaded from a system $tclock$ timer, and T is required the length of time.

The ladder diagram in Figure 4 can be converted by several ways: into one block by composing all rungs 1–3 into one transfer set; or into two blocks (rung pairs 1–2 and 3, or 1 and 2–3); or into three blocks by expressing each rung as one transfer set; eventually into more blocks, if some instructions will be emulated by stand-alone transfer sets.

In all cases, the call of the timer subroutine is inserted only in front the *block of converted instructions*, among which is a *timer* instruction — we denote this block by *BCIT* in this subsection.

The values of $x1$ and $x2$ outputs depends on the state of DN bit before calling the timer subroutine and by contrast, $x3$ value depends on the state of DN bit after its call. Therefore, we must also satisfy this property inside of BCIT.

If all output bits of a timer are read only after executing the timer instruction inside BCIT in all cases, the timer instruction is replaced only by $\{\hat{\tau_1}[\![f_{reg}]\!]\}$ transfer set, where τ_1 is a unique variable of the address utilized by the timer. It will be the input argument of the timer subroutine. [3]

Otherwise, if some timer outputs are also read inside BCIT before the timer instruction, the timer instruction is replaced by $\{\hat{\tau_1}[\![f_{reg}]\!], \hat{\tau_{1e}}[\![1]\!]\}$ transfer set, where τ_{1e} is a unique auxiliary variable initialized to zero by adding $\{\hat{\tau_{1e}}[\![0]\!]\}$ operation before BCIT, and the addresses of all timer bits accessed inside of this BCIT are replaced by multiplexing according to the state of τ_{1e}. For instance, DN bit will be replaced by the operation:

$$((\tau_{1e} \wedge DN) \vee (\neg\tau_{1e} \wedge DN0))$$

Notice, that this is necessary only in BCIT. The switching is useless in all other converted blocks, because there will be always $\tau_{1e} = 0$. [4]

The same method must be performed for all operations with side-effects or unpredictable effects, as storing data into an indirect address. To assure their correct evaluation, they must be converted as separate blocks, if it is possible, otherwise the switching of their before-after content should be included. In extremely critical cases, such operations can be emulated as the step by the step procedure, i.e. as stand alone statements. Therefore, side effect PLC operations will always require human's supervising and their fully automatic conversion is an idea for further research.

4. EXAMPLE OF CONVERSION

Figure 4 depict a fragment of a PLC program for counting boxes with a simple time filter of input. The instruction list exported from SLC 5/03 processor is:

```
SOR XIC S:1/15 CLR N7:0   EOR
SOR XIC I:1.0/0
     BST  TON T4:0 0.01 90 0
     NXB XIC T4:0/DN   OSR B3:0/0
          ADD N7:0 1 N7:0
     BND
EOR
```

---

[3] Each address of timer instruction must be utilized only once times in any PLC program.

[4] Reading timer bits before executing timer instruction usually means an ineffective coding because we add the delay of one scan. Such operations are normally utilized only in few special cases, 2nd rung in Figure 4 illustrates one of them — it represents a pulse generator.

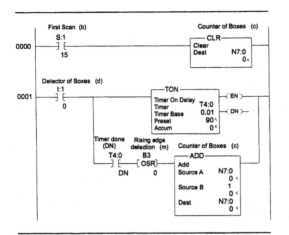

Fig. 5. Counter of boxes

If the detector I:1.0/0 gives signal longer than 0.9 second then N7:0 counter is incremented. The counter is initialized when PLC performs the first pass (bit S:1/15) after switching to run mode.

We first create PLC storage by assigning memory addresses to variables:

$$b \Leftarrow S{:}1/15, \quad c \Leftarrow N7{:}0,$$
$$d \Leftarrow I{:}1.0/0, \quad m \Leftarrow B3/0, \text{ and } DN \Leftarrow T4{:}0/DN,$$

to which we also add PLC flag register f_{reg} and PLC boolean evaluation stack e_1, e_2, whose depth is enough for this program. The last variable τ_1, corresponds to the input condition on the timer T4:0. It yields $S = \{b, c, d, m, DN, \tau_1, f_{reg}, e_1, e_2\}$ storage. Because our program does not access timer bits before evaluation of time instruction, we utilize simple emulation of timer and create transfer sets for the used instructions:

$$\text{SOR, EOR} \Rightarrow \{\hat{f_{reg}}[\![1]\!], \hat{e_1}[\![0]\!], \hat{e_2}[\![0]\!]\}$$

$$\text{BST} \quad \Rightarrow \{\hat{e_1}[\![0]\!], \hat{e_2}[\![f_{reg}]\!]\}$$
$$\text{NXB} \quad \Rightarrow \{\hat{f_{reg}}[\![e_2]\!], \hat{e_1}[\![f_{reg} \vee e_1]\!]\}$$
$$\text{BND} \quad \Rightarrow \{\hat{f_{reg}}[\![f_{reg} \vee e_1]\!], \hat{e_1}[\![0]\!], \hat{e_2}[\![0]\!]\}$$

$$\text{XIC x} \quad \Rightarrow \{\hat{f_{reg}}[\![f_{reg} \wedge x]\!]\}$$
$$\text{OSR m} \quad \Rightarrow \{\hat{f_{reg}}[\![f_{reg} \wedge \neg m]\!], \hat{m}[\![f_{reg}]\!]\}$$
$$\text{TON} \quad \Rightarrow \{\hat{\tau_1}[\![f_{reg}]\!]\}$$

$$\text{CLR c} \quad \Rightarrow \{\hat{c}[\![f_{reg}?0:c]\!]\}$$
$$\text{ADD c 1 c} \Rightarrow \{\hat{c}[\![f_{reg}?c+1:c]\!]\}$$

If we compose each rung separately, we obtain after removing uninterested f_{reg}, e_1 and e_2 that are initialized by EOR instruction:

$$\widehat{R}_1 = \{\hat{c}[\![b?0:c]\!]\}$$
$$\widehat{R}_2 = \left\{ \begin{array}{l} \hat{c}[\![d \wedge DN \wedge \neg m?c+1:c]\!], \\ \hat{\tau_1}[\![d]\!], \hat{m}[\![d \wedge DN]\!] \end{array} \right\}$$

The transfer sets correspond to C language statements where the timer subroutine is called with d argument, i.e., $\hat{\tau_1}[\![d]\!]$ transfer set:

```
if(b) c=0;            // 1st block
ExecuteTimer(d);      // 2nd block (BCIT)
m_temp = d && DN;
if(d && DN && !m) c_temp=c+1;
m=m_temp; c=c_temp
```

Finally we test composing the both rung into one block. It yields:

$$\widehat{R}_2 \circledcirc \widehat{R}_1$$
$$= \left\{ \begin{array}{l} \hat{c}\left[\!\!\left[\begin{array}{l} d \wedge DN \wedge \neg m ? (b?1:c+1) \\ \qquad\qquad\quad : (b?0:c) \end{array} \right]\!\!\right], \\ \hat{t_e}[\![d]\!], \hat{m}[\![d \wedge DN]\!] \end{array} \right\}$$

This program evidently leads to longer source code, so the separate conversion is more efficient.

5. CONCLUSION

The presented method converts a PLC program to programming statements, which can be used either for a low cost emulation of the program or as auxiliary tool when a debugged PLC program is moved to another cheaper hardware, for instance to PC computer.

There are many possibility for future research and many lines of further development suggest themselves, for example optimization of producing final code to reduce requirements for temporary variables. An interesting approach also represents utilizing some faster analogy of the described PLC emulation (for instance with a partial aid of binary decision diagrams) for dynamic verification of PLC programs or for searching their state space.

REFERENCES

Andersen, Henrik Reif and Henrik Hulgaard (1997). Boolean expression diagrams. In: *LICS, IEEE Symposium on Logic in Computer Science*.

Freund, E., A. Hypki, F. Heinze and R. Bauer (2001). COSIMIR PLC - 3D simulation of PLC programs. In: *Proceedings of the 6th IFAC Symposium on Cost Oriented Automation, Berlin*.

Hulgaard, Henrik, Poul Frederick Williams and Henrik Reif Andersen (1999). Equivalence checking of combinational circuits using boolean expression diagrams. *IEEE Transactions of Computer-Aided Design, 18(7)*.

IEC (1993). *International Standard 1131, Programmable Controllers. Part 3: Programming Languages*. International Electrotechnical Commission.

Šusta, Richard (2003). Verification of PLC Programs. PhD thesis. CTU-FEE Prague. avail. at http://dce.felk.cvut.cz/susta/.

IMPLEMENTATION OF MULTILOOP CONTROL FOR THE PLANT CONTROL SYSTEM IN IMPROVED KOREAN STANDARD NUCLEAR POWER PLANT (KSNP+)

Joon-Kon Kim, Hwan-Yong Jung, Soo-Won Kim, Sang-Kook Chung, and Il-Nam Choe

Korea Power Engineering Co., Inc.
360-9 Mabuk-Ri, Gusong-Eup, Yongin-Si, Kyonggi-Do, 449-713, KOREA

Abstract : As the performance and capability of the microprocessor adopted to the process control system has been improved, the number of plant equipment that can be operated by single microprocessor-based controller has been increased. Also, the licensing requirement such as the software verification and validation (V/V) requirement for the digital technology in the nuclear industry has been established. Shin-Kori 1&2 Nuclear Power Plants (SKN 1&2), the first project of KSNP+, adopt the multiloop control design in order to take advantage of the multitasking capability of the microprocessor based system. In this control design, equipment are grouped and segmented with a careful consideration of process system design, interaction, and operation. The grouped and segmented equipment are controlled by one controller. The identical redundant controller is designed as the backup of the controller with fault-tolerance of the control system. The multiloop control design reduces the plant construction cost without degrading the operational reliability of the plant control system. *Copyright © 2004 IFAC*

Keywords: Multiloop Control, Plant Control System (PCS), Segmentation, Redundancy, KSNP+, Shin-Kori 1&2 Nuclear Power Plants

1. INTRODUCTION

KSNP+ is an improved plant model of the Korea Standard Nuclear Power Plant (KSNP) developed by Korea Hydro & Nuclear Power Co., Ltd (KHNP) and Korea Power Engineering Company (KOPEC). SKN 1&2, which are the first nuclear power plant project implementing KSNP+ design concept, are scheduled to start commercial operation in 2009. The PCS of KSNP+ is a major I&C system which performs the control function of both discrete and analog loops in the plant process system. Instrumentation and control system for KSNP+ has various improved design features compared to KSNP design.

As the performance and capability of the microprocessor based PCS has been improved, the number of plant equipment that can be operated by single microprocessor-based controller has been increased. Also, the licensing requirement such as the software verification and validation (V/V) requirement for the digital technology in the nuclear industry has been established.

SKN 1&2 adopts the multiloop control design instead of the dedicated single loop-control design used in the previous KSNP in order to take advantage of the multitasking capability of the microprocessor based system.

In the dedicated single loop-control design, single controller performs the control function of single equipment. On the other hand, in the multiloop controller design, single controller performs the controlling function of multiple equipment. The purpose of this design implementation is to achieve a highly reliable, safe and economical PCS.

2. MULTILOOP CONTROL CONFIGURATION

The basic loop configuration of the PCS in KSNP is shown in the figure 1. The field I/O signal is connected to the dedicated I/O module in the system cabinet or in the field multiplexer cabinet. This signal is conditioned and transferred to the loop-dedicated controller card in the system cabinet for the processing, command-generation, creating the control and/or monitoring signal, etc. The command signal for the field equipment generated in the controller is transferred to the field equipment through the I/O module. The hardware and software of the controller card is designed based on the microprocessor technologies. And these technologies have been rapidly developed since the microprocessor based PCS was applied to the control system in KSNP.

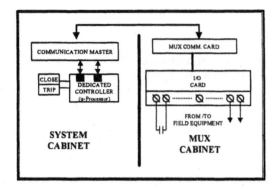

Figure 1. PCS Loop Configuration Diagram of KSNP

The multiloop control capability is provided by powerful microprocessors and this control capability is widely used in industrial power plants. However, the safety constraints of nuclear power plants have restricted wide usage of multiloop control capability for equipment control. The traditional dedicated control method has been applied to the plant control system instead, even though a digital control system has become highly reliable.

The possible concern in using the multiloop control for the control system is the weakness for the failure of many controlled loops at the same time when a controller fails. In order to address this concern, SKN 1&2 have adopted a methodology of the redundant controller. The identical redundant controller is designed as the backup of the controller. When a controller fails, the redundant controller automatically performs the control function without any interruption. Each of these controllers is provided with its own dedicated communication controller for interfacing with the data communication bus in the system so that the system can maintain fault-tolerance through automatic wrapping of fault tolerant communication even if either the primary or backup controller may fail.

The functionality and economy of the system can be enhanced, if the equipment of nuclear power plant are relevantly grouped based on the their functions and relationship in the process system, by using the multiloop control capability of the digital control system. In a multiloop controller design, the equipment are grouped and segmented with a consideration of process system design, interaction, and operation, and each group is assigned to a separate and independent controller. The controller is designed to perform the control function of the grouped equipment using the multitasking capability of the controller.

Each controller is truly independent of other controllers such that failures, maintenance or testing on one controller do not affect multiple plant functions. Plant equipment are assigned to a

controller in a manner that preserve the redundancy and independence that exist within the mechanical process system being controlled. This functionally segmented grouping is to prevent a failure of one controller from affecting more than one group of equipment.

In case of splitting of any one group due to the limitation of a controller capability with respect to the number of associated inputs/outputs, controller response time, etc., more than one controller is provided.

3. FEATURES OF THE PCS HARDWARE CONFIGURATION OF SKN 1&2 AND PREVIOUS PROJECTS

Table 1 shows the comparison of PCS configuration for KSNP+ and KSNP.

CLASSIFICATION		KSNP	KSNP+ (SKN 1&2)
Controller	Safety/Non-Safety System	Dedicated Single Loop Controller	Multioop Controller
Redundancy	Safety/Non-Safety System	Not Applied	Applied
Control Switch Module	Process System	Dedicated Switch Module on the Conventional Control Board	Dedicated Switch Module on the Conventional Control Board
	RW System	Dedicated Switch Module on the Conventional Control Panel	Soft Control via the OIS
Control Logic	Safety System	Separated Controller and I/O Cards	Controller Cards with Integrated I/O Functions
	Non-Safety System	Separated Controller and I/O Cards	Group Controllers with Intelligent I/O Cards
Controller Failure Effect	Safety System	No Actuation Possible	Controller Cards Independent from Group Controller
	Non-safety System	No Actuation Possible	Manual Actuation Possible
Maintenance/ Diagnostics	Safety/Non-Safety System	Engineered Diagnostic Features in the Controller SW	Comprehensive Integrated Diagnostic Features in the Controller and the Control & I/O Cards

Table 1. PCS Configuration Comparison between KSNP and KSNP+

In the previous project's PCS, dedicated single controller performed the control function for only one field equipment or only one device. Dedicated I/O card was used for interfacing the signal between the processor and field equipment. Therefore, the number of the control card and I/O card is decided approximately identical to the number of the field equipment and instrument. The redundant controller and redundant I/O module were not applied to the PCS in the previous plant except for the equipment, which are important to enhance the plant availability. (i.e. main steam isolation valves, main feed water isolation valves, reactor coolant pumps, and 13.8kV/4.16kV switch gear power circuit breakers)

The PCS for SKN 1&2 is designed based on the multiloop controller methodology following the design concept of KSNP+. One controller controls a group of equipment/devices. The secondary multiloop controller is designed for the backup of the primary multiloop controller. The dedicated I/O module is installed in the field multiplexer cabinet or in the system cabinet to transfer the signal from/to the controller and field equipment. And it is the same case as the previous system.

Regarding the Man-Machine Interface (MMI), the advanced feature of Soft Control has been adopted for Radwaste (RW) System. The control function for RW System has been performed by the operation of individual control switch module on the conventional control panel in the previous project. In SKN 1&2, the operator can operate RW system by the Soft Control. The Soft Control means that the operation of the system is executed by manipulating keyboard and mouse connected to the Operator Interface System (OIS). The operator can monitor the RW system through the Large Display Unit (LDU), which is a part of OIS. The LDU shows the system diagram, equipment status, process parameter value, and alarm status of the RW process system.

4. TELEPERM XP AND TELEPERM XS SYSTEM FOR SKN 1&2

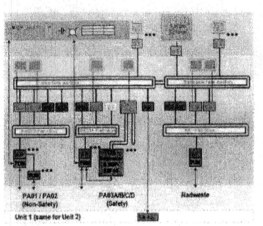

Figure 2.1 Overview Diagram of the PCS for SKN 1&2

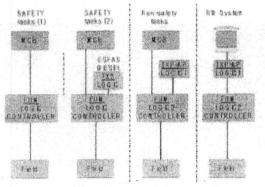

Figure 2.2 Different Structure for Different Task

The TELEPERM XP (hereinafter called as XP) and TELEPERM XS (hereinafter called as XS) supplied by Framatome ANP have been selected as the PCS of SKN 1&2. Figure 2 shows the system overview diagram for SKN 1&2 PCS.

The XP system consists of Process Operation and Monitoring System (OM), Engineering System (ES),

Diagnostic System (DS), Automation System (AS), Commissioning Tool and Plant Bus.

The XP system performs the control and instrumentation function for safety system and non-safety system except for the engineered safety feature actuation system (ESFAS) and the emergency diesel generator load sequencer, which are processed in XS system.

The multiloop controller is practically implemented in Function Module (FUM) and Automation Processor (AP) that are main components of Automation System. The basic configuration of Automation system is shown in Figure 3.

Figure 3. Basic Configuration of Automation System

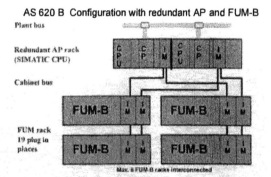

In the safety system, the FUM in the safety cabinet receives and conditions the field input signals, and performs the controlling function. The output signal to the field devices is transmitted from the FUM module. The process information is communicated through plant bus via the AP. The signal generated in a FUM is hardwired to other FUM for transferring the interlock signal.

In non-safety system, the FUM in the field multiplexer receives and transmits the I/O signal from/to the field device. The I/O signal between multiplexer and the system cabinet is communicated via the fiber optic cable. The FUM in the system cabinet receives the signal from the FUM in the multiplexer, performs function of controlling, converting the signal type, and transmits the output signal to the FUM in the multiplexer.

The FUM module also performs input and output (I/O) function in itself. All central components and the FUM are capable of operating in redundancy. Cabinet bus is also designed in redundancy including FUM interfaces.

The continuous closed loop control is implemented in FUM 280 module. And the discrete control loop is implemented in FUM 210 module. Since FUM is the module that includes I/O capability and signal processing capability simultaneously, the quantity of controlled loop in one FUM is restricted due to the number of I/O channel in the FUM. However, the I/O signals are connected to both primary module and secondary module, and the reliability of the I/O function is enhanced. If the I/O of the primary module fails, the I/O of the secondary module takes over the signal without interruption. Therefore the FUM designated as secondary controller continues to control the equipment.

The AP consists of CPU, communication processor (CP), and interface module (IM) for connection to the interface module in the FUM. The AP is also designed as redundancy. The system software and the user programs are executed in the AP.

The AP performs the central discrete control and continuous closed loop control functions and calculations as well as protection and monitoring. Data are transmitted via the AP and the communications processor to the system components linked to the systems (OM, DS, ES and other APs) for further processing. The other function of AP is the plant bus interface control by which the information between the plant bus and FUMs/APs is communicated.

Signals between APs are generally interchanged through the bus system. There are two main bus systems implemented in the architecture:

- Plant Bus
- Terminal Bus

The Plant Bus system interconnects the processors of the AS for signal interchange purpose. It also connects the OM with the AS. The Terminal Bus facilitates communication of the Operator Terminals with the servers of the OM.

When the AP boots up, the complete self-test is carried out. During operation, one section of the test program is executed in short time windows (test slices). The test program runs in the background, independently of the other software.
Faults will be clearly detected in the master AP before changeover to the standby AP is initiated. This is achieved by means of continuous self-tests in both APs that are integrated into the system software as follows :

- CPU: by testing operations, timers, interrupt masks and through cycle-time monitoring
- EPROM, RAM: by comparing the memory contents of the two automation processors and through checksum tests
- I/O bus: Cables and interface modules are checked for wire-breaks and short-circuits
- Page addressing: for incorrect acknowledgement on calling the interface number
- Parallel links: for interrupted paths and short-circuits in the dual-port RAM.

Both processors are connected to each other via the redundancy link. Both APs perform the same tasks at the same time, but one is the master and the other is in standby mode. In the event of a malfunction being detected in the master AP, changeover to the standby AP takes place, and the faulty AP goes from the operating state into the stop-state. In the event of a malfunction being detected in the standby AP, the master remains in the operating state and the standby

AP goes into the stop-state.

One FUM modules has two bus interfaces. Two FUM modules can be connected in parallel to one field device.

Due to this design, the system has a high degree of fault tolerance: Any failure in the application processors, modules, communication lines cannot bring the system to stop doing its task without any interruption. Therefore, in the fully redundant design, the system can handle failure in the application processor, failure in the communication lines, failure of a FUM, or even both failures at the same time. For example, one FUM module is faulty and additionally one communication line fails and the process still continues without interruption. Figure 4 shows the redundancy concept.

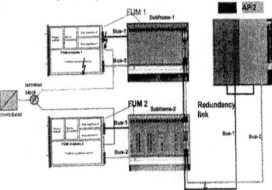

Figure 4. Redundancy of Loop Controller and Communication Controller

In Figure 4, AP-1 is the master controller and AP-2 is the reserve controller in the normal operation. FUM-1 was initiated, and supplies the transducer signal to AP-1 through Bus-1. When FUM-1 and AP-1 are disturbed at the same time, AP-2 detects the failure of AP-1 and the control function of AP-1 is switched over to AP-2 without any upset to the process and function. AP-2 performs the control function as a master controller through Bus-2. FUM-2 supplies the transducer signal to AP-2 through Bus-2 until AP-1 will be repaired. When the malfunction has been rectified in the faulty component and the equipment has been switched on again, the operating The AP automatically supplies all data, programs and signal states for the affected AP via the redundancy link. The AP is then available as a standby AP once more. Throughout the fault rectification process, the master without being affected executes all automation procedures.

The busses use Ethernet technology, which has a market share in excess of 80%. The transmission medium of all communication busses outside cabinets is fiber optic. The medium access method is Carrier Sense Multiple Access with Collision Detection (CSMA/CD) with full duplex support. The communication busses use the switching technology with 100Mbits per second, which is available in the SIMATIC Net since 1998. These features enhance the bus performance by more than the factor 50 in comparison with Industrial Ethernet. Messages are transported significantly faster via Fast Ethernet at 100 Mbits/s (IEEE 802.3u) and therefore only reserve the bus for an extremely short time. Full Duplex prevents collisions: The data throughput increases considerably because the usual message repeats are avoided. Between two stations, data can be sent and received simultaneously. The data throughput of a Full Duplex Fast Ethernet connection therefore increases to 200 Mbits/s.

Switching enables parallel communication: Load decoupling is achieved by subdividing a network into several segments using a switch. By filtering the data traffic on the basis of the Ethernet (MAC) address of the terminal equipment ensures that local data traffic remains local and only data destined for another part of the network are forwarded by the switch. In each individual segment, local data communication is possible independently of the other segments. Throughout the network, therefore, several messages can be transported at the same time. The performance gain lies in the fact that several messages can be sent in parallel. Wrong data packages will be detected by check sum method and thereafter eliminated.

Autosensing enables the stations involved to negotiate and agree the transmission speed for transmission between them before the transmission of the first data packet: 10 Mbps or 100 Mbps; Full Duplex or Half Duplex. It is also possible to deactivate auto-negotiation to ensure a specific transmission speed.

All communication busses are fault tolerant. To achieve this feature, the virtual ring technology applies. Virtual ring technology is used instead of double bus technology to achieve a redundant ring structure. So physically there is one bus line connected to a ring. SIMATIC NET uses a specially developed redundancy control process. This means that reconfiguring the network into a functional network infrastructure takes just a fraction of a second. In a 100 Mbps ring with 50 switch modules (OSM) the network can be reconfigured after an error (cable breakage or switch failure) in less than 0.3 seconds. One of the optical switch modules (OSM) is dedicated as redundancy manager (RM) and opens the ring in normal operation. Any OSM in the net can be assigned as RM. In case of a failure of a switch or of a physical fault of the fiber optic cable this RM closes the ring and data transmission is possible again to all participants of the ring. The terminal devices are unaffected by the changes in the network, and no logical connections are cleared. This guarantees control over processes and applications at all times. In addition to implementing high-speed media redundancy in the 100 Mbit/s ring, OSM also have the requisite functionality to enable high-speed redundant interfacing of rings or network segments. With two OSMs any number of structured network segments can be connected.

There exist two main groups of busses: the terminal bus and the plant busses. Due to this main structure, the communication controller network with the devices near to field is completely decoupled from the information system network and therefore enables the decoupling of the different communication sequences. Each Plant Segment has its own plant bus to ensure the highest possible independence. There is no control data exchange to the safety trains by using the plant bus. All signals to safety trains are hard-wired. Only communication data are passing the plant bus (except the RW System, which uses the soft control feature). For each plant segment all data are stored in a short term archives and simultaneously copied in a long term archives. Doing so no data will be lost and data retrieval is possible at any time either from the short term archives or the long term archives. Data transmission is only performed when data changes occur (event triggered). This reduces the traffic on the busses enormous compared with deterministic data transmission. In spite of the fact that each plant segment has its own data base, the access is possible through the information system network (Terminal Bus) which is a common bus for all plant segments. The participants on the terminal bus are not dedicated to a special plant segment therefore archived data can be retrieved from any terminal where the operator logged in has the access right to operate the system.

5. ECONOMICAL AND RELIABLE SYSTEM DESIGN OF THE PCS FOR KSNP+ (SKN 1&2)

In case of the PCS for KSNP, which is designed based on the dedicated single loop controller concept, the total number of control card was the same as the total number of control loops.

On the other hand, the total number of control card in the PCS for KSNP+ (SKN 1&2) is less than the total number of the controlled loops. The total number is reduced to approximately 75% of the previous project.

The card quantity of KSNP+ is reduced for various reasons. The major reason is the reduction of number of control card and I/O card due to the implementation of the multiloop controller to PCS.

The reduction of quantity for cards results in the saving the cost and time for the plant engineering and construction, which includes equipment installation, field wiring, start-up operation.

The reduction of the control card quantity may result in the decrease in the accident of the system failure during the plant operation. Thus, the reliability and the operability of the control system in the plant are expected to enhance.

The reliability of the multiloop controller could be calculated by the Mean Time between Failure (MTBF). The MTBF of the each component of PCS is obtained from the manufacturer. The MTBF of the loop, which is composed of multiloop controller and the associated circuit, is calculated as the following equation:

$$1/MTBF_{LOOP} = 1/MTBF_{PROCESSOR} + 1/MTBF_{INPUT\ MODULE} + 1/MTBF_{OUTPUT\ MODULE} + 1/MTBF_{RACK}$$

Figure 5 shows the typical correlation between the number of equipment per controller and MTBF based on the data from an industry manufacturer for the other projects. The MTBF for SKN 1&2 PCS will be available soon.

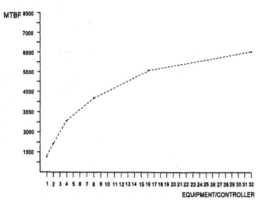

Figure 5. MTBF vs. Equip./Controller

According to the data shown on the figure 5, the MTBF increases as the number of controlled equipment per controller increases. The reliability would be increased as the number of equipment per loop increases. However, a failure of the controller with a lot of equipment assigned would render them inoperable at the same time. Therefore, an optimal segmentation might be needed to assign feasible number of equipment per controller according to the power plant system design philosophy and the supplier's recommendation based on their field experience.

6.   CONCLUSION

As stated above, the multiloop controller with redundant controller in the PCS maintains fault-tolerance of the control system and provides a significant increase in MTBF while reducing plant capital requirements simultaneously.

This approach is also expected to improve the economy and availability compared to the approach implementing the conventional dedicated loop controller. Some of the advantages include a significant improvement of the plant capacity factor while simultaneously reducing the plant construction and Operational and Maintenance (O&M) cost by reducing the hardware quantity.

The multiloop controller configuration of the PCS for SKN 1&2 Project provides a unique combination of advantages compared to the control system design in the previous project.

REFERENCES

SKN 1&2 Project (2002). Segmentation Guideline for Multiloop Controller (9-700-J485-001), Korea Power Engineering Company, Korea.

SKN 1&2 Project (2002). Purchase Specification for PCS (9-771-J212), Korea Power Engineering Company, Korea.

I&C Overall Analysis & Design Report (2002). Chapter 3 Multi-Loop Controller Segmentation (N-740-EJD460-001), Korea Power Engineering Company, Korea,.

I&C System for Improved KSNP (KSNP+) in Korea (2003). Byung-Soo Cho, Hwan-Yong Jung, and Sang-Kook Chung, IFAC Symposium on Power Plants & Power Systems Control,

TELEPERM XP(2002). Catalog LT600, Siemens AG, May

SIMATIC NET(1999). Industrial Ethernet, White Paper

IEEE 379 Application of Single-Failure Criterion to Nuclear Power Generating station Safety System

IEEE 603 Standard Criteria for Safety System for Nuclear Power Generating Station.

IEEE 7-4.3.2 Standard Criteria for Digital Computers in Safety System of Nuclear Power Generating Station

NRC Regulatory Guide 1.152 Criteria for Digital Computer in Safety Systems of Nuclear Power Plants

ELSEVIER
IFAC
PUBLICATIONS
www.elsevier.com/locate/ifac

IMPROVED OPERATION OF PUSHER REHEATING FURNACES BY USING A SUPERVISORY-PLUS -REGULATORY CONTROL

Zh.A. Icev¹, M.J. Stankovski¹, T.D. Kolemishevska-Gugulovska¹, J. Zhao², and G.M. Dimirovski*³

¹ SS Cyril and Methodius University, Faculty of Electrical Engineering
ASE Institute, P.O.B. 574, MK-1000 Skopje, Republic of Macedonia

² Northeastern University, School of Information Science and Engineering
Institute of Control Theory, Shenyang, Liaoning, 110004, P.R. of China

³ Dogus University, Faculty of Engineering, Computer Engineering Department
Acibadem, Zeamet Sk. 21, Kadikoy, TR-34722 Istanbul, Republic of Turkey
E-mails: z_icev@etf.ukim.edu.mk ; zhaojun305@sina.com; gdimirovski@dogus.edu.tr

Abstract: A systems engineering approach to integrated control and supervision for applications to industrial multi-zone furnaces has been elaborated and tested in a case study of three-zone 25 MW RZS pusher furnace at Skopje Steelworks. The integrated control and supervision design is based on combined use of general predictive control optimization of set-points, at the upper level, and steady-state decoupling classical two-term laws, at the lower executive control level. For practical engineering and maintenance reasons, digital implementations are sought within standard computer process control platform. *Copyright © 2004 IFAC*

Keywords: Generalized predictive control; integrated control and supervision; optimization; time -delay processes; set-point control.

1. INTRODUCTION

Systems engineering based essentially on control and optimization methods, by itself, it is a compromising quest to meet some main objective criteria while satisfying unavoidable engineering constraints and requirements. In addition, systems philosophy framework in which original real-world systems and their conceptual models and mathematical model approximations should be observed along with the comprehension of dynamical processes in the real world (below the speed of light) constituting a non-separable inter-play of the three fundamental natural quantities of energy, matter and information (Dimirovski and co-authors, 2001) where energy and matter are carrying the information needed.

Process designs and plant constructions of industrial furnaces have been subject to both scientific and technological research for long time (Ivanoff, 1934). Nonetheless, their control and supervision have never seized to be topics of extensive research due to the complexity of energy conversion and transfer processes in high-power, multi-zone furnaces (Rihne and Tucker, 1991). Industrial pusher furnaces in steel mills (the schematic is omitted) have high installed power and are operated by means of gas/oil fuel in a

difficult field environment and require considerable maintenance support. Practice of industrial steel-mill pusher furnaces has indeed demonstrated that the temperature regulatory level is capable of maintaining process temperature variables close to the command set-points provided separate executive controls stabilize the pressure, when fuel burners function properly. The operational pace is imposed by production scheduling of the slab rolling plant behind the furnace. Main disturbances come along with the line-speed and flow-rate of heated mass as well as the hot gas and air flows inside the reheat processing space, albeit varying slab/ingot dimensions and grades in lower-grade steel processing also act as disturbances. Thus furnace economic operation is rather delicate. The overall control task is to drive the slab heating process to the desired thermodynamic equilibrium and regulate it as well as to maintain the required temperature profile throughout the furnace. Therefore, a two- or even three-level control system architecture and hybrid controls (Yang and Lu, 1988; Zhao, 2001; Dimirovski and co-authors, 2001) ought to be employed for operating furnace plant systems.

The rest of this paper is written as follows. In Section 2, a rather short reference to the most relevant background research is given. Section 3 gives an outline presentation of the approach and solving techniques applied. In Section 4, a brief discussion on the relevant results is presented. Conclusions and references are given thereafter.

2. SOME OF THE RELEVANT BACKGROUND RESEARCH

The economic reasons, in particular the energy cost (fuel consumption) per product unit, and operational restrictions make necessary the plant operation optimization. Practice and research have shown three possible alternatives to resolve this optimization via combined set-point and overall control tasks: (i) an optimally designed controller to replace the regulatory level in performing both tasks (e.g., see Lu and Williams, 1983); (ii) an optimization expert system is combined with executive regulatory controls to perform both tasks (e.g., Yang and Lu, 1988); (iii) a supervisory set-point optimization control is introduced while retaining the regulatory level (e.g., Icev and Dimirovski, 2003) even if classical designs (see Ogata, 2002) are employed.

Practice has demonstrated that the first alternative hardly leads to feasible implementation designs, and the second one requires a costly development of an appropriate heuristic expert system. Hence the third alternative was found (Icev and Dimirovski, 2003) worth exploring via the generalized predictive control theory (Tsang and Clarke, 1988) for practical reasons of cost-effective operation and maintenance service.

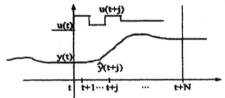

Fig. I. The conceptualization of model based predictive control theory (Clarke and co-authors, 1987).

Model-based predictive control algorithms and architectures (see Figures I-II) have been indeed successfully applied to many industrial processes, since the operational and economical criteria may be incorporated using an objective function to calculate the next control actions (Richalet and co-authors, 1978; Richalet, 1992; Clarke and co-authors 1987, 1988; 1992, 1994; Garcia and co-authors, 1989). In their very essence, these control strategies rely on optimisation of the future process behaviour with respect to future values of the executive control (process manipulated) variables, and their extension to use non-linear process models is motivated by improved quality of the prediction of inputs and outputs (Allgower and co-authors, 1999). This way, non-linear predictive controls are achieved albeit these do not have a specific control strategy of their own; rather these utilise the one associated with linear predictive control. That is, the optimisation procedure is applied on a suitable objective function, with constraints possibly, to carry out

control design (Camaco and Bordons, 1998).

3. SOLVING APPROACH FOR SUPERVISORY-PLUS-REGULATORY CONTROL

The adopted approach to derive a feasible system engineering solution to the problem of concern (see Figures II-III) may be summarized as given below. Firstly, a good design of the regulatory level is carried out (see Section 4 as well), which is to provide executive controls (vector u) to the supervisory level for controlling measured process variables (vector y) while compensating the non-measured varying disturbances (vector e). And secondly, an optimized design of the supervisory level (via minimizing a suitable performance objective function J) to determine the optimal set-points (vector r) so that the relevant constraints are satisfied, which are accounted for in the performance objective. (In principle, should it be available, an external reference trajectory - vector w, which is not needed in our case study - can be included.)

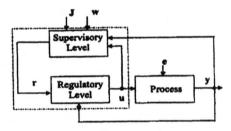

Fig. II. The overall architecture of a two-level supervisory-plus-regulatory control system.

In an industrial furnace operation, besides the limits on the controlled process variables, practical operating characteristics of actuators also impose constraints on the amplitude and the slew rate of the manipulated process variables. Hence, the inequality constraints to be observed are:

$$y_{\min} \leq \hat{y}(t+j) \leq y_{\max}, \ j=1,...,N_y; \quad (1)$$

$$u_{\min} \leq u(t+i-1) \leq u_{\max}, \ i=1,...,N_u; \quad (2)$$

$$\Delta u_{\min} \leq \Delta u(t+i-1) \leq \Delta u_{\max}, \ i=1,...,N_u. \quad (3)$$

Notice that the constraints are imposed on $\hat{y}(t+j)$, that is the $j-th$ step ahead prediction of the controlled process variables. In turn, via optimizing a suitable objective function subject to equality and inequality constraints (Luenberger, 1984) the overall control strategy is a hybrid one. The supervisory level dynamically optimises the system, while stability is preserved with certain conditions satisfied (Zhao, 2001). Since the regulatory level confines plant system transients locally, the resulting optimum may be a global one.

Real steel-mill industrial processes (such as at Skopje Steelworks) are known to be characterized by non-stationary stochastic disturbances. Therefore, as argued by Clarke (1987, 1988, 1992), the class of controlled autoregressive and moving-average process model (constant sampling period, equidistant time-shifting) is most appropriate. The resulting controlled autoregressive and moving-average process models, given the identified furnace operating process models, with the physical parameters T_s, τ_{ii}, τ_{av}, K_{ii} (Section 4), have the form

$$A_{ii}(q^{-1})y_i(t) = B_{ii}(q^{-1})u_i(t) + e_{ij}(t)/\Delta \, , \quad (4)$$

$$A_{ii}(q^{-1}) = 1 - \exp(-T_s/\tau_{av})q^{-1}, \quad (5)$$

$$B_{ii}(q^{-1}) = K_{ii}(1 - \exp(-T_s/\tau_{av}))q^{-(1+d)}, \, (6)$$

with $d = \left[\tau_{ii}/T_s\right]$, where now subscript i indicates the main channel ii of the regulatory control level (i.e., $ii = 1$, 2, 3); q^{-1} the backward shift operator (e.g. $q^{-1}v(t) = v(t-1)$) and $\Delta = 1 - q^{-1}$, where $e_{ij}(t)$ represents a zero mean white noise.

The regulatory control level of a well elaborated design implies known fixed control algorithms (e.g., SS-decoupling PI). Thus the respective process model for the supervisory level can be found by using the above class (4) as follows:

$$A_{rc}(q^{-1})u(t) = B_{rc}(q^{-1})r(t) + B_{ry}(q^{-1})y(t) \quad (7)$$

with appropriate polynomials $A_{rc}(q^{-1})$, $B_{rc}(q^{-1})$ and $B_{ry}(q^{-1})$. Since the designed regulatory level consists of PI controllers "$K_{Pii} + K_{Iii}/s$" (acting on error signals "$r_{ii}(s) - y_{ii}(s) \equiv r_i(s) - y_i(s)$" with K_{Pii} and K_{Iii} designs to produce controls "$u_{ii}(s) \equiv u_i(s)$"), the derived equations are:

$$A_{rc}^i(q^{-1})u_i(t) = B_{rc}^i(q^{-1})r_i(t) + B_{ry}^i(q^{-1})y(t), (8)$$

$$A_{rc}^i(q^{-1}) = 1 - q^{-1}, \quad (9)$$

$$B_{rc}^i(q^{-1}) = (0.5T_sK_{Iii} + K_{Pii}) + (0.5T_sK_{Iii} - K_{Pii})q^{-1} \quad (10)$$

$$B_{ry}^i(q^{-1}) = -(0.5T_sK_{Iii} + K_{Pii}) - (0.5T_sK_{Iii} - K_{Pii})q^{-1}. \quad (11)$$

Given the engineering construction of the furnace and the operating point with $U_{midrange}$, the inequality constraints on magnitude variations of the executive control variables were adopted; $\pm 20\%$ of the maximum actuator battery flow rates. These constrains along with actuator robustness allowed for neglecting the constraints on control increments. Furthermore, with regard to computational effort involved, the optimization variables are reduced to $\hat{y}(t+1)$, ..., $\hat{y}(t+N_y)$; $u(t)$, $u(t+1)$, ..., $u(t+N_u)$, $r(t)$, $r(t+1)$, ..., $r(t+N_y)$ on the grounds of the process model and recalling that the expectation of noise disturbances at time $t+j$ is $E\{e(t+j)\} = 0$.

The objective function $J = F(\hat{y}, \hat{u}, \Delta\hat{u})$, representing the optimization criterion at the supervisory level, has to account for the dynamical behaviour over the prediction-horizons. The objective function class used in (Tsang and Clarke, 1988; Camaco, 1993; De Prada and Valentin, 1996; Camaco and Burdons, 1998) is well justified with respect to both controlled and controlling variables, and was adopted in our work too (see Camaco and Burdons, 1998). The optimum predictive control action is to be obtained by minimising the objective function $J = F(\hat{y}, \hat{u}, \Delta\hat{u})$ subject to the equality (8)-(11) and the inequality (1)-(3) constraints. For this purpose, prediction of the control variables is calculated as a function of past values of the input controls and measured outputs and of a horizon of future control actions. Thus as known from Luneberger (1984), given the inequality constraints, only a numerical indirect solution to optimisation can be sought via introducing a Lagrangian function resulting in typical Kuhn-Tucker conditions. By employing the appropriate Lagrangian function yielded the solution formulae given bellow. (The entire derivation was available to the anonymous reviewers of draft version).

For each thermal process path, the energy based quadratic objective function in its simplest case (with weighting coefficients omitted) to be minimized is

$$F_i = \hat{y}^2(t+1) + \hat{y}^2(t+2) + \hat{y}^2(t+3) + u^2(t) + u^2(t+1) + u^2(t+2) + u^2(t+3). \quad (12)$$

The equality constraints associated with the controlled temperatures, $j = 1, ..., 3$, are

$$y_i(t+j) - \exp(-T_s/\tau_{av})y(t+j-1) - K_{ii}(1 - \exp(-T_s/\tau_{av}))u(t+j-3) = 0. \quad (13)$$

The $j = 1, ..., 3$ equality constraints associated with the executive controls (or manipulated process variables) - the fuel flow rates - lead to the following equations:

$$u(t+i-1) - u(t+i-2) - (0.5T_sK_{Iii} + K_{Pii})\delta r(t+i-1) - (0.5T_sK_{Iii} - K_{Pii})\delta r(t+i-2) + (0.5T_sK_{Iii} + K_{Pii})y(t+i-1) + (0.5T_sK_{Iii} - K_{Pii})y(t+i-2) = 0. \quad (14)$$

The sets of Kuhn-Tucker conditions for the objective function, with regard to Eq. (12) and to the inequality and the equality constraints, after a rather lengthy derivation finally become:

$$2y(t+j) + \lambda_j - \exp(-T_s/\tau_{av})\lambda_{j+1} + (0.5T_sK_{Iii} + K_{Pii})\mu_{j+1} + (0.5T_sK_{Iii} - K_{Pii})\mu_{j+2} = 0, \, j = 1, 2; \quad (15)$$

$$2y(t+3) + \lambda_3 + (0.5T_sK_{Iii} + K_{Pii})\mu_4 = 0, \, j = 1, 2; \quad (16)$$

$$2u(t) + \mu_1 - \mu_2 - K_{ii}(1 - \exp(-T_s/\tau_{av}))\lambda_3 -$$

$$-v_1^{u_{min}} = 0, \; j = 1,2; \qquad (17)$$

$$2u(t+i) + \mu_{i+1} - \mu_{i+2} - v_i^{u_{min}} = 0, \; j = 1,2; \; (18)$$

$$2u(t) + \mu_1 - \mu_2 - K_{ii}(1 - \exp(-T_s/\tau_{av}))\lambda_3 - \\ -v_1^{u_{max}} = 0, \; j = 1,2; \qquad (19)$$

$$2u(t+i) + \mu_{i+1} - \mu_{i+2} - v_i^{u_{max}} = 0, \; j = 1,2; \; (20)$$

$$2u(t+3) + \mu_4 = 0, \; j = 1,2; \qquad (21)$$

$$-(0.5T_s K_{Iii} + K_{Pii})\mu_1 = 0, \; j = 1,2; \quad (22)$$

$$-(0.5T_s K_{Iii} + K_{Pii})\mu_{j+1} + \\ -(0.5T_s K_{Iii} - K_{Pii})\mu_{j+2} = 0, \; j = 1,2; \quad (23)$$

$$-(0.5T_s K_{Iii} - K_{Pii})\mu_4 = 0, \; j = 1,2; \quad (24)$$

$$(y(t+j) - \exp(-T_s/\tau_{av})y(t+j-1) - \\ -K_{ii}(1 - \exp(-T_s/\tau_{av}))u(t+j-3))\lambda_j = 0, \\ j = 1,...,3; \qquad (25)$$

$$(u(t+i-1) - u(t+i-2) - \\ -(0.5T_s K_{Iii} + K_{Pii})\delta r(t+i-1) - \\ -(0.5T_s K_{Iii} - K_{Pii})\delta r(t+i-2) - \\ +(0.5T_s K_{Iii} + K_{Pii})y(t+i-1) - \\ +(0.5T_s K_{Iii} - K_{Pii})y(t+i-2))\mu_i = 0, \\ i = 1,...,4; \qquad (26)$$

$$-u(t+i-1)v_i^{u_{min}} = 0, \; i = 1,...,4; \quad (27)$$

$$-u(t+i-1)v_i^{u_{max}} = 0, \; i = 1,...,4; \quad (28)$$

$$v_i^{u_{min}} \geq 0, \; v_i^{u_{max}} \geq 0, \; i = 1,...,4. \quad (29)$$

The optimal solutions Eqs. (30)-(31) for the executive control signals $\delta r(t)$, generated by the supervisory predictive controller, can be obtained by taking $\lambda_1 \neq 0$, $\lambda_2 \neq 0$, $\lambda_3 \neq 0$, and $\mu_1 = 0, ..., \mu_4 = 0$. When the constraints are active as considered above and no changing of executive controls is allowed for (nonlinear programming), the time-varying set-points are determined as follows:

$$\delta r_i^{NLin}(t) = -(1/(0.5T_s K_{Iii} + K_{Pii}))(u(t-1) + \\ +(0.5T_s K_{Iii} - K_{Pii})(r(t-1) - y(t-1)) - \\ -(0.5T_s K_{Iii} + K_{Pii})y(t)), i = 1,...,3, (30)$$

where now subscript i indicates the main channel ii of the regulatory control level (i.e., ii =1, 2, 3). When the inequality constraints are not active (the respective Langranean multipliers are non-existent) and further

changing of executive controls allowed, the above optimization problem becomes a linear one, and the time-varying set-points are found as:

$$\delta r_i^{Lin}(t) = -(1/(0.5T_s K_{Iii} + K_{Pii}))(u(t-1) + \\ +(0.5T_s K_{Iii} - K_{Pii})(r(t-1) - y(t-1)) - \\ -(0.5T_s K_{Iii} + K_{Pii})y(t)) - \\ -\frac{K_{ii}(1 - \exp(-T_s/\tau_{av})}{(0.5T_s K_{Iii} + K_{Pii})(1 + K_{ii}^2(1 - \exp^2(-T_s/\tau_{av}))} \times \\ \times(\exp^3(-T_s/\tau_{av})y(t) + \\ +\exp^2(-T_s/\tau_{av})(0.5T_s K_{Iii} - K_{Pii})u(t-2) + \\ +\exp(-T_s/\tau_{av})(0.5T_s K_{Iii} - K_{Pii})u(t-1)) \; (31)$$

4. RESULTS AND DISCUSSION

The case-study RZS furnace at Skopje Steelworks has three zones, total size 25x12x8 m, maximum installed 28MW and normal operating power 25 MW Only some of the results could have been given here. The furnace is operated in a steady-state regime at a given pusher pace depending on slab/ingot size and other metallurgical specifications (beyond our expertise and not of concern in here) as shown by the respective characteristic curves in Figures III-IV for the standard heat processing at 1050-1250 °C identified. From the control point of view, thermal process in this 3zone furnace is a 3x3 (N_{inp} x N_{out}) multivariable process for both the regulatory and the supervisory control level. (Notice, temperatures in slab centre are not amenable for real-time measurements, hence reconstructed off-line using a special model and Kalman filtering because).

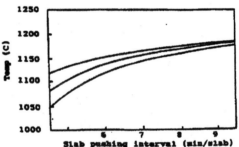

Fig. III. Operating steady-state characteristics desired: slab temperatures T_t, T_c, T_b as functions of pusher pace; upper curve at slab top surface T_t, lower curve T_b - bottom surface, and middle curve T_c – slab centre.

For high-power large industrial furnaces in operating conditions the input signals had be confined in the ranges from (0.03-0.05) to (0.10-0.15) of the maximum magnitude of executive controls. With a choice of 5-10% of the maximum allowed input signal and the records of a number of step as well as PRBS responses for every input, respectively, via known theory (Ogata, 2002; Ljung, 1999, and MATLAB-Systems Identification Toolbox), a family of Ziegler-Nichols and Küpfmüller-Strejc models and the respective non-parametric dynamical models in terms of k-time sequence matrices of the matrix weighting function have been identified.

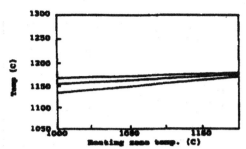

Fig. IV. Operating steady-state characteristics: slab temperatures T_t, T_c, T_b as functions of furnace heating zone temperature for the standard processing regime 1050-1250 °C (following the overall control system redesign).

Local process dynamics (K = $\Delta\Theta/\Delta Q$) is

$$G_{ij}(s) = K_{ij} \exp(-\tau_{ij}s)/(T_1+1)(T_2 s+1) \,. \ (32)$$

with ranges of values for all parameters. Thereafter the discrete-time representations in shift-operator q^{-1} have been identified, respectively. Besides, value ranges for the main attributes of slab heating dynamics have been identified: steady-state (SS) gains K_{ij} (in [°C/MJ/min]) and pure time-delays τ_{ij} (in [min]) within each of the process transfer paths (i,j), giving rise to an average pure time delay τ_{av}.

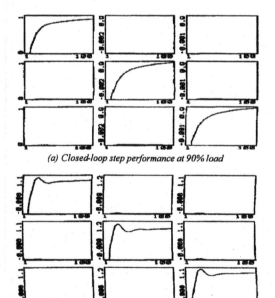

(a) Closed-loop step performance at 90% load

(b) Closed-loop step performance at 60% load

Fig. V. Achieved control performance of the designed regulatory level alone (noise is filtered out).

With regard to the main process channels (i,i) (i.e., transducer-actuator paths of the equalizing, lower, and

upper zones), the values of SS gains range within 7.25-8.95, 4.10-6.50, and 1.50-1.70, while time-delays within 1.50-1.70, 2.10-3.90, and 3.90-7.20, respectively, depending on the actual power load. Note that these variations account for operating process uncertainties too. The (i,i) static I/O characteristics appeared not entirely symmetric showing dynamical weighting sequences were not entirely symmetric either. Hence, re-design study on transducer-actuator collocation was carried out too.

Thus, given the complete knowledge of identified process dynamical and steady-state operations at the typical operating loads the analytical equations were derived for solving both the regulatory steady-state decoupling PI and the optimizing predictive reference controls (Icev and Dimirovski, 2003). The typical dynamical performance of the regulatory control level, achieved via the design carried out for the nominal operation with 80-90% loads, is depicted in Figure V for a range of loads 60-90%. As seen, the regulatory level fulfils the pre-conditions for the operating performance of the supervisory control.

The $y_i(0)$ =1150 °C implies non-zero initial values of the executive control variables $u_i(0) = U_{midrange}$ representing the mid-range control values to which controls $u_{ii}^t = \delta u_{ii}(t)$, $u_{ii}^{t+j} = \delta u_{ii}(t+j)$ due to the supervisory controller are added. The process noise vector $e(t)$ was assumed to have the same stochastic characteristics in every channel and approximated as a weak white noise with σ_e^2 =10. The residing horizon for the supervisory predictive control can be determined by observing the operating pure time delay phenomena in the furnace and the relationship between pure delay and inertial times $\delta = \tau_{ij}/(\tau_{ij}+T_{ij})$ and $\eta = \tau_{ij}/T_{ij}$ taking into consideration the adopted sampling period $T_s = 1$ [min] given the integer value of $d = \lfloor \tau_{ij}/T_s \rfloor$.

The resulting overall control system performance is illustrated by the sample results for the heating zones depicted in Figures VI-VIII. Since τ_{ij} is approximately the same for the main channels (i,i) of the upper and lower zones and not equal to the one of the equalizing zone, a common average τ_{av} was adopted and used throughout. In turn, it appeared that the residing horizon should have minimum $H_r = 3$ and maximum $H_r = 5$ sampling periods. At the desired operating points $y_{ii}(0)$ =1150 °C, the supervisory control level is acting as to maintain the current references $r_{ii}(t)$, $t \ge 0$: whenever executive controls satisfy the constraints and hence observes no active constraints, then controls are not identical to the regulatory components and therefore $\Delta u_{ii}(t) \ne 0$ and $u_{ii} = \delta u_{ii}(t)$; whereas whenever is acting so as to observe active constrains present and the current reference commands $r_{ii}(t)$, $t \ge 0$ not fixed, then $\Delta r_{ii}(t) \ne 0$ and $r_{ii} = \delta r_{ii}(t)$.

Fig. VI. Normal operating regime of heating zones T_{zone} = 1150 °C and 90% power (initial condition $r(0)=y_{ss}(0)=1150$ °C): A time history of the controlled reference to the regulatory control level for a sudden change of +50 °C at the initial time instant.

Fig. VII. Normal operating regime T_{zone} = 1150 °C and 90% power (initial condition $y_{ss}(0)=1150$ °C): A time history of the varying executive control under the supervisory controlled reference after the initial change.

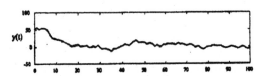

Fig. VIII. Normal operating regime T_{zone} = 1150 °C and 90% power (initial condition $y_{ss}(0)=1150$ °C): A time history of the varying regulated temperature under the supervisory controlled reference after the initial change.

5. CONCLUSION

An application case study on improving the operating performance of a 25MW pusher furnace, which is equipped with a standard computer process control system and aimed at reheat processing of low-grade steel slabs primarily, and ingots too, depending on market demand and supply. The improved furnace operation is achieved via control systems redesign introducing a combined predictive supervisory and PI-SS-decoupling regulatory control strategy (a know-how technological property) has been presented. In turn, an enhanced real-time process control ensuring about 4% cost saving for fuels has been achieved. The alternative of using computational intelligence techniques in a fuzzy-neural predictive control strategy as a prerequisite for furnace expert control is topic of a future research paper.

Acknowledgement: Authors express their gratitude to the anonymous reviewers for the respective remarks, which helped considerably to write up this final version.

REFERENCES

Allgower, F., T. Badgwell, J. Qin, J. Rawlings, and S. Wright (1999). Nonlinear Predictive Control and Moving Horizon Estimation - An Introductory Overview. In: *Advances in Control: Highlights of ECC'99* (P. M. Frank (Ed)), pp. 391-449. Springer-Verlag, London.

Camacho, E., and C. Bordons (1998), *Model Predictive Control*. Springer-Verlag, London.

Clarke, D., C. Mohtadi, and P. Tuffs (1987). Generalized predictive control (Pts. I and II), *Automatica*, **23**, no. 2, pp. 137-160.

De Prada, C., and A. Valentin (1996). Set point optimization in multivariable constrained predictive control. In: *Proceedings of the 13th World Congress of the International Federation of Automatic* Control. San Francisco, pp. 351-356. Pergamon Press, Oxford.

Demircioglu, R., and D. Clarke (1992). CGPC with guaranteed stability properties. *IEE Proceedings Pt. D Control Theory and Applications*, **139**, no. 4, pp. 371-380.

Dimirovski, G. M., A. Dourado, E. Ikonen, U. Kortela, J. Pico, B. Ribeiro, M. J. Stankovski, and E. Tulunay (2001). Learning Control of Thermal Systems. In: *Control of Complex Systems* (K.J. Astroem, P. Albertos, M. Blanke, A. Isidori, W. Schaufelberger, and R. Sanz (Eds)), Chapter 14, pp. 317-337. Springer-Verlag, London.

Garcia, C., D. Prett, and M. Morari (1989). Model predictive control: theory and practice - a survey. *Automatica*, **25**, no. 3, pp. 335-348.

Icev, Z. A., and G. M. Dimirovski (2003), Study of a Two-Level Supervisory-plus-Regulatory Control System for RZS Pusher Furnace Using Model Predictive Control Strategies and Fuzzy-System Models. *FEE-ASE Technical Report GPC-TSC-01/03*. SS Cyril and Methodius University, Skopje (in Macedonian).

Ivanoff, A. (1934), "Theoretical foundations of the automatic regulation of temperature." *J. Inst. Fuel*, **7**, pp. 117-130; discussion on pp. 130-138.

Ljung, L. (1999), *System Identification: Theory for the User* (2nd ed.). Prentice Hall PTR, Upper Saddle River NJ

Lu, Y. Z., and T. J. Williams (1983). Computer control strategies of optimal state feedback methods for the control of steel mill soaking pits. *ISS Transactions*, **2**, pp. 35-43.

Luenberger, D. G. (1984), *Linear and Nonlinear Programming*. Addison-Wesley, Reading, MA.

Ogata, K. (2002). *Modern Control Engineering* (4th edition). Prentice-Hall, Upper Saddle River, NJ.

Rhine, J., and R. Tucker (1991), *Modelling of Gas-fired Furnaces and Boilers*. McGraw-Hill, New York.

Richalet, J., A. Rault, J. Testud, and J., Papan (1978). Model predictive heuristic control: applications to industrial processes. *Automatica*, **14**, pp. 413- 428.

Richalet, J. (1992). Observations on model based predictive control. *Control Engineering*, pp. 39-41, August.

Tsang, T., and D. Clarke (1988). Generalised predictive control with input constraints. *IEE Proceedings Pt. D Control Theory and Applications*, **135**, no. 6, pp. 451-460.

Yang, Y. Y., and Y. Z. Lu (1988). Dynamic model based optimization control for slab reheating furnaces. *Computers in Industry*, **10**, pp. 11-20.

Zhao, J. (1993). A solution to the nonlinear optimal control problems. In: *Proceedings of the 32nd IEEE Conference on Decision and Control*, vol. 4, pp. 2999-3002. The IEEE, New York.

Zhao, J. (2001). Hybrid Control for Global Stabilization of a Class of Systems, in *Advanced Topics in Nonlinear Control Systems*, Chapter 4, pp. 129-160. World Scientific, Singapore.

ELSEVIER

IFAC
PUBLICATIONS
www.elsevier.com/locate/ifac

REMOTE CONTROL OF REAL AND VIRTUAL WELDING ROBOTS FOR LEARNING

Wilhelm Bruns°, Kai Schäfer°°

°*Bremen University, Germany*
Research Center for Work, Environment and Technology (artec)
bruns@artec.uni-bremen.de
°° *Bendit Innovative Interfaces GmbH, Germany*
kai.Schaefer@bendit-interfaces.com

Abstract: An application of a Mixed Reality concept to Internet-Based Robotics will be presented. It will be argued how a coupling of real and virtual, local and remote robot systems may support a cost oriented training and education in context related robotics. This application is related to *Hyper-Bonds*, our unified concept to describe complex effort/flow driven automation systems distributed over real and virtual worlds. It allows selected materialization of parts of the system in reality and is functionally connected to a simulation model. *Copyright © 2004 IFAC*

Keywords: mechatronics, internet-based robotics, simulation, mixed reality

1. INTRODUCTION

Blending real and virtual realities in a mixed reality environment (Ohta & Tamura, 1999) is becoming more and more popular. The extension of this blending towards internet-based distributed environments may offer some interesting possibilities for low cost automation, but also rises several problems. Remote control of robots using augmented reality is a well established field of control theory and practice (Milgram et al, 1995). However a systematic concept to integrate these components of automation into a broader context and its use for situated learning is still missing. We argue for a certain theoretical framework and demonstrate an application of using a remote melting robot for teaching robotics in a process oriented way.

In some previous work, the concept of complex objects was introduced (Bruns, 1999) being objects with a real concrete part coupled to various virtual representations (simulation, animation, symbolic) by means of grasp- or image-recognition. This coupling introduces the possibility to build and change real systems and synchronously generate their functional representatives. Simulation may be carried out with the virtual model and compared with the desired behavior of the real system, Figures 1 and 2.

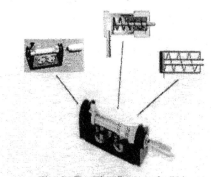

Fig. 1: Complex Pneumatic Object

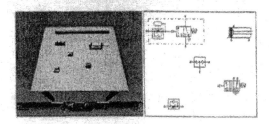

Fig. 2: Two virtual Representations

As long as one stays in physical modeling, it is only the (not trivial) problem to merge continuous and discrete behavior representations. Several simulators support this type of hybrid modeling (for more details see Mostermann, 1997). Two powerful theories approaching each other from different edges are Petri-Nets for discrete behaviors and Bond-Graphs (Paynter, 1961; Karnopp, 1995; van Amerongen, 2001) for continuous flows. Both concepts have in common that they are graph-oriented and therefore may open up a broad area of possibilities to apply methods of theoretical informatics, namely graph-transformation and -replacement methods. Secondly, they support unified views on physical phenomena, a feature more and more important in times of emerging mechatronic systems. Thirdly, they support a certain intuitive engineering point of view (Cellier et al, 1995).

As soon as we come to a mixture of physical and virtual worlds however, be it locally or distributed, we have to face not only hybrid physical connections but also the transfer between energy, signal flow phenomena and information processing. No general concept can be seen yet. However some steps in this direction can be presented.

In a prototype, it could be demonstrated how, by using complex objects of type *conveyor belt*, *container* and *tool-machine*, a system configuration could be built with real concrete models synchronously generating the topology of a virtual representation. Furthermore the desired behavior of the automation system could be demonstrated by concrete hand-movements and signs. From this demonstration, Petri-Net fragments were generated to serve as building blocks for the control algorithm driving a Programmable Logic Controller (PLC) responsible for the control of a real automation system by means of sensors and actuators (Figure 3) (Bruns, 1999).

Fig. 3: Specifying Behavior by Demonstration

This concept has been further extended by bi-directional links between the virtual and the real model, being able to sense and generate various relevant physical continuous effort and flow phenomena via universal connections: **Hyper-Bonds** (Bruns, 2003). The reason, to call this connection Hyper-Bond is because it aims, similar to Bond-

Graphs to a unified interface concept for various physical flow phenomena. However, there is no direct translation of bond-graphs into hyper-bonds. Bond-Graphs describe energy flows, whereas hyper-bonds provide the interface between energy, signal and information flows. Mostermann (1997) describes well difficulties related to cutting a bond-graph into two parts. Therefore the following description is more a conceptual than a detailed implementation view.

An interesting application problem is the remote control of a robot system, which is part of an automation system that is remotely designed, implemented, tested, controled and serviced. Several authors report about remotely controlled mixed reality robots (Milgram et al, 1995, Milgram & Ballantyne, 1997) and related design problems (Milgram & Colquhoun, 1999). Their focus is on a general taxonomy of mixed reality and for the resulting man-machine interface, strongly emphasising the visual representation of tremote reality.

Our vision is, to have an interface concept to be able to design a system and successively construct it locally in virtuality and reality and materialise it at a remote place, not only grasping through the Internet but transfer all physical phenomena through the network.

2. SYSTEMS AND BOUNDARIES

A system behaviour may be studied by cutting it at well defined boundaries and replacing the external influences by some observable and measurable relevant variables, reducing the investigation to the internal dynamics of the rest (figure 4). In work oriented systems design we may use this principle to cut a system into two parts, one non-relevant for man-machine interaction or ergonomics, and the other one important and relevant (safety, performance, human-skills). Certain well known aspects of a system can be represented in a formal way by algorithms in the computer, others to be investigated in more human related way are represented in reality, but coupled to a dynamic surrounding. This would allow completely new forms of easy experimental systems design.

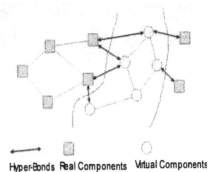

Hyper-Bonds Real Components Virtual Components

Fig. 4. Boundaries cutting a system

In order to provide arbitrary boundary conditions, we must have a mechanism to generate and sense phenomena (Fig. 5). We implemented a coarse prototype for electrics (voltage and current) and pneumatics (pressure and volume-flow) and demonstrated its successful integration into a virtual construction- and simulation-environment, Fig. 6 (Bruns, 2003). We used pressure valves connected to a pressure source to generate air flow and sensors to measure the pressure for a pneumatic-hyper-bond, and electrical switches connected to a voltage source to generate electric-current flow and sensors to recognize a high or low voltage level at the interface for an electrical-hyper-bond. Connections between real and simulation parts of a system are well known from hardware in the loop tests, however it is new, to provide this possibility in a flexible user-centred way.

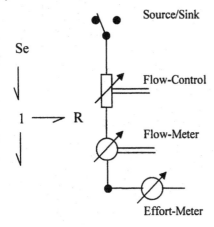

Fig. 5. Abstract mechanism to sense and generate effort and flow (Hyper-Bond)

3. APPLICATIONS

Our concept of Hyper-Bonds is being applied in a learning environment for electro-pneumatics, where students can work on complex systems, freely switching between virtuality and reality[1]. At present, it is not yet possible to cut a real complex system (Fig. 7) at **arbitrary** boundaries and project one part into virtuality and keep the rest in reality, both being connected by a general interface. Or take the other direction: cut a virtual model of the production system in Fig. 7 and materialize it stepwise into a real production. However, in principle it could be demonstrated. Fig. 6 shows a virtual model of the distribution station (a mover driven by pneumatic pressure and electrical signals Fig. 8) as part of a larger modular production system (MPS), Fig. 7. Electrical signals and pressure for the virtual model are generated from the real modeling desk,

[1] EU-IST Project DERIVE (Distributed Real and Virtual Learning Environment for Mechatronics and Tele-Service

transferred into virtuality by a pressure hyper-bond (left connector row) and electric hyper-bond (right connector row). As this automation process (sequence of distribution-station with mover, test - station and manufacturing-station with round table, drilling and testing) is a slow discrete event driven process, there are only minor real-time synchronization problems. This changes however for continuous robot control.

Especially if one is interested in a work-process-oriented focus, it is of high value to use authentic situations, see and feel the complex context, be able to select certain interesting aspects of a system, put them as real components on a laboratory desk, but still have them connected to and integrated into the overall system. As the virtual model can be distributed over different locations (however with some restrictions in time behavior), one has the possibility to have one complex virtual system materialized in parts at one location and in parts at other locations. This opens up completely new perspectives for distributed task oriented experiences and co-operation within groups.

Fig. 6. Virtual and real Part of a System connected by Hyper-Bonds

For systems development our concept may support an incremental implementation and testing of complex devices. For service and maintenance it would support the stepwise investigation and repair of disintegrated parts.

Fig. 7. Modular Production System in Reality

Fig. 8. Distribution Station as real Part of an MPS

One further aim is to combine this concept with real continuous robot processes like welding in a remote way. In a cooperation between several industrial and academic partners[2] we demonstrated the non-critical interactive control and visual observation of a welding robot via internet and its combination with a universal robot simulator COSIMIR (Freund et al, 2002), Fig. 9-10. Entering the Internet Webpage via Internet Explorer, the user has the possibility to program, simulate, control the real robot and observe the resulting process by video image capture.

Fig. 9: Internet Access to Remote Welding

http://www.bendit-interfaces.com/Demos/RobWeld/Demo/default.html

This is a possibility for vocational schools not having direct access to the real process or a production line. However it still has to be verified, what the benefits of a remote real but not graspable system may be over pure simulation.

One reason for real systems, even if they are not graspable but situated at a remote side, is the possibility to validate and adjust the virtual dynamic model (bond-graph) of the physical components by and with their real counterpart. Having the possibility to selectively switch between real and virtual components would be of some benefit.

Fig. 10: Simulation of melding Robot

Fig. 11: Video of real process

Another reason supporting our approach is the necessary communication and cooperation between real persons at the process location and remote experts or learning students. This important mode of collaboration can be experienced with the system. The demonstration may be used as an entrance to distributed collaboration in virtual and real environments not only relevant for mechatronic systems design but also for maintenance and control of complex automation systems. A third reason having robots remotely controlled for learning applications, is the desire to configure remote laboratories of the above type (Fig. 5) without 24-hours onside human help.

However, to reach a state where the theoretical unified concept of bond-graphs, its implementation in a universal mixed reality analog/digital interface and the necessary network quality of service is available, much work still has to be done (Hirche & Buss 2003; Melchiorri 2003). Some low-cost steps in this direction will be presented at the conference (Yoo & Bruns 2004), Fig. 12-13.

---

[2] Festo, K-ROBOTIX, Inst. Robotics Research, ARTEC, Bendit

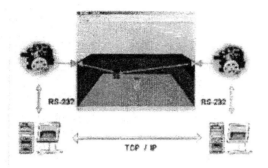

Fig. 12. Distributed Mixed Reality model

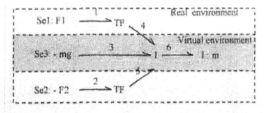

Fig. 13. Bond-graph model of mass transportation with force-feedback

REFERENCES

Amerongen, J. van (2001): Modelling, Simulation and Controller Design for Mechatronic Systems with 20-sim. Proc. 1st IFAC conference on Mechatronic Systems, pp. 831-836, Darmstadt, Germany

Brave, S., H. Ishii, A. Dahley (1998). Tangible Interfaces for remote Collaboration and Communication. Proc. of CSCW '98, Nov. 14-18

Bruns, F. W. (1999). Complex Construction Kits for coupled real and virtual engineering Workplaces. In: N. A. Streitz et al (Ed): Cooperative Buildings. Lecture Notes in Computer Science 1670. Springer, Berlin, pp 55-68

Bruns, F. W. (2003): Hyper-Bonds. Human Skills Oriented Systems Design with Mixed Reality. Proc. of 6th IFAC Symp. on Automated Systems based on Human Skill and Knowledge. Göteborg

Cellier, E. F., H. Elmqvist, M. Otter, (1995). Modeling from Physical Principle. In: The Control Handbook, (W. S. Levine (Ed)), CRC Press, Boca Raton, pp 99-108

Freund, E., Hypki,A., Pensky, D.H. (2002). Multimedia Robotics Teachware based on 3D Workcell Simulation System COSIMIR® Proc. 1st IEEE International Conference on Information Technology & Applications (ICITA2002), Bathurst, Australia

Hirche, S., Buss, M. (2003). Study of Teleoperation Using Realtime Communication Network Emulation. Proc. IEEE/ASME Int Conf. Advanced Intelligent Mechatronics, Kobe, Japan

Ishii, H., B. Ullmer (1997). Tangible Bits: Toward Seamless Interfaces between People, Bits and Atoms. CHI'97, Atlanta, Georgia

Karnopp, D. C., D.L. Margolis, R.C. Rosenberg (1990). System Dynamics – A unified Approach. John Wiley, New York

Melchiorri, C. (2003). Robotic Telemanipulation Systems: An Overview on Control Aspects. 7th IFAC Symposium on Robot Control, Wroclaw, Poland

Milgram, P., Rastogi, A., Grodski, J. J. (1995): Telerobotic control using augmented reality, Proc. IEEE Int'l Workshop on Robot-Human Communication (RO-MAN'95), pp.21-29

Milgram, P., Ballantyne, J. Real world teleoperation via virtual environment modeling. Proc. 7th Int. Conf. On Artificial Reality and Tele-Existence (ICAT'97), pp.1-9, 1997

Milgram, P., Colquhoun, H. (1999). A Taxonomy of Real and Virtual World Display Integration. In Ohta, Y., H. Tamura (Ed). Mixed Reality – Merging Real and Virtual Worlds. Tokyo, pp. 5-30

Mostermann, P. J. (1997). Hybrid dynamic systems: A hybrid Bond Graph Modeling Paradigm and its Application in Diagnosis. Dissertation, Vanderbilt University, Nashiville, Tennessee

Ohta, Y., H. Tamura (1999). Mixed Reality – Merging Real and Virtual Worlds. Tokyo

Paynter, H. M. (1961). Analysis and Design of Engineering Systems, MIT Press, Cambridge, MA

Yoo, Y.-H.; Bruns, W. (2004). Realtime collaborative mixed Reality Environment with Force Feedback. 7th IFAC Symposium on Cost Oriented Automation, 2004, Ottawa

http://www.derive.uni-bremen.de/
http://lab.artec.uni-bremen.de/
http://www.labfuture.net/
http://www.bendit-interfaces.de/

TELE-MAINTENANCE FOR IMPROVEMENT OF PERFORMANCES IN TPM AND RCM

Olivier Sénéchal *
Jean-Baptiste Léger **

* LAMIH, University of Valenciennes, France,
** Société PREDICT, Vandoeuvre, France

Abstract: Performance of production systems is firstly presented as the balanced obtaining of relevance, efficiency, effectiveness and effectivity. We justify then Total Productive Maintenance (TPM) and Reliability-Centered Maintenance (RCM) as leverage on performance, according to the principles of systemics. We show then that performance of production systems is improved if they are precise, rapid, robust and stable, and that these characteristics can be amplified with tele-maintenance. An application of a tele-maintenance tool (CASIP) in such a context is finally presented, and we document the consequences on performance. *Copyright © 2004 IFAC*

Keywords: Tele-maintenance, relevance, efficiency, effectiveness, effectivity, RCM, TPM.

1. INTRODUCTION

Methods for the interpretation and evaluation the performance of production systems have changed considerably over the last few decades. Formerly subject to mono-criterion assessments related to productivity, performance today must be evaluated throughout the entire product life cycle from a multi-criteria point of view, encompassing such diverse elements as productivity, flexibility, cost, time, quality, safety, societal and ecological standards. This evolution in the performance criteria means that both production system piloting instrumentation and the role of maintenance activity in the performance process need to be re-examined.

2. PRODUCTION SYSTEM

Basing us on the general system theory (Bertalanffy 1968), we consider "production system" is a special implementations of the system concept that:
- comprise a set of human, technical and financial resources,
- are introduced in a natural, economic, societal and political environment,
- work for the system's perpetuation, the good of its members, its customers, or society in general,
- produce material (goods) or immaterial (services) products,
- implement several processes (design, production, management, marketing,...),
- evolve in terms of composition (equipment, manpower, ...), organization, and activities.

According to this definition, a wide variety of entities can be called production systems—a workshop, a firm, an employment agency, a health service, an industrial group, for example. But what is "performance", given such varied kinds of systems?.

3. PERFORMANCE AND ASSOCIATED BASIC CONCEPTS

All too often, the focus of performance evaluation is on the results, despite the fact that this focus is known to lead to inappropriate actions (Ahmad and Dhafr, 2002). Numerous papers talk about performance in terms of effectiveness and efficiency. Some use the terms without ever defining them precisely (Martorell, *et al.*, 2002); others use them synonymously, though clearly linked to different decision-making levels (efficiency associated to internal and operational decisions; effectiveness associated to strategic and external decisions) (Yasin, *et al.*, 1999).

We suggest applying the following distinctions currently encountered in cost management (Mentzer and Konrad 1991).

Relevance is the articulation between the objectives and the means. Evaluating relevance requires answering the question: 'Will the proposed means permit the stated objectives to be attained?'. Answering this question is fundamental for the system design phase or the investment phase, because it can help prevent expensive over sizing and insure the feasibility of the project.

Efficiency is, here as in the domain of software capability/maturity models (Jiang et al., 2004), the articulation between the means and the results: 'Given the means implemented, are the results sufficient?'. This question is asked primarily during the exploitation phase of the production system. If the response is not satisfactory, decisions about system control (the operation and maintenance of technical resources) or about management (human resources) must be made.

Effectiveness is the articulation between the results and the objectives: 'Do the results achieved meet the stated objectives?' System effectiveness is very often evaluated in terms of quality indicators. If the degree of effectiveness is not satisfactory, it is possible to act on the internal system organization and the various accessible adjustment parameters.

Effectivity is the articulation between the objectives, the means and the results, evaluated in terms of the finality of the system : "Does the final outcome justify the expenditure of sufficient means to obtain satisfactory results for the stated objectives?" A negative response to this question can challenge the very existence of the production system.

Our vision of performance control is illustrated by the following image:

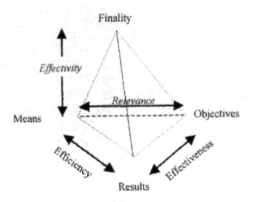

Fig. 1. . The performance tetrahedron for production systems

We imagine the control of production system performance as a regular tetrahedron in which the relationships (relevance, efficiency, effectiveness and effectivity) between the basic system concepts (finality, objectives, results and means) are expressed as equivalent distances between points. (See Fig. 1.).

By definition, maintenance does not question the finality of the systems to which it is applied. We therefore consider maintenance as a leverage on the triangle constituting the base of the tetrahedron (relevance, efficiency, effectiveness).

4. A CYBERNETIC LOOP FOR PILOTING PRODUCTION SYSTEMS

Since Norbert Wiener (1961) first used the word "cybernetics" (from the Greek "kubernétiké"), system theory has gradually evolved to include research into this area. Wiener developed many of his cybernetics proposals around essential concepts related to the definition of piloting: control, communication, feedback and isomorphism.

One of the fundamental principles of cybernetics is that all systems are subject to a disorganizing disturbing loop, which can either demonstrate antagonistic or friendly reinforcement of the state of balance. Systems also include a reorganizing loop charged with maintaining the state of the system so that its goals can be achieved.

In this article, we propose a generic model of the reorganizing loop that, in our opinion, is applicable to any type of industrial activity, and to any decision-making level (see Fig. 2) (Burlat, et al., 2003).

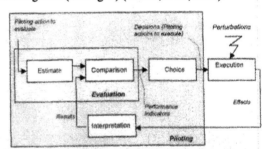

Fig. 2. Generic model of the reorganizing loop (Burlat, et al., 2003).

We insist on the fact that for production systems, where human being is omnipresent, the feedback of the reorganizing loop is not limited to measurement, but is based on interpretations compared to a global vision or frame of reference.

This generic model is valid during the production system's exploitation phase (for a posteriori piloting), as well as during its design phase (for a priori piloting) (Sénéchal and Tahon, 1997).

Still, in both situations, there is one fundamental tool without which closed loop piloting is impossible: the performance indicator (Berrah, et al., 2000), that must allow the evaluation of one or more of the characteristics described in section III (relevance, efficiency, effectiveness and effectivity).

In the following sections, we will demonstrate the relevance of associating TPM and RCM in order to obtain satisfactory levels of performance according to the principles of cybernetics introduced above.

5. ASSOCIATING TOTAL PRODUCTIVE MAINTENANCE (TPM) AND RELIABILITY-CENTERED MAINTENANCE (RCM) TO IMPROVE PERFORMANCE

5.1 Total Productive Maintenance

Born in Japan in the Seventies and widely diffused by the Japanese Institute of Plant Maintenance, Total Productive Maintenance (TPM) is one of the the fundamental "just-in-time" strategies. It is defined as a total system of productive maintenance whose objective is to obtain the best productivity possible (Nakajima, 1988).

In order to meet this objective, TPM uses a performance indicator called "Overall Equipment Effectiveness (OEE)":

$$OEE = \text{Availability} \times \text{Performance Rate} \times \text{Quality Rate} \quad (1)$$

In order to improve availability, performance or quality, TPM works to eliminate a set of losses:
- Equipment Downtime
- Engineering Adjustments
- Minor Stoppages
- Unplanned breakdowns
- Time spent making defective products
- Waste

Although it undeniably has a positively influence on the four principal dimensions of manufacturing performance (cost, quality, delivery and flexibility) (McKone, *et al.*, 2001), we believe that TPM is, first and foremost, consecrated to efficiency. Whether OEE is increased by reducing breakdowns, lost production, settings, or by increasing capacity and labor productivity, the consequence is the same: better results from fixed means, from both a quantitative and qualitative points of view. However, improving the OEE without taking strategic objectives or the economic context into account can be dangerous. For example, increasing OEE in bad economic conditions has no relevance if the market is not able to absorb the additional production. Therefore, sometimes an improved OEE entails a redefinition of the means.

This redefinition can be done using Reliability-Centered Maintenance (RCM).

5.2. Reliability-Centered Maintenance

RCM consists of conducting expensive maintenance tasks only on materials whose failure modes have important consequences for safety, operational availability and/or costs (MSG-3, 2001). The level of importance is evaluated in terms of the concept "criticality", which we consider to be the principal performance indicator in RCM.

Depending on the level of analysis and the level of progression, RCM applies this concept of criticality not only to potential failures, but also to all the equipment studied during the preliminary equipment selection phase. The equipment is selected according to defined critical paths for production, quality and safety, and through use of criticality matrices.

Thus, RCM is a strategy that targets the relevance of both the production tools and maintenance tools. The concept of "gravity", used along with "frequency" and "detectability" to determine "criticality", can be evaluated by taking economic conditions into account. For this reason, we see TPM and RCM as complementary rather than mutually exclusive strategies. Implemented in a real piloting approach in which it is possible to verify that results fulfill the objectives (for example, through the correct use of a data processing system), the joint use of TPM and RCM provides absolute performance leverage.

In keeping with our cybernetic approach, the characteristics introduced in the next section are those which should be maintained at high levels both for production systems and for automated equipment. As we will demonstrate further on, these characteristics can be maintained via tele-maintenance.

6. PRECISION, RAPIDITY, ROBUSTNESS AND STABILITY OF PRODUCTION SYSTEMS

Automated equipment is considered precise if the difference between the command and the obtained result (error) is low. There are two types of precision: static (when $t \rightarrow \infty$) and dynamic (when the command is variable and $t \rightarrow 0$).

This definition is completely applicable to production systems, the command being the stated objective (in terms of turnover, production volume, upturn, OEE or MTBF) and the disturbance being an external event contributing to the disorganizing loop mentioned in section IV (failure of a supplier, strike, fire, …).

Rapidity is the ability of the system to reduce the time between when the order is given and when the results are obtained. Rapidity is particularly important during the exploitation phase of the production system. For example, the mean down time (MDT) of a piece of equipment, which has an influence on OEE, depends closely on the rapidity with which data (orders, error detections, technical information, or commercial information) are transmitted.

In industrial applications, some characteristics of automated systems fluctuate, notably due to the ageing of system components, sensors and actuators. Compensatory arrangements must be made to preserve the ability of production systems to attain a given performance level, or in other words, to keep them robust. Maintenance is obviously the industrial activity that ensures the robustness of production systems.

A system is said to be stable if, for a limited input, it produces a limited output (stability BIBO : Bounded Input Bounded Output). Stability is a fundamental characteristic in risk management, where maintenance can be particularly useful for avoiding or limiting the consequences of a failure.

Section VII relates the effects of tele-maintenance on the characteristics described in the preceding paragraphs.

7. EFFECTS OF TELE-MAINTENANCE ON PERFORMANCE

7.1. The co-operative effects

The efficiency and effectiveness achieved using TPM are due to the capacity of driving operators to detect not only present failures, but also those trends that could lead to future failures. In complementary plans, the diagnoses necessary for avoiding or curing some failures come from experts who are rarely close to the process. Given that availability depends closely on both MTBF (Mean Time Between Failure), which can be increased by a better failure forecast, and MTTR (Mean Time To Repair), which can be reduced by a faster diagnosis, increasing cooperation between driving operators and experts would obviously help to improve availability, thus also improving OEE. The "co-operative" effect of tele-maintenance contributes here to the rapidity of the production system.

To find solutions for very complex failure cases, the advice of several experts can be necessary. The process of consultation and confrontation between experts can help to find the cause of a failure that one individual working alone could not find. This process could be facilitated by using the communication, formalization, coordination and knowledge management functions provided by tele-maintenance tools. Using the tools in this way would contribute to the robustness and the stability of the production system, as well as to the precision of the maintenance processes.

7.2. Examples of possible functionalities

Innovative functionalities CASIP is a distributed software platform making it possible to implement Remote Monitoring, Remote Diagnosis and Tele-maintenance for multi-site companies. In order to encourage cooperation and telemaintenance methods,

the CASIP modeling methodology is supported by a common repository (Léger and Morel, 2001).

This innovative software can detect malfunctions, make diagnoses based on recorded cases of risky situations, and predict future evolutions in the system. To do this, it must access three analysis levels:
- a reactive level, connected to the communication networks, for real time data acquisition, processing and storage;
- a proactive level, based on shop-floor data and its processing as well as expert algorithms and the information they produce, all stored in a data base (e.g. ORACLE, SQLServer, Sybase, Access...) for more global safety processing;
- a remote level for distant access over the Internet to/from anywhere.

CASIP obtains most of its information (about symptoms, deterioration, and causes) from sensor measurements and operations reports.

The database contains much useful information, formalized in terms of knowledge to make it more synthetic and more easily shared between users (experts, technicians, operators). The knowledge database system uses case-based reasoning (Aha, 1991) techniques, that allows it to learn new cases.

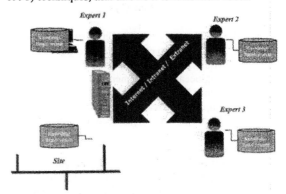

Fig. 3. . Co-operative maintenance plateform

The integration of the TCP/IP protocol allows CASIP to communicate with both the Enterprise Management (ERP) and the Operational levels (MES), while integrating the OPC standard (OLE for Process Control) permits direct communication with the industrial process through fieldbuses: Fieldbus Foundation, Profibus, Interbus-S, Devicenet, WorldFip, Fipway...

This opens the way to a large range of communication possibilities involving PLCs, sensors and distributed Input/Output. An independent exchange also ensures robustness and autonomy in the event of a Supervisory Control failure.

XML (eXtended Markup Language) has been used to obtain an ergonomic Man-Machine Interface with ample portability: for example, the CASIP interactive software for Expert Diagnosis can be used on an Internet Browser.

7.3. Contribution to RCM

CASIP uses a knowledge-based system to structure know-how (Coad, 1992). This knowledge is exploited to detect faults, diagnose hazardous situations and forecast the effect of events.

In this way and in order to implement RCM, this knowledge-based system is exploited to identify the critical equipment.

The performance indicators used are both quantitative (Criticality, Statistics, Probabilities...) and qualitative (accessibility for maintenance, availability of spare parts).

In order to precisely analyze causes and to study the impact of consequences on production performance (Mean Down Time, Quality, Environment...), the knowledge-based system is structured by applying the three complementary methods used in RCM: HAZOP (Lawley, 1974), FMECA (De Kleer and Williams, 1987), and FTA (Fault Tree Analysis).

Using these methods allows all the dependencies of critical equipment to be clearly defined (the critical path of RCM).

The experts can then propose either preventive solutions during the preliminary equipment selection phase, or corrective or ameliorative solutions for existing systems.

With existing systems, CASIP can be interfaced with CMMS (Computerized Maintenance Management System) in order to implement a part of the knowledge-based system. This allows the data from work-orders to be exploited to elaborate RCM indicators. For example, taking "gravity" into account can permit real economic conditions to be considered.

In order to solve complex cases of failure, the CASIP knowledge-based system is accessible and can be shared via networks. Each expert can access the data at every step, creating, modifying and adding comments.

7.4. Contribution to TPM

In order to solve complex cases of failure, the CASIP knowledge-based system is accessible and can be shared via networks. Each expert can access the data at every step, creating, modifying and adding comments.

The formalized RCM requirement level mentioned above allows each process to be drawn in detail in order to specify the necessary safety algorithms/procedures and to implement them within the Operational Platform.

In this way, CASIP covers three main domains (Léger, et al., 1999) of action:
- real-time, online monitoring of system degradation and failure,
- diagnosis aid based on localizing and identifying causes,
- prognosis concerning deviant behavior.

Each of these domains contributes to improving process availability, and thus OEE, by providing:
- better remote monitoring and improved failure detection,
- better diagnoses due to cooperation between CASIP, the experts and the maintenance operators, thus reducing Mean Time to Repair (MTTR),
- better operational reliability using prediction to avoid failures.

Due to its real-time and online characteristics, CASIP is able to rapidly detect sudden problems that can impact on performance, which is "Just-in-Time-Maintenance".

In addition, its prognosis capabilities permit progressive degradations to be monitored and the time to failure to be forecast, which is "Proactive Maintenance". The rapidity with which it system can detect problems contributes to the improvement of Mean Up Time.

Prognosis allows the effects of each failure or degradation to be identified, their criticality to be measured and their risk to be evaluated, as well as contributing to the stability of the production system, which is Risk Management. This contributes to the improvement of the OEE and Risk Management.

In order to rapidly implement a Tele-maintenance system, the CASIP toolbox integrates proactive Maintenance models that can be applied to different equipment: dead time, response time, functioning time, trend analysis... These mathematical models are applied to any equipment, to any variable and provide TPM indicators in real time. For example, the "dead time" model is applied by connecting the set point and the position measurement with a user friendly interface, and without programming code we can access to the dead time indicator and to a trend analysis in order to monitor the drift behavior of the equipment. This indicator is specially interesting for proactive maintenance.

Due to the online capability, the performances indicators can take the very detailed data coming from the equipment into account. In this way, TPM becomes more efficient because experts can analyze each part of the OEE indicator: time to detect, time to diagnose, time to find spare parts, time to repair, time to put into service, time to obtain nominal functioning...

7.5. Performance improvement

By using the previously described tool, the return on the investment is measured in the time saved when studying and developing a Tele-maintenance system. Indeed, CASIP supports the entire production system life-cycle, without information loss, without planning failures, without double analysis, without rewriting... the time saved varies from 30% to 50%.

The knowledge adapted during operations is instantaneously updated in the RCM analysis. The time needed to revise this knowledge is nearly null.

8. CONCLUSION AND PROSPECTS

In this paper, we introduced a systemic view of production system performance and its control. Taking into account the human dimension of such systems, we showed that tele-maintenance improves precision, rapidity, robustness and stability, and amplifies the positive effects on performance produced by associating TPM to RCM. More precisely, tele-maintenance as a co-operative tool federates a variety of engineering methods and takes expert knowledge and skills concerning plant degradation into account. In this way, it supports the cooperative processes necessary for the improvement of prognosis, detection and diagnosis, and reduces the time needed for corrective interventions.

These effects of tele-maintenance have been demonstrated in real cases applying the CASIP tool. The next step of our research is to proceed to a quantitative/analytic demonstration, using statistical methods. The second step is to focus on applying data mining to tele-maintenance, which should particularly improve the production system robustness in non-referenced risk situations.

We consider that these new possibilities for increasing and integrating competences, and for juxtaposing different points of view, can contribute to the overall improvement of operating safety.

REFERENCES

Aha D., (1991), Case-Based Reasoning Algorithms, DARPA, *Case-Based Reasoning Workshop*, pp 147-58

Ahmad M. and Dhafr N. (2002), Establishing and improving manufacturing performance measures, in *Robotics and Computer Integrated Manufacturing*, 18, pp 171-176.

Berrah L., Mauris G., Haurat A., and Foulloy L., (2000), Global vision and performance indicators for an industrial improvement approach, in *Computers in Industry* 43, pp 211-225.

Bertalanffy F (1968), *General System Theory*, George Braziller.

Brown, F., M.G. Harris and A.N. Other (1994). Name of paper. In: *Name of book in italics or underlined* (Name(s) of editor(s). (Ed)), page numbers. Publisher, Place of publication.

Burlat P., Marcon E., Sénéchal O., Dupas R., and Berrah L. (2003), Démarches d'évaluation et de pilotage de la performance, *in Ouvrage collectif GRP « Evaluation des performances des systèmes de Production »*, Traité IC2 Hermès Paris, pp. 49-77.

Coad P., (1992), Object-oriented patterns, *Communications of the ACM*, 35, pp 152-59.

De Kleer J., and Williams B.C., (1987), Diagnosing Multiple Faults, in *Artificial Intelligence* 32, pp 97-130.

Jiang J., Klein G., Hwang H.G., Huang J., and Hung S.Y. (2004), An exploration of the relationship between software development process maturity and project performance, in *Information & Management* 41, pp 279–288.

Lawley H.G., (1974), Operability Studies and Hazard Analysis, in *Chemical Engineering Progress* 70, pp 105-116.

Leger J.B. and Morel G., (2001), Integration of Maintenance in the Enterprise: towards an Enterprise Modelling Based Framework compliant with Proactive Maintenance Strategy, in *Production Planning and Control*, 12, pp 176-187.

Leger J.B., Iung B., and Morel G, (1999), Integrated Design of Prognosis, Diagnosis and Monitoring Processes for Proactive Maintenance of Manufacturing Systems. *IEEE Systems, Man, and Cybernetics Conference SMC'99*

Maintenance Program Development Document MSG-3, (2001), Maintenance Steering Group – 3 Task Force, Air Transport Association (ATA) of America.

Martorell S., Sanchez A., Carlos D., and Serradell V. (2002), Comparing effectiveness and efficiency in technical specifications and maintenance optimization, in *Reliability Engineering and System Safety* 77 pp 281–289.

McKone K. E., Schroeder R.G., and Cuab K.O., (2001), The impact of total productive maintenance practices on manufacturing performance, in *Journal of Operations Management* 19, pp 39-58.

Mentzer J.T. and Konrad B.P. (1991), An efficiency/effectiveness production-distribution system: Models and methods, in *Business Logistics* 21 , pp 33–62.

Nakajima S. (1988) Introduction to TPM. Productivity Press, Cambridge, MA.

Sénéchal O. and Tahon. C. (1997), :A modelling approach for production costing and continuous improvement of manufacturing processes , in *Production Planning and Control*, Taylor & Francis, 8, pp 731-742.

Wiener N. (1961), Cybernetics or Control and Communication in the Animal and the Machine, Herman et Cie, The Technology Press, John Wiley and Sons, The MIT Press.

Yasin M., Czuchry A. J., Dorsch J. and Small M. (1999), In search of an optimal cost of quality: an integrated framework of operational efficiency and strategic effectiveness, in *Journal of Engineering and Technology Management* 16, pp 171-189.

www.elsevier.com/locate/ifac

TOWARDS THE DEFINITION OF STANDARDISED MACHINE TOOL MONITORING SYSTEM

A.-L. Gehin[1], M. Staroswiecki[2]

[1]*LAGIS UMR 8021*, [2]*Ecole Polytechnique Universitaire de Lille*
Bâtiment Polytech'Lille - Cité Scientifique
59655 VILLENEUVE D'ASCQ cedex - France
anne-lise.gehin@polytech-lille.fr

Abstract: We present in this paper a part of the solutions we proposed in the context of the Thematic Network IDMAP[1] to develop standardised monitoring system for machine tools. The proposed approach rests on an inner description of the functions of a monitoring system completed by an external model which specifies the provided services and their organisation. It allows the definition of rules from which the interoperability and the interchangeability of monitoring systems could be satisfied. *Copyright © 2004 IFAC*

Keywords: monitoring system, machine tool, standardisation, interchangeability, interoperability.

1. INTRODUCTION

Significant progress in improving and optimising machining processes generally can be achieved by the implementation of on-line and real-time monitoring, diagnosis and control devices, measuring and processing signals acquired from one or several sources of the machining process.

Monitoring devices have been developed for several years but they are still not widespread in industry due to the lack of standardisation. The development of field bus systems and digital networks has made a progress in automation of machine and cutting processes due to the integration of standardised hardware interfaces and protocols for sensors, actuators and other devices.

But at the same time, monitoring and diagnosis methods have not grown up into standardisation in the same way. Functions and systems are working in special fitting to each application. There is no standards in function architecture and methodologies. The producers integrate to their systems specific techniques and methods and there is no compatibility between systems of different producers. That is the reason why we propose to describe a monitoring system with a generic model which rests on two parts:

1. the functional analysis gives a representation of the inner functions of the monitoring system by describing on which procedures it rests,

2. the external model describes the set of the services provided by the monitoring system and their organisation.

---

[1]IDMAP: Intelligent Devices for the on-line and real time monitoring, diagnosis and control of MAchining Processes – Contract Brite – EURAM Nr. BRRT-CT97-5045

The descriptions are given for a monitoring system taken outside any specific application context. They aim to contribute at the definition of rules from which

- the interoperability and the interchangeability of monitoring systems for machine tools could be satisfied by checking:
  - the consistency of the handled data and the service organisation (interoperability),
  - the presence, in the replacement equipment, of the data and services required by the application (interchangeability),

- interoperability language, standardised for all the components (monitor, sensor, actuator ...) could be formulated.

2. FUNCTIONAL ANALYSIS OF A MONITORING DEVICE

A monitoring device senses and processes signals from one or several sources of the machining process. A part of the sensors may be internal to the monitor, another part may be connected to the monitor via a communication support which can take different forms such as a direct link or a field bus.

According to the results of the procedures applied to the sensed signals, users can supervise and control the machining process in a consistent manner. Note, that users may be human operators or automation components such as actuators, regulators ....

Figure 1 summarises the main functions of a monitoring system. The function "to input" creates a raw data base (RDB). The internal sensors of the monitoring system are the hardware support of this functions. Consequently, the RDB is essentially constituted of signal values measured on the machining process.

The function "to validate" operates on the RDB content in order to:
- detect and isolate (and, when possible, diagnose) system faults,
- prove the consistency of the raw data by suppressing or replacing erroneous data by their most probable estimation.

Fault Detection and Isolation (FDI) algorithms are the support of the function "to validate" (Iserman, 1984; Patton, et al., 2000). They create a validated data base (VDB)

The function "to elaborate" uses the content of the VDB to generate extra information such as:
- identification of sensor, actuator or process fault,
- virtual sensors which correspond to data produced by algorithms from measured data (for example, elaboration of the acceleration signal from the measured speed signal, elaboration of statistical value such as mean

value of N sample, maximal value on a given temporal window ...)
- data structure such as (data+date), trends, histories elaborated from data directly accessible or virtual data.

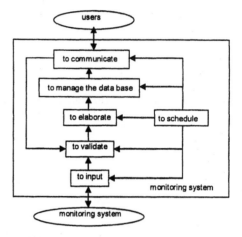

Figure 1: Functional architecture of a monitoring system

The function "to manage the data base" stores all the parameters required by the validation and elaboration algorithms and all the data produced by the functions "to validate" and "to elaborate".

The function "to communicate" implements the interface between the users and the monitoring system. This communication is bidirectional. In input, users send service requests, in output they receive the results of the service running. A user request may be a write request to initialise a configuration parameter, for example, or a read request to access a part of the monitor data base. Note, that the parameters of a write request may be validated as the raw data sensed on the process.

The function "to schedule" generates activation signals to start the execution of the different procedures implemented behind the different functions.

By describing each of these functions with a formalism such as SADT (I.G.L. Technology, 1988) which is a modular and hierarchical approach allowing to specify the input, output, control data and supports of the functions, a generic internal model of a monitoring system can be obtained (Gehin and Staroswiecki, 2001). This internal model is a first step to standardisation by allowing to cheek that the functionalities provided by two monitors of different providers are implemented in the same manner. Moreover, it can be a guide for the designer to specify, for a given system, information such as:

- the list of sensed physical value
- how each sensed data is validated
- the list of elaborated data
- the algorithm used to generate elaborated data
- ...

228

This description allows the user to know exactly the possibilities of its monitoring system and to evaluate the adequacy of the monitoring system to the application requirements. Moreover for a given monitoring system, the user is able to evaluate the (positive or negative) consequences of component (sensor for example) addition or removal.

3. EXTERNAL MODEL OF A MONITORING DEVICE

From an external viewpoint, a monitoring system provides services to users. Preoccupations of the users are not to know how an internal function is implemented but are to easily access system services and more generally, to know what they can obtain from the monitor and how they have to process to obtain it.

An answer to these questions can not be given to each specific monitoring system. To homogenise the use of monitoring systems some standardised models have to be defined. They have to integrate both communication and application aspects to satisfy the interoperability property which insures that a cooperation is possible between monitor, sensors and other automation components even if the providers of these components are different.

Field buses development has generated works to define standardised communication between components. We propose to complete those works by an external modelling adapted to monitoring system for machine tool.

3.1. Information representation

For communication networks, the International Standard Organisation (I.S.O.) has defined the O.S.I. (Open System Interconnection) reference model (ISO 7498, 1984). The O.S.I. model splits the global protocol which defines the communication rules between a set of heterogeneous components into 7 levels. For each level, the provided services are described.

The rules for information representation and for message structuring (syntax, name management ...) are defined by the level 7 protocols. The definition of these rules is essential to make sure that the exchanged data are understood by each component and that a service requested by a component from another one is correctly run.

3.2. The interoperability language : different approaches

The interoperability property makes sure that a cooperation is possible between components of different providers. The 4-20 mA standard used to represent a [0%-100%] interval is an example of a norm satisfying the intercommunication principle since it defines not only the transmission system but an information representation formalism too. The interoperability for this example, is satisfied only if each component correctly interprets the value expressed between 0 and 100 %.

The interchangeability concept is often associated with interoperability. The interchangeability property insures that a component can be replaced by another one having the same function but coming from another provider. The replacing component must provide the same services, produce the same data and represent them with the same formalism.

These two aspects of interoperability and interchangeability are essential because the users do not want to redefine the software of their application at each time they replace, add or remove a component. This necessity is well understood by field bus and intelligent component designers. That is why, different formalisms are proposed to satisfy interoperability and interchangeability requirements. Let's cite two complementary approaches :

- the Device Description concept consists in the definition of a standardized language to describe the components connected to a field bus,
- the Function Blocks description defines an equipment as a set of functionally autonomous entities.

3.2.1. Device Description approach. A Device Description (DD) is a formal description of the data and operating procedures for a field device, including commands, menus and display formats. The component behaviour with the communication system is completely described. This description can be directly integrated to the component at the application layer level .

As an example of DD let us mention the definition of Device Profiles for CANopen or DeviceNet, of Device Models for SDS, of a Device Description Language for HART. CANopen, DeviceNet and SDS are three application layers defined for the CAN field bus. CANopen is proposed by a users consortium, DeviceNet by Allen Bradley of Rockwell Automation and SDS by Honeywell. The Device Description Language (DDL) defined by the Rosemount society for the HART (Highway Addressable Remote Transducer) has been reused by Fisher Control, Siemens and Yokogawa in the context of the I.S.P. (Interoperable System Project).

The main entities the DD describes are variables, commands, methods and menus. Every accessible variable in the device is included such as process measurements, derived values, and internal parameters such as range, sensor type, choice of linearisation, materials. For each variable, the DD may specify, among other things, the data type (for example: integer, floating point, alphanumeric, enumerated), how it should be displayed, a name for the display to an operator, any associated units, an help text, perhaps describing the meaning of the variable or how to use it. For each command, the data structure of the command and its response and the meaning of any command response status bits may be specified.

3.2.2. The function block approach. Device descriptions take the form of device description vocabularies. The function block approach structures the devices into functional autonomous entities. For each function block, the performed function (for example, measurement, validation, communication), the associated data and the method adopted to manage the events, are defined. This approach aims at satisfying the interchangeability, i.e. to assure that for two components of the same type (for example, two differential pressure sensors), aimed at the same application class, the provided functionalities and data are the same.

The device description and the function block approaches are complementary. Device descriptions favour the interoperability and function blocks the interchangeability. Nevertheless, two extra notions have to be added to precise the service organisation and the service running conditions. Indeed, for consistent operating behaviour reasons, services have to be organised into coherent subsets taking into account the situations in which they can be run and by which users they can be run. Moreover, a service is run not only if it has been requested by an authorised user in a given situation, but also the resources required for its running have to be non faulty too.

A system such as a machine tool where sensors, actuators and monitors have to cooperate will behave correctly only if each component can know the organisation and the running conditions of the services of the devices with which it cooperates. The external model developed in the next part integrates these two requirements.

3.3. The proposed model

Notion of service .From an external viewpoint, a monitoring system provides services to users (trend of process signal, alarm when a physical value is out of range ...) (Staroswiecki et Bayart, 1996). A service s_i is described by a 6-uple:

$$s_i = \{cons_i, prod_i, proc_i, rqst_i, ena_i, res_i\}$$

where:
- *$cons_i$* are variables the service consumes,
- *$prod_i$* are variables the service produces,
- *$proc_i$* are the procedures applied on the consumed variables to generate the produced variables,
- *$rqst_i$* is the request associated with the service.

It is issued by the user,
- *ena_i* is the enabling condition processed by the monitor. Associated with the request, it specifies the running conditions of the service. A clock signal to sample an analog signal, an information to not provide a measurement service when the sensor is faulty or disconnected are examples of enabling conditions,
- *res_i* are hardware or software resources required for the service realisation.

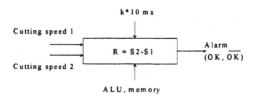

Figure 2 : example of service.

Figure 2 illustrates an example of an alarm generation service. Every 10 ms, an alarm is generated or not according to the value of a residual calculated from the value of the cutting speed measured by two different means (an external one and an internal one for example).

Versions of a service. Since a service needs consumed variables and hardware resources, its running will be nominal at a given time only if :

- the consumed variables are present with proper characteristics (for example they really represent the process at the current time : their time stamp is correct)

- the hardware resources are in good running.

Unfortunately, this is not always the case, and it might happen that some of the required resources are faulty or that some of the required variables are not usable. Fault tolerant devices offer different versions of the same service (nominal as well as degraded ones) (Gehin, *et al.* 1997). Each version is characterized by the set of the resources and consumed variables it needs. In order to introduce such possibilities, the external model describes each service as an ordered list of versions. In response to a request, the component runs the first version of the list whose resources are non faulty and whose consumed variables are all present with proper characteristics. The first item in the list corresponds to the nominal behavior of the component, the other ones are degraded versions and the service is unavailable only when no version of the list can be run.

Let's take an example to illustrate the versions notion. Suppose that three sensors S1, S2, S3 allow to know a physical value V. The different versions of the V estimation service are shown by Table 1. The first version is the nominal one. Its running requires the two resources S1 and S2. If S2 is unavailable, for example because the signal is too noisy, the V estimation service can be run according to a degraded version by estimating V only from S1 (degraded 1a). Another degraded version at the same level can be run if the resource S1 is unavailable (degraded 1b). If S1 and S2 are both unavailable, an even more degraded version consists in evaluating V from S3 (we assume that the accuracy of the result is in this case worse).

C	Versions	Available resources	Procedure
0	Nominal	S1, S2	V=(S1+S2)/2
1	Degraded 1a	S1	V=S1
	Degraded 1b	S2	V=S2
2	Degraded 2	S3	V=S3

Table1 : Example of multiple versions for a same service.

Service organisation. A priori, requests might be addressed to the monitor at any moment. However for coherence reasons, the monitor must not run incompatible services at a same time (initialisation of the sample frequency and periodic measurement services for example). That's why the set of services is split into coherent subsets called User Operating Modes (USOM). A USOM corresponds to a set of services that the system or the operators may request in a given situation of the component life cycle (Gehin and Bayart, 1995). USOM are designed taking into account safety conditions, precedence constraints, etc. Off operation, configuration, automatic, test, … are USOM examples.

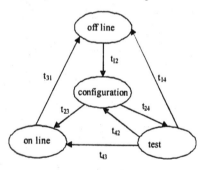

Figure 3 : A USOM deterministic automata

At each time, the monitor is in a current USOM. The only services it will accept to run are those which belong to that USOM. In other words, any request for another service will be automatically rejected. Obviously the list of the services of each USOM has to contain a "Set USOM" request, otherwise it would be impossible to leave the current USOM. This request has to indicate the destination USOM (the origin being, of course, the current one). However, it is clear that any <origin, destination> pair cannot be

allowed, for security or operating consistency. As an example, the request "set automatic USOM" should be rejected as long as the initialization parameters have not been fixed (writing these parameters is one of the services of the "configuration" USOM). For that reason, the conditions which allow to move from one USOM to another one have to be specified. The definition of the set of USOM and of the transition conditions can be described by a deterministic automaton (see figure 3).

CONCLUSION

The functional analysis which allows to give a representation of the inner functions of a monitoring system can be used as a design tool or to specify and describe an existing monitor. It allows to check that a service is implemented in the same manner in monitors from different providers. The external approach completes the inner representation by specifying the list of the monitoring system services and their organisation into coherent subsets named USOM. Associated with a standardised vocabulary, it allows a request sent by a component to another one, to be understood and run if the operating conditions of the receiver allow it.

The Device Descriptions and the decomposition into functional blocks are approaches essentially oriented towards standardised description of the produced and consumed variables and of the service provided. They have been defined to ease the communication between field bus connected components. They can be used to normalise components of a machine tool such as sensors, actuators and monitoring systems. If we add a codification of the service organisation and of the service running conditions we have all the required elements to make sure that components (sensors, actuators, monitors) of different providers can correctly cooperate in a same application.

By fixing design rules, the proposed modelling should allow to obtain safe and dependable machine tools even if they are not provided with a proprietary instrumentation.

REFERENCES

Gehin, A.-L. and M. Bayart (1995) – Operating Modes of a Distributed Intelligent Automated Production System – *IEEE International Conference on Systems, Man and Cybernetics IEEE SMC'95*, pp. 1333-1338, October 22-27, 1995 – Vancouver (Canada)

Gehin, A.-L., M. Bayart and M. Staroswiecki (1997), Faulty Resources Management in Automation Systems, *IEEE SMC'97*, Orlando (USA), October 12-15, 1997

Gehin, A.-L. and M. Staroswiecki (2001)- Models for machine tool component standardisation - *Contribution to the Final Technical Report of the*

Project: "Intelligent Devices for the On-Line and Real-Time Monitoring, Diagnosis and Control of Machining Processes" of the Thematic Network in Brite-Euram III - Contract-No: BRRT-CT97-5045

I.G.L. Technology (1988) - *SADT un langage pour communiquer* - Eyrolles.

Iserman, R. (1984) - Detection based on Modelling and Estimation methods - A Survey. *Automatica*, **vol. 20**, pp. 387-404., 1984

ISO 7498 (1984), Information processing systems – Open Systems Interconnection-OSI – *Basic Reference Model International Standard Organization*

Patton, R. J., Frank P. M., Clark R. N. (2000) – *Issues of Fault Diagnosis for Dynamic Systems* – Springer

Staroswiecki, M. and M. Bayart (1996). Models and languages for the interoperability of smart instruments. *Automatica*, **vol. 32**, pp. 859-873.

ELSEVIER
IFAC
PUBLICATIONS
www.elsevier.com/locate/ifac

PROGNOSIS-BASED MAINTENANCE DECISION-MAKING FOR INDUSTRIAL PROCESS PERFORMANCE OPTIMISATION

A. Muller, M.C. Suhner, B. Iung, G. Morel

Centre de Recherche en Automatique de Nancy (CRAN), CNRS UMR 7039
Université Henri Poincaré - Nancy 1
Faculté des Sciences – BP 239
54506 VANDOEUVRE-LES-NANCY Cedex – FRANCE
alexandre.muller, marie-christine.suhner@esstin.uhp-nancy.fr;
benoit.iung, gerard.morel@cran.uhp-nancy.fr

Abstract: Among the processes for management of product life cycle, maintenance is one of the key Enterprise domains to improve business objective, product quality and service while keeping the production system dependable. It leads to deploy maintenance strategy more efficient vs. proactive mainly based on component and functional degradation prognosis process able to set-up adapted maintenance decision making to better master its direct and indirect costs. In this paper is discussed a methodology supporting prognosis process modelling which is based on the combination both of probabilistic approach for degradation mechanism modelling and of event one for degradation dynamical monitoring. This methodology is experimented on manufacturing component and deployed within e-maintenance architecture. *Copyright © 2004 IFAC*

Keywords: Maintenance, e-maintenance, Decision-Making, Prognosis, Degradation, Bayesian Network.

1. INTRODUCTION

With today's increasing demand on productivity, safety, product quality and customer satisfying, but also the decreasing of profit margins, (Kutucuoglu, *et al.*, 2002) consider Maintenance as a key process of the Enterprise. Indeed maintenance plays a critical role for global performance fulfilment (effectiveness, efficiency, relevance) by leading to the reduction of **inefficiencies** (optimisation of the economic, technical, ecological, ... criteria). Nevertheless the correlations between these criteria even their contradictions renders the problem of maintenance deployment highly complicated, especially for complex systems (Lee, 1998). Frequent maintenance actions increase the system reliability and availability but the resulting costs are also increased. So, how to select and deploy the best maintenance policy both technically and financially justified, is one challenge to be solved.

Information technology can play a crucial role in the support of this challenge to be able to take the right maintenance decision, at the right time and at the right place for optimising the global plant and product performances. This optimisation goal requires mainly:

- To consider maintenance as an integral part of the overall Enterprise business strategy: holistic

approach for global framework (Léger and Morel, 2001).

- To operate maintenance, as far as possible, in **proactive way** (Iung, 2003) to be able to **anticipate** the critical scenarios related to dependability, productivity and/or economic aspects.

This anticipation action which characterises a predictive maintenance strategy is structured on monitoring, diagnosis, prognosis and decision making modules which can be supported through new Internet-based E-maintenance architecture (refer to IMS Centre http://www.imscenter.net).

Among these modules, the prognosis process is often considered as the Achilles heel. Most of the existing prognosis methods are component-oriented and without a real formalisation in the modelling. It is now not sufficient to face the expected performance optimisation with regard to the complexity of the plant where the degradation and deviation modelling is really difficult due to the economic or stochastic degradation dependencies among components. So our challenge consists in developing a new method based on realistic assumptions for prognosis modelling and implementation that is applied not only at the component level but also at the function or application ones.

This paper is organised as follows: Section 2 outlines the prognosis process in general completed in section 3 by an overview on maintenance strategy decision-making. Section 4 describes the most relevant aspects of the methodology proposed and section 5 provides an application of it to a real case study. Finally some conclusions and future developments are presented

2. OUTLINE OF THE PROGNOSIS PROCESS

According to the OSA-CBM architecture (http://www.osacbm.org), the prognosis functional scope consists of the prediction of future component health considering its present health and its current degradation/failure progression. (Byington, et al., 2003) classified the prognostic methods as being associated with approaches such as Experience-Based prognostics, Evolutionary/Statistical Trending Prognostics, Artificial Intelligence Based Prognostics, ... To be efficient these methods are often used in combination as we propose in our work while seeking to use this prognosis methods either only on the component level but rather on the function and application levels (function is supported by several components). In general, a component is subjected to degradation phenomenon such as ageing, wear, and erosion. Thus, a function is subjected to the consequences of degradations occurred with all of its components. The degradation mechanism obtained is complex but the function state can be determined from the combination of the states of each component, which support the function. So, the methodology developed for the prognosis process modelling has to be generic whatever the level considered (component, function, system) but the degradation attributes and the meaning of degraded functioning mode will be dedicated to each level.

For developing prediction, the prognosis process is always based on the two following steps:

- *Identification of the current state of the item (component, function, system)*: It is necessary to observe the item through direct or indirect indicators and then to characterise the degraded state in which it is at the current time.

- *Forecasting the future state of the item*: From the degraded current state, it is necessary to determine in which way the item is evolving and which state will be reached if nothing is done.

This last step leads to identify the input information for maintenance decision-making procedure. Indeed if the future estimated situation is a state considered as "safe" or "successful" to fulfil the application goal no maintenance action should be planned. The system will evolve towards an acceptable degraded state and it can derive under control. In contradictory way, if the degradation degree can be the cause of risk or insufficient performance, a maintenance strategy must be started immediately or at closed time.

3. OUTLINE OF THE MAINTENANCE STRATEGY DECISION-MAKING

From the current and estimated situations, the most efficient maintenance strategy has to be selected to keep the system/product in control. This selection concerns decision-making problems, which can be classified, as proposed by (Dekker and Scarf, 1998), according to the time scale involved and/or the level at which maintenance decision will be developed. In result, the optimum maintenance policy is defined by (Weintsein and Chung, 1999) as the maintenance policy that minimises the sum of the total costs for performance of preventive and emergency activities; and the financial consequences of interruptions to the production plan for the organisation. These activities affects **indicators** (for performance measurement) such as the useful life length of the maintained equipment, number of failures, ..., maintenance direct and indirect costs (Al-Najjar and Alsyouf, 2003) where:

- Direct costs (labour, material, investment,...) usually appear in the maintenance budget.

- Indirect costs are all the expenses that are indirectly related to maintenance. These could be related to lost profit due to missing production during planned and unplanned stoppages, loss of customers, bad reputation, loss of market share, ...

Taking the "best" maintenance decision consist therefore in comparing different alternatives such as corrective, condition-based, predictive, ... (Wang, 2002) affecting for each of them the indicators with different values taking into account the system/product situation expected after maintenance action development. Each alternative assessment uses different sources of data, is dependent of the points of view adopted (e.g. safety should be privileged rather than the availability) and is given with a certain degree of confidence. It leads to place the maintenance strategy resolution as a multi-criteria decision-making problem.

4. METHODOLOGY PROPOSED FOR PROGNOSIS PROCESS MODELLING

To be able to deploy the maintenance decision-making in a good way, we already contributed to the formalisation of the prognosis process by joining different mechanisms and sources of data through probabilistic and event models. It leads in operational phase, to combine the two models to implement the prognosis process. The main added value of this methodology is to develop degradation models the most realistic as possible by initiating them through return of experience and by improving them through operation data extracted from the on-line architecture implementation. This methodology is constructed on three steps which the theoretical mechanisms are more detailed in (Muller, et al., 2004a) and (Muller, et al., 2004b): Process & Flow-based approach,

Elaboration of the probabilistic network, Design of the monitoring architecture.

4.1. Process & Flow-based approach

First, the industrial system to be considered is modelled through functional analysis by using process approach. One process is broken-up into several sub-processes, and the same mechanism is applied for the sub-processes until the expected level of abstraction is reached (e.g. component). The proposed definition of a process relies on the four concepts interpreted as follow: *Support* which implements the process, the *Goal* which represents the purpose of the process, the *Function* which is a desired action and the *Behaviour* which explains how a system does what it is intended to do. The process behaviour is described by the causal relationships between its input flows, the support, and its output flows (causality mechanism). Each flow is composed of objects, and the whole is defined by attributes. The deterioration of process behaviour (performance loss) is due to the support degradation according to at least one functioning mode and it led to modify some output flows and objects attributes. Indeed the concept of flow-based performance evaluation means that the process performance can be directly measured on the produced flows and more precisely, on the value of its attributes. Moreover as a process is linked to another one by the flows it produces or consumes, the propagation mechanism could be used to propagate the effect of a component's degradation on the whole system and to simulate the deterioration process through time (causal and temporal inference rules). To computerise this modelling phase, the MEGA Process tool (http://www.mega.com) was selected based on its process representation mapping capacity and its flexibility of use.

4.2. Elaboration of the probabilistic network

The second step consists in transforming the previous process model into a probabilistic one that represents the causal relationships and temporal degradations in a unified way. The model goal is to support the mechanisms of inference which can propagate, in the future, the process degradation and determine at any time the impact of this degradation on the whole system. The probabilistic model is developed by means of Bayesian Network (BN) (Jensen, 1996) which is a directed acyclic graph defined by a set of nodes and of edges which characterize conditional dependencies among the variables. The relationship quantification is defined by a CPT (Conditional Probability Table) which contains probabilities $p_{i,j}$ of a child node being in a state "i" given the states "j" of its parents. The temporal degradation mechanisms are also integrated to the probabilistic model by means of dynamic nodes: Dynamic Bayesian Network (DBN) (Dean and Kanazawa, 1989). A dynamic variable $X(k)$ is represented by two nodes linked with an arc which explains the temporal probabilistic dependence (Weber and Jouffe, 2003). The CPT $P[X(k+1) / X(k)]$ quantifies the temporal relation between the state of the variable at time k and those at time k+1.

The translation of the Process & Flow-based model into a DBN is described in three stages:

1 - <u>Nodes and states</u>. The BN nodes represent the supports and the flows of processes. A support is associated to several nodes, which correspond to its potential degradation modes. A flow contains as many nodes as its related attributes. The basic idea is to identify each flow attribute with a "static" variable A and each degradation mode with a "dynamic" variable. A support state SP is defined as a reachable combination of the states of its degradation modes $Mi(k)$. A state of $Mi(k)$ is defined either by its physical meaning (e.g. oxidation level) or by the consequence of the degraded state on the process performance (e.g. % loss of conductivity). In the same way, a flow state F is defined as a reachable combination of the states of its q attributes Ai (e.g. water flow is characterized by its volume flow rate, pressure). The states of attributes correspond to the process performance allocation fixed by the company.

2 - <u>Structure</u>. The processes are linked together by their common flow (link between produced and consumed flows) and so, the model structure consists in specifying a set of edges which represents the causal relationships between the inputs flows, the support, and the outputs flows.

3. <u>Network's parameters</u>. Each node has a conditional probability distribution (defined in the CPT) that quantifies either causal or temporal relationships. Root nodes have only prior probability distributions. For causal relationship, a conditional probability is affected to each instance of the output flow attributes considering each configuration of input flows and support. For a temporal relationship, the elicitation of conditional probability $p_{i,j}$ defines the deterioration stochastic processes. For a temporal and causal relation the parameters $p_{i,j}$ are learning from operational database if there is enough used data or elicited by expert's knowledge in terms of subjective estimates.

4.3. Design of the monitoring architecture

The design of the monitoring architecture is deducted from probabilistic model: each observable variable X is associated to indicator I_x that can be considered as a substitution of the deterioration level (figure 1). I_x is a discrete variable whose states correspond to the nominal, degraded and failing states of a BN variable. The values of these indicators are refreshed with the current status of the shop floor data and controlled in real time by means of algorithm. An adapted algorithm starts an alarm as soon as a variable I_x exceeds a threshold α. In this case information about the variable states is translated as an evidence (i.e. an assignment of values to a set of variables) for the probabilistic model. If incertitude on the observation is developed, the evidence only updates the prior probability values. These events are displaying and storing in an operational experience database adapted to the refinement of the probabilistic model's parameters. This database allows verifying the coherence (and updating) between the rates of transition fixed a priori and the real rates observed.

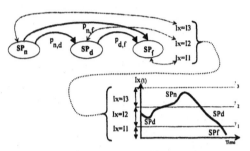

Fig. 1. Combination of probabilistic and event approaches

5. CASE STUDY

In order to demonstrate the feasibility and the added value of the methodology proposed, two experimentations are in progress. The first one developed in (Muller, et al., 2004b) is well invested and concern a platform called IMS-CRAN in which the physical process is a regulation process. The second one, developed in this paper is only started and concern a new platform called TELMA for unwinding of metal bobbin (sheet steel cutting out). It is composed of manufacturing components such as pneumatic cylinder, chuck, marking system.

At middle term, our scope is to test and evaluate, on these two platforms, the deployment of (e)-maintenance optimised services (from the methodologies proposed). TELMA (and IMS-CRAN) is characterized by a high level with new information and communication technologies such as the distribution and communication of a technical form of intelligence directly into the components (Iung, 2003). In addition to the physical part and to the software technical intelligence part, the platform is also composed of an e-maintenance architecture. This last one supports the structure required for e-prognosis process deployment (figure 2) by means of the CASIP architecture (http://www.predict.fr) for the event approach and by the BayesiaLab software (http://www.bayesia.com) for the probabilistic one. The server SAM of CASIP is connected with the shop-floor data acquisition system (OPC server) to the communication networks in order to acquire and process in real time the data coming from the field components. The monitoring tasks extracts information required to determine the system behaviour: Refreshing and calculus of the indicators I_x; the detection of an event related to degradation situation is considered as a symptom or an alert which can start a request for prognosis. All the events

and processing results are stored in an Oracle database allowing also to be fully open for integration towards CMMS tool (EMPACix) and remote "user" (remote client SAM).

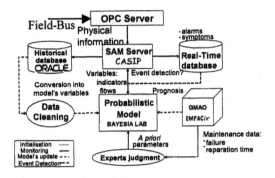

Fig. 2. E-prognosis architecture

On TELMA, the example developed below concerns mainly the subset that the finality is to lock/unlock the axis supporting the metal bobbin (locking action allows to change the bobbin in safety condition). The subset is composed of 4 components: a PLC that is shared with other systems, a distributor, a cylinder and a mandrel (in contact with the axis). In this example, the physical laws used in the models elaboration are voluntary simplified in order to minimise the variable number and the model size to give the methodology more understandable.

5.1. Process & flow-based approach.

The sub-process "To lock/unlock the bobbin axis" is fulfilled by operating a mandrel unit action and consequently applying (or not) a locking effort on the bobbin axis (figure 3a).

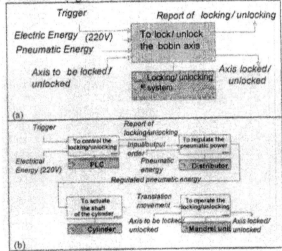

Fig. 3. Process-diagrams "To lock/unlock the bobbin axis"

The sub-process is itself broken up into four sub-processes in relation to the 4 components support (figure 3b). Only the development of the e-prognosis module related to the sub-process "To actuate the shaft of the cylinder" is explained in this section.

This process is defined by its: *Support:* Cylinder; *Goal:* To actuate the shaft of the cylinder. The goal's achievement of the process is directly measured by attribute of the output flow (translation movement); *Function:* To move the shaft according to a translation movement until an expected position; *Behaviour:* Described by all relationships existing between the input flow (regulated pneumatic energy), the support (cylinder) and the output flow (translation movement).

5.2. *Elaboration of the probabilistic network.*

This second step aims at transforming the process & flow-based diagram into a Bayesian Network.

- *Nodes & States.* Cylinder is subjected to a loss of tightness mechanism modelled as a Markovian deterioration process. The effect of a tightness loss is a modification of the (input/output) pressure and airflow rate in the cylinder. Consequently, the translation movement is degraded according to its attributes deviation. The support "Cylinder" is represented by one dynamic variable $TL(k)$ (Tightness Loss) with four states: *OK* is the nominal state, D_1 and D_2 are the degraded states which do not entail the loss of the function and *HS* is the failing state representative of function lost.

The input energy flow "Pneumatic energy" is characterised by two static variables: the input pressure P and the air flow rate Fr provided by the distributor. The states of these variables are defined by the above process "To regulate the pneumatic power". By default, four states are associated to each variable. The output flow "Translation movement" is characterised by two static variables: the final static force F and the average speed S of the translation.

- *Edges.* The causal relationships lead to the definition of oriented arcs that link the "potential cause" variables (P, Fr and $TL(k+1)$) to "effect" variable (F and S). Theses relationships come directly from the physical laws that are applied on the cylinder. The force is calculated from the input pressure of the pneumatic energy and its possible decrease due to the loss of tightness. Thus, the structure of the model contains 4 classical arcs and one dynamic relationship (figure 4).

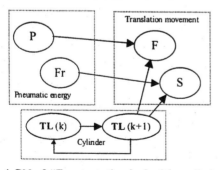

Fig. 4. BN of "To actuate the shaft of the cylinder"

- *Parameters of the model.* The prior probabilities attached to the input flow "Pneumatic energy" are

fixed by default in this case (actually, they would be generated by "To regulate the pneumatic power"). The definition of the CPT of F and S are based on the quantification of the causal relationships. It consists in the elicitation of a conditional probability p to each instance of the output flow attributes F, S in each configuration of (P, Fr and $TL(k+1)$). This step is based on two major rules: the value of F is determined thanks to the physical relation which links the pressure, its degradation by tightness loss and final static force supplied by a cylinder. In the same way, the value of S (which in fact is very difficult to calculate) is determined approximately by a relation between the input airflow rate and its degradation caused by tightness loss (assumption: the loading for the axis is constant). The CPT of $TL(k+1)$ is established thanks to the Markovian process of tightness loss. The conditional probabilities are equivalent to the transition rates λ (TL_{k+1}/TL_k), fixed *a priori* according to the operating knowledge available for similar cylinder (it is assumed that these parameters are time-invariant).

After configuration (design of the monitoring architecture), the monitoring process supported by CASIP platform executes continuous processing of **Pressure observations** which is the only indicator instrumented in the sub-system and used for refining the degradation model of the process "to regulate the pneumatic power". It is sequentially used also for initialising the BN model when the prognosis of the process "To actuate the shaft of the cylinder" has to be launched.

5.3. *Forecasting issued from prognosis process*

The prognosis process is started in order to evaluate the translation movement (F, S) in the future taking into account the initial conditions of P, Fr and TL. *Numerical application*

λ_1	10^{-2}	Cylinder : diameter 40mm
λ_2	$2.8.10^{-2}$	$P_{max} = 7$ bars
λ_{r1}	2.10^{-3}	$Fr_{max} = 20$ l/mn
λ_{r2}	5.10^{-2}	Static force loss = 10%
Fmin	620	Minimum static force (N)
Smin	2	Minimum Speed (cm/s)

Step 1: Starting. The monitoring process determines the states of the observable variable P at present time ($t=t_0$). At this time, the cylinder operate normally (no degradation). Therefore, the user can inject the observation of pressure variable state to the BN (figure 5) whereas the states of the others variables are unknown. In this example, the pressure is equal to 6 bars. This information is translated as the evidence Pr (P = '6 bars') = 1 to the variable P.

As for the cylinder, incertitude on the observation of the variable TL exists, the BN receives a soft evidence which updates the prior probability values and in this case affected the probability '0' to the failing state: Pr (TL='HS') = 0 and the probability 1/3 to the three other states.

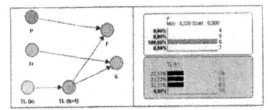

Fig. 5. Initialisation of the BN model (Bayesia tool)

Step 2: Inferences. Software BayesiaLab is used for implementing the causal and temporal inferences. Firstly, a causal inference is made to determine the probability of the non-observable nodes {Fr, F, S} giving the starting conditions. Then, from the observed situation at time t_0, the temporal inference computes the posterior probability $P[TL(t_0 + T)]$ of cylinder and consequently allows forecasting the probability $P[F(t_0 + T)]$ and $P[S(t_0 + T)]$ by a causal inference at time $t_0 + T$.

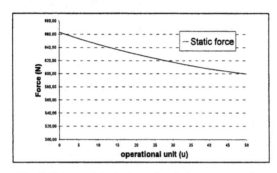

Fig. 6. Temporal evolution of static force expectation

With the same conditions (P and Fr are fixed), the performance of the process at time t_0 is evaluated by a static force $F_{t0} = 663N$ and a average speed $S_{t0} = 2.21$ cm/s while the prognosis at time $t_1 = t_0 + 50$ indicates an expectation for $F_{t1} = 600N$ and for $S_{t1} = 2.03$ cm/s (figure 6). This information represent the future estimated situation of this sub-process and can be propagated to the downstream sub-processes to calculate global process estimated situation and then to develop maintenance decision-making if necessary. For example, as F_{t1} is inferior to Fmin, this estimated situation is not satisfying in terms of performance, leading to develop a maintenance action before t_1 to keep F_t parameter as good.

6. CONCLUSION

The work presented in this paper gives an overview of how a methodology supporting prognosis process modelling based on the combination both of probabilistic and event approaches, is useful for maintenance strategy decision-making. Moreover experimentation on a manufacturing component is proposed to show the feasibility of this methodology and the prognosis process implementation on an e-maintenance architecture. On the basis of these first results, the experimentation context has to be extended to all the platform to develop abstraction complexity and then the results have to be compared with performances already obtained by conventional approaches to validate this methodology. In addition other future developments will be devoted in short term to integrate the possible maintenance alternatives to the model (decision nodes in BN) to concretely support maintenance strategy decision-making.

REFERENCES

Al-Najjar, B., Alsyouf, I. (2003). Selecting the most efficient maintenance approach using fuzzy multiple criteria decision making. *Int. Journal of Production Economics*, **84**, 85-100.

Byington, C., Watson, M., Roemer, M., Galie, T. (2003). Prognostic Enhancements to Gas Turbine Diagnostic Systems. In: *IEEE Aerospace Conference*, Big Sky, 2003.

Dean, T., Kanazawa, K. (1989). A model for reasoning about persistence and causation. *Computational Intelligence*, **5**, 142-160.

Dekker, R., Scarf, P.A. (1998). On the impact of optimisation models in maintenance decision making: the state of the art. *Reliability Engineering & System Safety*, **60**, 111-119.

Iung, B. (2003). From remote maintenance to MAS-Based E-maintenance of an industrial process. *IJIM*, **14/1**, 59-82.

Jensen, F. (1996). *An Introduction to Bayesian networks*. UCL Press, London (UK).

Kutucuoglu, K. Y., Hamali, J., Sharp, J.M., Enabling (2002). BPR in maintenance through a performance measurement system framework. *Inter. Journal of Flexible Manufacturing Systems*, **14**, 33-52.

Lee, J., (1998) Tele-service engineering in manufacturing: challenges and opportunities. *Int. Journal of Machine Tools&Manufacture*, **38**, 901-910.

Leger, J. B., Morel, G. (2001). Integration of maintenance in the enterprise towards an enterprise modelling-based framework compliant with proactive maintenance strategy. *Product Planning & Control*, **12/2**, 186–187.

Muller, A., Suhner, M.-C., Iung B. (2004a). Bayesian network based proactive maintenance. In: PSAM 7 / ESREL 4 International conference on probabilistic safety assessment and management, 14-18 June, Berlin – Germany

Muller, A., Suhner, M.-C., Iung B. (2004b). Probabilistic vs. dynamical prognosis process-based e-maintenance system, In: *11th IFAC INCOM04*, Salvador, Brazil, 2004.

Wang, H. (2002). A survey of maintenance policies of deteriorating systems. *European Journal of Operational Research*, **139**, 469-489.

Weber, P., Jouffe, L. (2003). Reliability modelling with dynamic bayesian networks. In: *5th IFAC Symposium on Fault Detection, Supervision and Safety of Technical Processes*, 57-62, Washington, USA.

Condition Monitoring Services for e-Maintenance

E. Hohwieler, R. Berger, C. Geisert

Fraunhofer Institut für Produktionsanlagen und Konstruktionstechnik, IPK
Pascalstraße 8-9
10587 Berlin, Germany
eckhard.hohwieler@ipk.fraunhofer.de, ralf.berger@ipk.fraunhofer.de,
claudio.geisert@ipk.fraunhofer.de

Abstract: Maintenance should follow quickly after the first appearance of problems, and be in accordance with the actual condition of the machine. This is made possible by monitoring the condition of machine parts via signals and sensors. The status of the system is ascertained based on the condition-values, load-values and the particulars of the situation. Main idea of the presented "Condition Monitoring Services" approach is to host and to provide electronic services for the analysis and prediction of machine health status using enhanced diagnostic algorithms and to deliver online assistance and additional information for related maintenance activities over the Internet. In this sense online services shall support condition based maintenance and help to reduce downtime caused by failure or degradation of components. Therefore a set of relevant machine devices and corresponding parameters and signals that indicate present machine status must be selected. The remaining lifetime of individual components can then be estimated. Using regular updates of the load, which are documented by the machine, the future progression of wear is predicted. With the result of classification of machine health status planning of maintenance and repair activities can be improved. The main requirement is the ability of the system to record and transfer relevant signal parameters via the Internet to the remote server for data evaluation. The results are stored in the machine producer's database and made available online to the customer via a web-browser.
Copyright © 2004 IFAC

Keywords: condition monitoring, online services, added value services, maintenance, prediction, Internet security

1. INTRODUCTION

New services and customer-care in the after-sales-area are becoming ever more decisive criteria differentiating firms in the field of machine and equipment manufacturing. New value-added services that accompany products are available to support the machine operator, improve accessibility and optimize use (Sihn and Graupner, 2001). Such services can be provided in different modes to the customer, use personal resources or run as automated services.

Fig. 1. Modes of service providing

In contrast to the already established remote diagnostics in case of machine failure, new online services can support the customer (Lee, 1998). This is not only for exceptional situations but includes also support in the daily production process during the whole life cycle of the machine (Hohwieler, 1998). In order to make the offering of such services to customers around the globe efficient, we draw on the most modern information and communication technology.

2. CONDITION MONITORING SERVICES

Main idea of the presented "e-Maintenance Services" approach is to host and to provide electronic services for the analysis and prediction of machine health status using enhanced diagnostic algorithms and to deliver online assistance and additional information for related maintenance activities over the Internet. In this sense online services shall support condition based maintenance and help to reduce downtime caused by failure or degradation of components. Therefore a set of relevant machine devices and corresponding parameters and signals that indicate present machine status must be selected. With the result of classification of machine health status planning of maintenance and repair activities can be improved.

The developed electronic service Condition Monitoring enables the customer to assess the health state of his machine in view of its wear and wastage. The information the customer receives after the assessment is given without interacting directly with an employee of the producer. The customer is able to regularly check the state of his machines by activating such a service and transferring recorded data to the machine producer (Ali, et al., 2002). Data evaluation is then automatically carried out by the service application running on the server and results are passed to the user who can access them via web browser. This service is based on recording reliability-relevant data of the machine and its components. It is also based on the evaluation of this transferred data in order to obtain information about wear and degradation with methods "status check" and "load monitoring".

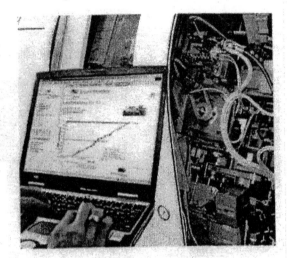

Fig. 2. Web access to remote monitoring service

Data is recorded by the numerical control and stored in a temporary file. The service can be activated through Internet access and a web browser. After the user has registered on the service-site and has been authorized, he can then choose a service and enter his machine's specifications. The service then requests the necessary data for transfer. The data analysis starts automatically on the manufacturer's server. The results are stored in a service-specific database. After finishing the evaluation, the user is informed about the new available results. Through his service access, he can look at the results or have it printed out.

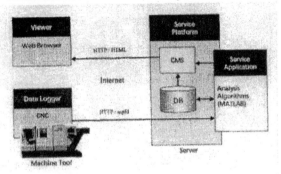

Fig. 3. Architecture of Web-Based Condition Monitoring

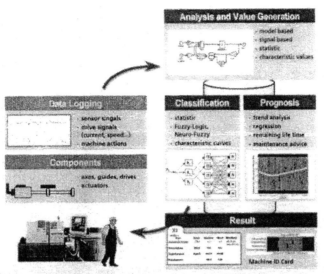

Fig. 4. Generating actual characteristic values with machine status check

In a concrete application, the functions and service for condition monitoring and generation of load profile for a turning machine were carried out. Data and signals available inside the numerical control of the machine are used as data source. This data is collected by a software component, which was specially made for this purpose and integrated in the control. The appropriate data is recorded regularly during the normal working operation of the machine in order to collect and determine relevant data about the life cycle and load profile of the machine or its components. Therefore drive current signals and speed signals for spindles and axes during first operation of the NC-program and the number of manufactured items are recorded for every work piece. Auxiliary machine commands that control a particular function can also be used as an information source for the evaluation. Additional information (i.e. machine type, serial number, date and time of first program run, the name of the NC-program, etc.) is stored in the data file produced. Derived values with properties such as operation performance, switching frequency of actuators, setting up frequency and work piece manufacturing time can be generated. The prepared data are stored in a profile card as profile values or charts.

Different type of use of machines at user's site and resulting individual load profiles lead to differing wear behavior and degradation during operation of the same machines at different customers. The fixed scheduling of maintenance intervals oriented at machine runtime does not take into account this load dependent wastage. Because of this, the condition monitoring service includes a continuous and automatic determination of physical characteristic values during the operation of a given system for generating and observing the load profile that will be used for the evaluation of the state of the system and components. The load profile and operation performance will then be used to estimate the load effect on indi-vidual components and the resulting lifetime remaining. In this way the future development of machine wear can be predicted by extrapolating from the load profile already recorded.

One component, that is most relevant for load depending wear and tear of spindles the bearings could be detected. The degradation process of a bearing is influenced by the number of revolutions within the permitted speed range. The expected remaining lifetime can than be estimated by

$$ t = K_B \cdot \frac{1}{n}, \tag{1} $$

where t is the remaining lifetime in hours, n is the rotation speed in % of the maximum rotation speed and K_B is a bearing specific constant.

On the other hand, the second parameter, that influences the degradation process of a bearing significantly is the transverse force, which is related to cutting forces. By cubing the ratio m of transverse force to rated transverse force it is possible to estimate the reducing factor R, that reduces the remaining life time to

$$ t_{new} = \frac{t}{R}. \tag{2} $$

Direct access of transverse forces via the numerical control is not possible without auxiliary sensor equipment. As a first approximation the drive current of the spindle motor was taken.

In addition to this load-based evaluation, a specific analysis of the individual components prone to degradation can be carried out. These components were identified to be the spindle and machine axis because

241

they are the parts most affected by forces during operation.

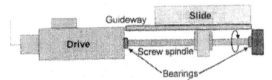

Fig. 5. Scheme of the electromechanical system feed axis.

The results indicate the amount of wear and can be used to recommend the appropriate service needed. Data recording is carried out during the diagnostic operation by a specifically designed test-NC-program under defined conditions. The signals used are current and speed signals of axis or spindle drives, which can be accessed directly from the CNC. To estimate the condition of a feed axis with controlled rotation speed, a mathematical linear model of this electromechanical unit (Figure 5) was built and verified. From this system of differential equations, which describes the process, characteristic diagnostic features can be generated. These features represent the physical parameters static and sliding friction and the moment of inertia. Drive current, rotation speed and acceleration derived from the rotation speed are used as input for the least squares method (Krüger, 1999; Uhlmann, et al., 1999). This approach can be used for parameter identification from an over-determined set of linear equations, built up on equation 3

$$\Theta \cdot \dot{\omega} = c_e \cdot i_D - \rho \cdot \omega - M_c , \qquad (3)$$

with

Θ : moment of inertia
ω : rotation speed
ρ : coefficient of sliding friction
c_e : motor constant
i_D : drive current
M_c : static friction

To prevent misinterpretation of the analyzed data caused by maloperation (e.g. choosing a machine to which the measured data don't belong) supplementary information like serial number, name of the NC-program, time and date, operating hours and the sampling time is added to the head of the transmitted file. A comparison of this information with machine specific data stored in a database is used to validate the correctness of the requested service task. The length of the measured time series is taken into account to ensure that the measurement was complete.

The identified parameters form so-called diagnostic features, which show typical behavior if conditions change. If a diagnostic feature exceeds a boundary value that was assigned by an expert a list of possible reasons (e.g. loss of tension of a drive belt, out-of-

balance of a spindle, loss of lubrication) and its countermeasures is generated using a classification algorithm. Because not all causal relationships are known and due to the presence of noise during data acquisition an accurate classification requires a wide pool of well known conditions and their corresponding diagnostic features. Therefore a continuous enlargement of this pool leads to better and refined results. The observation of trends is used to detect degradation before a failure occurs. With a linear regression over the last estimated values of the diagnostic features it is possible to predict and schedule expected maintenance tasks.

The generic structure of the service logic provides a combination of arbitrary algorithms for preprocessing and analysing data. To ensure flexibility and an easy and fast way for an upgrade of the implemented algorithms, MATLAB® – a commercial software for technical computing – has been integrated into the service logic (Figure 6). A separate MATLAB® instance is started by the service logic for each session. While the service logic is responsible for user guidance, specific service configuration and result representation, all actions in the context of data analysis are done by MATLAB® autonomously. This includes database requests and the generation of graphs.

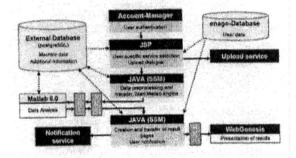

Fig. 6. Generic structure of the service logic

At the core of the service is a database, in which all service and user specific information (machine data, user data, algorithms, templates for the presentation in the content management system, etc.) is stored. Whether information is needed from or created by the service intelligence depends on the specific service task. It is also possible to send tables to different databases, which may even be located on distributed PCs. During the project a PostgreSQL-database was chosen, because of its platform independency and power. Moreover this sort of database has the advantage of being freeware and provides ODBC- and JDBC-interfaces. This fact becomes an important criterion, if different software applications that need access to the database are not able to support both types of drivers. The structure ensures an easy-to-do upgrade of algorithms and other features of the service, like changed boundary conditions or advanced signal processing methods.

3. MAINTENANCE ASSISTANCE

A list of the maintenance work to be done is generated for the given machines based on the current wear condition and predicted lifetime remaining as determined by condition monitoring. This can be looked up by the user in the service book, who can then interactively call-up the individual activities. The entries of the list contain short summaries about:

- the location on the machine where activity is to take place,

- a description of the work procedure to be taken,

- a statement of why the activity is required.

The specification concerning the location of activity is linked with detailed procedural instructions. The list contains routine maintenance procedures to be carried out as well as measures for repair generated because of recognized or predicted critical states of components. These measures are the core of the condition-based maintenance (Hohwieler and Berger, 2002). The purposes are specified along with the recommended measures to be taken. This might be demand for maintenance or repair due to load or accumulated running duration based on a diagnosed condition.

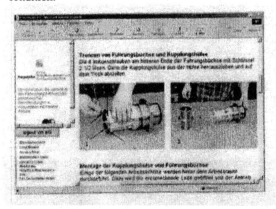

Fig. 7.   Example for Web Maintenance Assistance

By calling up an item of the To-Do-List in the "Service Book" the maintenance assistant service is started, which provides information about jobs, the procurement of spare parts and documents (Figure 7). If the user doesn't feel qualified enough, he can either initiate a service order to the manufacturer or book an Online-Qualification course for the given work and carry it out himself.

Even without the specific maintenance tip that the system generates, it is still possible to access the service provider with its prepared procedure-oriented electronic maintenance-, service- and machine documentation. Through this access, the user is offered the following information:

- Maintenance manuals,

- Service and diagnosis instructions,

- Circuit and wiring diagrams,

- Mechanical design drawings.

For this the individual documents listed are provided with users' rights, which allow the manufacturer to purposefully decide about the accessibility of the information. This is especially important when providing documents about the design information since it is to be treated extra carefully.

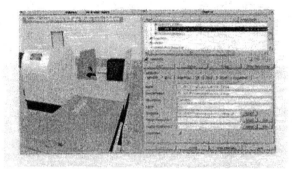

Fig. 8.   VR Training Scenario for Maintenance

E-Training offers adaptive qualification and training for operators, maintenance and service personnel. For the provision of training content, interactive, visual training scenarios are used in addition to simulation and virtual reality technologies. Tailored schooling and training via a service platform in the form of a virtual training center is a new service provided by e-Training, which can be offered by the system producer. For this, the manufacturer uses the information available in his database about the state of the machines and the qualification level of the user in order to put together a need-based offer from basic-modules.

An interactive simulation-based training system would be offered alongside multi-media machine documentation and computer supported learning programs (Figure 8). Through this, the user obtains an online-support with the preparation and carrying out of the maintenance work (Bluemel, et al., 2000).

4. SERVICE PLATFORM

The service portal serves as a platform supporting the entire interaction between producer and customer: from the signing of the contract and the providing of service content, to invoicing and billing (Berger and Hohwieler, 2003). It includes components for contract and operator management, encrypted transference and evaluation of recorded machine data, as well as modules for the displaying and reporting of results. The architecture is built around the service logic which is part of a service implementing the service functionality. A Web portal component is included to access user-individual information. For the implementation of the intended functionality the service logic can use the components integrated into the platform to get information about the user and the customer via API interface.

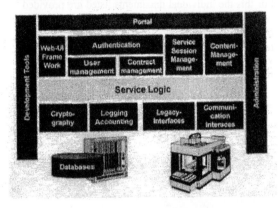

Fig. 9.   Components of the service platform

Since security plays an important role in service delivery and access to machines and services, a secure network topology was designed, which protects not only the hosted services but also the service clients. The authentication module works together with the user management and contract management components. It retrieves the authentication secret from the user management and checks if the contract is valid. If this is the case, a list of services can be retrieved from the contract management. In this way the portal component is able to provide the possibility to start the contracted services.

5. SUMMARY

Services that increase the accessibility and productivity for the customer have already begun to be developed. Among these are system-tailored services for status-oriented maintenance, as well as qualification and training with the use of simulation models.

Providing condition monitoring functionality and additional maintenance assistance by machine producers as web-based services via Internet will support users in manufacturing plants and shops. With the feedback and operation profiles gained through web-based Condition Monitoring machine producers can examine and learn more about stress behavior of machines and components and use this knowledge to increase reliability of future designed machines.

REFERENCES

Sihn, W. and T.-D. Graupner (2001). e-Industrial Services: Value-Added Services for the Producing Sector. In: *CIRP 34th International Seminar on Manufacturing Systems*, pp. 413-424 Athens, Greece.

Lee, J. (1998). Teleservice Engineering in Manufacturing: Challenges and Opportunities. In: *International Journal of Machine Tools & Manufacture*, 38, pp. 901-910.

Hohwieler, E. (1998). Internet-Zugang an Steuerungen für Produktionsunterstützung und Teleservice. In: *Innovative Produktionstechnik* (Krause, F.-L. (Ed)), pp. 505-518, Hanser, Munich.

Ali, S. A., Z. Chen, J. Lee and M. Koç (2002). Web enabled Device to Business (D2B™) platform for distributed and dynamic decision making systems. In: *Proceedings of 5th International Conference on Managing Innovations in Manufacturing (MIM)*, pp. 157 – 169, Milwaukee, Wisconsin, USA.

Krüger, J. (1999). Methoden zur Verbesserung der Fehlererkennung an Antriebsstrecken. Dissertation, TU-Berlin, ISBN: 3 816752683.

Uhlmann, E., E. Hohwieler and F. Becker (1999). New Structures and Methods for Control-Integrated Process Supervision. In: *Proceedings of the Second International Workshop on Intelligent Manufacturing Systems 1999*, pp. 809-815, Leuven, Belgium.

Hohwieler, E. and R. Berger (2002). E-Maintenance – Web-Based Services for Production Systems. In: *Proceedings of 5th International Conference on Managing Innovations in Manufacturing (MIM)*, pp. 215 – 224, Milwaukee, Wisconsin, USA.

Bluemel, E., A. Hintze, M. Schumann and S. Stuering, S. (2000). Using Virtual Prototypes for the Education of Maintenance and Service Personnel. In: *Proceedings of International Conference on Ship and Shipping Research (NAV)*, pp. 8.9.1-8.9.10, Venice.

Berger, R., and E. Hohwieler (2003). Service Platform for Web-based Services for Production Systems. In: *Proceedings of 36th CIRP International Seminar on Manufacturing Systems "Progress in Virtual Manufacturing Systems"*, pp. 209-213, Schriftenreihe Produktionstechnik, Band 29, Saarland University, Saarbrücken.

www.elsevier.com/locate/ifac

CLP-BASED PROJECT-DRIVEN MANUFACTURING

Zbigniew A. BANASZAK, Marek B. ZAREMBA

University of Information Technology Copernicus in Wrocław, Inowroclawska 56, 53-648 Wrocław, Poland
E-mail: z.banaszak@iizp.uz.zgora.pl
Département d'informatique et d'ingénierie, Université du Québec en Outaouais, Gatineau, QC, J8X 3X7 Canada
E-mail: marek.zaremba@uqo.ca

Abstract. This paper addresses decision-making support for project-driven design and manufacturing processes in small- and medium-size enterprises. The problem considered regards of finding of computationally effective approach aimed at scheduling of a new project subject to constraints imposed by a multi–project environment. In other words, we are looking for an answer whether a given production order specified by its cost and completion time can be accepted in a given manufacturing system specified by available production capability, i.e., the time-constrained resources availability. The problem belongs to a class of multi-mode case project scheduling problems, where the problem of finding a feasible solution is NP-complete. A CLP-based approach is proposed. Illustrative example of its application to a real-life company employing ILOG OPL Studio 3.6 software tool is presented. *Copyright © 2004 IFAC*

Key words: project management, small and medium size enterprises, constraint logic programming, task scheduling

1. INTRODUCTION

The field of project-oriented management of manufacturing systems is currently driven primarily by market forces. Some of the most challenging issues that arise in the domain of distributed manufacturing technology and management include manufacturability analysis, validation and evaluation of process plans, partnering in virtual enterprises, process design, and optimization of production plans and schedules. These issues are easily unified within a framework of a project-driven manufacturing concept which is focusing on small and medium size enterprises (SMEs), where products are manufactured based on make-to-order or build-to-order principle.

Most companies, particularly SME have to manage various projects, which share a pool of constrained resources, taking into account other objectives in addition to time. SMEs frequently manage various objectives at the same time. According to surveys, about 80% of companies have to deal with multiple projects, which relates to other data stating that about 90% of all projects occur in a multi-project context (Lova, Maroto., and Tormos, 2000). Since the project management belong to a class of NP-complete problems, new methods and techniques aimed at real-life constraints imposing on-line decision making are of great importance (Anaviisakow and Golany, 2003;

Wei, at al. 2002). Such methods enhance an on-line project management, and support a manager in the process of decision making, e.g. in the course of an evaluation whether a new project can be accepted for processing in a multi-project environment of a manufacturing system at hand. They can also be included into DSS tools integrated into standard project management software like MS Project or CA-Super Project. In this context it is worth to note that the currently available software tools allow pre-emption; however, they are not designed to cope with company production capability constraints in terms of resource and time availability. Moreover, they do not permit to consider production planning in a unified way to enable an integrated approach to such different tasks as production and transportation routings, production and batch sizing as well as tasks scheduling.

In that context, Constraint Logic Programming (CLP) languages, by employing the constraints propagation concept and by providing unified constraints specification, seem to be well suited for modelling of a company real-life and day-to-day decision-making (Bartak, 1999) process. The rest of the paper is organized as follows: Section 2 describes the modelling framework enabling to state the problem. A concept behind searching for a feasible project scheduling is then presented in Section 3. In Section 4, a case of the CLP-based approached usage to decision making in a

SME company is investigated. Conclusions are presented in Section 5.

2. PROBLEM FORMULATION

Constraint programming (CP) is a framework for solving combinatorial problems specified by pairs: <a set of variables with domains, a set of constraints restricting the possible combinations of the variables' values>. Constraint propagation, i.e., reference engine, is based on the idea of using constraints actively to prune the search space. The scope of propagation techniques, i.e. local consistency checking, is to reach a certain level of consistency in order to accelerate search procedures by drastically reducing the size of a search tree.

A constraint satisfaction problem can be stated as follows. Consider a set of n variables $x_1, x_2, ..., x_n$, their domains $d_1, d_2, ..., d_n$, and a set of constraints of this variables. Each constraint, i.e. an n-ary relation on $x_1, x_2, ..., x_n$ and can be treated as a subset of the Cartesian product $d_1 \times d_2 \times ... \times d_n$ of the domains. The solution is a value assignment of the variables such that all constraints are satisfied. In order to illustrate this kind of inference engine, let us consider a set of the following variables x, y, z and their domains $d_x = \{1,2,3,4,5\}$, $d_y = \{1,2,3,4,5\}$, $d_z = \{1,2,3,4,5\}$. Suppose the following constraints:

$\alpha: X \geq Y + 1$, $\quad \beta: Y \geq Z + 2$, $\quad \gamma: Z \leq X - Y$.

One of possible ways of constraint propagation is shown below:

$D_X = \{1,2,3,4,5\}$ $D_X = \{2,3,4,5\}$
$D_Y = \{1,2,3,4,5\}$ $\longrightarrow$ $D_Y = \{1,2,3,4\}$
$D_Z = \{1,2,3,4,5\}$ α $D_Z = \{1,2,3,4,5\}$

$D_X = \{2,3,4,5\}$ $D_X = \{2,3,4,5\}$
$D_Y = \{1,2,3,4\}$ $\longrightarrow$ $D_Y = \{3,4\}$
$D_Z = \{1,2,3,4,5\}$ β $D_Z = \{1,2,3,4,5\}$

$D_X = \{2,3,4,5\}$ $D_X = \{4,5\}$
$D_Y = \{3,4\}$ $\longrightarrow$ $D_Y = \{3,4\}$
$D_Z = \{1,2,3,4,5\}$ α $D_Z = \{1,2,3,4,5\}$

$D_X = \{2,3,4,5\}$ $D_X = \{4,5\}$
$D_Y = \{3,4\}$ $\longrightarrow$ $D_Y = \{3,4\}$
$D_Z = \{1,2,3,4,5\}$ γ $D_Z = \{1,2\}$.

The resulting set of feasible solutions consists of the following combinations of value assignment of the variables:

X, Y, Z   X, Y, Z   X, Y, Z   X, Y, Z
(3 , 2 , 1),   (5, 4, 1),   (5, 3, 1),   (5, 3, 2).

So, the task is to find a valuation of the variables satisfying all the constraints, i.e., a feasible valuation.

Note, that the constraints are not restricted to linear equalities and/or inequalities. They can express arbitrary mathematical or logical formula as well as bind variables with different non-numerical event domains. Taking into account this advantage, the following problem is considered.

Consider a manufacturing system providing a given production capability while processing some other work orders. So, only a part of the production capability (specified by the time-restricted resource availability) is available for use in the system. A newly introduced production order is represented by an activity-on-node network, and specified by project duration, which is equivalent to a presumed completion time (the work order cycle) as well as a total project cost constraint. Each activity may be executed in one of a set of modes (system resources). Also, each activity cannot be preempted and the mode once selected cannot be changed.

The problem consists in finding a makespan-feasible schedule that fulfils the constraints imposed by precedence relations, by time-constrained resources availability, as well as the duration deadline. In other words, the question concerns the feasibility of the project schedule, i.e., does there exist a project schedule respecting the project duration deadline and cost limits?

3. PROJECT FLOW PROTOTYPING

The most frequently observed decision problem a producer is faced with concerns a question whether a capability of his manufacturing system can meet the constraints imposed by a given production order. It means, the question considered is: Whether the constraints imposed by a production order can be satisfied by an enterprise capability, i.e. whether the consumer's requirements can be balanced with producer's resources availability? A positive response to this question means there exists a detailed schedule of production flow containing the manufacturing and transportation routes, production and delivery batch sizes, etc.

The approach proposed can be seen as an alternative to the one based on a computer simulation. On one side, it allows to respond to the same question "what if", providing detailed plan of production flow if a balance holds. On the other side, it provides suggestions (e.g., how to change consumer requirements and/or a producer capability) supporting negotiation aimed at a production order acceptance. The core technology of CP is therefore hidden in constraint propagation combined with sophisticated search techniques (Bartak,

1999). Depending on the order, the constraints propagation take a number of backtracking steps and the efficiency of the applied searching strategy may differ dramatically (see Fig. 1).

Taking this into account, our objective is to develop a task oriented searching strategy, the implementation of which in a CLP based language could be successfully applied in project-driven SME companies, i.e., to make possible on-line decision making under real-life task sizes and constraints.

4. ILLUSTRATIVE EXAMPLE

Consider the real life case of a SME company producing mechanical components, which has received a new production order for the manufacturing of 74 carrying tubes. The time of execution of the order as requested by the customer is 5 days. The price that the customer can pay for the manufacturing of 1 piece is 35 cost units.

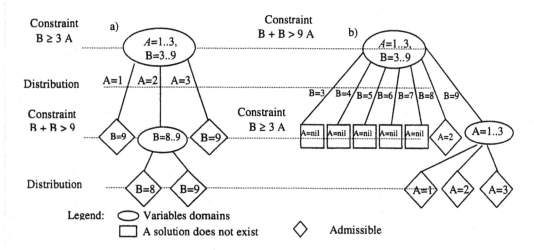

Fig. 1. Constraints propagation, a) $B + B > 9$ follows $B \geq 3 A$, b) $B \geq 3 A$ follows $B + B > 9$.

The carrying tube consists of 6 elements: external pipe, disk, pipe, holder, rivet and a plate. The plate and rivet are the elements that are purchased from a cooperating company and are accessible in the warehouse.

The company executes a basic production schedule. It has a production capacity surplus, which could be used if necessary for the need of further orders. It is assumed that there are known periods in which the specific resources remain available. Fig. 2 presents the available production capabilities of selected resources of the company in a specified period of time, e.g., 5

days. The dark fields denote periods when the resource is available, whereas the numbers in the fields signify the cost of the use of a given resource in a time unit.

The production order has been treated as a non standard product. Due to its uniqueness and a lack of a routine decision scheme, it may be treated as a project type order. In order to execute an analysis resulting in the adoption or rejection of the project, it is first necessary to specify a set of activities that will reflect the technological order of operations. Therefore, the network of activities following technological constraints is determined.

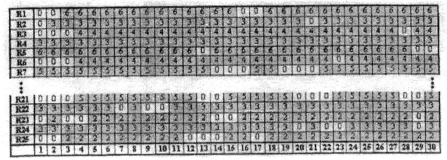

Fig. 2. Resource availability.

There are four means of transportation available in the company (R22, R23, R24, R25). The operation and transportation times as well as their allocations to the departments and particular workstations are presented in the Table 1, i.e., the transportation mean R22 is used in the course of the holder manufacturing, R23 in the course of the disk manufacturing, while R24 and R24 in the course of the pipe, and R25 the external pipe manufacturing. Buffers and warehouses are fully accessible and the transportation rate is unrestricted.

Table 1. Specification of activities and their executed times.

	Operation	Ten number of the technological operation	The number of the production department	Operation time	Predecessor	Resource
1.	**Operations of holder manufacturing**					
	Bendering	A1	520	5	A0	R2
	Transport operation	A2		10	A1	R22
	Lumbering	A3	403	55	A2	R1
	Transport operation	A4		5	A3	R22
	• • •	• • •	• • •	• • •	• • •	• • •
	Washing	A34	401	10	A12, A22, A26	R13
	Transport operation	A35		10	A34	R22
	Control	A36	401	20	A35	R19
	Transport operation	A41		10	A36	R22
2.	**Operations of disk manufacturing**					
	Cutting	A13	520	50	A0	R6
	Transport operation	A14		5	A13	R23
	Lumbering	A15	403	45	A14	R1
	• • •	• • •	• • •	• • •	• • •	• • •
	Transport operation	A37		5	A34	R23
	Control	A38	401	15	A37	R20
	Transport operation	A42		5	A38	R23
3.	**Operations of pipe manufacturing**					
	Cutting	A21	520	15	A0	R18
	Transport operation	A22		10	A21	R24
	Chamfering	A23	401	5	A22	R9
	• • •	• • •	• • •	• • •	• • •	• • •
	Transport operation	A43		10	A40	R25
4.	**Operations of external pipe manufacturing**					
	Transport operation	A27		10	A21	R25
	Rolling	A28	401	90	A27	R11
	Transport operation	A29		5	A28	R25
	• • •	• • •	• • •	• • •	• • •	• • •
	Grinding	A32	403	10	A31	R4
	Transport operation	A33		5	A32	R25
5.	**Welding**	A44	401	300	A41, A42, A43	R14
	Transport operation	A45		5	A44	R25
	• • •	• • •	• • •	• • •	• • •	• • •
	Grinding	A50	401	175	A49	R17
	Transport operation	A51		10	A50	R25
	Manual-tooling	A52	401	10	A51	R3

For the above input data, the responds to the following questions are sought: Does the company capability follows the cost and duration requirements of a given production order? If so, what is the relevant schedule of its execution? What are the alternatives, i.e., schedules guaranteeing a lower cost and/or a shorter work order completion? If not, which conditions should be released subject to producer-customer negotiations?

In order to respond to the above stated questions the ILOG OPL Studio 3.6 software package equipped with the Optimization Programming Language (OPL) has

been used. This package is a complete interactive modeling environment for rapid and efficient development of optimization models and applications (Ustin, 2001).

First of all the network activities, their operation times as well as production order completion time are declared. Function "maxduration" specifies the maximum time of the execution deadline for the 74 pieces of the carrying tubes. It was assumed that one time unit corresponds to 5 minutes. The execution time of an order, specified by the customer, is 5 days, i.e., 480 time units. (Fig. 3).

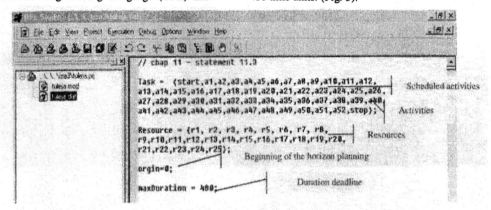

Fig. 3. Input data

Then, a variant of resource assignment is considered. As a result of resources allocation (taking into account resource availability, see Fig. 3), a schedule of the realization of a given production order is generated taking into account the time-limited accessibility of the resources (Fig. 4). The Gantt's chart obtained provides the starting time and the finishing time of the operations (activities), their durations as well as the work order completion time. The obtained makespan corresponding to the manufacturing of the 74 pieces of carrying tubes is 193 time units, i.e., 965 minutes.

Taking into account the time of the product designing, which takes 720 minutes, the total execution time of the production order is 1685 minutes (Fig.5). Since the deadline specified by the customer is 5 days, i.e. 2400 minutes, i.e., 480 time units (assuming that the production is executed in a one shift system) the time margin is 715 minutes.

It should be noted that the above result has been obtained in an on-line mode. Attempts to employ such task oriented software tools as MS PROJECT, ARENA, and IFS were unsuccessful (Banaszak, 2003).

Because the results obtained indicate that the production capabilities are sufficient for the realization

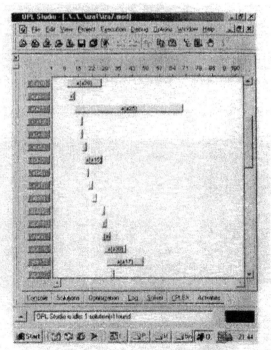

Fig. 4. Schedule of the work order activities.

of the production order within the specified time limit the next stage of evaluation is considered. The calculation of the total cost of execution of the production order has to take into account the cost of use of the production resources (Fig.2).

The maximum purchase price of the item specified by the customer is 35 cost units per piece. The values presented in Fig. 1 include departmental and general enterprise costs. The departmental costs include such components as: depreciation, gas, waste treatment, telephone bills, water, salaries, social insurance, equipment, and so on. The obtained cost of production order equals to 1706 cost units. That is because:

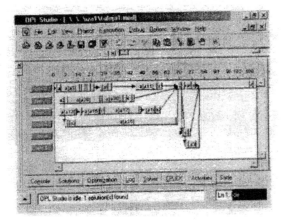

Fig. 5. The Gantt's chart of the production order

The cost of material for the execution of the specific parts is 443,2 cost units:
- pipe (2,28*74 =168,7 cost units),
- disk (1,16*74 = 85,8 cost units),
- holder (0,03*74 = 2,22 cost units),
- external pipe (2,52*74 = 186,5 cost units).

The cost of the rivet and the plate as purchased from the cooperating company is 170,2 cost units.

So, the total cost of the manufacturing of the 74 carrying tubes equals to 2319,4, i.e., 31,3 cost units per piece (1706 cost units + 443,2 cost units + 170,2 cost units). The maximum purchase price specified by the customer is 35 cost units per piece. Therefore, the margin of profit that the company may add to the price is 11,8%.

5. SUMMARY

A CLP – based modeling framework, which supports design of decision making systems aimed at finding the answer whether a given production order can be accepted for processing in an enterprise, is considered. The framework provides a good platform for checking consistency between the production order completion requirements and a workshop capability offered.

The CP methodology presented here is a promising alternative for commercially available tools based on other technologies, such as a class of ERP. Their application in solving a real-life problem is quite limited (Davir, et al, 2003; Lova, et al, 2000).

Also, the proposed approach can be considered as a contribution to project-driven production flow management applied in make-to-order companies as well as for prototyping of the virtual organization structures. That is especially important in the context of cheap and user-friendly decision support in SMEs. Further research is aimed at developing task oriented searching strategies, implementation of which will support the SME's decision making process.

REFERENCES

Anaviisakow S., and B. Golany (2003). Managing multi-project environments through constant work-in-process, *International Journal of Project Management*, Vol. 21, pp. 9-18.

Banaszak Z., I. Pisz (2003). Project-driven production flow management. *Project driven manufacturing*. WNT, Warsaw, pp. 54-71.

Barták R. (2003). Constraint-based scheduling: An introduction for newcomers, *Preprints. of the 7th IFAC Workshop on Intelligent Manufacturing Systems*, 6-8 April, 2003, Budapest, Hungary, pp.75-80.

Dvir D., Raz T., and A. Shehar (2003). An empirical analysis of the relationship between project planning and project success, *International Journal of Project Management*, Vol. 21, pp. 89-95.

Lova A., Maroto C., and P. Tormos (2000). A multicriteria heuristic method to improve resource allocation in multiproject scheduling, *European Journal of Operational Research*, Vol. 127, pp. 408-424.

Ustrin I. (2001). Getting Started: Modeling and Solving Combinatorial Problems with Constraint Programming using ILOG OPL Studio..

Wei C. C., Liu P.-H., and Y.-C. Tsai (2002) Resource-constrained project management using enhanced theory of constraint, *International Journal of Project Management*, Vol.20, No.2, pp. 561-567.

FUZZY REGRESSION IN FORECASTING AND MANAGEMENT OF SME INSURANCE BUSINESS

Georgi M. Dimirovski[1,2] and Cvetko J. Andreeski[3]

[1] *Dogus University, Faculty of Engineering, Department of Computer Engineering
Acibadem, Zeamet Sk. 21, Kadikoy, TR-34722 Istanbul, Republic of Turkey*
and
[2] *SS Cyril and Methodius University, Faculty of EE, Skopje, Rep. of Macedonia
Fax. +90-216-327-9631; E-mail: gdimirovski@dogus.edu.tr*

[3] *University St Clement Ohridski - Bitola, Faculty of Tourism and Hospitality Ohrid
"Kej Marsal Tito" 95, MK-6000 Ohrid, Republic of Macedonia
Fax: +389-96-262-147; E-mail: cipuslju@mt.net.mk*

Abstract: Insurance is an important activity of modern economies, and so are small and medium insurance companies. As inferred from developmental phases of economics as an axiomatic social science possibly, it has evolved into science that employs systems analysis and works with sophisticated mathematical tools. The gap between economic reality and predictions derived from sophisticated theories persists as intentions of using precise mathematical tools rises. The basic reason of uncertainty in economics is the impact of human decisions in creating and managing economic processes. People use imprecise predictions in the decision making process, hence more than just classical two-valued logic and classical theory of additive measures are needed. A possible solution satisfying these needs are decision support tools employing fuzzy-regression methodology. *Copyright © 2004 IFAC*

Keywords: Decision support; fuzzy regression analysis; insurance business; management; small and medium enterprises.

1. INTRODUCTION

Fuzzy logic as a system-theoretic and applied computing field (Klir and Yuan, 1995; Terano and co-authors, 1994), and even more so as a universal modeling methodology for complex uncertainty processes, was created largely due to fundamental results of Lotfi A. Zadeh (e.g., see his works 1965, 1973, 1975, 1978, 1979, 1980). Fuzzy logic and its applications have become popular via their applications in fields of engineering (Kaufman and Gupta, 1988; Zadeh, 1997, 1998) albeit the real potential of the fuzzy logic may well be even greater in social sciences, which deal with human-centered or "soft" systems (Dimirovski, 1994). This research work is related to applications in economics and finances, in particular (e.g., see Tanaka and co-authors, 1982; Heshmaty and Kendal, 1985; Andreeski, 2003; and references therein).

Economics, as part of social sciences, should have been one of the early targets for utilizing fuzzy set theory. However, the role of fuzzy set theory in economics was recognized much later than in many other areas. Nonetheless, British scientist Shackle has been a pioneer advocate in favour of fuzzy logic in economics (Kaufmann and Gupta, 1988). Since the late 1940s he has argued that probability theory, accepted in economics as the only mathematical tool for expressing uncertainty, was not meaningful for capturing the nature of the uncertainty in economics and human decision making (Shackle, 1961). Many applications of fuzzy set theory, which have been described in the literature since it emerged in 1965, are relevant to economics to various degrees.

Included in this category are fuzzy preference theory, fuzzy games, fuzzified methods for some problems of operations research such as fuzzy regression, and various problem areas of fuzzy decision making (Zimmermann, 1991). For indeed, as pointed out by Belman and Zadeh (1970), much of the real-world decision making takes place in an environment in which constraints, goals, and consequences of possible actions cannot be precisely known (Ma and co-authors, 1999). The experienced economists often express imprecise but highly relevant economic predictions in linguistic terms, which econometric specialists (Steward, 1991; Terrence, 1993) recognized too.

There are a number of sophisticated methods for economic time series analysis and forecasting (Box and co-authors, 1994). In this paper we use the theory of fuzzy regression in order to develop a decision support tool for making appropriate financial forecasting analysis in the insurance assets management (Andreeski, 2003). On the grounds of economic statistical studies in the Rep. of Macedonia (Gancevski, 2000), the exogenous variable data adopted is the average salary. The rest of this paper is written as follows. In Section 2, we present our case study on life insurance business in our country via statistical analysis methods (Box and co-authors, 1994; Terrence, 1993). Then Section 3 presents our novel research of this case study by using the fuzzy regression analysis (Kaufmann and Gupta, 1988), which makes-up the core of a decision support tool for managing small and medium enterprises in insurance business. Conclusions and references follow thereafter.

2. REGRESSION ANALYSIS OF THE LIFE INSURANCE PREMIUM IN R.M. FOR TIME PERIOD 1968-1998 YEAR

For this analysis, we made use of statistical data for the life insurance in the Republic of Macedonia (R.M.; a small developing country with a population of 2 million) from year 1968 till year 1998 by annual data. Presented data for all different type of insurance are taken from the insurance company "Q-be Makedonija", the privatized state insurance enterprise formerly controlled by the government. This company is the only insurance company that operates since 1961 in the R.M. It holds the greatest percent of clients and taken premium according to the statistical data, and moreover this company holds around 93% of insurance and reinsurance market in the R.M. (Gancevski, 2000). Statistical data are taken from the annual reports and relevant books of this company. The other statistical data are taken from the Statistical Yearbooks of the R. M.

The statistical regression analysis (Box and co-authors, 1994) gives the correlation between the life insurance demand and the average salary in R.M. for the same time period 1968-1998 year. In Table 1, there are given the statistical data for premium of life insurance in R.M. 1968-1998 and the average salary in the R.M. for the same period. There are given in this table the output of the regression model with interceptor and the values of the residuals. In the table of correlation coefficients we present the value of correlation between two different time series. One can see from these results that there is a high level of correlation between two analyzed time series.

The value of correlation found is 0.966 which suggest strong relationship between these two time series. We have constructed a regression model with interceptor. Output of this model is presented in Table 1 and the model, with its correspondent statistics are given in Table 2. From this table one can see that R-sqr parameter has value of 0.930 that implies good fit of the model what is expected in respect of great correlation between analyzed time series. The same conclusion can be made from the analysis of F statistics which has the value of 398.834. The critical value for F statistics for our model is 4.18 for 5% of significance and 7.60 for 1% of significance. The value of F statistics is much greater than two critical values, which implies that the model gives good fit of analyzed time series. The T statistics for the interceptor has small value, which indicates that we should create model without interceptor. Critical value of the t statistics for our model is 2.042 for 5% of significance and the value of T statistics for the interceptor has a value of 0.48, which implies acceptance of the hypothesis of rejecting the interceptor from the model. The value of Durbin-Watson statistics is 2.373 and this value is close to value 2, which implies that there is no serial correlation of first degree between the residuals of the model. Critical value of the Durbin-Watson statistics was found to be 1.26.

In Figure I, we present the correlation coefficients of the residual of regression model and the intervals of confidence. From Figure II, one can conclude that there are not significant values of the correlation coefficients of the residuals, which implies that the residuals indicate the series of white noise. The sum of the residuals is zero and that is another confirmation that they fulfill the constraints of normal distribution. In Figure II, we present comparative evolution movements of the following two series: data of life insurance in R.M. and the output of the regression model for this time series. In Figure II, also there can be seen the actual fit of the model.

The presented statistics has lead to conclusion that the amount of life insurance premium is strongly correlated with the amount of average salary in the R.M. Regression model that was found to fit best the time series is the model without an interceptor.

However, this model does not explain the whole structure of the analyzed time series; only 93-94% of it. In the sequel there are presented data for national gross product and income in the R.M. as well as their influence on the life insurance premium.

Table 1. Regression analysis on the life insurance business in the R.M.

Year	Life insurance premium in mil. Denars	Average salary in R.M.	Regression model with interceptor	Residuals
1968	10	750	61.912	-51.912
1969	12	841	63.469	-51.469
1970	15	989	66.002	-51.002
1971	20	1185	69.356	-49.356
1972	29	1416	73.309	-44.309
1973	42	1647	77.262	-35.262
1974	53	2120	85.356	-32.356
1975	82	2622	93.947	-11.947
1976	93	2970	99.902	-6.902
1977	125	3479	108.612	16.388
1978	133	4215	121.207	11.793
1979	199	5030	135.154	63.846
1980	218	5109	136.506	81.494
1981	297	8054	186.903	110.097
1982	378	10441	227.752	150.248
1983	490	13127	273.717	216.283
1984	677	17982	356.799	320.201
1985	1539	29613	555.838	983.162
1986	1726	59254	1063.077	662.923
1987	2152	119347	2091.435	60.565
1988	3228	320000	5525.164	-2297.16
1989	10926	568000	9769.132	1156.868
1990	14.842	3188	103.633	-88.791
1991	45.430	6270	156.374	-110.944
1992	52.667	635	59.944	-7.277
1993	1.817	3782	113.798	-111.981
1994	4.253	7754	181.770	-177.517
1995	8.243	8581	195.922	-187.679
1996	13.986	8817	199.960	-185.974
1997	33.781	9063	204.170	-170.389
1998	48.198	9394	209.835	-161.637

Table of correlation coefficients

	Life insurance	Average salary in R.M.
Life insurance	1.000	0.966
Average salary	0.966	1.000

Durbin-Watson statistics=2.373
Goldfeld-Quandt statistics=181.734

Furthermore, it can be seen from Table 2, the value of T statistics for the interceptor involvement found was 0.48, thus implying possible rejection of the interceptor from the model. In Table 3, the resulting outputs of regression analysis for modelling the life insurance time series without interceptor, as it was suggested from previous model with interceptor, are presented.

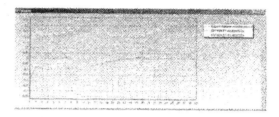

Fig. I. Results on the residual analysis (correlation coefficients).

Table 2. Regression with interceptor

Regressors	Values	Stand. dev.	T ratio
Interceptor	49.077	102.65	0.48
Average salary	0.017	0.00	19.97

s=538.973 R-sqr=0.932 R sqr(adj)=0.930

Analysis of variance:

Source	DF	Sum of sqr.	Aver. sqr	F
Regression	1	115858034.17	115858034.17	398.834
Sqr	29	8424269.46	290492.05	
Total	30	124282303.63		

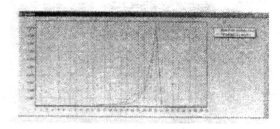

Fig. II. Graphical presentation of life insurance time series and the output of the model developed.

Table 3. Regression without interceptor

Regressor	Value	Standard dev.	T ratio
Average salary	0.017	0.00	21.63

s=531.998 R-sqr=0.940 R sqr(adj)=0.940

Variance analysis:

Source	DF	Sum sqr.	Aver. sqr	F
Regression	1	132365916.63	132365916.63	467.687
Sqr	30	8490668.50	283022.28	
Total	31	140856585.13		

Durbion Watson statistics=2.322
Goldfeld Quandt statistics=76.683

It is seen from Table 3 that the value of T statistics for this model is greater than the same of the previous model. Value of R-sqr is 0.940 and is greater then the same value of the previous model. These values indicate that this model is better the previous one. This model even better fits the time series, which is obvious from the values of R-sqr and F statistics. The value of the *Durbin-Watson* statistics is close to value of 2.0 and implies uncorrelated residuals for this model. The sum of residuals is equal to zero.

In Figure III, there are presented correlation coefficients and intervals of confidence for the residuals of the regression model without interceptor. It is apparent that there are no significant values of the residuals out of the confidence interval and there is no pattern of the residuals. Upon calculation of Q statistics for the residuals it was confirmed that the residuals are uncorrelated.

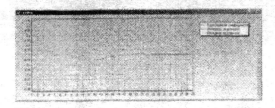

Fig. III. Residual analysis (correlation coefficients for the regression model without interceptor).

Fig. IV. Life insurance time series and output of the regression model without interceptor.

Figure IV depicts the comparative series of the life insurance and the regression model without interceptor.

3. APPLIED FUZZY REGRESSION ANALYSIS

Fuzzy regression analysis emerged in early 1980s with the works of Tanaka and co-authors (1982), of Heshmaty and Kandel (1985), and the Fuzzy-Delphi method due to Kaufmann and Gupta (1988). There are two motivations for developing fuzzy regression analysis. The first motivation results from the realization that is often not realistic to assume that a crisp function of a given form represents the relationship between the given variables. Fuzzy relation, even though less precise, seems intuitively more realistic. The second motivation resulted from

the fact that economic decision-making applications based on financial time series are inherently fuzzy (Terano and co-authors, 1994). Moreover, back-tracing observation analysis on life insurance business has demonstrated that the available data appear to be fuzzy too due to human interference and not entirely statistical (Andreeski, 2003).

We have used the time series of the case study on life insurance in order to explore the potential of the method of fuzzy regression analysis for a decision making support. We have adopted average salary in R.M as an exogenous variable in the analysis (Andreeski, 2003). All the calculations needed to get the results from linear programming are made with Matlab software. The linear fuzzy possibility theory (Dubois and Prade, 1991) in regression modelling with standard data in the insurance business has been used. To obtain practical results, they are sought with linear possibility regression model (Terrano and co-authors, 1994), using fuzzy numbers (Dubois and Prade, 1987) and Takagi-Sugeno-Kang fuzzy system (Takagi and Sugeno, 1985; Sugeno and Kang, 1988). It has been shown (Tanaka and coauthors, 1982) that solving this problem reduces to solving a linear programming problem with a model of the type

$$Y_i = A x_i. \tag{1}$$

Relationship (1) expresses the possibilistic linear function which gives the plausible value of the insurance premium paid in with respect of the average salary of potential clients. Variable x_i represents time series data of the average salary through years (1968-1998), and $A = A(\alpha, c)$ is the symmetrical fuzzy number to be determined in the course of fuzzy regression analysis.

The class of fuzzy numbers chosen is the one with symmetric triangular membership function. Then, following the theory of linear possibility regression model (Terrano and co-authors, 1994), the following LP problem is to be solved:

$$\min_{\alpha,c} J(c) = \sum c' |x_i|, \tag{2 a}$$

$$y_i \leq x_i^t \alpha + (1-h)c' |x_i|, \tag{2 b}$$

$$y_i \geq x_i^t \alpha - (1-h)c' |x_i|, \quad \alpha, c \geq 0. \tag{2 c}$$

Fuzzy coefficient A is determined by the degree to which the given data (y_i, x_i) is included in the inferred fuzzy number Y_i. More precisely,

$$\mu_{Y_i}(y_i) \geq h, \quad i = 1,...,N \text{ and } 0 \leq 0 \leq h \leq 1, \tag{3}$$

where h is the degree to which the available data set is included into the inferred fuzzy number.

This LP problem in our research was solved using h =0.5. Values of the fuzzy number A are α =0.0417 and c =0.0180; hence we can write A (0.0417, 0.0180). For the calculation we have chosen the large-scale method based on LIPSOL algorithm (Linear Interior Point Solver) and the number of iterations was 7. In Table 4, there are given the values of the lower bound and the upper bound as well as the most plausible value of Y_i. Of course, negative values should not be taken under consideration because the output cannot be negative.

Table 4. Lower bound, center and upper bound for
$$Y_i$$

Year	Center	Lower bound	Upper bound	μ
1968	31.275	-30.6	93.15	0.65
1969	35.0697	-34.3128	104.4522	0.66
1970	41.2413	-40.3512	122.8338	0.67
1971	49.4145	-48.348	147.177	0.69
1972	59.0472	-57.7728	175.8672	0.74
1973	68.6799	-67.1976	204.5574	0.80
1974	88.404	-86.496	263.304	0.79
1975	109.3374	-106.9776	325.6524	0.87
1976	123.849	-121.176	368.874	0.87
1977	145.0743	-141.9432	432.0918	0.93
1978	175.7655	-171.972	523.503	0.87
1979	209.751	-205.224	624.726	0.97
1980	213.0453	-208.4472	634.5378	0.98
1981	335.8518	-328.6032	1000.3068	0.94
1982	435.3897	-425.9928	1296.7722	0.93
1983	547.3959	-535.5816	1630.3734	0.94
1984	749.8494	-733.6656	2233.3644	0.95
1985	1234.862	-1208.2104	3677.9346	0.87
1986	2470.892	-2417.5632	7359.3468	0.84
1987	4976.77	-4869.3576	14822.897	0.71
1988	13344	-13056	39744	0.61
1989	23685.6	-23174.4	70545.6	0.72
1990	132.9396	-130.0704	395.9496	0.55
1991	261.459	-255.816	778.734	0.58
1992	26.4795	-25.908	78.867	0.50
1993	157.7094	-154.3056	469.7244	0.50
1994	323.3418	-316.3632	963.0468	0.50
1995	357.8277	-350.1048	1065.7602	0.50
1996	367.6689	-359.7336	1095.0714	0.51
1997	377.9271	-369.7704	1125.6246	0.53
1998	209.4862	-565.5188	984.4912	0.79

If we do consider results for the values of $\mu_i \geq 0.5$, then for values of 1968 we can have lower bound 0.3375 and upper bound 62.2125. The central value

is 10 with the value of μ =0.656162. Values closer to 0.5 are the so-called endpoints, and there are 4 of them in Table 4. These values appear to be during years 1992-1995. According to the obtained results, the years of optimally operating the life insurance business were years 1977-1985 when the value of μ is close to the possibly maximum value 1.0.

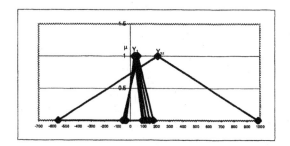

Fig. V. Typical inferred fuzzy number values and membership grades for the output $Y_i^* = A x_i^*$, the plausible value of the insurance premium.

The inferred fuzzy values for $Y_i^* = A x_i^*$ are shown in Figure V. All the essential comments have been given within the presented analysis above. Though, a couple of clarifying comments are still needed. The investigated problem was solved using h =0.5 for reasons of decision-making in insurance assets management on realistic observation grounds so that these lead to neither too optimistic nor pessimistic forecasts. Namely, in life insurance business seldom there are available enough data, in particular in a small developing country which is in economic crisis. Secondly, the focus was put on average salary because the statistical regression analysis has demonstrated its dominant role in the insurance market of the given society environment.

4. CONCLUSION

An applied fuzzy regression analysis on the case study of life insurance business in the Republic of Macedonia has been presented. Given the size and category of a developing country as well as the past of the existing insurance business, it is believed that these research findings may well serve as a prototype case of small and medium insurance enterprises.

Two reasons suggest creating and using the fuzzy regression model in insurance business where human impact is crucial. Firstly, the data analyzed with the regression can capture the underlying uncertainty character that is better defined with fuzzy numbers. And secondly, the obtained result via the fuzzy regression analysis can describe better the actual inherent conditions of the analyzed data although formal precision in the fuzzy model is lower than in the traditional statistical one. Therefore the fuzzy possibilistic approach possesses higher potential for choosing the most adequate optimum value of the premium, as well as possible choices of other values balancing the benefit of the company and/or of insured clients. Besides, an analysis survey by years can be carried out for the company's operation; i.e., the years during which the company did business most efficiently (the value h=0.5 taken) as well as years below and above the optimal threshold are readily available.

The fuzzy regression based model was shown to complement to the traditional regression analysis, which uses the traditional mathematical statistics approach. Hence it is worth investigating the alternative with non-symmetrical fuzzy numbers, the topic for a future paper. The other future research topic is on making use of random set description of possibility measure (de Cooman and Aeyels, 2000).

REFERENCES

Andreeski, C. J. (2003). Non-Classical System-Theoretic Approach to Decision and Control in Actuary Assets *Management* (*PhD dissertation*). SS Cyril and Methodius University, Skopje, MK.

Belman, R.E. and L.A. Zadeh (1970). Decision making in fuzzy environment. *Management Science*, 7, no. 4, pp. 141-164.

Box, G. E., G. M. Jenkins, G. C. Reinsel (1994). *Time Series Analysis, Forecasting and Control*. Prentice-Hall, Upper Saddle River, NJ.

De Cooman, G. and D. Aeyels (2000). A random set description of a possibility measure and its natural extension. *IEEE Trans. on Systems, Man and Cybernetics – Part A: Systems & Humans*, 30, no. 2, pp. 124-130.

Dimirovski, G. M. (1994), Mathematical Tools and Techniques of Fuzzy Systems and Fuzzy-logic for Decision and Control. *Technical Research Report SAAS-ULB/94-7*. Universite' Libre de Bruxelles, Bruxelles, BE.

Dubois, D. and H. Prade (1987). The mean value of a fuzzy number. *Fuzzy Sets and Systems*, 24, pp. 297-300.

Dubois, D. and H. Prade (1991). Fuzzy sets in approximate reasoning – Part I: Inference with possibility distributions. *Fuzzy Sets and Systems*, 40, pp. 143-202.

Heshmaty, B., and A. Kandel (1985). Fuzzy linear regression and its applications to forecasting in unecertain environment. *Fuzzy Sets and Systems*, 15, pp. 159-191.

Gancevski, P. (2000). *Theory and Practice of Insurance Business*. "Mikena", Bitola (in Macedonian).

Kaufman, A. and M. Gupta (1988). *Fuzzy Mathematical Models in Engineering and Management Science*. North-Holland, Amsterdam.

Klir, G. J. and Bo Yuan (1995), *Fuzzy Sets and Fuzzy Logic Theory and Applications*. Prentice Hall, Englewood Cliffs, NJ.

Ma, J., Z. Fan, and L.H. Huang (1999). A subjective and objective integrated approach to determining attribute weights. *European J. Operational Research*, 112, pp. 397-404.

Shackle, G.L.S. (1961). *Decision, Order and Time in Human Affairs*. Cambridge University Press, Cambridge, UK.

Steward, J. (1991). *Econometrics*. Philip Allan Pub. Co., New York.

M. Sugeno and T. Kang (1988). Structure identification of fuzzy system. *Fuzzy Sets and Systems*, 28, pp. 15-33.

Takagi, T. and M. Sugeno (1985). Fuzzy identification of systems and its application to modeling and control. *IEEE Trans. on Systems, Man and Cybernetics*, 15, no. 2, pp. 116-132.

Tanaka, H., S. Uejima, K. Asai (1982). Linear regression analysis with fuzzy model. *IEEE Trans. on Systems, Man and Cybernetics*, 12, no. 6, pp. 903-907.

Terano, T., K. Asai, and M. Sugeno (1994). *Applied Fuzzy Systems*. Academic Press, Cambridge, MA.

Terrence, M. C. (1993). *The Econometric Modelling of Financial Time Series*. Cambridge University Press, Cambridge, UK.

Zadeh, L. A. (1965). Fuzzy sets. *Information and Control*, 8, pp. 338-353.

Zadeh, L. A. (1973). Outline of a new approach to the analysis of complex systems and decision processes. *IEEE Trans. on Systems, Man and Cybernetics*, 3, no. 1, pp. 28-44.

Zadeh, L. A. (1975). The concept of linguistic variable and its application to approximate reasoning. Parts I, II & III. *Information Science*, 8, pp. 199-249, pp. 301-375; 9, pp. 47-80.

Zadeh, L. A. (1978). Fuzzy sets as a basis for a theory of possibility. *Fuzzy Sets and Systems*, 1, pp. 3-28.

Zadeh, L. A. (1980). Inference in fuzzy logic. *Proceedings of the IEEE*, 68, pp. 124-131.

Zadeh, L. A. (1997). Toward a theory of fuzzy information granulation and its centrality in human reasoning and fuzzy logic. *Fuzzy Sets and Systems*, 90, pp. 11-127.

Zadeh, L. A. (1998). Roles of Soft Computing and Fuzzy Logic in the Conception, Design and Deployment of Information/Intelligent Systems. In: O. Kaynak, L. A. Zadeh, B. Turksen, and I. J. Rudas (Eds.) *Computational Intelligence: Soft Computing and Fuzzy-Neuro Integration with Applications*, Pt. 1, Chapter 1, pp. 1-9, NATO ASI Series F Vol. 162. Springer-Verlag, Berlin.

Zimmermann, H. J. (1991). *Fuzzy Sets, Decision Making and Expert Systems*. Kluwer, Boston.

ELSEVIER
IFAC
PUBLICATIONS
www.elsevier.com/locate/ifac

COMBINATION OF IS/ICT AND CONSTRAINT MANAGEMENT IN OPTIMISATION OF DECISION MAKING PROCESSES

Josef Basl

Prague University of Economics, Prague, Czech Republic

Abstract: The paper deals with cost oriented decision process of the IS/ICT innovation. The innovation map is described like core element of the complex approach. The innovation dimensions and levels defined in the map are used for monitoring and decision making of the effective investment into IS/ICT. The whole approach is based on innovation and constraint management. The experience from the application is in the conclusion of the paper. *Copyright © 2004 IFAC*

Keywords: innovation management, process management, constraint management, information system, information technology, enterprise information system

1. INTRODUCTION

The current period brings to individuals, firms and national economics new opportunities they have not had before. The changes are different not only according to their quantity but to the unpredictability of their development and influence.

The situation is completely different and the traditional known rules of decision making are not very useful. These rules were invented and codified in the past based on the experience from the former world. It is quite naturally that this changed situation is asking for new rules including changes in behavior and attitude of people.

Many researches try to bring own solution and help. One good example is the identification of the need of the process oriented organization that has replaced the former function oriented organization since nineties. The main advantage of the process oriented organization is customer and aim orientation based on the potential of IS/ICT tools.

It is just area of IS/ICT (information systems and information and communication technology) that is connected with high cost paid within the last decade by companies. It is also the field of interest of the author.

The author has dealt with it for the last five years and has tried to combine the information and process management with the constraint management with aim to bring business benefit and reduce cost.

2. COST ORIENTATED INNOVATION OF IS/ICT

2.1 Key potential of is/ict

The reduction of cost and optimization of investment into the IS/ICT is very attractive for the research. The companies have invested enormous money into the implementation of the software systems like the ERP (Enterprise Resource Planning,). Now they pay for the software packages and connected services like the SCM (Supply Chain Management) or the CRM (Customer Relationship Management).

The special situation of IS/ICT has established in the Middle and East Europe. The economical and political changes forced the companies to change their information systems. The experience of the author is from this region and the article reflects its specific features.

The basic question "has such invested money been returned" was not actual at the beginning of 90's. At that time the first applications of ERP systems appeared in the Middle and East European countries. This question started being emphasized first after the year 2000. The reason for it was the big investment in companies connected with Y2K problem and then global economic situation at the beginning of this decade.

2.2 Known dificulties with the innovation of the IS/ICT

Generally since the beginning of their history the implementations of IS/ICT have been connected with the weak ability to keep implementations within the frame of planned budget and promised time. The complex efficiency was also not often reached.

For the better understanding of author's approach is suitable to summarize the reaction on the inefficiency of the IS/ICT discovered by researchers during 90's in America and the western Europe.

One group of researchers (Hammer, 1993; Ng Lee, 1999) identified the need for change and optimization of the business process. The other criticized the weakness of on the own software tools (Davenport, 1995). Third big effort was made in the direction of the creation of the metrics of delivered services (Molnar, 2000).

The necessity to change behavior of the IS/ICT users was identified in constraint management (Goldratt, 2001).

We can observe these several ways how to deals with problems of IS/IT but there is not only one complex solution. On the other hand the existence of these ideas has had influence on the approach described in this article especially processes and constraint management. The author decided to apply them on the key software category in a company – on the ERP systems.

2.3 Example of IS/ICT - the ERP systems

The ERP systems are good example for presentation of rapid development of IS/ICT and connection with cost and return of investment The history of the ERP began in 1990 and the current ERP systems area very comprehensive and covers not only the back-office but also functionality from front-office like CRM (Customer Relationship Management).

Interesting is that the majority of companies in many countries run their business on the ERP systems. This proportion is for example 98% of the „Top 100" in the Czech Republic.

The ERP is important IT service. From the research perspective the methodical integration of ERP is crucial. It is on the one hand integration within business processes and on the other it is integration with informatics processes.

The enterprise information systems (ERP) and their support of management and innovation of business processes represent the significant tool for improvement being more competitive

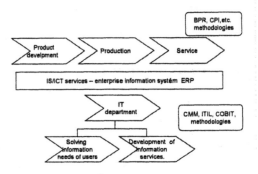

Fig. 1 - Key position of the ERP service

As we can see from the figure there are special methods for business process modeling and optimization. The example is BPR (Business Process Reengineering) and CPI (Continuous Process Improvement). There is also set of useful methods for IT process optimization like CMM (Capability Maturity Model), COBIT and the ITIL. On the other hand the similar description of the ERP like IS/ICT service is not sufficient but it would be useful for innovation of ERP.

It is important that the first wave of the ERP's in the Middle and East European countries from the beginning of 90's are near the end of their lifetime. The current situation is therefore much more complicated because of looking for effectiveness of implementation. More and more companies are looking for the cost optimization of the own implementation and later of the running of business processes and the system maintenance.

3. INNOVATION OF THE ERP SYSTEMS

3.1 Frame of approach

The innovation of the ERP needs new approaches. The author found the inspiration in the classical innovation process of the manufacturing technology. The author's research is oriented therefore on:

- methodical integration of ERP into the business processes support
- creation of model of ERP integration
- methodology for business and cost oriented ERP innovation

The whole complex author's approach is on the following figure. There are three main areas and approaches:

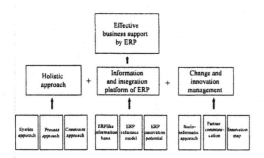

Fig. 2 – Frame of the approach

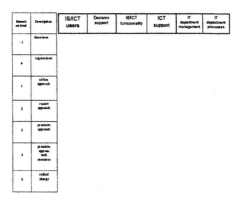

Fig. 3 – Level and dimensions in the innovation map

The approach comes out the description of company. This description is based on:
- system paradigm (input, output, system and subsystems),
- process paradigm (the system is set of processes with process owners, process goals, process effectiveness and efficiency).
- finally constraint paradigm where the crucial process (constraint) limiting the higher throughput is identified

The ERP system plays the key role in the described approach. It is mainly its ability for:
- integration and information platform within company
- key source of innovation
- innovation management base .

The last part is dedicated to management of change where communication, conflict solution and problem visualization are needed. The key part is the innovation map described in the following text.

3.2 Innovation map of the ERP

One of the output of author's research is the innovation map for IS/ICT. Its establishing was inspired by:
- the traditional approach to innovation known in late 50's in manufacturing, esp. in mechanical engineering. The innovation degrees are used are taken from this approach
- the traditional methods like capability maturity model and COBIT, e.g. approach known in IT management since late 80's, form the second part.

The both approaches, e.g. complex integration of ERP into business processes on the one hand and the innovation levels on the other are basis for the axes in the innovation table.

Two most important dimensions are taken from the three main levels:

Business process:
- user of IS/ICT (ERP)
- way of decision making

IS/ICT (ERP) tool:
- functionality of IS/ICT (ERP)
- IS/ICT support level

IT processes
- IS/ICT management)
- IS/ICT processes

All six dimensions are situated on the axis "y" and they can be changed in the real situation.

The second axis "x" represents the seven innovation levels:

The following figure is not for showing detail but all important relationships.

Innovation level	Description	IS/ICT user	Decision support	ERP functionality	ICT support	IT department management	IT department processes
-1	destruction	No training	None	Lost of functionality	Obsolence	None	Undescribed
0	regeneration	Skills keep at the same level	Reports	Functionality kept at the same level	Basic maintenance	Based on an internal needs	Only main business processes described
1	ad-hoc approach	Ad-hoc dispersion of knowledge	Ad-hoc SW support	Partial improvement of existing ERP	Renovation by own power	Ad-hoc according to user needs	Described due to other project
2	reactive approach	Basic dispersion of knowledge	Special query and reports	Adding of new functionality to existing ERP	Purchase of new components	Permanent according to user needs	Only main IT processes described
3	proactive approach	Planned dispersion of knowledge	Business intelligence	Upgrade of existing ERP	Purchase based on ICT plan	Based on plan	All IT processes described
4	proactive approac with measure	Benefit oriented dispersion of knowledge	BI with „cockpit" approach	Replace of existing ERP with a new one	Purchase based on ICT plan with business measures	Based on plan with measures	Measurable IT processes described
5	radical change	Radical staff retraining	Competitive intelligence	ERP based on a new concept	Outsorcing ASP	Based on methods (CMM, ITIL, Cobit, .)	Optimisation of IT processes

Fig. 4 – Full innovation map

3.3 Cost and benefit of optimisation in innovation of the ERP

The matrix in the innovation map covers 7 levels and 6 dimensions. They can help to describe the current level and the demand level in each dimension. The both sets of the values can be presented in the form of a common spider diagram:

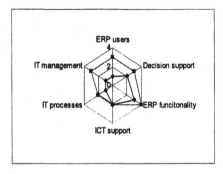

Fig. 5 –The innovation potential
of the investment

The diagram shows the innovation potential for improvement in different dimensions.

To get answer to the question in which direction is the investment most effective the application of the principles of Theory of Constraint (TOC) is useful. The TOC method can help with the optimisation of the decision. It means which improvement of dimension brings better results. It is when (using the traditional TOC terms):

- higher T (throughput), event. difference of T - ΔT is growing
- lower OE (operating expenses), event. difference of EO - ΔOE is decreasing

The main criteria function for good decision is (based on the throughput accounting):

$(\Delta T - \Delta OE)/ \Delta I$ – where ΔI is investment in the IS/ICT innovation.

Generally the whole methodology is applied via process decomposition of a company. The business processes are described first. The optimisation of them follows. The process owners have the key role for good decision. Their own motivation is important also. The adequate motivation of these persons that has to be in relation with the firm strategy, goals and politics are necessary conditions as well.

4. PRACTICAL APPLICATION OF THE APPROACH

The methodology has been applied till now in the three different types of organisation:

- manufacturing company
- utility company
- public sector organisation

The same set of dimensions was applied in all three cases as described in this article. The reason was to gain the platform for the cross comparison. The results showed the lowest level of actual situation in the public sector that is trying to apply management methods used in commercial firms. On the other hand there is the biggest potential for investment and improvement of current situation here.

The manufacturing company was on the symbolic second position. The level is higher here but the big potential is still here. This fact is very important as this sector creates the majority of GNP and export in the Czech Republic.

The analysis in the utility company showed the highest level of all dimension especially in used IS/ICT tools, application of processes and firm culture. These companies invested in the Czech Republic the highest sum into the innovation of IS/ICT during the last decade

5. CONCLUSION

The first experience from the application of the described approach shows very interesting outputs. Although all three cases are relatively different (in the sense of their core business, number of employees and customers) some common features of successful implementation was identified:

- the firms culture has to be on the high level
- the process of change is relatively slow and last long time
- the support of management is needed.
- the system of metrics is absolutely necessary during the project and at its final stage.

The current area for application of this approach is a group of small and medium enterprises.

REFERENCES

Basl, J., Skolud, B., Banaszak, Z. (2002): *Socioinfomatics approach to decision engineering based on the constraint management*, 9th ISPE conference, Cranfield, Great Britain, 2002, pp. 447-454

Hammer, M., Champy, J. (1993) : *Reengineering the Coorporation: Manifesto for Business Evolution*, Nicholas Brealsy, London, 1993

Danvenport, T.H., Bedra, M.C. (1995*): Managing Information about Processes*, Journal of Management Information Systems, 12, 1995, p. 57-80

Molnar, Z (2000).: *Efficiency of Information Systems*, Grada Publishing, Prague , 2000 (in Czech

Goldratt, E. (2001): *Neccesary but not Sufficient*, The Norton River Press, Great Barrington, 2001

Ng, J.K.C., Ip, W.H., Lee, T.C. (1999): *A paradigma for ERP and BPR integration*, International Journal of Production Research, 06/15/99, 9, 1999

ELSEVIER

IFAC
PUBLICATIONS
www.elsevier.com/locate/ifac

MODELLING AND SIMULATION IN MANAGING
OF PRODUCTION PROCESSES

Józef Matuszek, Dariusz Plinta

Department of Industrial Engineering, University of Bielsko-Biala
2 Willowa Street, 43-309 Bielsko-Biała, Poland
jmatuszek@ath.bielsko.pl, dplinta@ath.bielsko.pl

Abstract: Production of products and offering services are the target of all economic activities. However production is not realized in any method, but it is carefully planned with special regard to the effective utilization of resources. Modelling and simulation become substantial methods, which aid in production management. There are described examples of different simulation projects from production practice. *Copyright © 2004 IFAC*

Keywords: modelling, simulation, production process, processes management

1. INTRODUCTION

During production planning, it is not enough to determine how, what and when we should produce something but also to determine what, where and for whom we should produce it. Production planning is realized by starting with long-term task and short-term task planning and ending with planning of the realization of the operation tasks. Production planning can be understood as the creation of conditions for a fluent and effective run of production processes. We should at this stage take costs and realization time, which should be made optimal, into consideration (Bubenik and Plinta, 2003).

One of the main elements the effectiveness is the maximum utilization of the possessed production resources.

Development of production control aiding systems is a result of the targets, which were defined for the created systems. These targets were often quite different, but if they were the same they differed from each other by assigned priorities. Belonging to the most important targets of system aiding planning and managing of production are (Košturiak, *et al.*, 1996; Košturiak, *et al.*, 2000):

➢ satisfying customers needs,
➢ keeping to fixed time-terms,
➢ realization of technological requirements,
➢ maximum utilization of workplaces,
➢ minimization of realization time of production orders,
➢ decreasing of production costs,
➢ correct settlement of series size,
➢ minimalization of storage time,
➢ minimalization of defective products,
➢ minimalization of storage.

Activity of a company in free market economy conditions makes managers undertake more and more complex and complicated tasks. The consequence of this is the necessity of synchronisation of increasing quantities of the technological factors what brings looking for more effective methods of control of the production processes. Control of production planning is one of the most important tasks of a company. The target of these activities is to manufacture the products at the planned time. Furthermore, they have to fulfil the qualitative requirements and their manufacture costs should be as low as possible.

The introduction of computers into companies was accompanied with the development of software aiding the production planning and control. PPC (Production Planning and Control) systems for production planning at the operational level appeared just for this reason. Development of PPC systems is connected with new data processing technologies. At present, they are characterised by the possibility of integration with other computer systems and with module structure.

New tendencies within a companies organisational field, which also have an influence on computer systems, have a meaningful influence on the development of production planning and control. Among them, the most important are: Material Requirements Planning (MRP I), Manufacturing Resource Planning (MRP II), Enterprise Requirements Planning (ERP or MRP III), Just in Time (JIT), Kanban and Optimised Production Technology (OPT) (Bubenik and Plinta, 2003; Goldratt, 2001; Košturiak, *et al.*, 1996).

Furthermore, modelling and simulation becomes more and more substantial method, to aid a production management mainly in conditions of long-series production.

2. MODELLING AND SIMULATION

Computer technologies are the basic tool of accumulating and exchanging of information in contemporary enterprises. More and more commonly are used methods of artificial intelligence and expert systems there in management processes. Persons managing realized tasks in enterprise often ask themselves two questions: "Why?" and "What will be, when?", wanting to be the best, this means quickly, with lower costs and more effectively (Hromada, *et al.*, 1998).

The universal procedure used in methods of modelling and simulation, includes the following main stages (Gregor, *et al.*, 1998; Sadowski, 1998):

➤ problem's determination,
➤ model's building,
➤ preparation of input data,
➤ realization of experiments on the model through implementing of different input data and observation of the model's behaviour,
➤ verification and modification,
➤ analysis of simulation results,
➤ working out the activity programme.

The main criterion, at undertaking the decision about the utilisation of simulation in practice, are the advantages from its utilization – figure 1.

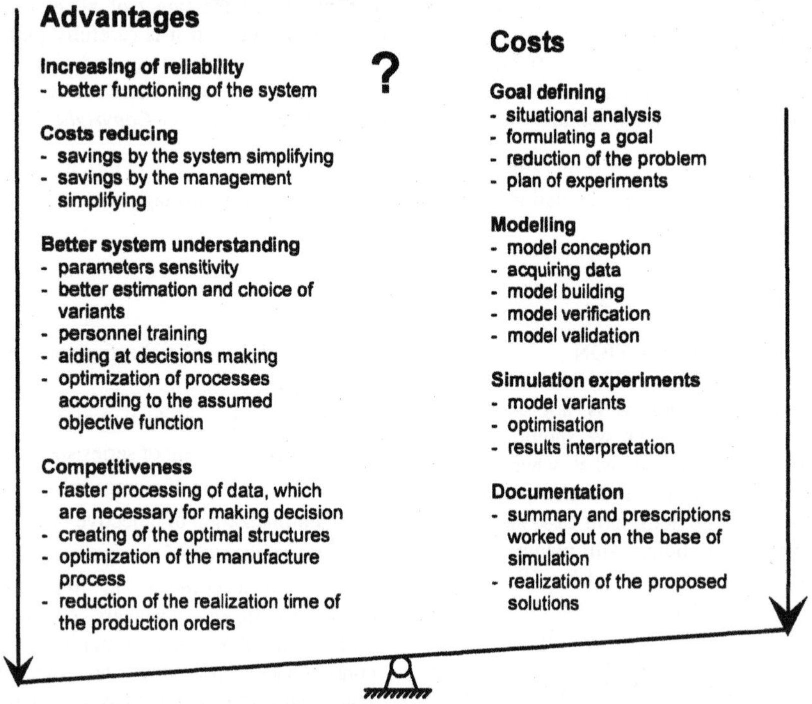

Advantages

Increasing of reliability
- better functioning of the system

Costs reducing
- savings by the system simplifying
- savings by the management simplifying

Better system understanding
- parameters sensitivity
- better estimation and choice of variants
- personnel training
- aiding at decisions making
- optimization of processes according to the assumed objective function

Competitiveness
- faster processing of data, which are necessary for making decision
- creating of the optimal structures
- optimization of the manufacture process
- reduction of the realization time of the production orders

Costs

Goal defining
- situational analysis
- formulating a goal
- reduction of the problem
- plan of experiments

Modelling
- model conception
- acquiring data
- model building
- model verification
- model validation

Simulation experiments
- model variants
- optimisation
- results interpretation

Documentation
- summary and prescriptions worked out on the base of simulation
- realization of the proposed solutions

Fig.1. Advantages contra costs in the modelling and simulation method (Košturiak, *et al.*, 2000).

At the beginning of any analysis it is necessary to define the input parameters of the analysed system and output parameters, which would be a result of the conducted analyses and on the basis of which an opinion about the analysed system can be established. Two cases can occur and it is difficult to find the optimum solution between them. Firstly, a simulation of system maintenance maybe conducted, in which the input parameters are simulated. Secondly, different collections of output data and the problem of finding the optimum are present.

During modelling and simulation there come into being many variations of proposed solutions. If it is not possible to examine all arrangements, study only those variants, which are chosen on the base of subjective opinion of researcher and his intuition and knowledge about the object of research should be chosen.

3. EXAMPLES OF MODELLING AND SIMULATION OF THE PRODUCTION PROCESS RUN

1ˢᵗ example – supply, manufacturing and sale

The analysis system of furniture production presents the following: the two-part office wardrobe with shifted doors and the desk with cabinet on circles.

The aim of the analyses was to test the present system functional ability and to test developmental possibilities.

In the first stage the production process was analysed. Different production programmes were compared. The duty of available resources was analysed and on the basis of test data a calculation of costs was accomplished. For different variants of possible changes, material and workplaces costs were compared. For example, in the next simulation models operations (cutting of panels, assembly) were chosen, which were realized in cooperation, through this shorter production process realization time was achieved.

It is possible to create the next variants of the analysed production system still comparing and estimating results from the conducted simulation.

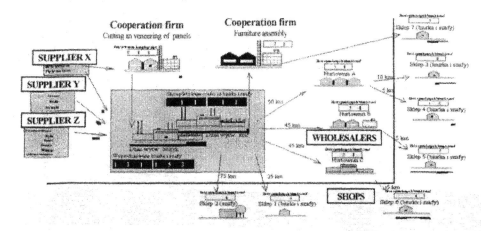

Fig.2. Simulation model – supply, manufacturing and sale.

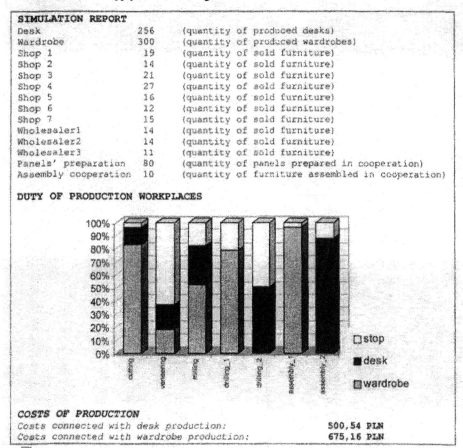

Fig.3. Part of the simulation report.

Results from the simulation made it possible to compare different variants of proposed changes in the analysed production system. They were compared from the point of view of time and cost of order realization. The possessed production resources make production increasing about 10% without necessity of additional financial expenditures possible. It is necessary to execute a part of the work in cooperation (cutting and veneering of panels and furniture assembly).

In figure 2 the variant of the simulation model with two cooperating firms is introduced: the firm which

cuts and veneers panels; and the firm which assemblies the compiled in our firm parts. This simulation model was worked out in ARENA packet. A fragment of the simulation report for this variant is presented in figure 3. Thanks to the simulation:
-   production abilities of the present system were checked,

-   influence of the proposed changes on: production ability of the company; cost of materials; manufacturing; transportation; and cooperation was estimated,
-   different directions of the company development in area of supply and sale were compared.

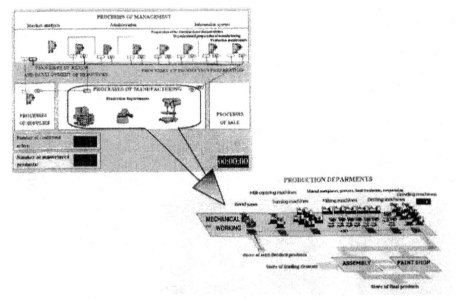

Fig. 4. Simulation model of the production process run.

2nd example – production process run

The second example is connected with the production process run, which is realized in the company where the toothed gears are manufactured (Plinta, 2001) – figure 4.

The present organizational structure did not fulfil the market's requirements. The production cycle was too long. Changes to improve the production processes realization were necessary. Proposals of the possible organizational changes were analysed by the modelling and simulation method. The simulation of realization of the same production order in different organizational conditions makes the comparison of the proposed solutions possible.

The simulations carried out allowed the possibilities for realization of the proposed production plan to check and to trace the utilization of the company's production resources more exactly. The simulation results allowed for the efficient finding of bottlenecks and indication of the unused elements in the production system.

There were also defined costs of activities, which made possible the analysis of other production orders. The way of calculation of the costs is given in figure 5.

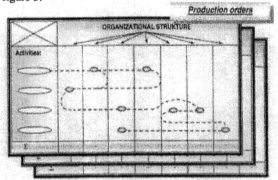

Fig. 6. Factory sheet of activities connected with production orders.

The activity based costing as opposed to the traditional calculation, demands the separation of single activities within the limits of each function of the company – figure 6, and determination the size of measurement of activity for each action.

Costs of researches and development	K_{br}
Costs of marketing activities	K_{mr}
Costs of designing	K_{pr}
Supply costs	K_z
Manufacturing costs	K_w
Costs of production protection	K_{op}
Selling costs	K_s
Service costs	K_{us}
Exploitation costs	K_e
Costs of product	K

Fig. 5. Calculation algorithm according to the costs of activities.

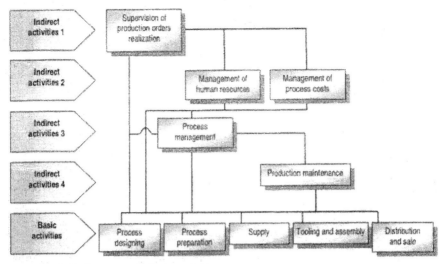

Fig.7. Settlement of activities' costs on the basic activities.

Having identified all activities they were grouped into indirect and basic activities. Next, the costs of indirect activities will be accounted into the basic activities according to figure 7.

Creating a simulation model of the analysed production system, we define the following stages of the production process. These stages most often match the activities, which were defined in the activity based reckoning. Regarding the costs in the modelling and simulation, enlarges effectiveness of the researches carried. Especially, creation of integrated model with the company computer system of the costs calculation, it brings the best effects.

3rd example – production subsystem

The third example is connected with the shaft-treating department (Plinta, 2001). This model embraces activities realized in particular workplaces in the manufacturing cell – figure 8.

Because of large quantity of different types of shafts (~70 types per month), before building the simulation model, they were divided into groups according to the criterion of constructional similarity. Additionally, for each group there was defined a representative, for whom the simulation was realised.

From the simulation of monthly production one obtains information about duty of workplaces, about manufacturing costs of various types of the shafts (Figure 9) and about the order realization time.

The modelling and simulation method with the analysis of costs enlarges the effectiveness of researches – designing or analysis of the system functioning. It gives us the possibility to design the system, which is close to an optimum in respect of costs and the time of tasks' realization.

Manufacture processes, machines

Op.	Name of operation	Workplace (quantity of machines)
1	Cutting on length	Band-saw (2)
2	Facing on length and centring	Mill-centring machine (1)
3	Turning	Turning machine DFS 450 NC (2)
4	Grinding	Centre-type grinder (4)
5	Slot milling	Sequence milling machine (3)
6	Grease conservation	Manual workplace

Simulation model

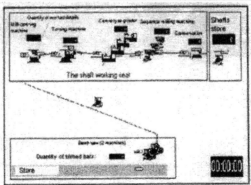

Fig. 8. Simulation model of the shaft-treating department.

Utilization of workplaces (simulation time: 168 hours)		
Workplace (quantity of machines)	Work time [hours]	Utilization [%]
Band-saw (2)	63.84	38
Mill-centring machine (1)	97.44	58
Turning machine (2)	164.64	98
Centre-type grinder (4)	90.64	54
Sequence milling machine (3)	57.12	34
Conservation (1)	6.77	4

Statement of expenses / piece						
Product (shaft)	Material costs [PLN]	Labour costs [PLN]	Storing costs [PLN]	Workplaces costs [PLN]	Indirect costs [PLN]	Sum of manufacture costs [PLN]
A	16,84	2,25	2,16	12,95	17,10	51,30
B	20,15	3,31	2,48	14,85	20,39	61,18
C	35,88	3,96	2,91	17,45	30,10	90,30
D	49,13	4,56	4,78	28,70	43,59	130,76
E	60,44	4,88	5,21	31,25	50,89	152,67
F	67,62	6,65	6,03	36,20	58,25	174,76
G	74,38	6,07	5,74	34,45	60,32	180,96
H	95,83	6,48	5,95	35,70	71,98	215,94
I	107,58	15,30	12,41	74,44	104,86	314,59
J	159,65	10,78	9,23	55,40	117,53	352,60

Fig.9. Example of simulation results.

The exact knowledge about the value of workplaces costs, costs of activities and minimalization of overheads are the basis for determination of the total production costs.

This production system can be still improved, conducting further simulations for various variants of the system (new machines, different production orders, various quantity of worked elements, different size of buffers, foreseen disturbances and possible breaks).

4. SUMMARY

The modelling and simulation is a universal technique, what inflicts, that it finds more and more wider usage in enterprises. It is more and more important aiding technique for designing of new production systems, and also aiding production management. It makes possible tracing of the production functioning and detection of bottlenecks points. Tracing of effects of proposed changes makes possible also. It makes selection of the best variant easier. It can take into account not only manufacturing process, but also supply and sale processes – the 1st example. For building of such complex model, it can approach in two ways: by creating of generalized model of the whole system, and then extend particular subsystems. On other side there is possible first to create models of subsystems, which will become joint later into one.

Described examples of projects from production practice show advantages of the modelling and simulation method in the dissolving of different problems connected with the managing of production processes. New production processes, realization of production orders, operating procedures, decision rules, organizational structures, transportation systems, information flow, etc., can be explored without disrupting ongoing operations.

This technique is very useful especially as extension of the traditional application in following areas:
- manufacturing analysis and improvement tool,
- investment decision support,
- tool of communication,
- explanation and educational tool.

REFERENCES

Bubenik, P. and Plinta D. (2003). *Solution of production planning and scheduling in modern manufacturing environment*. Scientific bulletin of The University of Bielsko-Biała, Bielsko-Biała, pp. 33-36.

Goldratt, E. (2001). *Neccesary but not Sufficient*. The Norton River Press, Great Barrington.

Gregor, M., Haluškova, M., Hromada, J., Košturiak, J. and Matuszek, J. (1998). *Simulation of manufacturing system*. Technical University in Lodz, Branch in Bielsko-Biała, Bielsko-Biała.

Hromada, J., Krajčovič, M., Gregor, M. (1998). *Computer simulation in inventory control area*. International conference – Production Engineering, Bielsko-Biała, pp. 65-68.

Košturiak, J., Gregor, M., Mičieta, B., Matuszek, J. (2000) *Projektovanie výrobných systémov pre 21 storočie*. University in Žilina, Žilina.

Košturiak, J., Gregor, M., Slamková, E., Chromjaková, F., Matuszek, J. (1996): *Method and Tools of the Enterprise Logistics*. Politechnika Łódzka Filia w Bielsku-Białej, Bielsko-Biała.

Plinta, D. (2001). *Modelling and simulation of the production processes in the conditions of the group working of machine elements*. Doctor's thesis, University of Žilina, Žilina.

Sadowski, R. P. (1998). *Simulation with ARENA*. McGraw-Hill, New York.

ELSEVIER

IFAC
PUBLICATIONS
www.elsevier.com/locate/ifac

OPTIMISATION METHODS IN THE MODELLING AND SIMULATION OF PRODUCTION SYSTEMS

Dariusz Plinta, Sławomir Kukla

Department of Industrial Engineering, University of Bielsko-Biala
2 Willowa Street, 43-309 Bielsko-Biała, Poland
dplinta@ath.bielsko.pl, skukla@ath.bielsko.pl

Abstract: In production run management the most proper solution is connected with larger numbers of criterions. This simulation project does not create a possibility for finding the optimum solution, but it is a tool to aid designers work and that enables quick testing on computer systems of different variants in production process runs. Within this simulation project is an example of the application of optimisation (the Yager's method) in the simulation project. *Copyright © 2004 IFAC*

Keywords: modelling, simulation, production system, production flow prototyping

1. ANALYSIS OF THE PRODUCTION PROCESS RUN

Efficient analysis of the production system requires fulfilment of two conditions:
- collecting and disposing of information about the production process,
- disposal of theory and technique assuring an optimum state.

The proper planning of particular stages of experimental research is necessary to find possible variants of solutions, and first of all proper settlement of estimation criteria.

If when during modelling and simulation there come into being many variations of proposed solutions, research of all possible combinations and all possible arrangements of the value of studied factors is very time-consuming. If it is not possible to examine all arrangements, study only those variants, which are chosen on the base of subjective opinion of researcher and his intuition and knowledge about the object of research should be chosen (Basl, *et al.*, 2002; Cleland and Ireland, 2002; McHugh, *et al.*, 1995).

Modelling of production systems is mainly based on a statical data base, that is registered information in the data processing system of a company. This information for example, norms of work time, material norms, information about possessed resources, costs of machine-hour, are seldom updated. These changes, which occur regularly and

which will occur more often in the future not often are taken into account. By the use of the modelling and simulation packet we can analyse the effects of changes which are to occur (Kelton, *et al.*, 1998; Montgomery, 1997). The procedure used in methods of modelling and simulation is presented in figure 1.

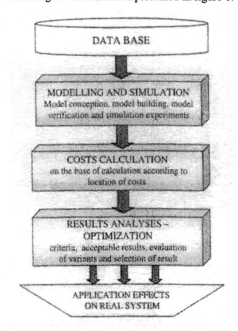

Fig.1. Management of production aided through modelling and simulation technique.

267

The estimation and section of the rational process run it is possible to execute by application of multi-criteria estimation tools using the subjective point criterions or criterions with fuzzy character (Yager and Filev, 1994).

2. ESTIMATION AND THE CHOICE OF THE VARIANT OF MANUFACTURING PROCESS RUN FOR REALIZATION

In the below example for estimation manufacturing process run structure, the proposed point-by-point method of assessment according to Yager is used (Płonka, 1998; Yager and Filev, 1994). Input data of this method are:
– number of criterions m,
– number of variants of manufacturing process run n,
– elements of matrix of individual criterions validity

$$B = \left[b_{ij} \right],$$

– elements of matrix $C = \left[c_{ij}(e) \right]$, which are the point estimate if the i-variant according to j-criterion, passed by p-expert.

Each expert is liable for the construction of the validity estimation matrix of criterions with the Saaty's method in comparing of pairs of received criterions (Saaty, 1980).

In the next step there is one summary matrix of validity criterions. For the summary matrix there was a search eigenvector Y, fulfilling the following matrix equation:

$$B\,Y = \lambda_{max}\,Y \qquad (1)$$

where:
  B - the summary validity matrix of criterions,
  Y - eigenvector,
  λ_{max} - the maximum eigenvalue of matrix B.

Eigenvector has many coordinates as there are accepted criterions, and these coordinates have to fulfil the following condition:

$$\sum_{j=1}^{m} y_j = m \qquad (2)$$

where:
  y_j - the j-coordinate of eigenvector Y.

Coordinates of the eigenvector, called weights, express the validity of the answering criterions (with higher value of weight, there is a higher value of criterion). Applying the involution method it determines the eigenvalue and the eigenvector. This task used the numeric procedure written in Pascal language.

The total estimations, which are normalized by averaging the estimation opinion given by particular experts.

$$c_{ij} = \frac{1}{p} \sum_{e=1}^{p} c_{ij}(e) \qquad (3)$$

where: p – number of experts.

Increased conduct depends on creation of the normalised decisions by involution of each component of the next normalised estimations to the even power suitable weight.

$$d_j = \sum_{j=1}^{m} c_{ij}{}^{y_j} / w_i \qquad (4)$$

After writing out of the formula is as follows:

$$d_1 = c_{11}^{y_1} / w_1 + c_{21}^{y_1} / w_2 + ... + c_{n1}^{y_1} / w_n$$
$$d_2 = c_{12}^{y_2} / w_1 + c_{22}^{y_2} / w_2 + ... + c_{n2}^{y_2} / w_n \qquad (5)$$
$$...$$
$$d_m = c_{1m}^{y_m} / w_1 + c_{2m}^{y_m} / w_2 + ... + c_{nm}^{y_m} / w_n$$

The result is the optimum decision, of which there is the rational run of manufacturing process which at best fulfils all received estimation criterions.

$$D = D_1 + D_2 + ... + D_n \qquad (6)$$

In the assumed method the decision of type of minimum is optimum. As the i-component of optimum decision, answering to the i-variant, it takes the smallest i-component from particular decisions $d_1, d_2, ..., d_m$.

$$D_i = \min_j c_{ij}^{y_j} \qquad (7)$$

The best variant of the production process run, under the assumed criterions, is the variant with the largest component in optimum decision.

$$D_{rac} = \max_i D_i \qquad (8)$$

The use of the point-by-point method of assessment according to Yager permits a simple and effective way to choose the rational manufacturing process run. With considering validity estimation of individual criterions and taking into account their weights in farther conduct, it is possible to estimate the particular variants of the process run and to rank it in order from the best to the worst. These workings will be helpful in planning the simulation experiments and the continuous improvement of production systems.

3. AN EXAMPLE OF MODELLING AND SIMULATION OF THE PRODUCTION PROCESS RUN

The analysis system of furniture production presents the following: the two-part office wardrobe with shifted doors and the desk with a cabinet on castors.
The necessary materials for producing the afore-mentioned articles were divided into three groups: elements of wardrobes, desks and parts common to desk and wardrobe. These materials are ordered together without defining there destination.

At the end of the manufacturing stage we receive packaged furniture, which is sent directly to shops and wholesalers.

The aim of the analyses was to test the present system functional ability and to test developmental possibilities of.

There were made the following assumptions:

- times of realization of production operations were determined,
- the planned size of production was determined,
- the demand of particular materials was defined,
- distances between suppliers and the firm were specified as well as the distance between firm and recipients (wholesalers and shops),
- transportation, times of transport and the unit cost (per km),
- potential firms were situated, which will cooperate in preparation of materials and assembly,
- parameters of simulation were defined (time of cycle; e.g. week or month and quantity of cycles).

The current production process was analysed in the first stage. The availability of resources analyse was done and on the basis of test data the calculation of costs was accomplished.

For example in the following simulation models chosen operations (cutting of panels, assembly) were met through cooperation and we achieved shorter lead time. The next stage analysed supply and sale processes taking earlier proposed changes into account. Table 1 includes the description of variants of simulation models.

Table 1 Variants of simulation models

Variant	Descriptions
1	The whole process is realized in the firm - all materials are compiled and processed in the company. Assembled wardrobes and desks are sent to wholesaler and shops.
2	Furniture panels and fibreboards will be bought after cutting them into suitable dimensions, directly from the supplier.
3	An agreement was made with the company which veneers edges of furniture' panels.
4	Assembly workshops were liquidated and prepared elements are assembled directly by the customers.
5	An agreement was made with the company that assembles wardrobes.
6	The production process was improved – an increase in production efficiency is at about 10% including the possibility for net distribution and finished pieces of furniture.

It is possible to create variants of the analysed production system by comparing and estimating the results from the simulation (table 2).

It was possible to compare the results form different variants of the proposed change with the analysed production system by comparison of three criterions: utilization degree of possessed resources, production costs and quantity of produced furniture. There was used the Yager's method.

Table 2 Results from simulation

Criterion of estimation	Variants:	1	2	3	4	5	6
Average production of workplaces connected with manufacturing [%]	Desks	23,1	22,0	21,9	24,4	25,4	28,8
	Wardrobes	39,0	38,5	38,5	43,5	44,1	47,0
Production costs [PLN]	Desks	543,41	524,00	525,50	521,05	510,20	500,54
	Wardrobes	701,10	695,00	692,30	685,25	680,34	675,16
Quantity of produced furniture [pieces]	Desks	225	235	237	235	230	256
	Wardrobes	260	275	280	284	280	300

The estimation of criterions and variants of production processes run of desks and wardrobes was entrusted to three experts: worker of the manufacturing section, worker of the production planning section and the cost-analysis section (figure 2).

The best results were achieved in variant 6 – tables 3 and 4. Owned production resources make production enlargement about 10% without necessity of additional financial expenditures possible. However it is necessary to realize part of work in cooperation (the cutting and veneering of panels and assembly of furniture pieces).

Expert 1

	k1	k2	k3
k1	1	1	2
k2	1	1	2
k3	0,5	0,5	1

Expert 2

	k1	k2	k3
k1	1	0,5	0,25
k2	2	1	0,5
k3	4	2	1

Expert 3

	k1	k2	k3
k1	1	4	0,333
k2	0,25	1	3
k3	3	0,3333	1

The summary matrix of the criterions validity

	k1	k2	k3
k1	1	1,8333	0,8611
k2	0,5455	1	1,8333
k3	1,1613	0,5455	1

Fig.2. The defined by particular experts estimation of criterions validity.

Table 3 The multi-criterion estimation of variants in case of production of desks

The co-ordinates of eigenvector:	$Y = \begin{bmatrix} 1,1554 \\ 0,9924 \\ 0,8523 \end{bmatrix}$
The function of optimum decision:	$D = 0,0917 / w_1 + 0,0469 / w_2 + 0,0469 / w_3 + 0,1173 / w_4 + 0,1530 / w_5 + 0,2522 / w_6$
Preferred solution:	Variant 6 with the largest value in optimum decision: **0,2522**

Table 4 The multi-criterion estimation of variants in case of production of wardrobes

Co-ordinates of the eigenvector:	$Y = \begin{bmatrix} 1,1554 \\ 0,9924 \\ 0,8523 \end{bmatrix}$
Function of the optimum decision:	$D = 0,0929 / w_1 + 0,0731 / w_2 + 0,0729 / w_3 + 0,1463 / w_4 + 0,1566 / w_5 + 0,2129 / w_6$
Preferred solution:	Variant 6 with the largest value in optimum decision: **0,2129**

4. CONCLUSIONS

Thanks to the simulation:
- production abilities of the present system were checked,
- influence of the proposed changes on production ability of the company; costs of materials; manufacturing; transportation and cooperation was estimated,
- different directions of the company development in area of supply and sale were compared,
- variants from point of view of three criteria – the duty of production workplaces, production costs and quantity of produced furniture was estimated.

Labour-consuming is a disadvantage of the modelling of such complex production systems. It is easy to make different changes later in the simulation model - to create different variants of the analysed production system. Checking different variants and section can bring significant advantages to a firm.

The example introduced and advantages of its use mentioned above its testify to the great usefulness of the modelling and simulation method in the dissolving of problems connected with the managing of production processes.

REFERENCES

Basl, J., Skolud, B. and Banaszak, Z. (2002). *Socioinfomatics approach to decision engineering based on the constraint management*, 9th ISPE conference, Cranfield, Great Britain, pp. 447-454.

Cleland, D. and Ireland, L. (2002). *Project management – strategies and implementation*. McGraw-Hill, New York.

Kelton, W., Sadowski, R. and Sadowski, D. (1998). *Simulation with Arena*. WCB/McGraw-Hill, Sewickley.

McHugh, P., et al. (1995). *Beyond Business Process Reengineering – Towards the Holonic Enterprise*. Willey, New York.

Montgomery, D. (1997). *Design and Analysis of experiments*. Wiley, New York.

Płonka S. (1998). *Metody oceny i wyboru optymalnej struktury procesu technologicznego*. Wydawnictwo Politechniki Łódzkiej Filii w Bielsku-Białej, Bielsko-Biała.

Saaty, T. L. (1980). *The Analytic hierarchy processes*. McGraw-Hill, New York.

Yager, R. and Filev, D. (1994). *Essentials of fuzzy modelling and control*. John Wiley & Sons, New York.

ELSEVIER

IFAC

PUBLICATIONS
www.elsevier.com/locate/ifac

DIMENSIONALITY REDUCTION OF PERTINENT FEATURES FOR EXPLORATORY ANALYSIS IN DECISION SUPPORT SYSTEMS

Marek B. Zaremba

Département d'informatique et d'ingénierie
Université du Québec en Outaouais
101, St-Jean-Bosco, Gatineau, Quebec, Canada
E-mail: zaremba@uqo.ca

Abstract: Efficient processing and interpretation of high-dimensional data sets, prevalent in a large number of problems related to engineering and management, has become an essential requirement that has to be addressed in the design and analysis of modern manufacturing systems. After a review of dimensionality reduction methods, this paper proposes the integration of reduced-dimensionality representations with sparse data filtering algorithms. The spatial filtering procedures implement an attention focusing mechanism that guides the user in locating the objects or data set segments most relevant to the user task. A non-linear dimensionality reduction method, capable of dealing with highly non-linear data distribution patterns, is investigated in more detail. Examples in visual inspection and robot navigation control are provided. *Copyright © 2004 IFAC.*

Keywords: Decision support systems, data mining, dimensionality reduction, process monitoring and diagnosis, image processing.

1. INTRODUCTION

The problem addressed in this paper relates to the issue of finding efficient solutions for problems involving the use of high dimensionality of data sets. The applications for those techniques can be found in such areas as diagnostics, condition monitoring, and maintenance of manufacturing processes and plants. In many problems related to process monitoring or decision support the information comes in the form of multidimensional signals and – in order to be efficiently process and interpreted - require some form of dimensionality reduction, since both the computational complexity of the problem and the result analysis and interpretation increase with the increase of its dimensionality. The goal of the dimensionality reduction procedure is to compress the signals in size and, even more important, to discover compact representations that would provide better insight into their variability, enhance their visualization, and optimize clustering and classification tasks. It should be noted that the optimality criteria for

different tasks can be distinct [Yin, 2002]. For example, dimensionality reduction for visualization should focus on preserving local inter-object structure in the data, whereas the classification tasks, in order to minimize the classification error, would call for better knowledge of underlying distributions.

The problem of extracting probability density functions and confidence intervals from industrial process data (steel mill NO_x emission, fluidised bed reactor, audiometric data) was discussed in [Chen et al., 2004]. Obtaining confidence bounds was investigated in [Martin and Morris, 1996]. One of sources of multidimensional data are remote sensors, such as satellite or airborne hyperspectral sensors. Results of environmental monitoring activities, using reduced-dimensionality hyperspectral imagery from a CASI (Compact Airborne Spectrographic Imager) was reported in (Lévesque et al., 2000). A review of issues related to analysis of hyperspectral data is given in (Landgrebe, 2002).

The important problem of dimensionality reduction has attracted attention of several groups of researchers, and resulted in a host of methods using approaches and techniques from different areas, such as artificial intelligence, optimal control, geometrical analysis, etc. Multivariate statistics methods have been increasingly employed in the practice of automation and manufacturing systems. A well-known and often used dimensionality reduction method is Principal Component Analysis [Jolliffe, 2002]. The method provides the optimum representation in a lower dimensional space, according to the criterion of the mean square reconstruction error. However, this method, along with other linear methods, cannot perform unfolding of a data manifold. Since the form of the manifold of the process data is rarely known, the presumption of the form amenable to Principal Component Analysis can be misleading.

This paper presents a nonlinear eigenvector method that preserves local neighbourhood within the underlying structure of the manifold. The application of the method is discussed in the context of detecting patterns of interest relevant to the user. The results of the dimensionality reduction procedure serve to define the input space for relevance analysis based on the principle of attention focusing [Malik and Perona, 1990; Stough and Brodlay, 2001]. The visual attention mechanism based on the relevance function provides several advantageous features in detecting objects. Due to the multi-scale model-based approach, it provides a mechanism for a quick localization of the objects of interest invariantly to object size and orientation. It also provides optimal segmentation in the sense of the maximum likelihood criterion. Another interesting feature of this approach is its capability to handle sparse data sets.

After presenting an eigenvector, nonlinear, local neighbourhood preserving dimensionality reduction technique, the paper further investigates the way to integrate the resulting low dimensionality representations with an attention focusing mechanism that guides the user in their selection of most pertinent portions of data. Examples of the analysis in diagnostics and mobile robot navigation are given.

2. DIMENSIONALITY REDUCTION METHODS

Dimensionality reduction can be defined as the process of mapping high dimensional objects to a lower dimensional representation. Formally, dimensionality reduction produces $\bar{x} = \varphi(x)$,

where $x \in \Re^n$, $\bar{x} = \Re^m$ and m < n. Mapping φ may be linear or nonlinear, and may be obtained using supervised or unsupervised learning procedures.

Apart from linear algorithms, such as Principal Component Analysis (PCA), there exist several algorithms for nonlinear dimensionality reduction. The main ones are:
- Multidimensional scaling (MDS),
- Sammon mapping,
- Topographic maps,
- Self-organizing maps (SOM),
- Generative topographic mapping (GTM),
- Principal curves and surfaces,
- Neural network methods,
- Isomap.
- etc.

In order to further use the resulting projection to a low-dimensional space as an input to the attention-focusing spatial filtering procedure discussed later in the paper, it is necessary that the dimensionality reduction algorithm maps high-dimensional data of unknown nonlinear distribution into a **single** global coordinate system of lower dimensionality. Therefore, methods such as MDS or Principal Component Analysis have to be excluded. In this paper, an eigenvector, nonlinear, local neighbourhood preserving dimensionality reduction technique is presented. The algorithm is based on the Local Linear Embedding approach [Roweis and Saul, 2000].

The algorithm consists in computing low dimensional vectors Y_i to which high dimensional data points X_i are mapped. After computing the neighbours of each data point, the weights W_{ij} are computed in such a way that they best reconstruct each data point from its neighbours. The weights W_{ij} denote the contribution of the jth data point to the ith reconstruction.

The reconstruction is based on minimizing the error cost function

$$\varepsilon(w) = \Sigma \left| X_i - \sum_j W_{ij} X_j \right|^2 \qquad (1)$$

and is subject to two constraints. First, each data point is reconstructed only from its neighbours, enforcing $W_{ij} = 0$ if X_j does not belong to the neighbour data set. Second, the rows of the weight matrix sum to one, i.e., $\Sigma_j W_{ij} = 1$. This constraint ensures that the geometric properties of the neighbourhood are invariant to rotations of the frame of reference.

The embedding cost function

$$\Phi(Y) = \Sigma \left| Y_i - \sum_j W_{ij} Y_j \right|^2 \qquad (2)$$

is based on locally linear reconstruction errors. In this algorithm, this is obtained by optimizing the coordinates Y_i while keeping the weights W_{ij} constant. In order to make the problem well posed, the optimization is performed subject to the following constraints: the low dimensional coordinates are centered on the origin, and the embedding vectors Y have unit covariance. The low-dimensional embedding space is defined by the bottom eigenvectors using the Rayleitz-Ritz theorem [Horn and Johnson, 1990].

Figure 1 depicts an output of the algorithm in the form of a two-dimensional distribution of a set of data representing textured objects. The input to the dimensionality reduction procedure consists of 49 data items of dimensionality $D = 1024$ (four-band images of the resolution 16×16). A clearly defined ordered nonlinear distribution is obtained.

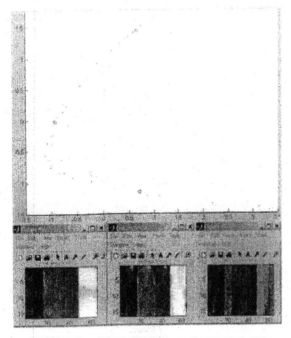

Fig. 1. Example of mapping of a set of 1024-dimensional data into a 2-D space.

The bottom portion of Fig. 1 shows an example of three data points selected from different regions of the distribution plot: the top (1), the left-hand side (2), and the bottom (3). The four vertical strips in the three example images correspond to the four spectral bands in the original image.

3. MULTI-MODALITY DATA RELEVANCE ANALYSIS

A number of engineering analysis and decision support problems can be defined in terms of the localization and detection of objects, patterns or data clusters of specific interest for the user. The solution to this type of problems is approached in this paper from the perspective of the theory of visual attention. Visual attention operators proved to be a time-effective solution to object detection problems, since they allow focusing data analysis only on a few particular regions of interest containing potential objects of interest. The object detection method discussed in this paper is based on an extended version of the attention operator called Image Relevance Function (IRF) proposed initially in (Palenichka and Zinterhof, 1996). The relevance function computation is based on a multi-

scale change detection simultaneously in space and time and center-point location (focus-of-interest) of those changes. The object of interest could be different real-world objects or their constituent parts or just regions (in space and time) possessing a property relevant to a data analysis task.

The relevance function is calculated by a spatial attention focusing operator capable to locate, identify (recognize), and concisely describe objects of interest present in a given data set (image). It involves four main object tokens (object constraints) as indicators for the presence of objects: object spatial contrast, object homogeneity, change in time, and object scale (size). An example of the relevance function operator with two object constraints, object spatial contrast and local homogeneity of object intensity, can be written as

$$R(i, j, S_k) = \left| c(i, j, S_k) \right| - \alpha \cdot \left| d(i, j, S_k) \right| \quad (3)$$

where $c(i,j,S_k)$ is the local object-to-background contrast estimated at kth scale, $d(i,j,S_k)$ is the local homogeneity measure of the object region estimated at kth scale, α is a constant coefficient that controls the contribution of the second (constraint) term. These constraints are calculated for maximal and minimal object sizes defined as scale S_k.

A specific form of (3) that function applied to the vector of the input data $g(i,j)$ can be defined as in (4), where $|O|$ and $|B|$ denote the number of points in object and background sub-regions $B(i,j)$ and $O(i,j)$,

$$R\{g(i, j)\}$$
$$= \left(\frac{1}{|O|} \sum_{(m,n) \in O(i,j)} g(m,n) - \frac{1}{|B|} \sum_{(m,n) \in B(i,j)} g(m,n) \right)^2$$
$$- \alpha \cdot \left(a - \frac{1}{|O|} \sum_{(m,n) \in O(i,j)} g(m,n) \right)^2$$
$$(4)$$

Vector $g(i,j)$ is often called *property map*, and can be computed according to a model of the object O. As an example, the following form of the property map was introduced in (Zaremba and Palenichka, 2002):

$$g_{i,j}(b_1, b_2, ..., b_T)$$
$$= \lambda \cdot y(i, j) + \sum_l \tau_l(b_1, b_2, ..., b_T) \cdot \varphi_l(i, j) \quad (5)$$

The map generates a piecewise constant model of objects to be segmented in the initial multi-dimensional image by defining a function $g_{i,j}(b_1, b_2, ..., b_T)$, which is an implicit function of components of a multidimensional vector $[b_1, b_2, ..., b_T]$ at point (i,j). Vector $\tau_l(b_1, b_2, ..., b_T)$ is a vector of constant values of image segments corresponding to objects belonging to class l, the indicator function $\varphi_l(i,j)$ is a binary map of objects of the lth class. The indicator function $\varphi_l(i,j)$ is equal to zero in the whole image plane except for the points belonging to objects of interest in the lth class. Term λ is the perturbation level, i.e., the standard deviation of the white noise term $\lambda \cdot y(i,j)$. This term can be interpreted as intensity trend (e.g., a significant slope in intensity) or random noise in the general case.

The positions of local maximum values of the multi-scale IRF coincide with location points of the objects of interest or their constituent parts in the data set domain A:

$$(i_f, j_f)_l = \max_{(i,j) \in A} \max_k \left\{ R[(g(i,j), S_k], (i,j) \notin \Gamma_l \right\} \quad (6)$$

Calculation of the relevance function and its local maxima can be performed using fast iterative procedures and multi-scale matched filters.

4. INTEGRATION WITH RELEVANCE ANALYSIS

Integration of the multi-modality data relevance analysis with the dimensionality reduction pre-processing phase can be obtained in two stages:

1) Transformation of vectors $[b_1, b_2, ..., b_T]$ to a lower dimensionality representation $[d_1, d_2, ..., d_L]$, where $L < T$.
2) Calculation of the property map $g(i,j)$ from the reduced dimensionality data set.

The first stage can be done using one of the methods discussed in Section 2. Analysis of the form of the distribution of the original data set is an important factor. If the distribution is highly nonlinear, especially when the data set is "warped", the Local Linear Embedding method offers significant advantages.

Calculation of the property map using (5) may not always be possible because of the unavailability of binary map $\varphi_l(i,j)$ in $\Re^L$. Therefore, we propose a weighted norm that works on a set of N example data $\{\mathbf{d}_c\}$. Given a set of examples relating to the relevant object class, the property map $g(i,j)$ is obtained as

$$\sum_{c=1}^{N} w_c d_c(i,j) N(d_c, d) \quad (7)$$

where $N(d_c, d)$ is a topological neighborhood relation between an example vector and the current data d, which may be defined in several ways. In a typical choice for the neighborhood function

$$N(d_c, d) = \exp(-|d - d_c|^2 / 2\sigma^2) \quad (8)$$

it falls off with distance r between the data units. Thus nearby units receive similar updates, enabling the topological information to be fed into the property map. The optional weights w_c permit to assign higher importance to the examples that are closer to the example cluster centres. In our experiments, the clustering was performed using the Self-Organizing Maps technique (Kohonen, et al., 1996).

5. APPLICATION EXAMPLES

5.1. Machine vision

Figure 1 depicts the results of the application of the method presented in this paper to defect detection on a photoreceptor surface.

a)

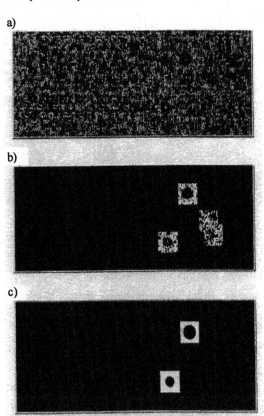

b)

c)

Fig. 2.   Defect detection results in two steps, localization (b) and verification (c), for a photoreceptor surface image (a).

The original data are in a form of a multi-feature vector. A grey-scale image of the photoreceptor surface is shown in Fig. 2 a). The candidate defect areas obtained as a result of the application of the relevance function procedure are depicted in Fig. 2 b). The resulting defects, after a verification procedure, are shown in Fig. 2 c).

5.2. Navigation planning and control

Vision systems have found numerous applications in a variety of image-based control schemes. An efficient use of visual information for navigation purposes in the area of autonomous robot systems was discussed by Talluri, R. and Aggarval (1996). Applications of image-based servoing in 2.5-D and 3-D setups were presented in (Malis and

Chaumette, 2000; Wilson et al., 1996). The problem of path planning in image space was dealt with in (Cowan and Koditschek, 2000). An approach consisting in coupling path planning in image space and image-based control, and taking into account constraints in the realized trajectories was discussed in (Mezouar and Chaumette, 2002). As shown below, the image relevance function operator preceded by a dimensionality reduction procedure can be used for both path planning in image space and vision-based navigation of a mobile robot system.

Path planning

In the case of multi-dimensional image data, such as multi-spectral satellite imagery, a pixel with image coordinates (i,j) of a multi-spectral image is a vector $\boldsymbol{b}_{i,j} = [b_1, b_2, ..., b_T]$ of T intensity values corresponding to T wavelength bands. In the example presented in this paper, images were obtained from the Landsat-7 satellite that uses the Enhanced Thematic Mapper Plus (ETM+) scanning radiometer instrument which produces seven bands of reflected energy and one band of emitted energy. Since one of the eight ETM+ channels is a panchromatic channel, in Landsat-7 imagery $T = 7$. The objective of the path planning procedure was to obtain a path for navigation in water basins. The first stage of the satellite image processing was to result in extracting the image of the water body. Application of dimensionality reduction methods showed that two bands, blue and near-infra-red produce the precise water image. Hence, the input data for the calculation of the property map are of dimensionality $L = 2$.

The task of planning a path between a pair of points can be defined as finding first, a piece-wise linear curve connecting successive maxima of the IRF function, and second, smoothing the curve as a Figure 3 shows an example of a path obtained in this way.

a)

b)

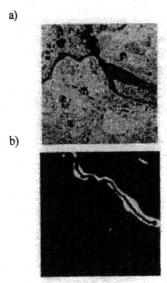

Fig. 3. Navigation planning. Maxima of IRF (a); navigation path (b).

Local maximum values of IRF are presented as circles – structuring elements - superimposed on the original satellite photo (Fig. 3 a). The calculated path is superimposed on the intensity image of IFR.

Navigation

The navigation algorithm that applies the IRF approach is based on the potential field method (Khatib, 1986) and adapted to the image-based discrete space operations by using a reactive control scheme proposed by Zaremba and Porada (1998) and defined in a equivalent to a raster image. The robot displacement vector, i.e., a planar vector **d** tangent to the robot trajectory at the current robot position P, is a linear combination of an attraction force **U** directed toward a target point P_{target} and a repulsive force **V** associated with the local configuration of the obstacles:

$$\mathbf{d} = a\mathbf{V} + \beta\mathbf{U} \qquad (9)$$

The above reactive navigation control algorithm is applied to the image obtained by thresholding the IRF function at different levels. The obstacle area is defined by the points for which $R\{f(i,j)\} < H$, where H is a variable threshold. The value of H depends on the size of the mobile robot and the precision of the navigation trajectory with respect to the optimal path. The changes of the value of H are also induced dynamically by the dead-end situations, i.e., when the robot is no longer able to move toward the target. In such cases the value of H decreases until the size of the structuring element S reaches a maximum value S_{max}. The effect of the variations of H is illustrated in Fig. 4.

a)   b)

c)   d)

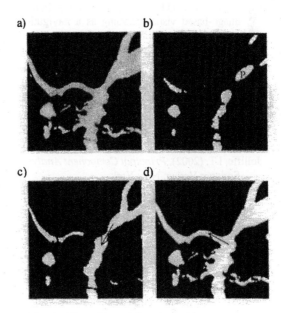

Fig. 4. Obstacle configuration.
a)   IRF shown as input gray-scale image,
b) - d)   secure navigation zones obtained from IRF at three different thresholds: $H_b > H_c > H_d$.

6. CONCLUSIONS

A new technology of combined nonlinear dimensionality reduction technique with multi-modality data analysis for such purposes as process monitoring, diagnosis, and object detection is presented. Since the analysis applies a generic operator for attention focusing, it allows for an easy integration of different object-relevant constraints for custom-based operator design. The main functional features of the technology:

- Fast detection (localization and identification) of objects of interest.
- Integration of object temporal changes and object motion detection with object spatial properties in different data modalities.
- Concise and multi-scale object shape description in the form of object skeletons amenable to graph-theoretical analysis.

Non-linear dimensionality reduction techniques permit to extend the relevance analysis to multidimensional data sets. An integrated software tool for dimensionality reduction and multi-modal data relevance analysis – FeatureEx-RA (Feature Extraction and Relevance Analysis) – is currently under development.

ACKNOWLEDGMENTS

The support of NSERC and CoRIMedia consortium is gratefully acknowledged. The author also wishes to thank Dr. R. Palenichka for providing access to the photoreceptor data.

REFERENCES

Cowan, N.J. and D.E. Koditschek (2000). Planar image-based visual servoing as a navigation problem. *Proc. IEEE Int. Conf. Robotics and Automation*, San Francisco, CA, 611-617.

Horn, R.A. and C.R. Johnson (1990). Matrix Analysis, Cambridge University Press, Cambridge.

Chen, Q., U. Kruger and A. Leung (2004). Regularized Kernel Density Estimation for Clustered Process Data. *Control Engineering Practice*, 12(3), pp. 267-274.

Jolliffe, I.T. (2002). *Principal Component Analysis.* Springer Verlag, New York.

Khatib, O. (1986). Real time obstacle avoidance for manipulators and mobile robots. *Int. J. Robotics Research*, 5(1), 90-98.

Kohonen, T., E. Oja, O. Simula, A. Visa and J. Kangas (1996). Engineering applications of the self-organizing map. *Proceedings of the IEEE*, 84, pp. 1358-1384.

Landgrebe, D. (2002). Hyperspectral Image Data Analysis as a High Dimensional Signal Processing Problem. *IEEE Signal Processing Magazine*, 19(1), pp. 17-28.

Lévesque J., K. Staenz and T. Szeredi (2000). The Impact of Spectral Band Characteristics on Unmixing of CASI Data for Monitoring Mine Tailing site Rehabilitation, *Canadian J. of Remote Sensing*, 26(3), pp. 231-240.

Malik, J. and P. Perona (1990). Preattentive Texture Discrimination with Early Vision Mechanisms, J. *Optical Society of America*, 7(2), pp. 933-932.

Malis, E. and F. Chaumette (2000). 2.5-D visual servoing with respect to unknown objects through a new estimation scheme of camera displacement. *Int. J. Comput. Vision*, 37(1), 79-97.

Martin, E.B. and A.J. Morris (1996). Non-parametric Confidence bounds for process performance monitoring charts. Journal of Process Control, 6(6), pp. 349-358.

Mezouar, J. and F. Chaumette (2002). Path planning for robust image-based control. *IEEE Trans. on Robotics and Automation*. 18(4), 534-549.

Palenichka, R.M. and P. Zinterhof (1996). A fast structure-adaptive evaluation of bcal features in images, *Pattern Recognition*, 29(9), pp. 1495-1505.

Palenichka, R. M., P. Zinterhof, and M. Volgin (2000). Detection of Local Objects in Images with Textured Background by Using Multi-Scale Relevance Function. *Proc. SPIE Int. Annual Symposium*, San Diego, USA, 4121, pp. 158-170.

Roweis, S.T. and L.K. Saul (2000). Nonlinear Dimensionality Reduction by Locally Linear Embedding. Science, 290, pp. 2324-2326.

Stough, T.M. and C.E. Brodley (2001). Focusing Attention on Objects of Interest Using Multiple Matched Filters, *IEEE Trans. Image Processing*, 10(3), pp. 419-426.

Talluri, R. and Aggarval, J.K., (1996). Mobile robot self-location using model-image feature correspondence, *IEEE Trans. Robotics and Automation*, RA-12, 63-77.

Wilson, W.J., C.C.W. Hulls and G.S. Bell (1996). Relative end-effector control using Cartesian position-based visual servoing. *IEEE Trans. Robotics and Automation*, 12, 684-696.

Yin, H. (2002). ViSOM-A Novel Method for Multivariate Data Projection and Structure Visualization. *IEEE Trans. on Neural Networks*, 13(1), pp. 237-243.

Zaremba, M.B. and E. Porada (1998). A Reactive Neuromorphic Controller for Local Robot Navigation. *Journal of Intelligent & Robotic Systems*, 23(2-4), 129-146.

Zaremba, M.B. and R.M. Palenichka (2002). Probabilistic morphological modelling of hydrographic networks from satellite imagery using self-organizing maps. *Control and Cybernetics*, 31(2), pp. 343-369.

ELSEVIER

IFAC

PUBLICATIONS

www.elsevier.com/locate/ifac

DESIGN METHODOLOGY FOR AN ADAPTIVE ATM SWITCH

Ilham Benyahia (1) & Philippe Kouame (2)

(1) Université du Québec en Outaouais, ilham.benyahia @uqo.ca
(2) Human Resources Development Canada, philippe.kouame@hrdc-drhc.gc.ca

Abstract : An ATM (Asynchronous Transfer Mode) switch has to process a variety of services that require different quality of service (QoS) and resources. Uniform switch processing may thus lead to network performance degradation. Adapting the switch processing to each traffic service is an important key to optimizing network resources. We present here a methodology for designing an ATM switch capable of adapting its processing to traffic behavior. Based on simulation techniques a number of experiments were carried out on different traffic behaviors to identify the most suitable switch processing. *Copyright © 2004 IFAC*

Keywords : ATM adaptive switching, traffic behavior, traffic control, cell loss, cell delay, ATM performance network, leaky bucket algorithm, dual leaky bucket algorithm.

1. INTRODUCTION

Multimedia traffic is becoming the most important form of communications for personal and industrial network applications. A single application now routinely combines different categories of traffic behavior and services including data, sound and video. This type of traffic integrates different behaviors that are compatible with ATM traffic services such as CBR (Constant Bit Rate), rt VBR (real-time Variable Bit Rate), and UBR (Unspecified Bit Rate). These categories of traffic behavior can be modeled by different probability distributions of traffic packet (cell) inter-arrival. We emphasize the importance of assigning specific processing characteristics to each category of service and traffic behavior to avoid network performance degradation or wasting of resources. Existing research interests on ATM traffic are limited to the definition of new services or performance

evaluation for a control strategy on a given traffic service (Pung and Bajrach, 1999; Kim, *et al.*, 1999).

Our goal here is to consider an ATM switch design that can adapt to each traffic service requirement during its operation. We present in this paper an approach for a dynamic control processing where strategies suitable for a given traffic behavior are chosen from rules defined through a set of experiments. This paper is organized as follows. In Section 2 we enumerate examples of different categories of ATM traffic services and control strategies. In Section 3 we introduce our dynamic modeling of an ATM switching system. It will be used as a tool for defining optimal operations of an ATM switch. The validation of the defined model is based on simulation techniques described in Section 4. A conclusion and future directions are presented in Section 5.

2. CONTROL REQUIREMENTS FOR ATM TRAFFIC SERVICES

ATM technology is suitable for multimedia applications however a variety of traffic service requirements must be satisfied. These services vary from the least tolerant such as the rt-VBR (Real-time Variable Bit Rate) service which is sensitive to any cell delay variation to UBR (Unspecified Bit Rate) which utilizes the remainder amount of bandwidth. Each category of service is characterized by specific parameters which are part of the traffic cell overheads describing the specific requirements and performance measurements (Rathgeb, 1990). For an ATM switch there are two main schemes for processing problems that may lead to a degradation of quality of service, namely flow and congestion control strategies.

Traffic control strategies are concerned with establishing the above cell traffic parameters and enforcing them. Thus, they are concerned with congestion avoidance. Different control strategies are defined as feedback mechanism, UPC (Usage Parameter Control), virtual scheduling algorithms (Goyal, *et al.*) sustainable cell rate algorithm. Examples of sophisticated congestion control are the leaky bucket and dual leaky bucket (Handel, *et al, 1994.*). These mechanisms are based on the definition of a capacity representing the amount of data which can be transmitted without an overflow. If the capacity is exceeded, the cells are lost. These algorithms necessitate the definition of operation parameters as finite capacity, a bucket that drains at a continuous rate of one unit per unit of time and whose content is increased by a given period. These control strategies are also related to parameters associated with the source of the traffic such as cell inter-arrival and the upper bound of a connection. Due to the global aspects of their parameters these control strategies are considered here to illustrate a complete study on switching function in order to define a design approach. The ABR (Available Bit Rate) service defines the minimum engagement necessary for a required quality of service and it requires a high degree of integrity. This service is rather sensitive to cell loss. It will be considered here for the first experiments to generate the most appropriate control strategy.

ATM switches have to deal with different traffic behavior and must satisfy different requirements to guarantee network performance. Thus, a control strategy has to be selected according to each context defined by the traffic behavior and the network requirements. We aim here to design a switching system such that its operation will lead to an optimization of the ATM network. Our design is based on a static model and a dynamic one representing operation scenarios to be simulated and analyzed to find associations between traffic behavior and control strategies that optimize the network performance. Thus the design phase will produce a rule base with such associations to be triggered at any traffic behavior change.

In the next section we briefly present our model of a switching system to be implemented and tested in order to find the most suitable combinations of traffic behavior and control strategies.

3. MODELING ADAPTIVE ATM SWITCH AND ITS ENVIRONMENT

Our design methodology incorporates both static and dynamic descriptions of the switching system. The dynamic model is based on the interaction of static model components and their response to a given traffic behavior. A set of scenarios representing given traffic behaviors during a specific period of time is defined. These scenarios are then used to identify the best strategies for network congestion control processing to minimize cell loss. Our approach uses a simulation study to collect data on which to base this choice. The components of our static model are the source, the switch and the traffic queue. Let us briefly define each switching component.

- The source component
This component defines the traffic behavior which is modeled by cell inter-arrival (Pitts and Schormans, 1996). Different parameters contribute to finalize characteristics of the traffic such as peak cell rate, sustainable cell rate and the maximum size of the burst (Pitts and Schormans, 1996). The peak cell rate defines the upper bound of the traffic. The sustainable cell rate defines an upper boundary in an ATM

connection calculated for a given period and the maximum size of the burst is the maximum number of cells sent to the peak cell rate.

- The switch component

This component is defined by its behavior based on a library of control strategies. These strategies can be chosen according to the traffic behavior and network load which requires either congestion or flux control.

- The component queue

This component is characterized by its size and the format of cells that compose the traffic. This format results from the service specification parameters. Examples of such parameters are those representing the quality of service.

We present here a brief description of the switching system operating scenario (see Figure 1). When the traffic source generates ATM cells, they are transmitted to the switch component which triggers the selected control strategy, then transmits these cells to the queue to be stored before their transmission to another network component.

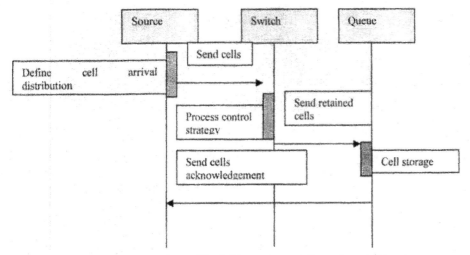

Fig. 1: Dynamic modeling of a switching system

4. SWITCH MODEL IMPLEMENTATION AND EXPERIMENTS

The ATM switch design presented in the above section is implemented and tested using the object-oriented framework simjava which is based on a discrete event simulation process (http://www.dcs.ed.ac.uk/home/simjava/simjava-1.0). The switch component interacts with its environment defined by the traffic source and the queue. The architecture test-bed is made up of the three presented components namely the source, the queue and the ATM switch. The traffic behavior experiments are limited here to different inter-arrival distributions such as Poisson, Exponential and Deterministic. Simjava integrates in the class Sim_event the appropriate functions associated with these distributions with probabilities of the occurrence of events. The congestion control strategies implemented in the

ATM switch are leaky bucket and dual leaky bucket. Finally, the queue component implements the mathematical M/M/1 model. When the queue receives a number of cells that exceeds its size, the surplus cells are lost. The component queue restores its space after each acknowledgement For each experiment we generate the same number of cells to be transmitted from the component source. For experiment results presented here, we focus on ABR service and we consider its particular performance requirement namely the minimization of cell loss. Each experiment is defined by input parameters such as packet size, buffer occupancy, the bucket token size and mean service time. For the same control strategy we consider different experiments with different cell inter-arrival distribution probability. The output parameters are defined by burst duration. The performance measurements used to take decisions are the total number of cells received by the switch and the number of lost cells. We illustrate in Table 1 examples on output experiment parameters.

The current experiment results prove that the combination of Poisson distribution and dual leaky bucket strategy minimizes cell loss better than other experiment combinations. However, when the ATM switch is operating with the leaky bucket strategy and a new connection is accepted for a short time duration with a new traffic behavior, it is better to maintain this strategy since the difference between the two control strategies is less significant than the processing overhead needed for strategy interchange. More details on experiment results can be found in (Kouame, 2001). A key element of our design methodology is the generation of a rule base which associates scenarios with congestion control strategies enabling the switch component to adapt its behavior to the current scenario.

Cell inter-arrival probability distribution	Percentage of lost cells
Poisson	11 %
Exponential	36 %
Deterministic	25 %

Table 1 Results of dual leaky bucket experiments

5. CONCLUSION

ATM technology is well suited for multimedia applications, which require a combination of traffic services. However, an ATM switch must respond to a variety of traffic requirements since using a uniform processing strategy to deal with different traffic services would risk degradation of network performance and could waste resources. In this paper we have presented a mechanism for designing an ATM switch capable of adapting to different traffic services. Our approach involves development of a test-bed to design a switching system. Our defined design methodology is based on experiments to evaluate the performance of each considered control strategy applied to a given traffic behavior (scenario). From these experiments we generate rules that associate each traffic behavior with the most suitable control strategy. The focus is on ABR service in the presented experiments since the main performance criterion is the

minimization of cell loss. The dual leaky bucket control strategy is very promising especially for Poisson inter-arrival distribution. More experiments are planned to finalize the ATM switch design by defining other control strategies to be applied for each cell inter-arrival probability distribution. We intend to build a library of traffic control strategies and to consider traffic requirements of other ATM services such as their performance parameters.

REFERENCES

Goyal, R., R. Jain, S. Fahmy, S. Narayanaswamy (..). Modeling Traffic Management. In: *ATM Networks with OPNET*.

Handel, R., M. N. Huber, S. Schroder (1994). ATM Networks. Addison Wesley.

Kim, D., Y. Cho, J. Par (1999). Analysis of Relative Rate Switch Algorithms for ABR Flow Control in ATM Networks. IEICE Trans. COMM, Volume E82-B No 1999.

Kouame, P. (2001). Conception d'un commutateur ATM dynamique. Report of DESS (Diplôme des Etudes Spécialisées en Informatique des télécommunications, Université du Québec en Outaouais.

Pitts, J. M. and J. A. Schormans (1996). Introduction to ATM Design and Performance: with Applications Analysis Software, John Wiley & Son Ltd.

Pung, H. K. and N. Bajrach, (1999). A Programmable ATM Multicast Service with Congestion Control, In: *IEICE/IEEE Joint Special Issue on Recent Progress.* ATM Technologies.

Rathgeb, E.P. (1990). Policing Mechanisms for ATM Networks: Modeling and Performance Comparison. In: Proceedings of the 7th International Tele-traffic Congress Seminar on "Broadband Technologies: Architectures, Applications, Control and Performance. Morristown, paper 10.1.

http://www.dcs.ed.ac.uk/home/simjava/simjava-1.0

www.elsevier.com/locate/ifac

PATHSET BASED CONDITIONING FOR TRANSIENT SIMULATION OF HIGHLY DEPENDABLE NETWORKS

Héctor Cancela*, Gerardo Rubino♣ and María E. Urquhart*

♦ *Instituto de Computación, Facultad de Ingeniería,*
Universidad de la República, Montevideo, Uruguay

♣ *IRISA/INRIA, Campus de Beaulieu 35042 Cedex, Rennes, France*

Abstract: Dependability modeling and evaluation is an important aspect of complex computer and telecommunication networks design. One of the most flexible tools is Markov chain modeling, which allows representing dynamic behavior and interactions between the different system components.

This work is centered on Monte Carlo estimation of transient dependability measures in networks modeled as Markovian systems, and particularly of the reliability $R(t)$, when the probability of system failure is a rare event. A variance reduction method, based on conditioning over events corresponding to the operation or non-operation of pathsets of the network, is presented. This method guarantees a better precision than standard Monte Carlo; the accuracy improvement is greater when t is small causing the failure events to be very rare in the considered interval. *Copyright © 2004 IFAC*

Keywords: network reliability, Monte Carlo simulation, Markov models

1. INTRODUCTION

Nowadays, dependability modeling and evaluation is an important aspect of complex computer and telecommunication networks' design. For this purpose, one of the most flexible tools is Markov chain modeling, which allows to represent dynamic behavior and interactions between the different network components (Marsan et al, 1988; Walrand and Varaiya, 1996). When the model of the system is small, or even moderately large, it also allows for exact numerical evaluation of the measures of interest. But as the number of components grows, the numerical methods are no longer feasible.

An alternative is the use of Monte Carlo simulation techniques, which give estimates (with a certain confidence) of the dependability of the system. Standard ("naive") Monte Carlo does not work well when the interesting events are rare (more precisely, when they occur with a very low probability), which is usually the case of nowadays very reliable systems. This leads to the use of variance reduction techniques, which can give a more accurate estimate using the same sample size. In the case of steady-state dependability measures, a family of importance sampling schemes (broadly called failure biasing schemes) has been proven to be appropriate (Cancela et al., 1996; Carrasco, 1991; Goyal et al. 1992; Heidelberger, 1995; Shahabuddin, 1994; Tuffin, 1999). These schemes do not work equally well when evaluating transient measures; in this case other techniques must be used, either alone or in combination with importance sampling.

This paper focuses on one important and widely used dependability measure, the reliability at time t, denoted $R(t)$; and proposes a Monte Carlo technique for its evaluation which guarantees variance reduction with respect to standard Monte Carlo. This technique is based on conditioning over events concerning the pathsets of the system, and is related to the family of recursive variance reduction methods, which have been used in static system reliability computation (Cancela and El Khadiri, 1995; Cancela and Urquhart, 2002).

The rest of this work is organized as follows. Section 2 briefly presents the class of systems under study (Markovian dependability models for multi-component systems). Their evaluation using standard Monte Carlo is discussed in Section 3. Section 4 describes the proposed technique. Section 5 presents an application example of the method, with some numerical results. Finally, Section 6 is devoted to the conclusions.

2. MODEL DESCRITION

We consider a model of a multicomponent network with failures and repairs, whose state space S can be divided into two disjoint sets U and D, where U is the set of the operational states of the system (the states where the network is performing according to the specifications), and D is the set of the failure states (where the network is not performing according to the specifications).

We denote by $K = \{1,2,...,K\}$ the set of the components of the network. We assume that the components are (i) either operational (or up), or (ii) unoperational (or down, that is, failed). An operational component k has failure rate λ_k. The system can include an arbitrary repair subsystem, to bring failed components back to operational state.

The system is represented (modeled) by a finite continuous time homogeneous and irreducible Markov chain $X = \{X_s, s \geq 0\}$. As said before, we denote by S the state space of X, and we suppose that the number of states is very large (that is, $1 << |S|$).

We assume that the initial state, denoted by x_0, is fixed. We also assume that all the components are up in that state, and that $x_0 \in U$. In this paper we will focus on the reliability $R(t)$ in $[0,t]$, defined as the probability that the network is operational during the interval $[0,t]$, that is:

$$R(t) = \mathbf{P}\{X_s \in U, \forall s \in [0,t]\}. \qquad (1)$$

A frequent situation is that the model (the stochastic process X) is quite complex and large (that is, $|S| >> 1$) and that the failed states are rare, that is, $\tau_D >> t$ with high probability. The size of the model may make difficult or impossible its exact numerical evaluation and the rarity of the interesting events can do the same with a naive Monte Carlo estimation, as discussed for example by Heidelberger (1995).

In the next section we present how to estimate the reliability $R(t)$ by a standard Monte Carlo simulation of chain X.

3. STANDARD MONTE CARLO METHOD

If we have a given system and a fixed time horizon t, the continuous operation of the network during $[0,t]$ is a random event and the reliability $R(t)$ of the network is its probability. If we denote by $\mathbf{1}_{(E)}$ the indicator function of an arbitrary event E, we are concerned by the expectation of the Bernoulli random variable $Y(t) = \mathbf{1}_{(X_s \in U, \forall s \in [0,t])}$ with value 1 if the network is operational during $[0,t]$, and with value 0 otherwise (if the network arrives at a failed state before time t); we have $R(t) = \mathbf{E}\{Y(t)\}$.

The unbiased standard Monte Carlo estimator of the reliability parameter $R(t)$ is a sample mean $\hat{Y}(t)$ based on the random variable $Y(t)$. More precisely,

$$\hat{Y}(t) = \frac{1}{N}\sum_{i=1}^{N} Y^{(i)} \qquad (2)$$

where N is a fixed sample size and $Y^{(1)},...,Y^{(N)}$ are independent and identically distributed random variables with same distribution as $Y(t)$.

If N is large enough, the central limit theorem can be applied to obtain a confidence interval. More precisely, a ε-level confidence interval is given by

$$\hat{Y}(t)\pm\xi\sqrt{\mathrm{Var}\left\{\hat{Y}(t)\right\}} = \hat{Y}(t)\pm\xi\sqrt{\mathrm{Var}\{Y(t)\}/N} \qquad (3)$$

where

$$\xi = G^{-1}(1-\varepsilon/2),\ G(x) = \frac{1}{\sqrt{2\pi}}\int_{-\infty}^{x} e^{\frac{-t^2}{2}}\,dt \qquad (4)$$

and

$$\mathrm{Var}\left\{\hat{Y}(t)\right\} = \mathrm{Var}\{Y(t)\}/N = (1-R(t))R(t)/N. \qquad (5)$$

To obtain a confidence interval with formula (3), and since the value of $\mathrm{Var}\{Y(t)\}$ is unknown, it is usually replaced by the following unbiased estimator:

$$\hat{V}_Y = \frac{1}{N-1}\sum_{i=1}^{N}\left(Y^{(i)} - \hat{Y}(t)\right)^2. \qquad (6)$$

The simulation algorithm consists of independently repeating N times the following experiment. The simulation starts at time 0 with the system in its initial state, and repeatedly finds the time of the next event (component failure or repair), updating the system state and advancing the time, until the chain reaches one state belonging to D or when the accumulated sojourn time in the different states is greater than t. In the first case the value of the sample is 0, in the second case it is 1. The estimation of $R(t)$ is the frequency of the observations where the system is operational up to time t.

The standard simulation is easy to implement. Unfortunately, it becomes extremely wasteful in execution time when the system is highly reliable, as the number of experiments must be very large in order to obtain some failed experiments and a reasonable relative error; a general reference on this subject is (Ross, 1996).

In the case of highly reliable networks, the problem is fundamentally caused by the fact that the sojourn time in the operational states of the system is usually longer than the considered interval $[0, t]$. Then, the chain X will stay in the operational states during all the interval $[0, t]$ for most of the samples; this means that a very large sample size is needed to obtain an acceptably accurate estimate of $R(t)$.

Next section presents a method which can be used to tackle this problem, by forcing the occurrence of some failures before t.

4. PATHSET BASED CONDITIONING

Following the notation in (Colbourn, 1987), we define a pathset as a set of components such that their operation ensures the correct operation of the network; in a coherent structure system, any superset of a pathset is also a pathset.

In this section we present a simple conditioning method based on the pathsets of the network. We will assume that there exists an efficient method to select a pathset of the system when the state of a certain number of components has been already fixed; this is true for most cases of practical interest.

We denote by T_i the time of the first failure of component i. As we are in a Markovian setting, the T_i are exponentially distributed random variables, with rates λ_i.

Given a certain pathset $C = \{i_1,...,i_{|C|}\}$, we can define the event

$$O_t(C) = \left\{ \forall i, \ 1 \le i \le |C|, \ T_i > t \right\} \qquad (7)$$

and the complementary event

$$F_t(C) = \left\{ \exists \ i, \ 1 \le i \le |C|, \ \text{s.t.} \ T_i \le t \right\} \qquad (8)$$

The first event corresponds to trajectories of the chain X where no element of pathset C fails before t; by definition of pathset, this implies that the network will be operational up to time t. The second event corresponds to trajectories where at least one element of pathset C fails; this is a necessary but not sufficient condition for a network failure.

We know that the r.v. T_i is exponentially distributed with rate λ_i. If we denote $\Lambda(C) = \sum_{i \in C} \lambda_i$, then we have $\mathbf{P}(O_t(C)) = e^{-\Lambda(C) t}$ and $\mathbf{P}(F_t(C)) = 1 - e^{-\Lambda(C) t}$.

Let us now define the event $Et = \{\forall s \in [0,t], \ X_s \in U\}$. The reliability of the system at time t is $R(t) = \mathbf{P}\{E_t\}$. The random variable $Y_t = \mathbf{1}_{(E_t)}$ satisfies $\mathbf{E}\{Y_t\} = R(t)$ leading to the standard estimation of $R(t)$ consisting of building N independent copies $Y_t^{(1)}, Y_t^{(2)}, ..., Y_t^{(N)}$ of Y_t and estimating $R(t)$ by the average $\sum_{n=1}^{N} Y_t^{(n)} / N$. In words, we simulate N independent trajectories of the model from 0 to t, we count how many times the chain remains in U during $[0,t]$ and then we divide by N.

Observe now that

$$\begin{aligned} R(t) = \mathbf{P}\{E_t\} &= \mathbf{P}\{E_t | O_t(C)\} \ \mathbf{P}\{O_t(C)\} \\ &+ \mathbf{P}\{E_t | F_t(C)\} \ \mathbf{P}\{F_t(C)\} \\ &= \mathbf{P}\{O_t(C)\} + \mathbf{P}\{E_t | F_t(C)\} \ \mathbf{P}\{F_t(C)\}. \end{aligned} \qquad (9)$$

Let us define now the random variable Z_t as follows:

$$\begin{aligned} Z_t &= \mathbf{P}\{O_t(C)\} + \mathbf{P}\{F_t(C)\} \ Y_t \\ &= e^{-\Lambda(C) t} + \left(1 - e^{-\Lambda(C) t}\right) Y_t. \end{aligned} \qquad (10)$$

We then have

$$R(t) = \mathbf{E} \left\{ Z_t | F_t(C) \right\}. \qquad (11)$$

This leads to a new estimator of $R(t)$ as follows. We build N independent trajectories of the chain from 0 to t but now, *in all trajectories at least one component of pathset C fails before t*. As before, we count in how many of these the chain remains in the set U of operational states, and then we divide by N. This is an estimator of the number $\mathbf{E}\{Y_t | F_t(C)\} = \mathbf{P}\{E_t | F_t(C)\}$. By substituting in the expression (9) we obtain the estimation of the reliability at t:

$$R(t) = \mathbf{P}\{O_t(C)\} + \mathbf{P}\{E_t | F_t(C)\} \mathbf{P}\{F_t(C)\} \qquad (12)$$

Let us look now at the variance $\text{Var}\{Z_t | F_t(C)\}$. We have

$$\text{Var} \left\{ Z_t | F_t(C) \right\} = \mathbf{P}\{F_t(C)\}^2 \ \text{Var} \left\{ Y_t | F_t(C) \right\}. \qquad (13)$$

Computing $\text{Var}\{Y_t | F_t(C)\}$, we obtain

$$\begin{aligned} \text{Var} \left\{ Y_t | F_t(C) \right\} &= \mathbf{P}\{E_t | F_t(C)\} \left(1 - \mathbf{P}\{E_t | F_t(C)\}\right) \\ &= \frac{R(t) - \mathbf{P}\{O_t(C)\}}{\mathbf{P}\{F_t(C)\}} \left(1 - \frac{R(t) - \mathbf{P}\{O_t(C)\}}{\mathbf{P}\{F_t(C)\}}\right). \end{aligned} \qquad (14)$$

and thus

$$\begin{aligned} \text{Var} \left\{ Z_t | F_t(C) \right\} &= (R(t) - \mathbf{P}\{O_t(C)\}) \ (1 - R(t)) \\ &< R(t) \ (1 - R(t)) = \text{Var} \left\{ Y_t \right\}. \end{aligned} \qquad (15)$$

This guarantees a reduction in variance over the standard Monte Carlo estimator. The technical point of this approach is the problem of sampling the trajectories of X conditioned on $F_t(C)$. This is easy to implement; when sampling random variables T_i with $I \in C$, instead of sampling from the original exponential distributions, we must sample from modified distributions that take into account this conditioning (see Appendix A).

Assume now that we build two disjoint pathsets C_1, C_2. Denote by $Z_t^{(1)}$ the process defined in (10) when the pathset C is C_1. Now, define

$$\begin{aligned} Z_t^{(2)} &= \mathbf{P}\{O_t(C_1) \cup O_t(C_2)\} \\ &+ \mathbf{P}\{F_t(C_1) \cap F_t(C_2)\} Y_t \\ &= e^{-(\Lambda(C_1) + \Lambda(C_2)) t} + \left[1 - e^{-(\Lambda(C_1) + \Lambda(C_2)) t}\right] Y_t. \end{aligned} \qquad (16)$$

We have, similarly as before,

$$R(t) = \mathbf{E} \left\{ Z_t^{(2)} | F_t(C_1) \cap F_t(C_1) \right\} \qquad (17)$$

The procedure to use this new process to build an estimator of $R(t)$ is as before. All we need to know is how to sample N trajectories of X conditioned on the event $F_t(C_1) \cap F_t(C_2)$ (this is described in Appendix A). If we compute the variance of this new estimator, we obtain

$$\begin{aligned} \text{Var} \left\{ Z_t^{(2)} | F_t(C_1) \cap F_t(C_2) \right\} &= \\ &= [R(t) - \mathbf{P}\{O_t(C_1)\} - \mathbf{P}\{O_t(C_2)\}] \ (1 - R(t)) \quad (18) \\ &< \text{Var} \left\{ Z_t^{(1)} | F_t(C_1) \right\} \end{aligned}$$

We see that the variance reduction is larger using two disjoint paths than only one. More generally, we can select K disjoint pathsets $C_1,...,C_K$ and define the corresponding process

$$Z_t^{(K)} = \mathbf{P}\left\{\bigcup_{k=1}^{K} O_t(C_k)\right\} + \mathbf{P}\left\{\bigcap_{k=1}^{K} F_t(C_k)\right\} Y_t$$
$$= e^{-\left(\sum_{k=1}^{K} \Lambda(C_k)\right) t} + \left[1 - e^{-\left(\sum_{k=1}^{K} \Lambda(C_k)\right) t}\right] Y_t. \qquad (19)$$

We have again

$$R(t) = \mathbf{E}\left\{Z_t^{(K)} \middle| \bigcap_{k=1}^{K} F_t(C_k)\right\} \qquad (20)$$

and

$$Var\left\{Z_t^{(K)} \middle| \bigcap_{k=1}^{K} F_t(C_k)\right\} =_t$$
$$= \left[R(t) - e^{-\left(\sum_{k=1}^{K} \Lambda(C_k)\right) t}\right](1 - R(t)). \qquad (21)$$

The more disjoints paths we use, the better estimator we get (since the variance reduction will be larger).

5. NUMERICAL RESULTS

In this section we show the application of the proposed method for $R(t)$ evaluation of a given example, and we compare its performance versus standard Monte Carlo.

The system under study has 30 components, 16 of type A and 14 of type B. Every component has the same failure rate $\lambda = 1$ failure/hour; there are two repairmen, one for each type of component; the repair rate for the components of type A is $\mu_A = 100$ repairs/hour, and for the components of type B is $\mu_B = 1000$ repairs/hour.

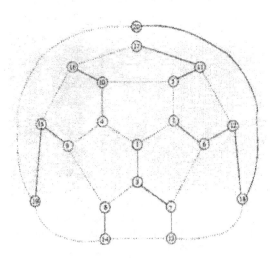

Fig. 1. The "dodecahedron" system.

The structure function of the system is given by the graph represented in Figure 1. The components of

the system are represented by the edges of the graph; the components of type A are shown by a dotted line, those of type B by a continuous line. The system is up if there is an operational path (a set of operational links) between the nodes 1 and 20. From the number of components, and the fact that they can fail independently, we can compute the size of the state space of this system, which is more than a billion states.

We have implemented the method based on pathsets as described in Section 4, and a standard Monte Carlo method. The experiments were done on a SUN SPARCstation 10 Model 602 computer.

Table 1 shows the simulation result for the unreliability $Q(t) = 1- R(t)$, computed for four different values of t: $t = 0.001, 0.01, 0.1$ and 1 hours. The sample size is 10^6 experiments for each different value of t. In the first column we indicate the values of parameter t. The column $\hat{W}(t)$ presents the estimated values of $Q(t)$ as computed by the proposed method. The third column contains an estimation of the accuracy improvement, given by the variances quotient (computed as the quotient of the variance of the standard Monte Carlo, $Var\{Y(t)\}=R(t)(1-R(t))$, and the estimation $\hat{V}\{W\}$ of the variance $Var\{W(t)\}$ of the proposed method). The following columns present the quotient of the execution times, and the relative efficiency (also called speedup) of the proposed variance reduction method relative to the standard Monte Carlo (computed as the product of the accuracy improvement by the time reduction).

Table 1: $Q(t)$ estimation: performance of the proposed method versus standard Monte Carlo.

	$\hat{W}(t)$	VR	TR	Speedup (VR×TR)
$t=0.001$	1.51×10^{-9}	0.99×10^{7}	2.65×10^{-1}	2.62×10^{6}
$t=0.01$	5.63×10^{-7}	1.11×10^{4}	2.73×10^{-1}	3.03×10^{3}
$t=0.1$	4.35×10^{-5}	2.53×10^{1}	3.17×10^{-1}	8.02×10^{0}
$t=1$	5.01×10^{-4}	1.05×10^{0}	7.71×10^{-1}	8.10×10^{-1}

$\hat{W}(t)$: estimated unreliability.

VR: variance reduction $Var\{Y(t)\} / \hat{V}\{W\}$.

TR: time reduction T_Y/T_W.

The proposed method has always a smaller variance than the standard Monte Carlo method; but this reduction depends on the value of t. For the smaller values, the accuracy improvement is very important, but it decreases when t grows. This result can be explained looking at the fact that the variance of W is bounded by $(R(t)-\mathbf{P}\{O_t(C)\})(1 - R(t))$ with $\mathbf{P}\{O_t(C)\} = \prod_{i\in C} \mathbf{P}\{T_i^{(1)} > t\} = \prod_{i\in C} e^{-\lambda_i t}$ which is a decreasing function on t; when t is small, so will be the variance, but when t grows, $\mathbf{P}\{O_t(C)\}$ goes to 0, and the bound approaches the value of the variance of the standard Monte Carlo method.

The execution time of the proposed method is a little bigger than the standard Monte Carlo one, specially, for the smaller values of t (as can be seen in Column 4); this can be explained by the necessity of finding the pathsets of the system and by the use of a more complex sampling procedure. The combined effect variance reduction-execution time (usually called relative efficiency) is presented at Column 5. We can see that in general the overall result is satisfactory, excepting the case where $t=1$, where the execution time penalization is more important than the variance reduction obtained. For smaller values of t, the efficiency improvement is very important, and allows to evaluate in a reasonable time systems which could not be evaluated by the standard Monte Carlo method.

6. CONCLUSIONS

In this work we have presented a new technique for the evaluation of transient reliability measures, based on the pathsets of the system. Its objective is to provide an efficient estimation tool in the context of rare events. The results obtained on a particular case show the interest of the method, presenting the accuracy improvements which can be obtained with a small computer time overhead. These improvements are especially important in the case of very reliable networks (i.e., very rare failure events), and time windows relatively small.

Future work includes the application of this method to evaluate other transient measures, such as instantaneous and expected availability. A naive application seems relatively straightforward, but further accuracy improvements may be possible by considering different pathsets at different states of the system. Another point of interest is to study the case of wider time windows, and how the method can be improved in this situation.

ACKNOWLEDGEMENT

This work is a result of project PAIR, funded by the INRIA (France).

APPENDIX A

We show how to derive the distribution function of an exponential random variable T_1 conditional to the event that the minimum of a set C of independent random variables T_1, T_2,... with respective parameters λ_1, λ_2, ..., is less than t, that is

$$\mathbf{P}\left\{T_1 \leq x \mid \min(T_1, T_2, \cdots) < t\right\} \quad (22)$$

The distribution function can then be used, for example with the inverse distribution function method, to generate samples for T_1.

We concentrate on the case where $C = \{T_1, T_2\}$; we denote $\Lambda(C) = \lambda_1 + \lambda_2$. If $C = \{T_1, T_2, \ldots, T_{|C|}\}$ with

$|C| > 2$, we can use the fact that $\min(T_1, T_2, \ldots) = \min(T_1, \min(T_2, T_3, \ldots))$, and $\min(T_2, \ldots, T_{|C|})$ is an exponential random variable T_2' with rate $(\lambda_2 + \ldots + \lambda_{|C|})$, giving again the case of only two variables in C.

Let us denote by V the random variable $\min(T_1, T_2)$. If $x \leq t$, we have

$$\mathbf{P}\left\{T_1 \leq x \mid V < t\right\} = \frac{\mathbf{P}\left\{T_1 \leq x, V < t\right\}}{\mathbf{P}\left\{V < t\right\}} =$$
$$= \frac{\mathbf{P}\left\{V < t \mid T_1 \leq x,\right\}\mathbf{P}\left\{T_1 \leq x\right\}}{\mathbf{P}\left\{V < t\right\}}. \quad (23)$$

Using the fact that if $T_1 \leq x$, and $x \leq t$, then $T_1 \leq t$, and $\mathbf{P}\{V < t \mid T_1 \leq x\} = 1$, we have

$$\mathbf{P}\left\{T_1 \leq x \mid V < t\right\} = \frac{1 - e^{-\lambda_1 x}}{1 - e^{-\Lambda(C) t}}. \quad (24)$$

If $x > t$, we have

$$\mathbf{P}\left\{T_1 \leq x \mid V < t\right\} = \frac{\mathbf{P}\left\{T_1 \leq x, V < t\right\}}{\mathbf{P}\left\{V < t\right\}} =$$
$$= \frac{\mathbf{P}\left\{T_1 \leq x, V < t, T_1 \leq t\right\} + \mathbf{P}\left\{T_1 \leq x, V < t, T_1 \leq t\right\}}{\mathbf{P}\left\{V < t\right\}}$$
$$= . \frac{\mathbf{P}\left\{T_1 \leq t\right\} + \mathbf{P}\left\{t < T_1 \leq x, T_2 \leq t\right\}}{\mathbf{P}\left\{V < t\right\}} \quad (25)$$
$$= . \frac{\mathbf{P}\left\{T_1 \leq t\right\} + \mathbf{P}\left\{t < T_1 \leq x\right\}\mathbf{P}\left\{T_2 \leq t\right\}}{\mathbf{P}\left\{V < t\right\}}$$
$$= \frac{1 - e^{-\lambda_1 t} + \left(e^{-\lambda_1 t} - e^{-\lambda_1 x}\right)\left(1 - e^{-\lambda_2 t}\right)}{1 - e^{-\Lambda(C) t}}.$$

This completes the derivation of the distribution function for T_1.

Once generated the sample for T_1, we need the distribution function for T_2 conditional to the generated value for T_1: $\mathbf{P}\left\{T_2 \leq x \mid V < t, T_1 = t_1\right\}$

If $t_1 \leq t$, we have that
$$\mathbf{P}\left\{T_2 \leq x \mid V < t, T_1 = t_1 \leq t\right\} = \mathbf{P}\left\{T_2 \leq x \mid T_1 = t_1\right\}$$
$$= \mathbf{P}\left\{T_2 \leq x\right\} = 1 - e^{-\lambda_2 x} \quad (26)$$
equal to the unconditional distribution of T_2. If $t_1 > t$, we have that
$$\mathbf{P}\left\{T_2 \leq x \mid V < t, T_1 = t_1 > t\right\} = \mathbf{P}\left\{T_2 \leq x \mid T_2 \leq t\right\}$$
$$= \frac{\mathbf{P}\left\{T_2 \leq x, T_2 \leq t\right\}}{\mathbf{P}\left\{T_2 \leq t\right\}}. \quad (27)$$

For $x \leq t$, this gives
$$\mathbf{P}\left\{T_2 \leq x \mid V < t, T_1 = t_1 > t\right\} = \frac{\mathbf{P}\left\{T_2 \leq x\right\}}{\mathbf{P}\left\{T_2 \leq t\right\}}$$
$$= \frac{1 - e^{-\lambda_2 x}}{1 - e^{-\lambda_2 t}}. \quad (28)$$

Finally, if $x > t$, we have
$$\mathbf{P}\left\{T_2 \leq x \mid V < t, T_1 = t_1 > t\right\} = \frac{\mathbf{P}\left\{T_2 \leq t\right\}}{\mathbf{P}\left\{T_2 \leq t\right\}} = 1. \quad (29)$$

This completes our derivation.

The preceding development directly concerns the case of a single pathset. In case of using K pathsets, the development is the same for each one. Since the pathsets are disjoints and the components independent, we simply apply the obtained distribution for the components of each one of the pathsets.

REFERENCES

Ajmone Marsan, M., Balbo, G. and Conte G. (1988). *Performance Models of Multiprocessor Systems.* The MIT Press.

Cancela, H. and El Khadiri, M.(1995). A recursive variance-reduction algorithm for estimating communication-network reliability. *IEEE Transactions on Reliability*, **44**(4):595-602.

Cancela, H., Rubino, G. and Tuffin, B. (1996). Fast Monte Carlo methods for evaluating highly dependable Markovian systems. In *2nd International Conference on Monte Carlo and Quasi-Monte Carlo Methods in Scientific* Computing, Salzburg, Austria.

Cancela, H. and Urquhart, M.E. (2002). RVR simulation techniques for residual connectedness network reliability evaluation. *IEEE Transactions on Computers*, **51**(4):439-443.

Carrasco, J. A. (1991). Failure distance based simulation of repairable fault tolerant systems. In *Proceedings of the 5th International Conference on Modelling Techniques and Tools for Computer Performance Evaluation*, pages 351-365.

Colbourn, C. J. (1987). *The Combinatorics of Network Reliability.* Oxford University Press, New York.

Goyal, A., Shahabuddin, P., Heidelberger, P., Nicola, V. F. and Glynn, P. W. (1992). A unified framework for simulating Markovian models of highly dependable systems. *IEEE Transactions on Computers*, **41**(1):36-51.

Heidelberger, P.(1995). Fast simulation of rare events in queueing and reliability models. *ACM Transactions on Modeling and Computer Simulation*, **54**(1):43-85.

Ross, S.M. (1996). *Simulation.* Academic Press.

Shahabuddin, P. (1994). Importance Sampling for the Simulation of Highly Reliable Markovian Systems. *Management Science*, **40**(3):333-352.

Tuffin, B.(1999). Bounded Normal Approximation in Highly Reliable Markovian Systems. *Journal of Applied Probability*, **36**(4):974-986.

Walrand, J. and Varaiya. P. (1996). *High Performance Communication Networks.* Morgan Kaufmann Publishers.

ELSEVIER
IFAC
PUBLICATIONS
www.elsevier.com/locate/ifac

FAST BROADCASTING WITH BYZANTINE FAULTS

Michel Paquette [*,1] Andrzej Pelc [*,2]

*Département d'informatique et d'ingénierie, Université
du Québec en Outaouais, Hull, Québec J8X 3X7, Canada.
E-mail:* michel.paquette@polymtl.ca, pelc@uqo.ca

Abstract: We construct and analyze a fast broadcasting algorithm working in the presence of Byzantine component faults. Such faults are particularly difficult to deal with, as faulty components may behave arbitrarily (even maliciously) as transmitters, by either blocking, rerouting, or altering transmitted messages in a way most detrimental to the broadcasting process. We assume that links and nodes of a communication network are subject to Byzantine failures, and that faults are distributed randomly and independently, with link failure probability p and node failure probability q, these parameters satisfying the inequality $(1 - p)^2(1-q) > 1/2$. A broadcasting algorithm, working in an n-node network, is called *almost safe* if the probability of its correctness is at least $1 - 1/n$, for sufficiently large n. Thus the robustness of the algorithm grows with the size of the network. Our main result is the design and analysis of an almost safe broadcasting algorithm working in time $O(\log^2 n)$ and using $O(n \log n)$ messages in n-node networks. The novelty of our algorithm is that it can cope with the most difficult type of faults, potentially affecting all components of the network (both its links and nodes), and that it is simultaneously robust and efficient. *Copyright © 2004 IFAC*

Keywords: fault tolerance, binary tree architectures, communication networks, distributed computer control systems, noisy channels, optimality, probabilistic models.

1. INTRODUCTION

As interconnection networks grow in size and complexity, they become increasingly vulnerable to component failures. Links and/or nodes of the network may fail, and these failures often result in delaying, blocking, or even distorting transmitted messages. It becomes important to design communication algorithms in such a way that the desired communication task be accomplished efficiently in spite of these faults, usually without knowing their location ahead of time. Such communication algorithms are called *fault-tolerant*.

No communication algorithm can work properly for all fault types and configurations. For example, if all links of a network are faulty, and faults result in permanently blocking all transmitted messages, there can be no hope of accomplishing any communication task. On the other hand, such massive failures are extremely rare in practice and they need not be of primary concern in algorithm design. Much more frequent are faults that damage a limited number of components. Hence an

[1] Supported in part by the M.Sc. CALDI Scholarship of the Research Chair in Distributed Computing of the Université du Québec en Outaouais.
[2] Supported in part by NSERC grant OGP 0008136 and by the Research Chair in Distributed Computing of the Université du Québec en Outaouais.

important goal is to design communication algorithms that work properly, under some assumptions bounding the number of possible faults.

We are concerned with one of the fundamental communication procedures called *broadcasting*. One node of the network, called the *source*, holds a message which must be transmitted to all other fault-free nodes. (It is assumed that the source is fault free: otherwise, no broadcasting is possible). There are two fundamental qualities which are demanded of a good fault-tolerant broadcasting algorithm. One of them is *robustness*, i.e., the ability of the algorithm to work correctly in spite of the presence of faults of some assumed type, which are distributed in unknown locations but according to some prescribed assumptions. The other one is *efficiency*: this requirement is similar as in the fault-free situation. In this work we adopt two primary measures of efficiency of a communication algorithm: its *time*, i.e., the number of time steps used by the algorithm, and its *cost*, i.e., the number of messages used in the communication process.

1.1 The model and terminology

The communication network is modeled as a simple undirected graph. We assume that the network is complete, i.e., all pairs of nodes are connected by communication links. However, our algorithm works for much sparser networks as well, in fact for some networks of logarithmic degree. Communication is synchronous: every elementary transmission is assumed to take one *time step*. We adopt one of the communication models most widely used in the literature, the so called *one-port* or *whispering* model (Hedetniemi, Hedetniemi and Liestman, 1988; Fraigniaud and Lazard, 1994). In one time step, each node can communicate with at most one other node, i.e., pairs of nodes communicating simultaneously must form a matching.

Since we are concerned with communication in the presence of faults, we must specify the nature and the assumed distribution of faults which our algorithms can tolerate. We work with the most general fault type: *Byzantine faults*. Such faults are particularly difficult to deal with, as faulty components may behave arbitrarily (even maliciously) as transmitters, by either blocking, rerouting, or altering transmitted messages in a way most detrimental to the broadcasting process. The main advantage of assuming such general type of faults is that an algorithm working under this assumption, also works correctly, and with the same or better efficiency, in the presence of any more benign type of failures, such as crash or fail-stop faults. Hence, designing an efficient and robust communication algorithm for Byzantine faults increases

its portability. Another issue to be specified is the distribution of faults. We work under the probabilistic distribution scenario, widely considered to accurately model realistic faulty environments (cf. surveys (Lee and Shin, 1994; Pelc, 1996)). Faults of links and nodes of the network are assumed to be distributed randomly and independently. with link failure probability p and node failure probability q. For our algorithm to work reliably, these constant parameters must satisfy the inequality $(1-p)^2(1-q) > 1/2$. It should be noted that this assumption is satisfied in most real cases: for example our algorithm can tolerate link and node failures occuring with probability up to 20%. Most realistic communication networks have much more reliable components, hence our algorithm will behave well in practice. It is assumed that the source of broadcasting is fault free: otherwise, no broadcasting is possible.

Since components of the network are subject to random failures, it is even possible that all components fail and hence, no broadcasting algorithm can work correctly at all times. Hence we seek algorithms that have high probability of correctness. A broadcasting algorithm, working in an n-node network, is called *almost safe*, if the probability of its correctness is at least $1 - 1/n$, for sufficiently large n. Thus the robustness of the algorithm grows with the size of the network, which is a desirable property, as networks encountered in real applications are often very large. For this reason, almost safe fault-tolerant communication algorithms have been widely studied in the literature (cf. the survey (Pelc, 1996)).

1.2 Our results

Since redundancy is a fundamental ingredient in all fault-tolerant communication algorithms, there exists a trade-off between robustness and efficiency of such algorithms. If we want a broadcasting algorithm to tolerate a lot of faults, especially Byzantine faults, messages must be sent along many alternate routes, and verified during the communication process, thus augmenting the running time and the cost of the algorithm. Our main result is the design and analysis of a broadcasting algorithm which is fast, uses relatively few messages, and has high probability of correctness. More specifically, we design an almost safe broadcasting algorithm working in time $O(\log^2 n)$ and using $O(n \log n)$ messages in n-node networks, for network components subject to Byzantine failures, under the assumption that link failure probability p and node failure probability q satisfy the inequality $(1-p)^2(1-q) > 1/2$.

Our algorithm is *oblivious*, in the sense that all transmissions are scheduled in advance: it is predetermined which pairs of nodes communicate in a given time step. The only thing that is decided during the algorithm execution, is the content of the messages, which depends on outcomes of a voting scheme used to mask erroneous transmissions. Such algorithms have the advantage of being easy to implement and of requiring small local memory in network nodes. It can be proved that every almost safe oblivious broadcasting algorithm must use $\Omega(n \log n)$ messages, hence our algorithm has cost of optimal order of magnitude. On the other hand, it is well known that any broadcasting algorithm, even working in a fault-free network, must have running time at least $\log n$. Hence the execution time of our algorithm exceeds the optimal one by at most a logarithmic factor.

1.3 Related work

Efficient broadcasting algorithms in various communication networks have been widely studied in the literature (see, e.g., surveys (Hedetniemi, Hedetniemi and Liestman, 1988; Fraigniaud and Lazard, 1994)). In particular, a lot of effort has been devoted to the study of broadcasting under the assumption that components of the network may be faulty (see the survey (Pelc, 1996)). Two commonly used fault models are the *bounded* model and the *probabilistic* model. In the bounded model, an upper bound k is imposed on the number of faulty components, and their worst-case location is assumed. Under this scenario, k-*tolerant* broadcasting is usually sought (see, e.g., (Gargano, 1992; Peleg and Schäffer, 1989)): the source message must reach all fault-free nodes provided that no more than k components (links, or nodes, or both, depending on the particular scenario) are faulty. In the probabilistic model, faults are assumed to occur randomly and independently of each other, with specified probabilities. Under this scenario, almost safe broadcasting is often sought, similarly as in the present paper (cf. (Berman, Diks and Pelc, 1997; Bienstock, 1988; Chlebus, Diks and Pelc, 1994; Diks and Pelc, 1992)).

Byzantine faults have been widely studied in the context of network communication, of the consensus problem in distributed computing, and of multiprocessor fault diagnosis (cf., e.g., (Lynch, 1996) and the survey (Barborak, Malek and Dahbura, 1993)). In (Bao, Igarashi and Katano, 1995), broadcasting in hypercube networks was studied under the assumption that either nodes or links (but not both) are subject to randomly distributed Byzantine faults. In (Berman, Diks and Pelc, 1997), a $O(\log n)$-time broadcasting algorithm was designed under the assumption that

links are subject to randomly distributed Byzantine faults but all nodes are fault free. On the other hand, in (Chlebus, Diks and Pelc, 1994) a $O(\log n)$-time broadcasting algorithm was proposed for randomly distributed link and node failures but only of *crash* type: such faults can only block messages but cannot distort them, which significantly facilitates the design of fault-tolerant broadcasting algorithms. To the best of our knowledge, our broadcasting algorithm is the first to cope with randomly distributed Byzantine faults of links and nodes at the same time.

2. THE BROADCASTING ALGORITHM

2.1 The underlying network

Although we work in the complete network, our broadcasting algorithm will in fact use very few links. This is a desirable feature, as sparse networks are easier to implement. We will construct an n-node network of logarithmic degree in which almost safe broadcasting is possible (cf. (Chlebus, Diks and Pelc, 1994; Berman, Diks and Pelc, 1997), where a similar network was used previously).

We use $\log n$ to denote the logarithm with base 2 and $\ln n$ to denote the logarithm with base e. For a positive constant c, we define an n-node network $G(n, c)$. Let $m = \lceil c \log n \rceil$ and $s = \lfloor n/m \rfloor$. For clarity of presentation, assume that n is a power of 2, m is odd and divides n, and $s = 2^{L+1} - 1$, for some integer $L \geq 0$. By definition, $L < \log n$. It is easy to modify the construction and the proof in the general case. Partition the set of all nodes into subsets $S_1, ..., S_s$, of size m, called *supernodes*. In every supernode S_i, enumerate nodes from 0 to $m - 1$. For any $i = 1, ..., s$ and any $j = 0, ..., m - 1$, assign label (i, j) to the jth node in the ith supernode. We assume that node $(1, 0)$ is the source of broadcasting. Arithmetic operations on the second integers forming labels are performed modulo m.

Arrange all supernodes into a complete binary tree T with $L+1$ levels $0, 1, ..., L$. Level 0 contains the root and level L contains leaves of tree T. The supernode S_1, called ROOT, is the root of T. For every $1 \leq i \leq \lfloor s/2 \rfloor$, S_{2i} is the left child of S_i and S_{2i+1} is the right child of S_i in the tree T. For every $1 < i \leq s$, supernode $S_{\lfloor i/2 \rfloor}$ is the parent of S_i. If a supernode is a parent or a child of another supernode, we say that these supernodes are adjacent in T.

The set of edges of the network $G(n, c)$ is defined as follows. If supernodes S_i and S_j are adjacent in T then there is an edge in $G(n, c)$ between any node in S_i and any node in S_j. Moreover, every pair of nodes in the supernode S_1 forming the root of T are joined by an edge. These are the only edges in $G(n, c)$. The graph $G(n, c)$ is called a *thick tree*. It should be noted that a thick tree is not a tree but has a tree-like structure which will be exploited in the design of our broadcasting algorithm. Clearly, a thick tree is a graph of maximum degree $O(\log n)$.

2.2 The overview of the algorithm

The idea of our algorithm is the following. First the source message is propagated to all nodes of ROOT, using all other nodes as intermediaries. Every node of ROOT gets $m - 1$ versions of the source message. Then it computes the message considered to be correct, by majority voting. Next, the source message is propagated down the thick tree, along its branches. Every supernode S_i, whose nodes already got the message, transmits first to its left child and then to its right child. This is done as follows. Every node of S_i sends its version of the source message to every node of the given child. After obtaining m versions of the source message from nodes of S_i, every node of the child of S_i computes the message considered to be correct, by majority voting. This is done until all nodes in all supernodes compute the version of the source message considered to be correct.

2.3 The description of the algorithm

We now proceed with a formal description of the broadcasting algorithm. For every node u, X_u denotes the content of register X in this node. In the beginning, the source message is held in register M of the source, i.e., it is $M_{(1,0)}$. For every node u, the message that it computes during the scheme execution, and considers to be the source message, is stored in its register M. Every node u outputs the final value of M_u as its version of the source message.

The elementary procedure SEND(u, X, v, Y) will be used for adjacent nodes u, v, for the following action: u sends X_u to v, and v assigns the received value to its register Y. If u, v, and the link joining u with v are fault free, the effect of SEND(u, X, v, Y) is the same as the assignment $Y_v := X_u$. Otherwise, an arbitrary value is assigned to Y_v.

Another elementary subroutine is VOTE(v, V, A), where V is a vector of values. This subroutine is performed locally by node v, and it serves to compute the correct version of the source message, with high probability. It outputs the majority value of all terms of a vector V of records held in v, and stores it in record A, if such a majority exists. Otherwise, it stores a default value.

We now describe the main procedures used in our broadcasting algorithm.

Procedure ROOT-Propagation spreads the source message among nodes of ROOT. After exchanging their versions of the source message, all nodes in ROOT vote on all received values, to compute the correct version of the source message, with high probability. Let $R_1,...,R_{m-2}$ be the partition of all edges in ROOT$\setminus\{(1,0)\}$ into pairwise disjoint perfect matchings, and let $R_i(v)$ denote the node matched with v by matching R_i.

---

procedure ROOT-Propagation
for $j := 1$ **to** $m - 1$ **do**
    SEND$((1,0), M, (1,j), W)$
for $i := 1$ **to** $m - 2$ **do**
    for all $j = 1, ..., m - 1$ **in parallel do**
        SEND$((1,j), W, R_i((1,j)), B[j])$
        $B[j]_{(1,j)} := W_{(1,j)}$
        //$R_i((1,j))$ stores the message coming
        //from $(1,j)$ in record $B[j]$. Also, every
        //node $(1,j)$ assigns the value of its record
        //W to the record $B[j]$ in its vector B.
for all $j = 1, ..., m - 1$ **in parallel do**
    VOTE$((1,j), B, M)$
    //Every node in ROOT computes the majority of
    //values from vector $B[1, ..., m - 1]$

Procedure GROUP-Transmission transmits the message from all nodes of one supernode to all nodes of another. Then all nodes in the receiving supernode vote on all received values, to compute the correct version of the source message, with high probability. Index addition in the procedure formulation is done modulo m.

---

procedure GROUP-Transmission (i, k)
for $r := 0$ **to** $m - 1$ **do**
    for all $j = 0, 1, ..., m - 1$ **in parallel do**
        SEND$((i,j), M, (k, j + r), B[j])$
    for all $j = 0, 1, ..., m - 1$ **in parallel do**
        VOTE$((k,j), B, M)$

Procedure TREE-Propagation (l) reliably transmits the source message from level l to level $l + 1$ of the thick tree.

---

procedure TREE-Propagation (l)
for all supernodes S_i on level l **in parallel do**
    GROUP-Transmission$(i, 2i)$; //to left child
    GROUP-Transmission$(i, 2i + 1)$; //to right child

Now Algorithm FBBF (Fast Broadcasting with Byzantine Faults) can be succinctly formulated as follows.

Algorithm FBBF
ROOT-Propagation
for $l := 0$ **to** $L - 1$ **do**
   TREE-Propagation(l).
---

3. THE ANALYSIS OF ALGORITHM FBBF

Let $q' = (1-p)^2(1-q)$ and $c = \frac{80\ln(2)q'}{(2q'-1)^2}$. By assumption, $q' > 1/2$. We consider Algorithm FBBF for the thick tree $G(n,c)$, for this value of c.

Proposition 3.1. Algorithm FBBF works in time $O(\log^2 n)$ and uses $O(n\log n)$ messages.

Proof. Since the size of every supernode is $O(\log n)$, the procedures ROOT-Propagation and GROUP-Transmission each take time $O(\log n)$. Since the number of levels in the thick tree is $O(\log n)$ as well, it follows that the execution time of Algorithm FBBF is $O(\log^2 n)$. In order to estimate the number of messages, observe that each pair of nodes, adjacent in the thick tree, exchange at most two messages. Since the maximum degree of the thick tree is $O(\log n)$, the number of edges in it is $O(n\log n)$, and consequently Algorithm FBBF uses $O(n\log n)$ messages.   □

It remains to prove the main result of this section, concerning the correctness of Algorithm FBBF. In the proof we will use the following probabilistic lemma known as Chernoff's bound (cf. (Hagerup and Rüb, 1989/90)). It will be needed to estimate the probability that the majority of a set of pairwise disjoint paths joining two nodes, are fault free. Since paths are pairwise disjoint, the events that they do not contain faulty components are independent, and Chernoff's bound can be used.

Lemma 3.1. Let X be the random variable denoting the number of successes in a Bernoulli series of length m with success probability q. Let $0 < \epsilon < 1$. Then $Prob(X \le (1-\epsilon)mq) \le e^{-\epsilon^2 mq/2}$.

Theorem 3.1. Algorithm FBBF is almost safe.

Proof. We define the following events:

- C is the event that, at the end of Algorithm FBBF, all fault-free nodes get the correct source message. We denote by p_C the probability of event C.
- Let x be a branch of the thick tree. B_x is the event that, at the end of Algorithm FBBF, all fault-free nodes in all supernodes in the branch x, get the correct source message. We denote by p_{B_x} the probability of event B_x.
- C_{ROOT} is the event that, at the end of procedure ROOT-Propagation, all fault-free nodes in ROOT get the correct source message. We

denote by $p_{C_{ROOT}}$ the probability of event C_{ROOT}.

- $C_{SOURCE,i}$, for $i = 1, ..., m - 1$, is the event that, at the end of procedure ROOT-Propagation, a fault-free node $(1, i)$ gets the correct source message. We denote by $p_{C_{SOURCE,i}}$ the probability of event $C_{SOURCE,i}$.
- Let supernode S_i be the parent of supernode S_k in the thick tree. CS_k is the event that, at the end of procedure GROUP-Transmission(i, k), all fault-free nodes in S_k get the correct source message. We denote by p_{CS_k} the conditional probability $Prob(CS_k|CS_i)$.

For any event E, we will use the notation $\overline{E}$ to denote the complement of E, and $\overline{p_E}$ to denote $1 - p_E$.

Consider the event $C_{SOURCE,i}$. Node $(1, i)$ is joined with the source by $m - 2$ pairwise disjoint paths of length 2, and one path of length 1 (the joining edge). The probability that such a fixed path of length 2 does not contain faulty components, is $q' = (1-p)^2(1-q)$. The probability for the path of length 1 is even larger: $1 - p$. For pairwise disjoint paths, these events are independent, hence Lemma 3.1 can be used to estimate the probability that the majority of these paths are free of faults. If this holds, event $C_{SOURCE,i}$ holds as well. Hence we get

$$\overline{p_{C_{SOURCE,i}}} \le e^{\frac{-(2q'-1)^2(m-1)q'}{8q'^2}} \le e^{\frac{-(2q'-1)^2 mq'/2}{8q'^2}}. \quad (1)$$

In view of $m = \lceil c\log n\rceil$ and $c = \frac{80\ln(2)q'}{(2q'-1)^2}$ we have

$$\overline{p_{C_{SOURCE,i}}} \le e^{\frac{5\ln(2)\log(n)(-(2q'-1)^2}{(2q'-1)^2/8q'}\frac{q'}{8q'^2}}. \quad (2)$$

$$= e^{-5\ln(n)} = 1/n^5. \quad (3)$$

We know that $\overline{p_{C_{ROOT}}} \le \sum_{i=1}^{m} \overline{p_{C_{SOURCE,i}}}$, since C_{ROOT} is the intersection of all events $C_{SOURCE,i}$, for fault-free nodes $(1, i)$. Hence $\overline{p_{C_{ROOT}}} \le m \cdot 1/n^5 \le 1/n^4$, for sufficiently large n.

Let supernode S_i be the parent of supernode S_k in the thick tree. Consider the event E_j that a single fixed fault-free node (k, j) in supernode S_k gets the correct source message at the end of procedure GROUP-Transmission(i, k). Suppose that all fault-free nodes in S_i got the correct source message prior to the execution of this procedure. Node (k, j) gets the source message through m disjoint paths, each consisting of one node from S_i and one joining link. The probability that both components in such a single path are fault free, is $(1-p)(1-q) > q'$. Hence the conditional probability $Prob(E_j|CS_i)$ is larger than $p_{C_{SOURCE,i}}$. Consequently, $Prob(\overline{E_j}|CS_i) \le 1/n^5$, which implies $\overline{p_{CS_k}} \le m \cdot 1/n^5 \le 1/n^4$, for sufficiently large n.

Fix a branch x of the thick tree. For sufficiently large n, we have

$$p_{B_x} = p_{C_{ROOT}} \cdot \prod_{i=1}^{L} p_{CS_i}. \tag{4}$$

$$p_{B_x} \geq (1 - \frac{1}{n^4})^{L+1} \geq (1 - \frac{1}{n^4})^n. \tag{5}$$

We have $(1-1/n^4)^{n^3} \to 1$ and $(1-1/n^2)^{n^2} \to 1/e$. Hence, for sufficiently large n,

$$(1 - 1/n^4)^{n^3} > (1 - 1/n^2)^{n^2}, \tag{6}$$

which implies

$$1 - 1/n^4 > (1 - 1/n^2)^{1/n}. \tag{7}$$

Thus we get $p_{B_x} \geq (1 - \frac{1}{n^4})^n \geq 1 - 1/n^2$. By definition of events C and B_x we have

$$\overline{C} \subset \bigcup \{\overline{B_x} : x \text{ is a branch in the thick tree}\}. \tag{8}$$

Since there are $2^L \leq n$ branches in the thick tree, we finally get, for sufficiently large n,

$$\overline{p_C} \leq n \cdot \frac{1}{n^2} = 1/n, \tag{9}$$

hence Algorithm FBBF is almost safe. $\square$

4. CONCLUSION

We proposed an almost safe broadcasting algorithm working in time $O(\log^2 n)$ and using $O(n \log n)$ messages in n-node networks, subject to Byzantine faults of components. The fact that the algorithm is almost safe guarantees that its reliability grows with the size of the network. The novelty of our algorithm is that it copes with the most difficult type of faults potentially affecting all components of the network: both its links and nodes. Our algorithm has the optimal cost complexity, and its running time differs from the lower bound only by a logarithmic factor. The question of whether there exists an almost safe broadcasting algorithm working in time $O(\log n)$ under our Byzantine fault scenario remains a challenging open problem suggested by our result.

REFERENCES

Bao, F., Y. Igarashi and K. Katano (1995). Broadcasting in hypercubes with randomly distributed byzantine faults. *Proc. WDAG'95*, **972**, pp. 215–229.

Barborak, M., M. Malek and A. Dahbura (1993). The consensus problem in fault-tolerant computing. *ACM Computing Surveys*, **25**, pp. 171–220.

Berman, P., K. Diks and A. Pelc (1997). Reliable broadcasting in logarithmic time with byzantine link failures. *Journal of Algorithms*, **22**, pp. 199–211.

Bienstock, D. (1988). Broadcasting with random faults. *Disc. Appl. Math.*, **20**, pp. 1–7.

Chlebus, B.S., K. Diks and A. Pelc (1994). Sparse networks supporting efficient reliable broadcasting. *Nordic Journal of Computing*, **1**, pp. 332–345.

Diks, K. and A. Pelc (1992). Almost safe gossiping in bounded degree networks. *SIAM J. Disc. Math.*, **5**, pp. 338–344.

Fraigniaud, P. and E. Lazard (1994). Methods, problems of communication in usual networks. *Disc. Appl. Math.*, **53**, pp. 79–133.

Gargano, L. (1992). Tighter bounds on fault-tolerant broadcasting, gossiping. *Networks*, **22**, pp. 469–486.

Hagerup, T. and C. Rüb (1989/90). A guided tour of chernoff bounds. *Inf. Proc. Letters*, **33**, pp. 305–308.

Hedetniemi, S.M., S.T. Hedetniemi and A.L. Liestman (1988). A survey of gossiping, broadcasting in communication networks. *Networks*, **18**, pp. 319–349.

Lee, S. and K.G. Shin (1994). Probabilistic diagnosis of multiprocessor systems. *ACM Computing Surveys*, **26**, pp. 121–139.

Lynch, N.A. (1996). *Distributed Algorithms*. Morgan Kaufmann Publ., Inc.. San Francisco.

Pelc, A. (1996). Fault-tolerant broadcasting, gossiping in communication networks. *Networks*, **28**, pp. 143–156.

Peleg, D. and A.A. Schäffer (1989). Time bounds on fault-tolerant broadcasting. *Networks*, **19**, pp. 803–822.

ELSEVIER

IFAC
PUBLICATIONS
www.elsevier.com/locate/ifac

Development of an interface between the IEEE 802.11b WLAN technology and the new mobile ad hoc subnetwork relay (SNR) technology

Bruno Ouellet, Gérard Nourry, Larbi Talbi

Département d'informatique et d'ingénierie
Université du Québec en Outaouais (UQO)
C.P. 1250 succ. B, Gatineau, QC, H3A 2N4

Abstract: *In this paper, we investigate the development of an interface between the two following ad hoc networking technologies: IEEE 802.11b and the subnetwork relay (SNR) technology, which is the only fully distributed multi hop wireless ad hoc networking technology available on the market today. The newly formed hybrid protocol could be adapted to both the civilian and military domains and could offer new wireless ad hoc applications in domains not explored yet. Three integration approaches are discussed and technically compared. The preferable approach, which consists of the SNR protocol above the physical layer of 802.11b, seems the most promising. The following work, that has already started, consists in building OpNet models of each approach in order to further explore and analyse their behaviour. Copyright © 2004 IFAC*

Keywords: WLAN, SNR, 802.11, ad hoc, wireless, interface.

1. INTRODUCTION

In the last decade, information technology has grown at a phenomenal pace. With advances in processor speed, miniaturization and power management, the wireless domain has become one of the key areas of interest for the consumer market. This has indirectly triggered a substantial R&D effort in many sectors of the communication industry and in academia, on wireless infrastructure networks and also on wireless ad hoc or independent networks. This last category of networks is seen as a logical extension to wired infrastructure network. Despite this effort, only one commercial fully distributed ad hoc multi-hop network seems to exist to date. Initially developed to meet military requirements in the HF, VHF and low UHF frequency bands, the subnetwork relay (SNR) technology is not limited to these bands. Its performance can be greatly increased and its applicability substantially broaden by the addition of higher speed and lower switching time bearers, such as those used in IEEE 802.11. The coupling of the SNR and IEEE 802.11 technologies would result in a technology that can be used to support a richer set of military applications as well as civilian applications. In this paper, we study the feasibility of integrating a wireless LAN technology such as IEEE 802.11b together with SNR in order to provide an efficient ad hoc, distributed, mobile networking technology that can be adapted to a wide variety of military or civilian deployment scenarios.

This paper first reviews key aspects of the sub network relay (SNR) technology (Section 2). In section 3, we briefly discuss the well-known IEEE 802.11b standard, which was the wireless LAN technology selected for integration with SNR in this study. Following these descriptions, the proposed integration approaches are described and this is followed by a section on the technical feasibility of each approach. The last section briefly covers future work, including ongoing efforts towards modelling each approach.

2. SUBNETWORK RELAY (SNR)

The subnetwork relay (SNR) technology concept was evolved within an international military working group in the early 00s. The aim of SNR was to reduce the reliance of current maritime communications on satellite communications, by exploiting line of sight (LOS) and extended LOS (ELOS) radio links. The desired technology needed to possess the following functional characteristics: fully distributed network architecture, self configuring and self-healing network protocol, support for roaming platforms, dynamic slot allocation, dynamic allocation of relay nodes, spatial slot reuse, reliable or unreliable link service, and lastly a parameterized design that would make it independent of the modems, radios and other equipment used at the physical layer.

The development work resulted in the design of the SNR protocol as detailed in (Labbé, 2002), {lab2,lab3,lab4,lab5,lab6}, and shown in Fig. 1. This figure represents two SNR nodes within a ship's LAN scenario. As can be seen, the SNR protocol is located just above the physical layer and covers the data link and network layer of the OSI model.

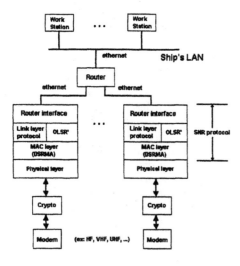

Fig. 1 Generic architecture of SNR

SNR is composed of five main components. The first one is the medium access control (MAC) sublayer, which is based on the Dynamic Slot Reservation Medium Access (DSRMA) protocol. DSRMA is a fully distributed TDMA MAC protocol (Lapic,2001). The second element is the link layer protocol, which establishes the interface between the other modules of SNR and the MAC sublayer. The link layer accepts outgoing data from the upper layers and from the routing sublayer and forwards it to the MAC sublayer. Upon reception of data from the MAC sublayer, it either relays it or extracts it and forwards it to the appropriate module. The third element of the SNR is the routing sublayer based on an adapted version of the optimised link state routing (OLSR) protocol. The OLSR module maintains the routing tables and insures that the routing information is readily available. The fourth element of SNR is the router interface. Its main function is to provide an interface between SNR and higher layer components such as a LAN router, as shown on Fig.1. The last element of SNR is the serial driver of the physical layer. It establishes synchronous communication between the OS and the physical connector of the selected physical layer.

Even though, the SNR protocol meets the specified requirements and supports most of the traffic classes

found on the Internet, its current implementation, on LOS and ELOS physical layers, would definitely benefit from the addition of a higher throughput bearer capable of short to medium range communication. In return, this adaptation to higher speed bearer would make SNR a technology of choice for a wider range of applications in both military and civilian markets, and would also represent a significant advancement in the wireless ad hoc networking R&D domain.

3. IEEE 802.11b

The IEEE 802.11b was selected for this experimentation following a detailed study of the current WLAN standards available. It was chosen, over HiperLAN, HomeRF and other 802.11 physical layers, mainly because 802.11b actually possesses the largest market share, which should grow substantially in the years ahead. 802.11b has also become the de facto standard in almost all military and civilian applications. In addition, the development of IEEE 802.11g, which will be fully compatible with 802.11b, makes it even more attractive.

As with all others 802 standards, 802.11b (IEEE-SA, 1997,1999) covers only the bottom two layers of the OSI model. It consists of a MAC sublayer, based on CSMA/CA, over a physical layer. Since the key characteristics of SNR needed to be preserved, an ad hoc network or IBSS topology was required. Therefore, the DCF function was retained, because it was the only one supporting a fully distributed scenario, and was adapted to optimise the efficiency of the resulting interface. The DCF function uses acknowledge frames, four different interframe spaces and the NAV in order to insure proper channel access management. In addition, in order to mitigate the hidden node problem, a RTS/CTS mechanism can be activated as required. Fig. 2 resumes the DCF with RTS/CTS mechanism activated.

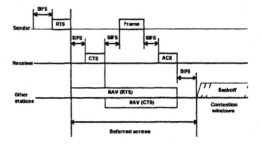

Fig. 2 DCF with RTS/CTS mechanism

In addition to the hidden node problem, it was found that since the 802.11 MAC has not been initially designed to work in an ad hoc multi-hop network, it experiences some serious problems due to the exposed node problem. In (Xu et al, 2002) the authors noted an instability problem, a severe unfairness problem and an incompatibility problem which deteriorate greatly the efficiency of the 802.11 wireless network.

Below the MAC sublayer, IEEE 802.11b supports not only the three physical layers (FHSS, DSSS and IR) at 1 and 2 Mbps available in 802.11, but also adds an enhanced DSSS physical layer functioning at 5.5 and 11 Mbps. For this experiment, we choose the 802.11b DSSS physical layer at 11 Mbps.

4. INTEGRATION APPROACHES

Once the detailed study of both SNR protocol and 802.11b standard was completed, we looked at possible ways to integrate these two wireless technologies together. Three approaches were retained for further examination. These approaches are detailed in the following subsections.

4.1 Approach 1: SNR less DSRMA over 802.11b

In this approach, showed in Fig. 3, we propose to modify SNR by taking away its MAC sublayer and by building an interface between the modified SNR protocol and 802.11b. Doing so, the link layer and the OLSR sublayer would also need to be modified to support the relaying function of the new protocol.

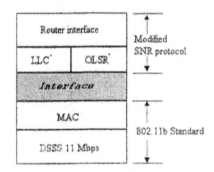

Fig. 3. SNR less DSRMA over 802.11b

4.2 Approach 2: SNR over 802.11b

The second approach consists of simply adding the SNR protocol stack over the 802.11b protocol stack and creating an interface between both stacks. Fig. 4 shows the concept of having two MAC sublayers one over the other (TDMA over CSMA/CA). For this approach we are talking about adaptation instead of modification because neither technologies are modified.

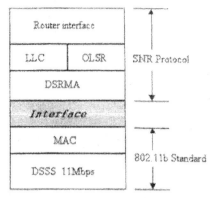

Fig. 4 SNR over 802.11b

4.3 Approach 3: SNR over physical layer of 802.11b

The third approach, showed in Fig. 5 is the simplest architecture so far. It consists of adding SNR protocol on top of the physical layer of 802.11b. No important modification is forecasted to either technology.

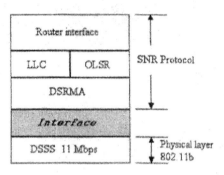

Fig. 5 SNR over PHY 802.11b

5. TECHNICAL EVALUATION

The evaluation of each approach begins with a functional analysis of the newly formed protocol stack, followed by an outline of the 802.11 network configuration that is recommended and, finally, an evaluation of the throughput efficiency of the new protocol, as per (Anastasi et al, 2003), is performed. For the purpose of comparing the throughput efficiency of the three approaches, it is assumed that all three approaches use the same 802.11 link layer and physical layer, using short PLCP frames. The efficiency is compared by looking at the overhead added to a 1500-byte frame from the moment it leaves the link layer up to its arrival at the physical layer.

5.1 Evaluation of Approach 1

For this approach, the 802.11b protocol remains unchanged. Its MAC sublayer will manage access to the medium within an IBSS scenario and its physical layer should also work as expected. For SNR, the changes are more important. With the removal of its MAC sublayer, the SNR protocol needs to be modified to include the ARQ functions and also both the neighbourhood discovery and the multipoint relay calculation functions need to be reactivated in OLSR. It is obvious that major modifications are required to support this approach.

On the configuration side, this approach will use an IBSS topology with DCF access. Acting as the only MAC sublayer, 802.11b needs all of its functionalities. As demonstrated in (Sheu et al, 2002), it is preferable to have both acknowledgement and RTS/CTS activated for this scenario. Fragmentation should be set as required and power management should be deactivated for our scenario where power is not an issue. Scanning and authentification settings are left to the discretion of the user requirements.

Under this configuration, the maximum throughput or efficiency will use the equation shown below, which represents the ratio between the time required to transmit the user data and the overall time the channel is busy due to the overhead needed.

$$Maximum\ Efficiency = {T_{payload}}\Big/{T_{All}}$$

where:

$T_{All}=$

$$DIFS + T_{AvBo} + (3xSIFS) + T_{RTS} + T_{CTS} + T_{ACK} + T_{Data}$$

$T_{payload}$ the time required to transmit the user data (1500 bytes) at 11 Mbps

$TDIFS = SIFS + (2X\ a\ SlotTime\ of\ 20\mu\ Sec)$

T_{AvBo} Average backoff time for the first contention window

$SIFS$ $10\mu Sec$ for 802.11b

T_{RTS} transmit time for a RTS frame,

T_{CTS} transmit time for a CTS frame,

T_{ACK} transmit time for a ACK frame,

T_{Data} transmit time for a data frame, includes PHY header, MAC header and FCS

With this formula, this approach offers a maximum efficiency of 55.3% for a load of 1500 bytes sent with the RTS/CTS procedure at 11 Mbps. This efficiency is expected to decrease significantly with the exposed node problem and also with the fact that the backoff timer rises with a power of two every time a collision occurs.

Even though the architecture of this approach is simple and 802.11b is not modified, this approach still requires important modifications to the layers above 802.11b. Even with the modifications, the maximum efficiency will still stand at around 50\% and should definitely decrease with collisions and with the interference from the exposed node.

5.2 Evaluation of Approach 2

The second approach requires no change to both protocols but they should be fine-tuned in order to coexist efficiently together. The architecture is loaded because two MAC sublayers are juxtaposed. Even if this adds processing time, it is still negligible for this experiment. In revenge, the juxtaposition brings a few benefits. For example, the DSRMA allows access to only one node at the time during each slot, which will undoubtedly limit contentions at the MAC sublayer found below. The only contention should occur from frames sent by the 802.11b MAC. With proper tuning, only the beacon and authentification frames could interfere. This situation will improve the network performance because any node will find a medium available immediately and the backoff procedure is then becoming almost useless.

Having TDMA over CSMA/CA should also address the hidden and exposed node problems. No RTS/CTS is required and broadcast frame, with no ACK, can be used to transmit all the information because 802.11b do not need to filter frames anymore. DSRMA, located above 802.11b MAC, does it for the protocol.

As per approach 1, the configuration of this approach is an IBSS using DCF access. As mentioned, broadcast frames should be used. No ACK frame will be used and the fragmentation function will come from the link layer found in SNR and will need to be adjusted to 802.11b limitations. The RTS/CTS procedure can be deactivated along with the fragmentation function in 802.11b. During affiliation, it is recommended to use passive scanning to limit transmission in someone else slot. For this approach, the corresponding efficiency formula is:

$$Maximum\ Efficiency = {T_{payload}}\Big/{DIFS + T_{Data}}$$

where:

$T_{payload}$ the time required to transmit the user data (1500 bytes) at 11 Mbps

$DIFS = SIFS + (2X\ a\ SlotTime\ of\ 20\square\ Sec)$

T_{Data} transmit time for a data frame, includes PHY header, MAC header and FCS

This simpler formula allows approach 2 efficiency to reach 82.2% for an identical load of 1500 bytes. Without modification on either 802.11b or SNR, this approach limits the developmental risk and offers a far better efficiency than the previous approach. In addition, both the exposed and hidden node problems are addressed and implementation components are readily available on the market.

5.3 Evaluation of Approach 3

The last approach structure is the simplest of all three. The SNR protocol fully controls the access to the medium and uses only the physical layer of 802.11b as a transceiver. This approach offers the fastest processing time with no concern about the exposed or hidden node problems. On the configuration aspect, the only points being noted are the respect by SNR of the frame length supported by the physical layer and the use of the short PLCP frame to enhance the throughput.

The maximum efficiency formula looks similar as the previous one, except that the DIFS time was taken out because the 802.11b MAC sublayer is not used here.

$$Maximum\ Efficiency = {T_{payload}} / {T_{Data}}$$

where:

$T_{payload}$ the time required to transmit the user data (1500 bytes) at 11 Mbps

T_{Data} transmit time for a data frame, includes PHY header, MAC header

Using this formula, the third approach becomes the more efficient at 87\% for a load of 1500 bytes transmitted at 11 Mbps.

To summarize the results of the technical evaluation process, Fig. 6 shows a comparison between all approaches based on a LLPDU (the load) size varying from 100 to 2000 bytes. From this graph, it is concluded that approach 3 is the most efficient of the three approaches, with approach 2 not very far behind.

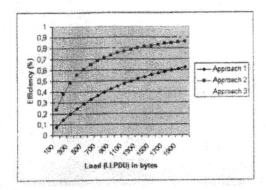

Fig. 6 Efficiency graph

6. CONCLUSION

In order to extend the technical analysis of the three approaches and to properly explore and quantify their differences, the next step is to create an OpNet model of each approach. Currently, OpNet models exist for both IEEE 802.11b and SNR protocol stacks. Looking at one approach at a time, the model of the interface will be built, tested and the results will be analysed. Each newly designed OpNet model will allow a complete study of the operational characteristics of the new corresponding hybrid protocol stack.

It is to be noted that encouraging preliminary simulation results have been obtained for the

approach 2 which is also directly confirming approach 3 results.

REFERENCES

Anastasi,G., E.Borgia, M.Conti, and E.Gregori (2003). IEEE 802.11 Ad Hoc Networks: Performance Measurements, 23rd International Conference on Distributed Computing Systems Workshops, pages 758-763.

IEEE-SA Standards Board (1997),Part 11: Wireless LAN Medium Access Control (MAC) and Physical Layer (PHY) specifications.

IEEE-SA Standards Board (1999), Part 11: Wireless LAN Medium Access Control (MAC) and Physical Layer (PHY) specifications: Higher-speed physical layer extension in the 2.4 GHz band.

Labbé, I.(2002), Subnetwork Relaying System (SNR) Link layer-MAC layer Interface specifications, Technical report, IPUnwired Canada.

Labbé, I.(2002), Subnetwork Relaying System (SNR) Link layer-OLSR interface specifications, Technical report, IP Unwired Canada.

Labbé, I. (2002), Subnetwork Relaying System (SNR) OLSR-MAC layer interface specifications, Technical report, IP Unwired Canada.

Labbé, I.(2002), Subnetwork Relaying System (SNR) OLSR protocol specifications, Technical report, IP Unwired Canada.

Labbé, I. and G. Nourry (2002), Subnetwork Relaying System (SNR) Link layer protocol specifications, Technical report, IP Unwired Canada.

Labbé, I. and G. Nourry (2002), Subnetwork Relaying System (SNR) MAC layer protocol specifications, Technical report, IP Unwired Canada.

Lapic, S. (2001), A Distributed TDMA Protocol for RF Networking, Technical report, Spawar Systems Center, USA.

Lapic, S. (2001), A Distributed TDMA Protocol for Wireless Networking, Technical report, Spawar Systems Center, USA.

Sheu, S.-T., T. Chen, J. Chen and F. Ye (2002), The Impact of RTS Threshold on IEEE 802.11 MAC Protocol, 9th International Conference on Parallel and Distributed Systems, pages 267-272.

Xu, S. and T. Saadawi (2002), Revealing the problems with 802.11 medium access control protocol in multi-hop wireless ad hoc networks, Computer Networks, 38,pages 531-548.

www.elsevier.com/locate/ifac

MINING AUTOMATION

Heinz-H. Erbe
Report on the Industrial Workshop A 1
(based on the presentation of the panelists)

Panelists:
John E. Udd, CANMET MMSL, Sudbury, Canada
J. Sasiadek, Carleton University, Ottawa, Canada

Abstract: Mining Automation is on demand in most of the countries with mining industry. The workshop discussed the strategies of automatic control of the mining equipment for transporting and processing the minerals. An important subject to consider is the supervision and maintenance: condition based maintenance should replace scheduled maintenance to saving cost when avoiding unnecessary changing of parts. The crucial point is indeed the collection of relevant data, and their transformation into meaningful information. The panelists gave an input through statements for the discussion with the audience.

Keywords: history of mining technology, remote control, unmanned autonomous transport, maintenance

1. A BRIEF HISTORY OF THE EVOLUTION OF MINING TECHNOLOGY

By the 20th Century it was the mining industry turn to be affected by the industrial revolution. Up to then, most mines had been small to medium-sized operations, often in narrow veins, often equipped with primitive equipment, and often operated with unskilled labour working in dark, damp, dank, dirty, and dangerous conditions. Images die hard. Because of this, this image still occupies a prominent place in what people today think of as being the mining industry.

In retrospect, one can imagine what it must have been like to work in a mine a century and a half ago:

• compressed air and the piston-type rock drill was introduced in the 1860s and 1870s.

Prior to that time, drilling was with hand steel, in which one person held and rotated the drill steel, while the second struck at the exposed end with a sledgehammer;

• nitroglycerine was introduced in North America in 1866, with dynamite coming later. Prior to that, black powder had been used to break rock. Before that, fire-setting was used. The most primitive implements were picks made of animal horns;

• motive power underground was by man power or animals pushing or pulling mine cars on rails;

• personal safety equipment was non-existent;

• illumination was by oil lamps of various kinds;

• ventilation was by natural convection induced by fires;

- child labour was common. Workers were generally unskilled with no hopes of being employable elsewhere;
- accidents were frequent and often fatal.

During the 20th Century both electricity and the internal combustion engine were introduced into mining. Until the first two decades of the last century, most Canadian underground mines were relatively shallow and were exploited using expensive and highly-selective methods of mining. Following the vein, or chasing the vein was common practice in the narrow-vein silver and gold mines of North America. It is an image that still lingers in the minds of many.

Starting early in the century, however, North Americans began to develop underground mines in large, massive, base metal ore bodies. The mines of the Sudbury Basin, the Sullivan Mine, at Kimberley, British Columbia, the Horne Mine, at Noranda, Quebec, and many others throughout the continent were all being developed starting in this period.

From a Canadian perspective some of the most important innovations in mining were: the conversion to underground mining at the King asbestos Mine, at Thetford Mines, in Quebecs Eastern Townships, in 1929; the introduction of block caving, at the same mine, in 1932; experimentation with furnace slag and pyrrhotite tailings, at the Horne Mine, in 1933; and the introduction of the cut-and-fill method to replace shrinkage stoping at the Falconbridge Mine, in 1935. With the latter came the introduction of sand and gravel as backfilling materials.

By 1948, INCO was using sand tailing fill at its Frood Mine, in Sudbury. In 1959-1960, Portland cement was being added to tailings fill at both Falconbridge and INCO. Where, previously, backfill had been used principally to provide a working floor in cut-and-fill stopes, it was now beginning to be used as an engineered mix which would provide not only a working floor but also support to the adjacent mine pillars. During this period the mining industry was making a transition from the high-cost highly-selective mining of yesteryear to the low-cost bulk-mining methods of today. The changes in backfill technology, and automation, made this possible.

Up until the 1950s, rail haulage was in common use in Canadian mines. The introduction of the Scooptram (or LHD, for

load-haul-dump unit) introduced tremendous flexibility into the design of underground mines. Underground ramps, by which equipment could be moved between stopes and mine levels, became common practice. By the 1980s, equipment was being operated through

Fig.1. Remote control of transport vehicles

remote control (Fig. 1)so that miners could be isolated from the more hazardous areas (in, particular, under the brows at drawpoints into large stopes). By the year 2000, tele-remote operation, with operators at far-distant locations from the machines they were operating, was being introduced. Today, mine-wide communications systems are common.

As the result of these advances the productivity of miners has increased dramatically. At the beginning of the 20th Century, the daily productivity of an underground hard rock miner was about 0.5 to 1 tonne. Today it is at least two orders of magnitude above that in many mines, and, in some it is in the 200 to 300 tons per day range. Thus, at the beginning of the 21st Century, the conditions in an underground mine are:

- unskilled labour has been replaced with skilled, highly-educated (and highly-paid) labour;
- the great majority of those working in a mine use high-technology;
- mines have become highly-automated workplaces;
- mining has become one of the safest of the heavy industries. It is far safer than fishing, forestry, or farming. For the past few years there have been about 20 fatalities annually in the entire nation (in a workforce of about 30,000).

2. THE INVOLVEMENT OF CANMET-MMSL

Today, three mining research programs of the Mining and Minerals Sciences Laboratories Division, of Natural Resources Canada are located in three regional laboratories - the third being at the Bell Corners Complex, in the western suburbs of Ottawa. The Underground Mine Environment Program is located in Sudbury, with some of its members in Bells

testing facility and a learning environment. In cooperation with la Commission scolaire de l'Or- et- des-Bois, it serves as a location for mine modular training for new miners. It is also the home for the mine hoist inspection service that was previously operated by the Quebec Mining Association. Also, in partnership with the Université du Québec en Abitibi-Témiscamingue, it is establishing an underground wireless communication laboratory.

At present, the focus of the laboratory are on

Fig. 2. Cross-section of deep mining

Corners, while the Mine Mechanization and Automation program is located exclusively at Val-d'Or. The third program, dedicated to Ground Control and Backfill, has its principal home in Bells Corners with some of its staff in the other locations (Fig.2).

Since its inception, in 1992, the Mine Mechanization and Automation Program has developed very close working relationships with the mining industry of western Québec. Located as it is at a mine site on the Cadillac Break, it is surrounded by many former narrow vein gold producers. The underground mine part of the laboratory serves as a real-mine

drilling and rock fragmentation, mine transportation, and underground mining communications.

One of the objectives of the work on rock fragmentation is the development of a selective method of thermal fragmentation that can be used in narrow vein mining. If successful, such a method would bring about important reductions in dilution, the volumes of the rock are ore to be handled, and, hopefully, mining costs. To date, a burner has been developed and tested in field applications on the surface. Studies are now underway of the implications of taking the method underground. These include studies of the emissions of the gases and dusts produced, and of those actions that are required for the certification of the technique for use underground.

Parallel studies are also being made on thermal fragmentation using an arc plasma torch system which is based on a sweeping movement, and of an electric pulse blasting technique which is based on electrohydraulic forming. The latter involves the transmission of a shock wave from a fluid to the surrounding rock as the result of a sudden discharge of stored electric energy through a wire in a fluid environment.

On a very practical note, our work in drilling technology has resulted in the application for a patent for a new rockdrill handle through which a 50% reduction in the harmful vibrations can be achieved. The design continues to be optimized, with commercialization being foreseen. We have also worked with a South African manufacturer of high-pressure rock drills, and have developed a portable unit for the production of high-pressure water close to the drill. We are also working on the development of a multi-functional drill carrier for use in narrow-vein mines. It is to be noted that the needs for improving the efficiencies and cost-effectiveness of small and medium-sized mines are every bit as acute as those for large operations.

During the past forty years, one of the main sources of motive power in Canadian mines has been the diesel engine. Unfortunately, some of the products of combustion, notably soot, are carcinogenic. At present, the solution has been to dilute the concentrations to beneath the levels which are permitted by mine health and safety regulations. As has been mentioned, the provision of ventilating air is expensive. It becomes very much more so as mines are worked at greater depths and as the permitted levels of concentrations are reduced by regulators.

As everyone knows, the costs of energy are increasing rapidly. Our love-affair with the internal combustion engine and our reliance on fossil fuels has the potential of creating a global energy crisis in the not-too-distant future. Today, the rate at which our global reserves of petroleum are being replenished has dropped to 40% of sales. Ontario has now run out of possible sources of hydroelectric power and must consider other alternatives. It is absolutely imperative that we must re-think our transportation infrastructure and how we should operate.

To that end our organization has been a proponent for the use of fuel cell technology in mining. Working with a number of partners, we have participated in the development of the worlds first fuel cell underground mine locomotive. The prototype underwent field trials at our Mine-Laboratoire, near Val-d'Or, and at the Campbell Mine, in Ontario. We are participating in the development of a hybrid scooptram and are considering possible applications of fuel cell technology to other underground mining equipment. The objective is to be able to offer the industry equipment which produces much less pollution and at an acceptable cost.

In the mines of the future, one of the requirements of management will be the knowledge of where every person and every piece of equipment is, at all times. The organizational and communications deficiencies of the past have led to intolerably long delays in getting work done - particularly in respect of getting communications through and in effecting repairs and in the deliveries of supervision, goods, parts, supplies, and electrical and mechanical services. The mine of the future must be wired.

To address some of these issues, our team is working on ways to improve the tracking of vehicles underground and on wireless communications technologies. We are doing both in collaboration with universities, cell phone manufacturers, various levels of government, and the industry.

Some Thoughts on the Future:

From a global perspective, if Canada's mining industry is to thrive and survive it must be both safe and competitive. Since we have no control over the global metals markets, our only option is to be able to sell our products, profitably, at prevailing prices. We must consider every single way in which we can possibly become less costly and more efficient.

At the same time we face the unique challenge that each and every one of our mining operations is unique in some respects. While ore-bodies and their settings may appear to be similar, there will unquestionably be some challenges and local conditions which are unique. Of all of our mining operations, the open pits on the surface are probably the most amenable to automation. The scale is usually quite large and communications are generally no problem. Many of these mines can be thought of as being rock factories. Somewhat more difficult are the tabular (or layered, or bedded) deposits at relatively flat inclinations, such as those in which many of our slat, gypsum, potash, and coal mines are developed. These too, are usually large, and possess the characteristic of being amenable to planar mechanization - similar in many respects to that in a large factory. A great many of these operations have been extensively mechanized for many years. Those that were not are now closed, either because of a lack of minerals to mine or costs.

Most difficult to automate is the class of mines known as metal mines. All of these have unique characteristics; either the size of the ore-body, the spatial attitude, the grade (or, value), the size, the location, the depth, the structural characteristics of the rock, and on, and on.

In this context, we must realize that some,

because of size, location, and the value of the product, will be amenable to automation. Others will present nearly impossible challenges. All will be somewhere in this spectrum.

The challenge to those who are developing goods and services for the industry is to respect this uniqueness and to be able to respond to it. Each operation will require a solution which is suited to its uniqueness. Our greatest need is for those breakthrough technologies which will permit us to do what is presently impossible.

3. AUTOMATION OF SCOOPING AND TRANSPORT

Autonomous underground transporting is in development. Figures 3 is a sketch of underground transportation a century ago, figure 4 shows an underground scooping machine and figure 5 development of automatic transport.

Figs. 4 and 3. Scooping today and years ago

4. CONCLUSIONS

By doing these things, Canadians have become known as the smartest miners in the world. In spite of our high standard of living and high wages, we have been able to maintain the competitive edge in a world marketplace which knows no boundaries between

Fig. 5. Model of an autonomous scooper

producers. To do so in the future will require more of the same innovation. Today, one of the great challenges for our mining industry is that the value of our dollar is rising in world markets, in which metals prices are commonly expressed in either U.S. dollars or British pounds sterling. With metals prices being dictated by the marketplace, we can only adjust to conditions by improving productivity. In doing so, however, we also unlock the potential of mineral deposits that are sub-economic at today prices. There are vast resources of low-grade copper-nickel areas in the Sudbury Basin, perhaps as much has been mined in the previous century, that will become more viable for production as unit productivity increases. It is in our nation interest to maximize the potential use of these resources.

Concurrently, the viability of any mining operation is inextricably linked to the rising or falling fortunes of the minerals which are produced from it. It has often been said (but much less so today) that good ore covers many excuses for poor practice. By taking every opportunity possible to implement every possible improvement to each and every one of Canadas mining operations, we will ensure our future as an international competitor.

As the industrial revolution has swept over the mining industry, it has been accompanied by both globalization and a concentration of ownership. As mining companies have sought greater efficiencies, there have been trends to focusing on the core business of the enterprise and divesting the inefficient and non-core components. Efficiencies are demanded at all levels of the corporation. The chances are that any non-productive part will not survive.

In the global mining industry the three major players today, Rio Tinto, BHP/Billiton, and Anglo American, are all multi-national corporations. In the global mining equipment industry there has been a corresponding concentration of ownership into two or three firms.

In the setting of the mining industry the challenge is to try to identify each and every area in which operations can be made more productive or in which savings can be achieved. Just as we, as people, must find efficiencies in the operations of our households, so, too, must industry seek the same in its operations. As one example, we have long been advocates of the concept of ventilation on demand in our mines (i.e. the delivery on ventilating air, when-needed, where it is needed). The cost of implementing this is high, but the savings in power that can be achieved are enormous. These savings will become much larger as we mine at greater depths and the costs of delivering the air increase.

The future is very much one which will necessitate the development of superb communications and control centres for mining operations.

DISCUSSIONS

Challenges of automation:
- Milling process of oars is already highly automated;
- The automation of scooping, transportation underground and at the surface is under development (remote control, tele-operation, autonomous vehicles);
- Maintenance organization is very important (condition monitoring, predictive maintenance).
- Developing unmanned autonomous vehicles scooping and transporting the oars underground; trend: using fuel-cells underground.

Trend: driverless transportation of oars.

ELSEVIER

IFAC
PUBLICATIONS
www.elsevier.com/locate/ifac

COST REDUCING ENGINEERING STRATEGIES IN THE AUTOMOTIVE INDUSTRY
Future trends in intelligent systems and sensors applied to the automotive industry and their impact on automation
and cost reduction issues in this industry

Heinz-H. Erbe
Report on the Industrial Workshop A 2
(based on the presentation of the panelists)

Panelists:
Denis Gingras, IMSI-Intelligent Materials and Systems Institute, Université de Sherbrooke, Canada
Ove E. Schuett, VP Americas, Delmia Corporation, Auburn Hill, USA
Peter Kopacek, Institute for Handling Devices and Robotics, TU Wien, Vienna, Austria

Abstract: The workshop discussed demands of the industry on technologies and
control strategies for reducing the cost of design, manufacturing and service for
products through an Integrated Product- and Process-Development, Virtual
Machining (digital factory), and e-Work. The panelists gave an input through
statements for the discussion with the audience. *Copyright © 2004 IFAC*

Keywords: smart devices, MEMS, vehicle guidance, telematics, e-manufacturing

1. INTRODUCTION

D. Gingras explained the intentions of the network of
centers of excellence AUTO 21 (Fig. 1), which aims
at developing the technologies and highly qualified

Fig. 1. Themes of AUTO 21

personal required to produce the automobile of the
XXI st century and support the Canadian automotive
industry. About 16 researchers are involved in this
activity. The University of Sherbrooke is on the
board of director of AUTO21 and is leading one of
its 6 research thematic dealing with intelligent in-
vehicle systems and sensors including telematics and
driver assistance systems. Seven researchers are
involved in research topics such as embedded
intelligent sensors and composite materials in
innovative infrastructures. For example several smart
bridges and overpasses have been constructed on
Quebec highways using technologies developed at
the University of Sherbrooke.

2. THE AREA OF INTELLIGENT SYSTEMS AND SENSORS (I2S)

Intelligence to systems (I2S) covers a broad range of
diverse technologies to make vehicles and driving
smarter (Fig.2). I2S refers to diverse technologies

including information processing, communications, telematics control, electronics and photonics. With the recent advances and breakthrough in areas such as information and computing technologies, in sensors and in displays, joining these technologies to our vehicles will make road transportation safer, easier, cleaner and cheaper. But these embedded technologies are also making the automotive manufacturing process more complex.

2.1. Smart sensors and actuators.

The basic definition is: transducers linked together with a microprocessors through a common bus. Smart sensors are now made of multiple chips. Micromachining and VLSI circuitry is merging. Smart sensors do more than just picking up and sending a signal, they have embedded algorithms that

Fig. 2. Technologies involved in intelligent vehicles

process and interpret data, communicate and self-calibrate over time. Integrating bus interface circuitry is a future development that will start at the module level and spread to the chip level.

2.2. To minimize production time and cost.

Devices are obtained by employing production proven techniques utilizing existing manufacturing equipment (Ex: Silicon IC industry being used for automotive sensors and actuators). Compatibility with the materials and operations in standard silicon IC production foundry (materials restricted to those used in standard IC process). Fabrication process partitioning can be used to solve compatibility problems. Packaging aspects account for the largest fraction of automotive device cost. One of the major capital investments lies in the equipment to perform automated packaging.
Sensors application examples are:
- remote tire pressure sensor
- frontal impact detection associated with airbag deployment
- high –g and low-g accelerometers for crash and side-impact detections
- angular rate motion sensors for improved braking, safety and navigational assist

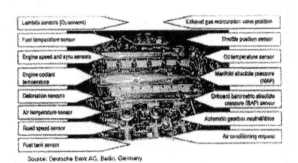

Source: Deutsche Bank AG, Berlin, Germany

Fig. 3. Input signals proved by MEMS/MST sensor applications in an ECU

2.3. MEMS (Microelectronic Mechanical Systems) as sensor concepts.

Piezoresistance: change of electrical resistance of a material in response to a mechanical strain (silicon piezoresistivve pressure sensors were the first automotive micromachined products – early 70s) .Electrochemical etching is currently the process of choice for micromachined pressure sensors). Capacitance variation : a moved or deformed plate with respect to a fixed plate generates a change in capacitance

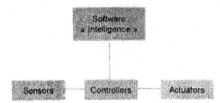

Fig. 4. Typical functional partitioning of automotive systems

Unlike ICs, MEMs type of sensors and actuators must interact with their environment and often utilize moving parts: this implies added complexity in the modeling, manufacturing and packaging processes (Fig. 3).

3. I2S FOR CONTROL AND MONITORING OF VEHICLE BEHAVIOR

Many new technologies are emerging to make vehicles behave more intelligently and exhibit a "smarter response" (Fig. 4). The application of advanced electronic technology to transportation vehicle performance has opened up many exciting applications as such as electronic suspensions, integrated chassis control systems, airbags active control systems, mission control, active noise control, thermal control systems, in-cylinder combustion control, advanced thermal control systems and active traction control. New technologies such as MEMS can be used for these kinds of applications. Furthermore, new integrated vehicle control systems (IVCS) can yield improved performance,

Cars are becoming smarter !

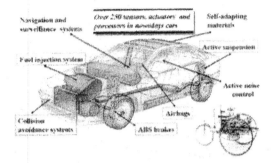

Fig. 5. Electronic support in modern cars

function and safety at a reduced cost. With such a concept, integration is taken beyond vehicle dynamics control, to incorporate occupant safety and comfort as well. Units such as Vehicle stability control (VCS) and Electrically Assisted Steering (EAS) constitute core technologies in vehicle dynamics control. The addition of an Active Roll Control to the VCS for example, can create significant stability and control improvements over conventional design (Figs. 5 and 6). Future generations of IVCS will be embedded in intelligent

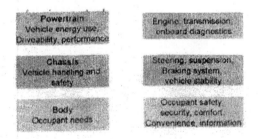

Fig. 6. Areas of system application

chassis modules and will enable control of all the primary subsystems that affect the dynamic behavior of the vehicle.

4. I2S FOR DRIVING ASSISTANCE AND AUTOMATION

These systems help drivers to process information, make decisions, and operate vehicles more effectively, efficiently, and with increased productivity. They exploit cameras, sensors and fast information processing sub-systems to provide an

interface between the driver and non-safety-critical in-vehicle controls and display systems. One of the current trends is the integration of consumer electronics and information services into the car. As on-board intelligent systems evolve, the industry will benefit from an open and standardized platform environment for automobiles. Up to now, the lack of standardization has severely limited manufacturers from sharing development costs and leveraging technological advances that could benefit all customers. Recently, an agreement was concluded between the Automotive Multimedia Interface Collaboration (AMIC) and the Consumer Electronics Association (CEA) to develop vehicle standards for the Intelligent Transportation Data Bus (IDB), a serial communication bus that creates an open, non proprietary standard architecture to allow multiple electronic devices to be installed easily, cost-effectively and safely in any vehicle. Development of new integrated display terminals will focus on the Human Machine Interface (HMI); individualizing functions for driver and passengers; and enhancing the vehicle's ability to serve as a multimedia command center. The technologies range from video monitoring and communications, drowsy-driver safeguards and voice, touch and fingerprint recognition controls.

5. I2S IN VEHICLE GUIDANCE, NAVIGATION AND TELEMATICS

Vehicle guidance, navigation and telematics focus on the collection and transmission of information on route guidance, traffic conditions and transit schedules for travelers before and during their trips.

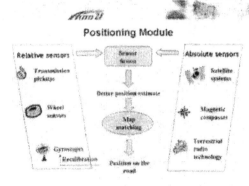

Fig. 7. Tools for Navigation assistance

The productivity of commercial, transit, and public safety fleets will be improved by using automated tracking, dispatch and weigh-in-motion systems that speed vehicles through much of the red tape associated with long route commerce. A new report from Strategy Analytics (In-Car Telematics Terminals Market 2000-2006) says 50% of new cars being sold in the US, western Europe and Japan will be telematics capable by year 2006. Emerging telematics applications such as e-mail, Internet, personal information management, e-commerce and remote access will see strong competition between in-car terminals and portable devices. Converging technology markets (for example GPS and inertial), combined with development of in-car voice synthesis and integration technology will drive telematics growth. There are two scenarios within telematics, with the first one being the Independent Vehicle Concept which puts a smart vehicle in the existing infrastructure. In-vehicle technology lets the vehicle operate automatically with on-board sensors and computers. The vehicle can use data from roadside systems but does not depend on infrastructure support. The second scenario is the Vehicles Cooperative Concept whereby smart vehicles communicate with each other, although not with the infrastructure. With on-board radar, vision, and other sensors, these equipped vehicles will be able to communicate with each other and coordinate their driving operations, thereby achieving best throughput and safety (Figs. 7, 8 and 9).

Vehicles are becoming more and more complex: automation is required. The traffic is higher and vehicle environment is denser: fast and accurate sensing is required to extend human perception. These technologies have yet to be integrated to create a fully intelligent vehicle that works cooperatively with the driver. The designer of an intelligent vehicle must integrate disparate technologies and systems to create a coherent machine that complements the

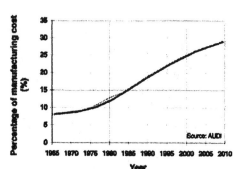

Fig. 9. Electronic modules in automotive manufacturing cost

posed by a lack of interoperability among computers, software and manufacturing hardware systems within the automobile industry prompted the Automotive Industry Action Group (AIAG) to create the Collaborative Engineering and Product Development Steering Committee in late 2000. The time-to-market (TTM) for integrating advanced electronic solutions into the automotive environment is challenging due to the legacy validation environment and the tendency for automotive developers to take a conservative approach to new technology adaptation. EMS (electronic manufacturing services) technology development groups, which have taken a ground-up approach to new business and technology models, may provide a fresh perspective into potential solutions to this challenge.

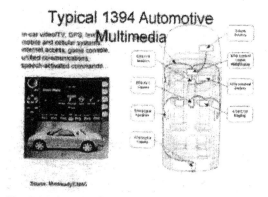

Fig. 8. Multimedia for Navigation and Entertainment

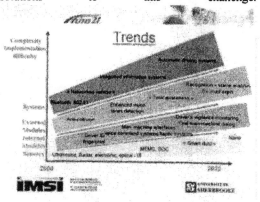

Fig. 10. Trends of future developments

human driver. These technologies also provide the intelligent link between travelers, vehicles, and infrastructure. Considering its network of SMBs involved in information, electronic and photonic technologies, Canada is an appropriate ground to develop the future automotive industry related to the intelligent sensors and systems sector, whereas it is likely to see the smooth integration of these technologies, to make a coherent smart vehicle in the US and other large automotive manufacturing centers (Fig. 10).

As outsourcing continues to increase, the lack of tight integration between trading partners has become a significant problem. Concern about the obstacles

Traditional serialized development processes and manufacturing collaboration was the cornerstone of product development life cycles in traditional OEMs in the automotive markets. As OEMs divested manufacturing and now technology development, a new breed of process was formed to more efficiently handle multidisciplinary markets, products and solutions. To a large extent, the commercial and consumer markets have seen the results of these new efforts which, in turn, have competitively forced OEMs to recognize these new development methodologies. Although the automotive market has been slow to adapt new models, this is changing as the successes in other areas surface. The automotive and military/defense industries have taken a more conservative approach to implementing the latest electronics industry breakthroughs favoring tried-and-true technologies where size and speed are conveniences rather than requirements for success. As the convergence of electronics continues to gain competitive advantage for the automotive industry, more and more automotive electronics OEMs are prepared to benefit from state-of-the-art technologies before they become mainstream technologies.

Product development often requires a series of different skill sets, and growing all of these skills at one location is time intensive and cost prohibitive. State-of-the-art technology integration requires a close working relationship between the design and process development teams.

The automotive industry presents reliability, quality, and cost challenges that require extensive risk mitigation before state-of-the-art technologies can be employed in a new automotive electronics design. For some OEMs, the risk continues to overshadow the benefits of the EMS model. However, software will continue to increase as a percentage of the value of the automotive product.

The virtual organization has to be able to manage close to real-time updates as features are added. Configuration control as well as test configurations will need to be kept in synch. The systems used for configuration management will need to stretch across all of the virtual organization. Where the programming is performed will also be subject to change. The majority of parts today are masked parts. In the future the majority will move to either OTPs or Flash memory. While most parts today leave the electronics assembly process fully programmed, in the future more parts will complete their programming at later stages in the assembly process of the vehicle.

6. CONCLUSIONS

D. Gingras concluded, vehicles can be seen as very complex systems involving a large number of technologies, sensors and sub-systems. This added complexity has a major impact on the manufacturing process and will bring new challenges in automation and cost reduction issues. One of the key question is how these same technological innovations can be exploited to address these same issues not only at the design and manufacturing stages but also at the post-production levels, including distribution, retailing and after sales servicing and maintenance:

Due to increasing complexity of vehicles, car owners and small independent repair shops can no longer perform full maintenance and repair of their vehicles; Vehicles are becoming expensive blackboxes where users have less and less understanding of their internal functioning; More than a dozen microprocessors in a single car are becoming routine;

Exclusive proprietary software and hardware diagnosis tools are becoming available only to OEMs distributors; shortage of shops.

6.1 Requirements.

All system components must be:
- Low in cost
- Manufacturable in very high volume
- Low in weight and small in size
- Robust
- Very reliable and with very high signal to noise ratio
- The added-value of a car will be more and more in its embedded intelligence at all levels materials-sensors-systems.
- The building of tomorrow's cars requires strong transdisciplinary R&D and the convergence of several economic, environmental and social factors.
- Remember : a careless integration of intelligence in vehicles may lead to lethal consequences !
- The very essence of mobility adds degrees of difficulty in implementing intelligence in vehicles
- This added intelligence will change radically the end-users behavior and the way the various automotive sectors will evolve (ex. aftermarket sector)
- One of the biggest challenge will be to bring this intelligence at a very low cost with a high level of reliability.
- Automotive components require the ruggedness of military parts at the price of consumer products.

REFERENCES

mentioned by D.Gingras:

Proceedings, Convergence 2002 Conference , SAE, Detroit, USA

Proceedings, Convergence 2003 Conference , SAE, Detroit, USA

Proceedings, Intelligent Transportation Systems (ITS) World congress 2004, Madrid, Spain

7. MANUFACTURING COST

O. Schuett discussed the cost aspects in today manufacturing of cars. One strategy is to design and then develop the manufacturing of the designed car. Another strategy is to look from the manufacturing site and develop the design. The best way would be to integrate product and process development. The figures 11 to 14 explain the presentation.

Reduction of Manufacturing Cost

1) Focus on Mfg. Engineering (only)

Plan	Launch

- **Target Area :**
  - Provide better Tools to Mfg. Engineers
- **Business Value :**
  - Reduction in Planning Time by x%
    30% – 40%)
  - Better Production System

Fig. 11.

Reduction of Manufacturing Cost

2) Interface to Design   *Integrate to Design*

- **Target Area :**
  - Provide better Tools to Mfg. Engineers
  - Propagate Change faster
- **Business Value :**
  - Reduction in Planning Time by 2x%
  - Better Production System
  - DFM

Fig. 12.

Reduction of Manufacturing Cost

3) Integrate / Interface to Design and Shopfloor

- **Target Area :**
  - Provide better Tools to Mfg. Engineers
  - Propagate Change faster
  - Production Control
- **Business Value :**
  - Reduction in Planning Time by 4x%
  - Optimal Production System

Fig. 13

Digital Manufacturing : PPR Model

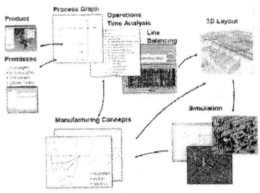

Fig. 14.

DISCUSSION

P.Kopacek:

Remarks on the production of cars; Today we have no intelligent robots at the shop floor, programming is still necessary; This will change soon for autonomous robots (robot football games are for-runners of this development; Developing effective Human-machine/robot collaboration at the shop floor is a challenge;

O.Schuett:

Production philosophy in Europe/USA: design and then it will be manufactured in all circumstances;
Production philosophy in Japan: design from the manufacturing point of view; Reducing manufacturing cost through:

-Better tools for mfg-engineers
-Reduction of planning time
-Better production systems
Example: Boeing 7E7

D. Gingras:

Electronic parts and systems in a car: dependable?
Do we get an over -electronizing?
Are cars becoming more driver friendly through drive by wire?
Trend: no more individual cars or trucks in the road, instead communication between nearly all cars in a road will become vital; IC-technology will make this possible, but what happens with the driver?
Smart, intelligent devices (MEMS, etc) are becoming available at the market; Cost aspects: design, production, maintenance, recycling, service; Possible bottleneck: repair and maintenance service is behind the car development; Therefore dependability of cars/trucks becomes important.

AUTHOR INDEX

Title/Year of publication	Editor(s)	ISBN
2002 continued		
Periodic Control Systems (W)	Bittanti & Colaneri	0 08 043682 X
Modeling and Control in Environmental Issues (W)	Sano, Nishioka & Tamura	0 08 043909 8
Computer Applications in Biotechnology (C)	Dochain & Perrier	0 08 043681 1
Time Delay Systems (W)	Gu, Abdallah & Niculescu	0 08 044004 5
Control Applications in Post-Harvest and Processing Technology (W)	Seo & Oshita	0 08 043557 2
Intelligent Assembly and Disassembly (W)	Kopacek, Pereira & Noe	0 08 043908 X
Adaptation and Learning in Control and Signal Processing (W)	Bittanti	0 08 043683 8
New Technologies for Computer Control (C)	Verbruggen, Chan & Vingerhoeds	0 08 043700 1
Internet Based Control Education (W)	Dormido & Morilla	0 08 043984 5
Intelligent Autonomous Vehicles (S)	Asama & Inoue	0 08 043899 7
2003		
Proceedings of the 15th IFAC World Congress 2002 (CD + 21 vols)	Camacho, Basanez & de la Puente	008 044184 X
Modeling and Control of Economic Systems (S)	Neck	0 08 043858 X
Mechatronic Systems (C)	Tomizuka	0 08 044197 1
Programmable Devices and Systems (W)	Srovnal & Vlcek	0 08 044130 0
Real Time Programming (W)	Colnaric, Adamski & Wegrzyn	0 08 044203 X
Lagrangian and Hamiltonian Methods in Nonlinear Control (W)	Astolfi, Gordillo & van der Schaft	0 08 044278 1
Intelligent Control Systems and Signal Processing (C)	Ruano, Ruano & Fleming	0 08 044088 6
Guidance and Control of Underwater Vehicles (W)	Roberts, Sutton & Allen	0 08 044202 1
Analysis and Design of Hybrid Systems (C)	Engell, Gueguen & Zaytoon	0 08 044094 0
Intelligent Manufacturing Systems (W)	Kadar, Monostori & Morel	0 08 044289 7
Control Applications of Optimization (W)	Gyurkovics & Bars	0 08 044074 6
Fieldbus Systems and Their Applications (C)	Dietrich, Neumann & Thomesse	0 08 044247 1
Intelligent Components and Instruments for Control Applications (S)	Almeida	0 08 044010 X
Modelling and Control in Biomedical Systems (S)	Feng & Carson	0 08 044159 9
2004		
Advances in Control Education (S)	Lindfors	0 08 043559 9
Robust Control Design (S)	Bittanti & Colaneri	0 08 044012 6
Fault Detection, Supervision and Safety of Technical Processes (S)	Staroswiecki & Wu	0 08 044011 8
Technology and International Stability (W)	Kopacek & Stapleton	0 08 044290 0
System Identification (SYSID 2003) (S)	Van den Hof, Wahlberg & Weiland	0 08 043709 5
Control Systems Design (C)	Kozak & Huba	0 08 044175 0
Robot Control (S)	Duleba & Sasiadek	0 08 044009 6
Time Delay Systems (W)	Garcia	0 08 044238 2
Control in Transportation Systems (S)	Tsugawa & Aoki	0 08 0440592
Manoeuvring and Control of Marine Craft (C)	Batlle & Blanke	0 08 044033 9
Power Plants and Power Systems Control (S)	Lee & Shin	0 08 044210 2
Automated Systems Based on Human Skill and Knowledge (S)	Stahre & Martensson	0 08 044291 9
Automatic Systems for Building the Infrastructure in Developing Countries (Knowledge and Technology Transfer) (W)	Dimirovski & Istefanopulos	0 08 044204 8
Intelligent Assembly and Disassembly (W)	Borangiu & Kopacek	0 08 044065 7
New Technologies for Automation of the Metallurgical Industry (W)	Wei Wang	0 08 044170 X
Advanced Control of Chemical Processes (S)	Allgöwer & Gao	008 044144 0

Printed and bound by CPI Group (UK) Ltd, Croydon, CR0 4YY

03/10/2024

01040310-0019